T0251055

MICROBIOLOGY
FOR **MINERALS,** **METALS, MATERIALS** AND THE **ENVIRONMENT**

MICROBIOLOGY FOR **MINERALS, METALS, MATERIALS** AND THE **ENVIRONMENT**

ABHILASH
B. D. PANDEY
K. A. NATARAJAN

CRC Press
Taylor & Francis Group
Boca Raton London New York

CRC Press is an imprint of the
Taylor & Francis Group, an **informa** business

CRC Press
Taylor & Francis Group
6000 Broken Sound Parkway NW, Suite 300
Boca Raton, FL 33487-2742

First issued in paperback 2017

© 2015 by Taylor & Francis Group, LLC
CRC Press is an imprint of Taylor & Francis Group, an Informa business

No claim to original U.S. Government works

ISBN-13: 978-1-138-74878-1 (hbk)
ISBN-13: 978-1-4822-5729-8 (pbk)

This book contains information obtained from authentic and highly regarded sources. Reasonable efforts have been made to publish reliable data and information, but the author and publisher cannot assume responsibility for the validity of all materials or the consequences of their use. The authors and publishers have attempted to trace the copyright holders of all material reproduced in this publication and apologize to copyright holders if permission to publish in this form has not been obtained. If any copyright material has not been acknowledged please write and let us know so we may rectify in any future reprint.

Except as permitted under U.S. Copyright Law, no part of this book may be reprinted, reproduced, transmitted, or utilized in any form by any electronic, mechanical, or other means, now known or hereafter invented, including photocopying, microfilming, and recording, or in any information storage or retrieval system, without written permission from the publishers.

For permission to photocopy or use material electronically from this work, please access www.copyright.com (http://www.copyright.com/) or contact the Copyright Clearance Center, Inc. (CCC), 222 Rosewood Drive, Danvers, MA 01923, 978-750-8400. CCC is a not-for-profit organization that provides licenses and registration for a variety of users. For organizations that have been granted a photocopy license by the CCC, a separate system of payment has been arranged.

Trademark Notice: Product or corporate names may be trademarks or registered trademarks, and are used only for identification and explanation without intent to infringe.

Visit the Taylor & Francis Web site at
http://www.taylorandfrancis.com

and the CRC Press Web site at
http://www.crcpress.com

Contents

Preface

Microbiology for Minerals, Metals, Materials and the Environment is a generic book aimed at analysing and discussing the symbiotic relationship between microbiology and materials through an interdisciplinary approach. Minerals, metals and all other materials exhibit a close interlinking with different environmental microbes in different ways. For example, biogenesis and biomineralisation are now established to explain the role of indigenous microorganisms in the occurrence, transformation and transport of several metals and minerals across the Earth's crust as well as ocean depths. The biology–materials cycle in nature involves various biomediated processes such as materials processing, metals extraction from ore minerals, microbially influenced mobilisation and corrosion of metals and materials, and environmental degradation as well as bioremediation. Although the role of microorganisms in mineral dissolution from naturally occurring ore deposits was known centuries ago, relevance and applications of biology–materials interfaces in a truly interdisciplinary fashion were scientifically established for only 30–40 years. Several microorganisms and microbiological concepts find applications in different facets of metals and materials processing such as biogenesis and biomineralisation, bioleaching of ores, metal-containing wastes and secondary resources, biomineral beneficiation, desulphurisation of fossil fuels, microbially influenced metallic corrosion, and microbial aspects of environmental processing, such as acid mine drainage and biosorption and bioremediation.

A major objective of this book is to present the status and developments in various aspects of the above-mentioned themes. The book is aimed at a readership that includes teachers, students, researchers, scientists and engineers specialising in mining, metallurgy, materials processing, coal processing, corrosion engineering, secondary metals recovery, environmental pollution and protection. In this respect, personnel from the mining, metallurgical as well as materials processing industries would find the contents of this book very interesting and practically relevant. Each chapter is written by one or more experts from academic and R&D institutions as well as industries.

The book can be divided into thematic areas of applications of microbiology in mineral, metal, material and environment sectors. The chapters illustrated in the book cover different aspects such as biohydrometallurgy, biomineralisation, bioleaching, biobeneficiaton, biosynthesis and bioremediation. The basics and historical aspects of biohydrometallurgy, biomineralisation and biogenesis of commercially important ore minerals, microbially induced mineral flotation and flocculation processes are extensively illustrated. Other aspects that are covered include the applications of microbes for metal extraction (including mechanisms and methods) from primary ores/minerals and mining wastes, biomining and related concepts of microbial diversity and various operations, and molecular biology of microbes involved in such systems (extremophiles). Selected chapters on specific systems include the importance and exploitation of microbes in the processing of enargite, uraninite and coal. The current industrial scenario of biohydrometallurgy is critically highlighted. The dissolution of minerals to metals may finally lead to the recovery/synthesis of products. Material synthesis by microbial intervention is depicted in detail highlighting the various mechanisms involved and essential role of microbes. Apart from minerals and ores, the recovery of metals from secondary resources/wastes such as e-wastes and urban mine wastes is also presented with respect to their types, composition, mechanisms of biodissolution and other allied aspects. Biogeochemistry of metals in the environment is an added attraction. The role of microbes in remediating (solid and liquid) wastes containing organic and inorganic contaminants as well as acid mine drainage are also presented in a few designated chapters. This also includes the role of microbes in influencing metallic corrosion. It would thus be useful to understand the linkage of microbiology with minerals, metals and materials, and grasp the importance of bacteria/ microbes in this sector, microbial intervention for waste clean-up, and the role of microbes in extreme environments, and finally to establish connectivity of a non-biologist to biologist with a difference and a microbiologist to process metallurgy in its entirety.

The editors are thankful to all the authors for their valuable contributions. They also express their sincere thanks to CRC Press for bringing out this timely publication.

Abhilash
B. D. Pandey
K. A. Natarajan

Editors

Dr. Abhilash is a scientist in the biohydro-metallurgy lab of the Waste Recycling and Utilisation Group of the Council of Scientific and Industrial Research-National Metallurgical Laboratory (CSIR-NML), Jamshedpur, India. He graduated in microbiology from Nagpur University and earned his post-graduate degree from Bangalore University specialising in mineral biotechnology. He earned his PhD (engineering) from Jadavpur University, India for his work titled 'Biohydrometallurgy of Indian Uranium Ores'. He worked as a lecturer in the Department of Biotechnology, SRN Adarsh College, Bangalore (affiliated to Bangalore University). Later on, Dr. Abhilash joined CSIR-NML as a project fellow and worked in the field of bioleaching of uranium from low grade Indian ores (a project under CSIR's Xth five-year plan) for 2 years. Later, he was selected as a junior scientist in the Metal Extraction and Forming Division of CSIR-NML to work in the area of biohydrometallurgy and other allied disciplines. He has been associated with various projects sponsored by the Government of India and industry under the designations of project leader and team member. In addition to these, he also served as an assistant professor in the Academy of Scientific and Innovative Research (AcSIR-NML).

He has been awarded three consecutive MISHRA awards for the best paper in extractive metallurgy by the Indian Institute of Mineral Engineers (IIME). He was also awarded the prestigious 'Young Scientist Award' at the 100th Indian Science Congress and Indian Nuclear Society for his work on uranium bioleaching. He was bestowed an award for his work on metal–microbe interactions by the Association of Microbiologists, India. He is a Fellow of the Society of Applied Biotechnology, India. He is also a member of various national and international professional bodies including the Engineering Section Committee of the Indian Science Congress Association.

Dr. Abhilash has published over 50 papers (out of which 20 are dedicated to uranium biohydrometallurgy) in SCI (Science Citation Index) journals and in National/International Conference Proceedings. He has presented over 30 papers under the oral/poster category at various national and international symposia/colloquia in India and abroad. He has authored the book *Chromium(III) Biosorption by Fungus*, Lambert Academic Publishing House, GmBH, and edited two compendium volumes. He is on the editorial board of international journals and also is a reviewer for prestigious journals in his field of expertise. He has visited countries such as Egypt, Australia, Russia and China concerning ongoing projects, participated in conferences and fostered international collaboration. He has delivered keynote lectures in India, and also has guided a large number of students with their master's thesis. He is also actively engaged in interaction with school students and teachers to promote biotechnology education.

His areas of interest include microbial technology, environmental microbiology, bioleaching, bioremediation, biosynthesis, microbial corrosion and waste management (metal scraps, effluents, e-waste, etc.).

Dr. B. D. Pandey is currently the chief scientist and head of the Metal Extraction and Forming Division, Council of Scientific and Industrial Research-National Metallurgical Laboratory (CSIR-NML), Jamshedpur, and a professor at the Academy of Scientific & Innovative Research (AcSIR)-NML, India. He earned his PhD in metal extraction and separation by solvent extraction and metal complexation from the Banaras Hindu University, Varanasi, India. While working at CSIR-NML for over 32 years, he has contributed significantly in the area of extraction of non-ferrous metals from both primary and secondary resources, including the extraction of rare earths. He is a life member of professional bodies such as the Indian Institute of Metals and Indian Institute of Mineral Engineers (IIME) (India). He has received several medals and awards, such as the Best Mineral Engineer Award of IIME and Distinguished Researcher Award (International Conference on Non-ferrous Metals, India). He has led very active collaboration programs with the Korea Institute of Geo-science and Mineral Resources (KIGAM), Daejeon (as invited scientist during 2010–2011) under the Brain-pool scheme of Korea Government, Russian Institutes (IGIC, Moscow and Science Centre of Siberian Branch, RAS, Krasnoyarsk) and Paris Tech., France. He is on the editorial board of international journals such as *Hydrometallurgy* and *Metals & Materials International*, and has been the editor-in-chief of the *International Journal of Non-ferrous Metallurgy*. His areas of research include chemical beneficiation, hydro- and biohydro-metallurgy and bioremediation, sulphation roasting and waste processing, recycling and environmental management. He has published over 130 research papers in leading international journals in the above-mentioned areas and more than 85 papers in international conference proceedings. He is involved in teaching and managing the affairs of the Academic Committee as chairman, AcSIR-NML, Jamshedpur.

Dr. K. A. Natarajan is currently a NASI (National Academy of Sciences) senior scientist—Platinum Jubilee Fellow and emeritus professor in the Department of Materials Engineering, Indian Institute of Science, Bangalore, Karnataka, India. He earned his MS and PhD specialising in mineral beneficiation and hydrometallurgy from the University of Minnesota, USA. The Indian Institute of Science, Bangalore conferred on him the degree of Doctor of Science in 1992 for his pioneering research contributions in minerals bioprocessing. He is a Fellow of the Indian Academy of Sciences, Indian National Academy of Engineering and the National Academy of Sciences (India). He has received several medals and awards such as the National Metallurgist Award by the Ministry of Mines, Government of India; National Mineral Award by the Ministry of Mines, Government of India; Alumni Award of Excellence in Engineering Research by the Indian Institute of Science, Bangalore; Kamani Gold Medal of the Indian Institute of Metals; Hindustan Zinc Gold Medal and the Biotech Product and Process Development and Commercialization Award by the Department of Biotechnology, Government of India. He is on the editorial board of several international journals in the area of mineral processing. His areas of research include mineral processing, hydrometallurgy, minerals bioprocessing, corrosion engineering and environmental control. He has published over 300 research papers in leading international journals in the above areas. He was the chairman of the Department of Metallurgy, Indian Institute of Science, Bangalore during 1999–2004.

Contributors

Abhilash
Metal Extraction and Forming
    Division
CSIR-National Metallurgical
    Laboratory
Jharkhand, India

Ata Akcil
Department of Mining
    Engineering
Mineral Processing Division
    (Mineral-Metal Recovery and
    Recycling Research Group)
Suleyman Demirel University
Isparta, Turkey

Alessia Amato
Department of Life and
    Environmental Sciences
Università Politecnica delle
    Marche
Ancona, Italy

Francesca Beolchini
Department of Life and
    Environmental Sciences
Università Politecnica delle
    Marche
Ancona, Italy

Abhilasha Bharadwaj
Chemical and Biomolecular
    Engineering
National University of Singapore
Singapore

Rahul Bharadwaj
Department of Biotechnology
Indian Institute of Technology
West Bengal, India

Jian Chen
Institute of Food Quality and
    Safety
Jiangsu Academy of Agricultural
    Sciences
Jiangsu, China

Shailesh R. Dave
Department of Microbiology and
    Biotechnology
School of Sciences
Gujarat University
Gujarat, India

Wang Dianzuo
National Engineering Laboratory of
    Biohydrometallurgy
General Research Institute for
    Nonferrous Metals
Beijing, China

and

School of Minerals Processing and
    Bioengineering
Central South University
Changsha, China

Gjergj Dodbiba
Department of Systems Innovation
Graduate School of Engineering
The University of Tokyo
Tokyo, Japan

Viviana Fonti
Department of Life and
    Environmental Sciences
Università Politecnica delle Marche
Ancona, Italy

Toyohisa Fujita
Department of Systems Innovation
Graduate School of Engineering
The University of Tokyo
Tokyo, Japan

Chandra Sekhar Gahan
Department of Microbiology
Central University of Rajasthan
Rajasthan, India

Anirban Ghosh
Metal Extraction and Forming
   Division
CSIR-National Metallurgical
   Laboratory
Jharkhand, India

Suchismita Ghosh
Department of Biological
   Sciences
Kent State University
Kent, Ohio

Tingyue Gu
Department of Chemical and
   Biomolecular Engineering
Institute for Corrosion and
   Multiphase Technology
Ohio University
Athens, Ohio

Yu. L. Gurevich
Krasnoyarsk Science Centre
Siberian Branch of Russian
   Academy of Science
Krasnoyarsk, Russia

Raman Gurusamy
Department of Biotechnology
School of Life Sciences
Pondicherry University
Puducherry, India

Fengxiang X. Han
Department of Chemistry and
   Biochemistry
Jackson State University
Jackson, Mississippi

Eric D. van Hullebusch
Université Paris-Est
Laboratoire Géomatériaux et
   Environnement
Champs-sur-Marne, France

Sadia Ilyas
Mineral Resources Research
   Division
Korea Institute of Geoscience and
   Mineral Resources (KIGAM)
Daejeon, Republic of South Korea

Wen Jiankang
National Engineering Laboratory of
   Biohydrometallurgy
General Research Institute for
   Nonferrous Metals
Beijing, China

G. Patricia Johnston
Department of Biological Sciences
Kent State University
Kent, Ohio

Sufia K. Kazy
Department of Biotechnology
National Institute of Technology
West Bengal, India

V. P. Ladygina
Krasnoyarsk Science Centre
Siberian Branch of Russian
   Academy of Science
Krasnoyarsk, Russia

Jae-Chun Lee
Mineral Resources Research
    Division
Korea Institute of Geoscience and
    Mineral Resources (KIGAM)
Daejeon, Republic of South Korea

Laura G. Leff
Department of Biological Sciences
Kent State University
Kent, Ohio

Yingchao Li
Department of Chemical and
    Biomolecular Engineering
Institute for Corrosion and
    Multiphase Technology
Ohio University
Athens, Ohio

Amy L. Lindenberger
Department of Chemical and
    Biomolecular Engineering
Institute for Corrosion and
    Multiphase Technology
Ohio University
Athens, Ohio

Barada Kanta Mishra
Bioresources Engineering
    Department
CSIR-Institute of Minerals and
    Materials Technology (IMMT)
and
Academy of Scientific and
    Innovative Research
Odisha, India

Srabani Mishra
Bioresources Engineering Department
CSIR-Institute of Minerals and
    Materials Technology (IMMT)
and
Academy of Scientific and
    Innovative Research
Odisha, India

Umaballav Mohaptra
North Orissa University (NOU)
Odisha, India

Gayathri Natarajan
Chemical and Biomolecular
    Engineering
National University of Singapore
Singapore

K. A. Natarajan
Department of Materials Engineering
Indian Institute of Science
Karnataka, India

Dhanasekar Naresh Niranjan
Department of Biotechnology
School of Life Sciences
Pondicherry University
Puducherry, India

Sandeep Panda
Bioresources Engineering Department
CSIR-Institute of Minerals and
    Materials Technology (IMMT)
and
North Orissa University (NOU)
Odisha, India

B. D. Pandey
Metal Extraction and Forming
    Division
CSIR-National Metallurgical
    Laboratory
Jharkhand, India

Bhargav C. Patel
Institute of Forensic Science
Gujarat Forensic Sciences University
Gujarat, India

Jayakumar Pathma
Department of Biotechnology
School of Life Sciences
Pondicherry University
Puducherry, India

Dhiraj Paul
Department of Biotechnology
Indian Institute of Technology
West Bengal, India

Yoan Pechaud
Université Paris-Est
Laboratoire Géomatériaux et
    Environnement
Champs-sur-Marne, France

Josiane Ponou
Department of Systems Innovation
Graduate School of Engineering
The University of Tokyo
Tokyo, Japan

Nilotpala Pradhan
Bioresources Engineering
    Department
CSIR-Institute of Minerals and
    Materials Technology (IMMT)
and
Academy of Scientific and
    Innovative Research
Odisha, India

Thulasya Ramanathan
Chemical and Biomolecular
    Engineering
National University of Singapore
Singapore

Ruan Renman
Institute of Process Engineering
Chinese Academy of Sciences
Beijing, China

Alescia Roberto
Department of Biological
    Sciences
Kent State University
Kent, Ohio

Natarajan Sakthivel
Department of Biotechnology
School of Life Sciences
Pondicherry University
Puducherry, India

Pinaki Sar
Department of Biotechnology
Indian Institute of Technology
West Bengal, India

Angana Sarkar
Department of Biotechnology
Indian Institute of Technology
West Bengal, India

Keiko Sasaki
Department of Earth Resources
    Engineering
Kyushu University
Fukuoka, Japan

Sujata
Metal Extraction and Forming
    Division
CSIR-National Metallurgical
    Laboratory
Jharkhand, India

Lala Behari Sukla
Bioresources Engineering
    Department
CSIR-Institute of Minerals and
    Materials Technology (IMMT)
and
Academy of Scientific and
    Innovative Research
Odisha, India

Paul B. Tchounwou
Department of Chemistry and
    Biochemistry
Jackson State University
Jackson, Mississippi

M. I. Teremova
Krasnoyarsk Science Centre
Siberian Branch of Russian
    Academy of Science
Krasnoyarsk, Russia

Yen-Peng Ting
Chemical and Biomolecular
    Engineering
National University of Singapore
Singapore

Devayani R. Tipre
Department of Microbiology and
    Biotechnology
School of Sciences
Gujarat University
Gujarat, India

Liu Xingyu
National Engineering Laboratory of
    Biohydrometallurgy
General Research Institute for
    Nonferrous Metals
Beijing, China

Dake Xu
Institute of Metal Research
Chinese Academy of Sciences
Shenyang, China

Peiyu Zhang
Department of Chemical and
    Biomolecular Engineering
Institute for Corrosion and
    Multiphase Technology
Ohio University
Athens, Ohio

1

Biomineralisation and Microbially Induced Beneficiation

K. A. Natarajan

CONTENTS

1.1 Introduction

A faster depletion of high-grade ore deposits has necessitated the development of alternative efficient beneficiation processes to treat low-grade ores. The beneficial roles of microbes in the field of mineral processing starting from mining, beneficiation and metal extraction to efficient waste disposal have now been well recognised. Physicochemical methods of mineral beneficiation such as flotation, acid dissolution, magnetic and gravity separation are often energy-intensive, costly and environmentally unacceptable.

The utility of microorganisms and bioreagents in mineral beneficiation has been well demonstrated (Chandraprabha and Natarajan, 2010; Natarajan, NPTEL Course, 2013). For example, microbially induced flocculation or flotation of minerals, remediation of toxic chemicals discharged from mineral processing operations, degradation of cyanide and so forth have been proven as promising alternatives. Different from bioleaching, microbially induced beneficiation involves selective removal of undesirable mineral constituents from an ore through interaction with microorganisms or their metabolic products and enriching it with respect to the desired value minerals. Unlike conventional techniques, microbial processes would be more energy efficient, cost effective and environmentally benign. Conventional beneficiation techniques involve the use of various toxic reagents, which could be safely replaced by environment-friendly biological reactants.

Microbially induced mineral beneficiation utilises microbe–mineral interactions to bring about an efficient removal of undesirable constituents in an ore through

a. Selective dissolution or
b. Conferment of surface hydrophobicity or hydrophilicity to bring about selective flotation or flocculation.

Consequences of microbe–mineral interactions relevant to microbially induced mineral beneficiation are the following:

a. Adhesion to mineral substrates resulting in biofilm formation,
b. Bio-catalysed oxidation and reduction reactions and secretion of biopolymers.
c. Interaction of cells and bioreagents with ore matrix.
d. Surface chemical changes on minerals or dissolution.

Several types of autotrophic and heterotrophic bacteria, fungi, yeasts and algae as well as their metabolic products can be used in mineral beneficiation processes.

Microbiological, surface-chemical as well as chemical factors involved in microbially induced beneficiation are discussed in this chapter with special reference to clays, bauxite, iron ores and sulphide minerals. Biomineralisation and biogenetic aspects of clay, bauxite, iron and sulphide ore deposits are analysed to understand the role of indigenous microorganisms. The different types of native mining organisms with respect to their role in mineral beneficiation are identified.

Microbially induced mineral beneficiation processes are illustrated with respect to surface properties of bacterial cells, their adhesion behaviour towards minerals and the surface chemical changes on the minerals consequent to microbe–mineral interaction.

1.2 Microbiological Aspects of Cell Adhesion to Mineral Substrates

Since microbial adhesion to mineral substrates and the formation of biofilms are important events in biobeneficiation processes, it is essential to understand the basic microbiological characteristics of different microorganisms. The structure and architecture of the bacterial cell wall play a prominent role in adhesion to mineral surfaces and resulting consequences. The mechanisms and consequences of adhesion differ depending on the type of the organism (Chandraprabha and Natarajan, 2010; Natarajan, NPTEL Course).

Bacteria are classified into two categories with reference to cell wall structure, namely, Gram-positive and Gram-negative. The chemical composition and structural features of cell walls are different in these two types. Gram-positive bacteria possess a well-defined, rigid outer cell wall, 15–30 nm thick, and an inner, closely held, cell-limiting plasma membrane. The wall constitutes 15%–30% dry weight of the cell. A major polymeric component is peptidoglycan and it may also contain one or more secondary polymers such as teichoic and teichuronic acids. On the contrary, the cell walls of Gram-negative bacteria contain a more general architecture having an outer membrane placed above a thin peptidoglycan layer. The bilayer membranes contain proteins, phospholipids, and lipopolysaccharides and also separate the external environment from the periplasm. In between the outer and the plasma membranes, a gel-like matrix (the periplasm) exists in the periplasmic space. The plasma membrane and the cell wall (outer membrane, peptidoglycan layer and periplasm) together constitute the Gram-negative envelope. In bacterial cells, the functional groups of polymers confer surface charges. The bacterial adhesion mechanisms are governed by surface charges and the nature of membrane polymers.

Bacterial cells possess a net negative surface charge at neutral pH due to the presence of peptidoglycan and teichoic acids. In this respect, microorganisms can be considered as living colloidal particles. The cells acquire charge through the ionisation of surface polymeric groups, which is pH dependent. Competition between electrical and chemical forces may control surface charge neutralisation. An electrical double layer is established at the mineral–solution–bacteria interface. Similar to mineral particles, the surface chemical properties of microorganisms can be characterised by zeta potential and the isoelectric point (IEP). Increasing solution ionic strength may favour bacterial adhesion to a mineral substrate. Bacterial cell dispersion and flocculation are also governed by electrical forces as in the case of mineral particles. The DLVO theory can be applied to explain the mineral–cell–solution interfaces. Bacterial surface hydrophobicity is also controlled by its cell membrane polymers. Hydrophobicity is conferred by hydrophobic molecules such as lipids present on cell surfaces. Hydrophobic bacterial cells are known to adhere to surfaces due to repulsion from the polar water molecule. Electrostatic repulsive forces decrease higher surface hydrophobicity. Since the bacterial surface charge enhances the possibilities for polar interactions with water molecules, the more charged the cell surfaces, the less hydrophobic they become. A decrease in the cell negative charge may then enhance cell surface hydrophobicity. Cells having higher hydrophobicity and lower electrophoretic mobility are more adherent. Bacterial adhesion mechanisms are controlled by cell surface charges and hydrophobicity. The chemical composition and architecture of the cell wall outer layers play a significant role. The surface proteins and polysaccharides may be involved in bacterial adhesion to minerals. Many bacterial cells possess both hydrophilic and hydrophobic surface regions. Cell surface hydrophobicity can be modified depending on the environmental conditions. Since interfaces offer a better environment for bacterial nutrition and growth, bacterial adhesion at mineral–solution interfaces is necessitated under environmental conditions.

Organic and inorganic reagents generated by bacterial metabolism are useful in mineral beneficiation. Different types of mineral acids, fatty acids, polymers and chelating agents are generated by bacteria, fungi and yeasts. A large variety of fatty acids, polysaccharides and proteins produced by microorganisms have been tested in the flotation of various ores. The microbial products can function as flotation collectors, depressants or dispersants.

1.3 Biomineralisation and Biobeneficiation of Clays

Clay formations portray a unique subsurface microbial habitat. Microorganisms are often trapped during deposition of clay layers (Shelobolina et al., 2005). Clays are biogenic and the plasticity of clays is due to microbial activity. Clay minerals are biologically altered and iron-rich diagenetic minerals in the form of iron hydroxide nanocrystals and biogenic clays are deposited around bacterial cells. Iron is present in cream-colored kaolins as ferric oxides and hydroxides (hematite and goethite) and as a structural replacement in kaolinite. In grey kaolins, which have not yet been oxidised, iron is present mainly as its sulphide (pyrite).

Clay minerals are abundant in soils and sediments. Together with microorganisms, they provide catalytic surfaces in sedimentary environments, which are important in many biogeochemical cycles. Microbial clay mineral oxidation and reduction can occur at temperatures and pH values existing in soils. Similar to iron oxide minerals, organic and electron transfer agents enhance the bioavailability of clay-bound ferric ions for reduction. Geochemical evidence attests to the fact that ferric iron in clay minerals can be rapidly reduced. Iron-rich clay minerals serve as the primary electron acceptor available for microbial ferric reduction. Various types of aerobic and anaerobic bacterial species have been isolated from different clay deposits such as smectite, hard kaolin, soft grey kaolin and soft tan kaolin. The abundance of organic matter in clay samples suggested indirect iron oxidation by heterotrophic bacteria. There were significantly higher numbers of iron oxidisers than ferric-reducers (Shelobolina et al., 2005).

Another group of microorganisms that could influence the iron chemistry of clays (Shelobolina et al., 2005) is aerobic heterotrophic microorganisms producing ferric-specific chelating agents called siderophores. Iron is an essential micro-nutrient for many microorganisms. In oxygen-rich neutrophilic environments, in the absence of organic or inorganic chelators, iron availability is limited by the solubility of its hydroxides. In response to low iron availability, aerobic microorganisms generate siderophores for iron mobilisation. An example of a siderophore-producing bacterium is *Pseudomonas mendocina*, which can gather smaller concentrations of iron from kaolinite. Bacterial siderophores can increase the release of iron, silicon and aluminium from kaolinite. Aerobic heterotrophs have been located in large numbers in several kaolin samples.

Microbial iron mineral transformation and mobilisation in kaolin clays are schematically illustrated below:

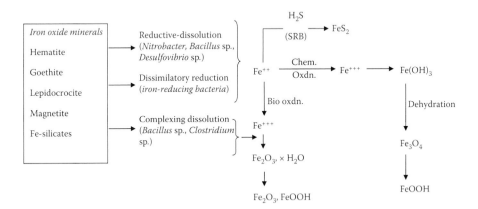

Microbial activity can result in the following changes in the composition of iron minerals:

a. Quantity of magnetically separated magnetite significantly increases.
b. Newer thermodynamically unstable ferric hydroxides such as ferrihydrite and lepidocrocite are synthesised.
c. Formation of iron sulphides (pyrite).

d. Changes in morphology, structure and composition of thermodynamically stable goethite and hematite. For example, acicular goethite and chemical hematite.

Consequences of microbial adhesion on clay minerals need to be understood.
Electrokinetic properties of clays are modified by microorganisms and their
metabolic products. Significant surface chemical changes are thus brought
about on clay minerals by bacterial interaction. The surface-chemical properties
of clays such as hydrophobicity, flocculation and dispersion can be controlled
through microbial interaction. The dispersion and flocculation of clays can
be controlled by interaction with proteins, amino acids and polysaccharides,
which are bacterial metabolic products. The formation of polymeric hydrous
oxides of iron and aluminium due to microbial weathering result in changes in
electrokinetic properties. The possible role of extracellular slime layers in bacterial adhesion need to be understood. Microbial cells readily attach (adhere)
onto clay particles through electrostatic and chemical interaction mechanisms.

The need for clay beneficiation in the light of the mineralogical changes
due to biomineralisation is analysed below.

Kaolin clays contain coarse and fine impurities that need to be removed.
Coarse impurities, more than half the crude volume of clays consist of
quartz, feldspar, mica or tourmaline. Sedimentary kaolin contains titania
mineral impurities such as anatase and rutile. Titania minerals contain lat-
tice-substituted iron imparting a yellowish brown color. Along with titania,
significant amounts of iron oxides and iron sulphides also occur in clays.
Iron oxide minerals such as hematite and goethite impart a dull yellowish
brown color. Iron sulphide minerals such as marcasite and pyrite occur in
kaolins containing organic matter, which impart a gray colour. Even in small
amounts, iron impurities reduce the brightness and whiteness of raw materials derived from kaolin. Bright, white pigments are generally preferred over
dull, yellow pigments manufactured from kaolin clays. Also, kaolin used in
paper and ceramic industries needs be whitened through iron removal.

Most of the high-quality clay deposits occurring on the top layers of the
Earth's crust have already been mined. The naturally whiter clays contained
the more easily removable hematite and/or goethite, and most of their dark
organic matter had already been oxidised. We are now forced to mine deeper
regions of clay deposits where the as-yet un-oxidised kaolin is grayer, containing more organic matter and iron sulphides.

Various methods of beneficiation are practised for iron removal from clays.
Physico-chemical methods of clay beneficiation for removal of impurities
such as iron and titanium oxides are the following:

a. High intensity superconducting magnetic separation.
b. Froth flotation.
c. Chemical treatment using hydrosulphides to reduce and solubilise
iron oxides.

d. Wet high intensity magnetic separation.

e. Iron leaching by oxalic acid.

f. Beneficiation and dewatering (flocculation, deflocculation).

g. Selective flocculation, reductive leaching and dewatering.

All the above abiotic methods are energy-intensive, costly and not environment-friendly. A biotechnological alternative to clay beneficiation would prove to be cost-effective, energy-efficient and environment-friendly (Styriakova and Styriak, 2000; Groudev, 2001; Kostka et al., 2002; Lee et al., 2002; Mandal and Banerjee, 2004; Gao et al., 2010; Ajayi and Adefila, 2012).

Development of a flexible biotechnology for kaolin beneficiation, either through microbial ferric reduction or ferrous oxidation, can be considered depending on the iron mineralogy. For commercial quality kaolin, ferric-reduction of hematitic and goethitic iron by using indigenous ferric-reducing bacteria would enhance the quality.

Selective bioflocculation can be considered to remove titania. In some selective flocculation methods, kaolin containing titania (and other fine impurities) is interacted with flocculants to promote settling of titania impurities, leaving the product kaolin to be recovered from the supernatant in a dispersed form.

Clay-slimes are generated during the beneficiation of phosphate and potash ores. The use of microorganisms for dewatering such clay-slimes and enhancing settling rates would prove beneficial. Biopolymer-producing microbes can flocculate phosphatic clay-slimes. Dilute phosphatic clay-slimes can be flocculated with indigenous organisms. Polymers isolated from *Leuconostoc mesenteroides* and *Xanthomonas* sp. were found to flocculate phosphatic clay-slimes.

Microorganisms also play an important role in the dissolution of silicate structures. *Bacillus* spp. are active and can remove both free and bound iron occurring in kaolin. Biodegradation of iron oxyhydroxides and partial destruction of mica structure can be brought about by microbial interaction.

Microbiological reducing and solubilising agents can also be used. Biotechnological methods for leaching kaolin using fungi such as *Aspergillus niger* can also be considered.

Iron removal from quartz sands, kaolins and clays can be achieved in a number of ways such as:

a. The use of bacterial and fungal metabolites at high temperatures.

b. *Bacillus cereus* isolated from kaolin deposits.

c. Fungal strains of the type *Aspergillus*, *Penicillium* and *Paecilomyces* spp.

d. High amounts of oxalic acid produced by *A. niger*.

Passive 'in situ' beneficiation methods at mine sites also become possible. Stimulation of activity of natural microflora present in clay deposits through nutrient additions is an 'in situ' biobeneficiation approach. Bacterial growth and production of organic acids decrease pH leading to transformation of iron minerals in clay deposits.

Microbial removal of pyrite from low-grade clay using sulphur and iron-oxidising bacteria such as *Acidithiobacillus ferrooxidans* and *Acidithiobacillus thiooxidans* has been demonstrated. About 82%–90% of the pyrite was removed in 5–12 days for pulp densities up to 70%. With the refined clay no red colour due to the presence of pyrite was developed after firing and its whiteness was similar to that of high-grade clay.

1.3.1 Microbially Influenced Iron Removal from Some Indian Clays

Iron is present in several Indian clays in the form of Fe_2O_3 and FeS_2 which need to be removed to achieve whiteness and desired qualities for use in ceramic, paper, paint and other industries. Using cell cultures and acidic metabolites of *A. thiooxidans* and *Bacillus* sp. the efficient removal of iron from two types of clay samples (one containing iron as hematite, another as pyrite) was demonstrated. Microbially influenced selective flocculation–dispersion as well as reductive dissolution of iron oxides from clays were also demonstrated successfully.

A few examples are illustrated below:

China clay samples containing pyritic iron having the following chemical composition were used to demonstrate biobeneficiation.

SiO_2	44%
Al_2O_3	32%
Fe_2O_3	5.2%
TiO_2	2.3%
CaO	0.3%
LOI	20%

Note: Brightness (as received) = 45–46.

Kaolinite was the major mineral with small amounts of pyrite and quartz.

The above china clay deposits are rather unique in the sense that iron is present as a sulphide (pyrite). Due to weathering of the sulphides, the clay samples in contact with water generated acidity and equilibrated at a pH of 2–2.5. Indigenously occurring acidophilic chemolithotrophs such as *Acidithiobacillus* were isolated from the clay samples. For this purpose, clay samples were inoculated into 9 K medium and Basal medium containing sulphur at a pH of 2 to promote the growth of *A. ferrooxidans*, *A. thiooxidans* and also *Leptospirillum ferrooxidans*. Variations of cell population along with changes in concentrations of ferric and sulphate ions were monitored as a

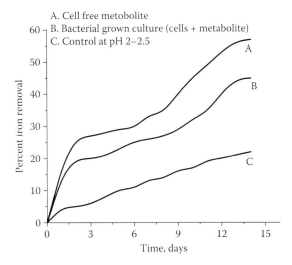

FIGURE 1.1
Iron removal from pyritic clays through microbial dissolution.

function of time. Several subculturings were carried out to attain higher cell populations of isolated acidophiles.

All the above isolated acidophilic bacteria could be used to remove the iron (as pyrite) from the clay samples. Among the clay isolates, *A. ferrooxidans* and *A. thiooxidans* were used to study iron removal. Typical results are illustrated below:

a. With cultures of *A. ferrooxidans* grown for different periods of subculturing up to 70%–80% of iron could be removed after a period of two to three weeks.

b. Similar results were obtained with cultures of *L. ferrooxidans*.

Iron removal from clays using different acidic metabolites from *A. thiooxidans* are illustrated in Figures 1.1 and 1.2.

Bioleaching of iron (pyrite) using metabolites (pH 0.5) of *A. thiooxidans* is illustrated in Table 1.1.

Brightness measurements of samples treated with pH 0.5 metabolite of *A. thiooxidans* are given in Table 1.2.

1.3.2 Microbially Induced Selective Flocculation of Hematite from Kaolinite

Microbially induced selective flocculation of hematite from kaolinite was also demonstrated with respect to interaction with *B. subtilis* (Poorni and Natarajan, 2013, 2014). The growth of bacterial cells in the presence of kaolinite resulted

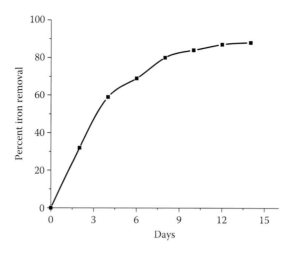

FIGURE 1.2
Pyrite removal from pre-acid washed clay fines with metabolite of *A. thiooxidans*.

in enhanced production of extracellular proteins whereas that of hematite promoted significant secretion of exopolysaccharides. Bacterial cells were successfully adapted to grow in the presence of both of these minerals and the advantage of using hematite-grown and kaolinite-grown cells and their metabolic products in the selective flocculation of hematite and effective dispersion of kaolinite was demonstrated. Bacterial cells and extracellular polysaccharides exhibited higher surface affinity towards hematite, rendering it hydrophilic, while significant protein adsorption on kaolinite led to enhanced

TABLE 1.1

Pyritic Iron Removal Using *A. thiooxidans*

	Interaction Time (h)	Percent Iron Removed
Growing culture	1	55
	12	68
	24	99
Cell free extract	1	50
	12	75
	24	98

TABLE 1.2

Brightness of Clay Samples

	Brightness (% ISO)
Untreated pyritic kaolin clay	58
12 h interaction	66
24 h interaction	73

TABLE 1.3

Selective Bioflocculation for Hematite–Kaolinite Separation

	Percent Settled			
			Mixture	
Conditions	Kaolinite Only	Hematite Only	Kaolinite	Hematite
Without bacterial interaction	40	60	40	65
Bacterial cells adapted to hematite	10	95	10	90
Kaolin-adapted bacterial cells	05	70	10	60
Bioproteins from hematite-adapted cells	45	80	25	82
Bioproteins from kaolinite-adapted cells	10	75	10	80
Exopolysaccharides from hematite-adapted cells	04	95	20	95
Exopolysaccharides from kaolinite-adapted cells	08	90	20	75

Source: Adapted from Poorni, S., Natarajan, K.A., *International Journal of Mineral Processing* 125; 2013: 92–100.

surface hydrophobicity. Bacterial interaction promoted selective flocculation of only hematite, while kaolinite was fully dispersed. Mineral-specific proteins were generated on growing *B. subtilis* in the presence of kaolinite.

Typical selective bioflocculation results are illustrated in Table 1.3.

At neutral pH, maximum settling (flocculation) for hematite was achieved using exopolysaccharides extracted from hematite-grown cells. A maximum dispersion of kaolinite was achieved after interaction with kaolinite-adapted cells and proteins and polysaccharides separated from adapted cells. Higher-surface affinity of kaolinite for proteins and of polysaccharides for hematite is responsible for the above observed flocculation–dispersion behaviour.

An efficient separation of hematite from kaolin samples could be obtained using cells of *B. subtilis* preadapted to kaolinite as well as the proteins and polysaccharides extracted from such cell cultures. The developed bioprocess could be compared with existing chemical alternatives such as selective flocculation and flotation using toxic chemicals such as amines and polyacrylamides. Efficient separation of iron oxides from clays is not possible through such methods including high-intensity magnetic separation and selective flocculation–dispersion. A bioprocess, on the other hand, would be more process efficient, cost effective and environment friendly.

1.4 Biomineralisation and Biobeneficiation of Bauxite

Significant chemical and biological weathering of silicates can occur in bauxite deposits. Strongly oxidised conditions to gradually reduced environments are encountered during the bioevolution of bauxites. Bauxite deposits contain water, organic carbon, sulphur and iron that can be used as energy sources by indigenous microbes. The role of microorganisms in the formation and evolution of bauxite has now been well established (Laskou and Eliopoulos, 2007, 2013).

Biomineralisation from oxidised to reduced environments involves the following stages:

a. Rock weathering due to excretion of microbial metabolites (organic and inorganic acids, ligands)
b. Microbial weathering of Al-silicates release Al, Fe and Si followed by precipitation of biogenic carbonates, sulphides and iron oxides
c. Aerobic and anaerobic microbes participate in sulphate formation and sulphide precipitation.

Bacterial consortia occurring in bauxite deposits dissolve primary rock-forming minerals and serve as nucleation sites for the precipitation of secondary minerals. For example, the presence of various types of microorganisms in gray-colored bauxites has been well documented. Many metals can be enzymatically enriched and dispersed by organisms (Papassiopi et al., 2010).

Two stages of the transformation of bauxite ores can be envisaged:

a. Reduction of ferric to ferrous leading to the formation of pyrite in a variety of forms and size.
b. Subsequent pyrite oxidation and formation of goethite.

Various morphological forms of pyrite could be identified commonly associated with fossilised bacteria in grey-red bauxites. The coexistence of fine-grained spherical pyrite and pyrite pseudomorphs after iron oxides suggest that pyrite is the product of bacterial activity and has formed from iron oxides. Bacteria facilitate nucleation and growth of minerals.

The role of microorganisms in bauxite mineral formation can be understood in relation to the following:

a. Bacterial metabolism involving redox reaction with sulphur and metals such as iron and aluminium.
b. Bacterial communities catalyse mineral formation since they derive energy for growth through selective oxidation–reduction of metals and serve as nucleation sites for precipitation of secondary minerals.
c. Both aerobes and anaerobes participate in the above biomineralisation processes.

To demonstrate the role of microorganisms in bauxite mineralisation, studies were carried out using Indian west coast bauxites. Indigenous microbes from several deposits were isolated, identified and their role in mineralisation analysed (Phalguni et al., 1996).

Typical microorganisms isolated from some Indian west coast bauxite deposits are listed in Table 1.4.

Various autotrophic and heterotrophic bacteria and fungi isolated from west coast Indian bauxite ore deposits were found to participate in the formation and concentration of bauxite. *A. ferrooxidans* generates sulphuric acid and the pH changes cause weathering of the aluminosilicates and precipitation of iron oxyhydroxides. Biological and chemical weathering causes mobilisation of aluminium by the settling of alumina in swamps, where it precipitates as aluminohydroxides. *Bacillus* spp. mediate the release of alkaline metals, such as sodium, potassium and calcium from the bauxite deposits. Magnesium and iron from bauxite also can be released by microbial action. Aluminium and silica remain in the bed leading to bauxite mineralisation. Fungi such as *Cladosporium* are alumina solubilisers and iron reducers. *Pseudomonas* anaerobically reduce ferric oxides, while *Bacillus coagulans* and *Paenibacillus polymyxa*, through the excretion of polysaccharides, flocculate iron oxides, alumina and calcite. Biogenic origin of the bauxite mineralisation thus becomes evident.

TABLE 1.4

Mine Isolates and Their Functions

	Isolates	Salient Characteristics Relevant to Biomineralisation
Fungi	*Aspergillus niger* *Aspergillus* spp.	Soil fungus, utilising carbohydrate sources from degrading plant waste, secretes organic acids such as citric, oxalic and gluconic acids.
	Cladosporium *Cladosporoides*	Fungus capable of reducing iron and dissolving aluminium.
Bacteria	*Bacillus* spp. *Bacillus circulans* *Paenibacillus polymyxa*	Bacteria secrete exopolysaccharides and organic acids such as acetic and citric acids and can grow anaerobically reducing ferric iron.
	Bacillus coagulans	Spore forming, Gram-positive bacteria, acidophilic, thermophile with optimum temperature of 35–45°C; and generate acid conditions of pH 4–5.
	Pseudomonas spp. *Pseudomonas putida* *Pseudomonas stutzeri* *Acidithiobacilli*	Capable of anaerobic ferric iron reduction. Autotrophic acidophile, capable of oxidising ferrous iron and precipitating iron oxyhydroxides. Also generates sulphuric acid and forms sulphates.
	Sulphate-reducing bacteria	Precipitates sulphide minerals.

Source: Adapted from Natarajan, K.A., Modak, J.M., Anand, P. *Minerals and Metallurgical Processing* 14(2); 1997: 47–53.

TABLE 1.5

Biobeneficiation of Bauxite Ore Using Mine Water
as Inoculum

Days	Percent Ca Removed	Percent Fe Removed
4	55	34
15	88	40

Source: Adapted from Natarajan, K.A., Modak, J.M., Anand, P. *Minerals and Metallurgical Processing* 14(2); 1997: 47–53.

Beneficiation of low alumina bauxites (<50% Al_2O_3) was studied in detail using mine-isolated native organisms (Phalguni et al., 1996; Modak et al., 1999; Vasan et al., 2011).

Initial screening tests were performed to establish the feasibility of using mine isolates in the removal of calcium and iron. Both these impurities as present in Indian western bauxites need to be removed to produce a commercially acceptable raw material for the abrasive and refractory manufacturers. Typical results are shown in Tables 1.5 and 1.6. The mine water contained a consortia of heterotrophs capable of removing both calcium and iron.

Isolates capable of efficient removal of calcium included the fungus, *A. niger* and a number of neutrophilic bacteria such as *B. cirulans, P. polymyxa* and *Pseudomonas* spp.

It becomes readily evident from the above discussions that many types of microorganisms indigenously present in bauxite deposits are capable of bringing about different reactions such as reduction of ferric ions, reductive dissolution of iron oxides and dissolution of calcium carbonate. It then becomes possible to use such mine-isolated microorganisms to achieve selective removal of calcite, iron oxides as well as silica and silicates (clays) from the bauxite ores.

Paenibacillus polymyxa, a neutrophilic, organic-acid-producing bacterium was found to be present in the mine waters and bauxite deposits. The bacteria, when grown in a Bromfield medium containing sucrose, ammonium sulphate, magnesium sulphate and yeast in the presence of calcium carbonate, were found to undergo calcium-dependent metabolism utilising calcium

TABLE 1.6

Beneficiation of Bauxite Ore Using Different Microorganisms Isolated from Indian West Coast Bauxite Mines

Isolates	Percent Ca Removed	Percent Fe Removed
Apergillus niger	78	15
Bacillus circulans	88	21
Paenibacillus polymyxa	86	23
Pseudomonas spp.	83	33

Source: Adapted from Natarajan, K.A., Modak, J.M., Anand, P. *Minerals and Metallurgical Processing* 14(2); 1997: 47–53.

to produce amylases to synthesise a dipicolinate. Profuse excretion of extra-cellular polysaccharides and organic chelating acids led to the dissolution of calcium carbonate.

The use of the above bacteria in the removal of iron oxides and calcium carbonate from low-grade bauxites was studied to establish an environmen-tally benign and cost-effective biobeneficiation process (Modak et al., 1999; Vasan et al., 2011).

To demonstrate calcium removal using *P. polymyxa,* a column bioreactor was designed which could be operated either in the fluidised-bed mode (for fine particles) or in a drain mode (coarser particles). Pulp density of the bauxite ground slurry could be controlled and fed from the top of the column reactor as total recycle slurry. Air vents on top facilitated aera-tion and spillage control. A uniform suspension of the particles using con-trolled air passage was ensured. Depending on bacterial growth cycles, two types of operations, namely, cascade and uncascade were carried out. Since the cell growth ceases after 24 h, the contact time was maintained at 24 h, when the slurry was drained off to a storage tank. After settling the suspended slurry, the leach liquor was decanted. Fresh bacterial cul-ture was added and the solids pumped back to the reactor. The entire operation was repeated 4–5 times as recycle mode in the cascade operation sequence. In the uncascade mode, no such reinoculation or recycle of the slurry was permitted. The entire operation was carried out continuously for 4–5 days at a desired pulp density. Higher calcium removal could be observed when the reactor was operated in the cascade mode and typical results are presented below:

a. More than about 90% of calcium removal was achieved in the cas-cade mode.

b. Bioleaching of calcium carbonate took place in two stages, namely, (i) a rapid indirect mode in the presence of acidic bacterial metabolites containing exopolysaccharides and (ii) a slower direct mode, involv-ing bacterial attachment to particles and enzymatic dissolution.

c. Typical compositions of the raw feed and leached residue are given in Table 1.7.

TABLE 1.7

Metallurgical Analysis of Ground Bauxite Ore and Bioleached Residues

	Percent Composition			
	Al_2O_3	Ca	Fe_2O_3	SiO_2
Ground bauxite ore	58.6	2.0	2.3	3
Biotreated residue	59	0.3	1.8	2

Source: Adapted from Modak, J.M., Vasan, S.S., Natarajan, K.A. *Minerals and Metallurgical Processing* 16; 1999: 6–12.

Calcium solubility in the bacterial metabolite was found to be very high even when compared to hydrochloric acid leaching.

The following mechanisms are proposed to explain the enhanced $CaCO_3$ removal through indirect mechanism.

a. Bacteria produce levan from sucrose (soluble capsule) which is a homopolysaccharide.

b. $CaCO_3$ solubility in culture increases from 70 mg/L in growth media to 300 mg/L in full-grown culture.

c. Fully grown bacterial culture contains organic acids such as formic acid, acetic acid, butyric acid and lactic acid.

d. Ca-organic acid salts have high solubility in water. Calcium forms stable complexes with many of these acids.

e. Compared to mineral acid, grown culture shows significantly higher calcium solubility.

f. Al_2O_3 solubility is very limited. Al-organic acid complexes are not known.

The mechanism of direct leaching of calcium carbonate on the other hand may involve:

a. Adhesion of cells to bauxite minerals.

b. Ca-dependent metabolism resulting in the formation of polysaccharides (in the capsule and biofilm).

c. Formation of soluble Ca–polysaccharide complexes resulting in dissolution of complex.

The following changes in calcium carbonate substrate due to adhesion of *P. polymyxa* could be observed:

a. Higher than 90% surface coverage in 24–30 h.

b. Presence of polysaccharides in the bacterial capsule.

c. Negative shift in zeta potential.

d. Decrease in surface hydrophobicity and enhanced settling rate in aqueous solutions.

The use of *P. polymyxa* in the dissolution and flotation beneficiation of bauxite ores has also been reported (Zhou et al., 2010).

The following observations are noteworthy:

a. pH change from 7 to about 4 (18 h) due to organic acid generation by bacterial growth.

TABLE 1.8

Calcium Removal from High Calcium
Bauxite by Selective Bioflocculation
Using Adapted Cells of *P. polymyxa*

Deslimings	Percent Ca Removal
1	57
2	63
3	67
4	71
5	73
6	79

b. Bacterial metabolites (attached bacteria + metabolites) complexed iron, silica and calcium.

c. Flotation of alumino-silicates (kaolinite) and dissolution of iron hydroxides could be achieved through bacterial interaction.

Yet another biological approach to remove calcium and iron from bauxites is through selective bioflocculation (Deo, 1998). Unlike in biodissolution, flocculation involves only interfacial changes brought about by bacterial interaction and is therefore a very fast process. Typical results are given in Tables 1.8 and 1.9.

The use of *P. polymyxa* is thus very effective in the removal of both calcium and iron from bauxite ores. Silicate bacteria such as *B. mucilaginosus* and *B. circulans* can remove SiO_2 from bauxite. The screening of silicate bacteria and bioleaching of silicon from bauxite (*Desiliconisation*) using *B. mucilaginosus* to control Al_2O_3:SiO_2 ratio has been attempted. The isolated strains adapted (grown) in hematite, olivine and calcite showed different leaching capabilities. Bioremoval of iron from bauxite amounted to 92% using microbial isolates. An interaction with *P. polymyxa* released iron and silicon. Owing to bacterial adhesion and generation of acidic metabolites iron could

TABLE 1.9

Iron Removal from High Iron Bauxite by
Selective Bioflocculation Using Adapted
Cells of *P. polymyxa*

Deslimings	Percent Fe Removal
1	23
2	40
3	51
4	69
5	85
6	96

be efficiently leached out. Iron-reducing bacteria can also be effectively used to bring about reductive dissolution of iron oxides (Groudev and Groudeva, 1983; Groudev, 1987; Karavaiko et al., 1989).

Microbial removal of silicon from low-grade bauxite has been reported. Silicate bacteria isolated from rock samples and soil were used. Silicon leaching is attributed to the formation of mucilaginous capsules consisting of exopolysaccharide oxides (Groudev and Groudeva, 1983; Groudev, 1987; Karavaiko et al., 1989). Continuous leaching was carried out in bioreactors at room temperature with a residence period of 5–7 days. Typical leaching results indicated reduction of silica from about 25% to about 9% with enrichment of alumina from about 44% to 64%.

Bauxite beneficiation using 'silicate' bacteria (such as *Bacillus circulans*) has also been demonstrated. Up to 75% of silica could be removed.

In situ microorganisms also find application in environmental control with respect to red mud disposal. Microbiology and microbial ecology of red mud samples can be established. The use of bacteria to reduce red mud alkalinity has also been established. Acid producing and alkaline bacteria can be readily isolated from red mud storage piles. Native bacteria proliferation as well as inoculation from the outside can be practiced to bring down alkalinity of red mud. The genera of red-mud isolated microbes include *Bacillus, Lactobacillus, Micrococcus, Pseudomonas, Flavobacterium* and different fungi.

Reclamation of bauxite residues through afforestation has also been tried. Bioremediation of bauxite residues has been reported in Western Australia by adding organic substrate to stimulate the growth of indigenous microorganisms (Jones et al., 2012). Native organisms generate organic acids and neutralise red mud. Inoculation of appropriate microbes can also be done from the outside. Bioextraction/microbial uptake of trace metals from bauxite red mud also has been attempted (Jones et al., 2012).

Some examples are:

   a. Microbial selenate reduction.
   b. Bioleaching of heavy metals from red mud using *A. niger.*
   c. Microbial dispersion and flocculation to settle particles and water harvesting.

We have so far analysed two ore mineral systems, namely, clay and bauxite, both of which are basically aluminium silicates containing iron oxides, calcium carbonate and silica as impurities. Biomineralisation features of the above two mineral systems, namely, kaolin clay and bauxite appear similar and complimentary. From a beneficiation angle also, there are complimentary aspects pertaining to the removal of iron. A close correlation among the clay and bauxite systems could then be made with iron ore mineralisation and beneficiation.

For example, clay beneficiation deals with iron removal. Bauxite beneficiation dealt with iron oxide removal from alumina. In iron ore beneficiation, separation of alumina, silica and aluminium silicates from iron oxide becomes paramount. The role of different microorganisms such as yeasts, anaerobic bacteria (SRB) and two *Bacillus* spp. on iron ore beneficiation with respect to silica and alumina removal is discussed below.

1.5 Biomineralisation and Biobeneficiation of Iron Ores

The following questions need to be answered in order to understand the relevance of iron ore biogenesis and biomineralisation with respect to beneficiation.

What types of microorganisms inhabit iron ore deposits?

Why they are there and what roles do they play?

What mineral-related biochemical and surface chemical reactions they can perform?

Microorganisms play a significant role in the natural iron cycle. Many iron-based redox reactions promote bacterial growth through iron dissolution and precipitation. Biogenic iron oxides occur as nanocrystals having different morphology and mineralogy. Direct bacterial nucleation processes can result in the formation of various iron minerals. Acidophilic and neutrophilic, aerobic bacteria promote oxidation of ferrous iron to ferric form, while anaerobic bacteria reduce ferric to ferrous ions. Iron oxides such as goethite, lepidocrocite, magnetite and hematite are found in sediments. Iron oxide particulates occur closely associated with bacterial cell walls having exopolymers. Iron-oxidising microbes can be isolated from iron-rich seepages. The formation of intracellular magnetite has been reported in magnetotactic bacteria. Banded iron formations represent biogenic iron mineralisation (Bonneville et al., 2004; Fortin and Langley, 2005; Gilbert et al., 2005; Roberts et al., 2006; Liu et al., 2006; Williams and Cloete, 2008; Arakaki et al., 2008; Natarajan, 2013).

The following aspects of biomineralisation are significant with respect to the identification of microorganisms relevant to iron ore beneficiation.

- Biotic and abiotic mechanisms are involved in the formation of iron ore minerals such as iron oxides, clays, silica, alumina, calcite and apatite.
- Biogenic iron oxides exist in close association with microorganisms that inhabit the ore body. For example, in natural sediments, iron oxide particulates occur in close vicinity of bacterial cell walls

containing secreted biopolymers and extracellular biogenic iron minerals. Iron oxidising and reducing bacteria are associated with biofilms formed on iron oxide minerals.

- A wide range of microorganisms existing under acidic to neutral pH as well as oxic and anoxic environments bring about oxidation and reduction of iron. A few examples are *Gallionella* sp., *Leptothrix* sp., *Acidithiobacillus*, *Leptospirillum ferrooxidans* and *Thermoplasmales* (archea).

- FeOOH sheaths are formed by *Lepthothrix* spp. in iron ore mines and generation of exopolysaccharides in the capsule is considered as a protection mechanism against metal toxicity.

- Extracellular polymers such as polysaccharides and bioproteins secreted by several iron bacteria would be useful in iron ore beneficiation. Examples are magnetotactic bacteria, *Bacillus* spp.

- Biogenic iron minerals carry biosignatures. Banded iron formation is a case in point. Nanocrystals of lepidocrocite on and away from the cell wall of *B. subtilis* have been identified.

- Diverse groups of magnetotactic bacteria (MTB) present in iron ore deposits synthesise intra- and intercellular magnetic minerals such as magnetite and magnetosomes.

- Microorganisms isolated from iron ore deposits will be useful in iron ore beneficiation (e.g. removal of phosphorous, alkalis, silica, clays and alumina). Since phosphorous promotes bacterial growth, iron ore particles having higher phosphorous contents were seen to be colonised by bacterial cells. Microbial mobilisation of phosphorous occurs in iron ore deposits. Microorganisms such as *Acidithiobacillus*, *Clavibacter* and *Aspergillus* isolated from iron ores may act as good phosphate solubilisers, since they generate inorganic and organic acids.

- *Shewanella oneidensis*, an iron-reducing bacterium produces mineral-specific proteins and exhibits a high affinity towards goethite under anaerobic conditions. *S. oneidenisis* can recognise goethite under anaerobic environments.

- Microbial ferric iron reduction by *Shewanella putrefaciens*, a facultative anaerobe would be useful in iron ore beneficiation since they attach preferentially to magnetite and ferrihydrite.

Isolating, characterising, and establishing the usefulness of native microorganisms in iron ore processing hold the key.

The advantages of a biotechnological approach to iron ore processing are many as listed below (Sarvamangala et al., 2012):

- Use indigenous organisms that are adapted to mine environments.
- Cost effective, energy efficient and environment friendly.

- Conventionally used costly toxic chemicals can be replaced by bio-degradable, biosynthesised reagents.
- Fast interfacial processes bring about rapid surface chemical changes on minerals
- Three approaches are possible:
  - Selective biodissolution
  - Selective microbially induced flotation
  - Microbially induced selective flocculation
- Diverse areas of applications such as dephosphorisation, desiliconi-sation, desulphurisation as well as silica, alumina and clay removal from iron ores.

Biobeneficiation of iron ores with respect to separation of silica, alumina, phosphorous (apatite) and clay (aluminium silicate) from hematite is illustrated below using four types of microorganisms, namely, *Saccharomyces cerevisiae* (yeast) SRB (*Desulfovibrio desulfuricans*), *Bacillus subtilis* and *Paenibacillus polymyxa*. All the tested organisms are closely associated with iron ore deposits. The developed biobeneficiation processes are specially suited to treat iron ore fines and slimes (Natarajan and Padukone, 2012).

1.5.1 Yeast-Mediated Separation of Quartz from Hematite

The utility of *S. cerevisiae* in the separation of silica from hematite has been investigated. *S. cerevisiae* is a unicellular and chemoorganotrophic eukaryote occurring associated with iron ore deposits.

Adsorption density of yeast cells was found to be significantly higher on hematite compared to quartz. The growth of yeast cells in the presence of quartz resulted in the generation of higher amounts of mineral-specific proteins, which exhibited higher surface affinity to quartz. Exopolysaccharides were increasingly generated when the yeast cells were adapted or grown in the presence of hematite, which exhibited a higher affinity towards yeast exo-polysaccharides (Natarajan and Padukone, 2011, 2012).

Yeast-mediated separation of quartz from hematite was demonstrated through microbially induced flotation and selective flocculation as illustrated in Table 1.10.

Unadapted yeast cells and their metabolites did not bring about significant flotation separation of quartz–hematite mixtures. Yeast cells grown in the presence of quartz and metabolic products separated from adapted cells promoted efficient selective flotation of quartz from hematite. Similarly, quartz-adapted yeast cells and metabolites promoted the dispersion of quartz particles while significantly flocculating hematite. Quartz can be effectively separated from hematite through selective bioflocculation, also using quartz-adapted yeast cells or metabolites.

TABLE 1.10

Quartz–Hematite Separation through Yeast-Induced Flotation

| | Percent Flotation | | | |
| | | | Mixture | |
Conditions	Hematite	Quartz	Hematite	Quartz
No biointeraction	8	12	14	15
Quartz-grown yeast cells	8	95	10	93
Hematite-grown yeast cells	10	12	9	14

Source: Adapted from Natarajan, K.A., Padukone, S.U. *Minerals and Metallurgical Processing* 29(2); 2012: 81–87.

1.5.2 Use of *B. subtilis* in Iron Ore Beneficiation

B. subtilis is a Gram-positive, neutrophilic, aerobic bacterium found associated with iron ore deposits. The use of *B. subtilis* for the separation of silica and alumina from hematite was investigated as discussed below (Sarvamangala and Natarajan, 2011):

Profuse and significant cell adhesion of *B. subtilis* on hematite were observed, unlike adhesion on quartz and corundum. When bacterial cells were grown in the presence of various iron ore minerals such as hematite, silica (quartz) and alumina (corundum), varying amounts of extracellular proteins and polysaccharides were secreted. The presence of quartz during bacterial growth promoted a higher secretion of proteins while that of hematite enhanced secretion of polysaccharides.

Bacterial interaction conferred increased surface hydrophobicity on quartz, while hematite was rendered more hydrophilic. Interaction with bacterial cells and cell-free metabolites significantly enhanced the flocculation (settling rate) of hematite at neutral pH, while quartz particles were increasingly dispersed. Dispersion of corundum was also facilitated by bacterial interaction, though not to the same extent as observed in quartz.

Flotation behavior of various minerals after interaction with *B. subtilis* is illustrated in Table 1.11.

TABLE 1.11

Flotation of Some Iron Minerals in Presence of *B. subtilis*

| | Percent Flotation | | |
	Control	With Cells	With Cell-Free Extract
Hematite	10	5	15
Corundum	12	70	30
Quartz	15	92	90

Source: Adapted form Sarvamangala, H., Natarajan, K.A. *International Journal Mineral Processing* 99; 2011: 70–77.

After interaction with *B. subtilis* cells and metabolites, hematite was found to be significantly depressed. Bacterial interaction however promoted the flotation of quartz and corundum. Efficient separation of quartz from hematite from their mixtures could be achieved after interaction with cells and metabolites of *B. subtilis*.

1.5.3 Biobeneficiation Using Anaerobic Bacteria

The use of an anaerobic SRB, namely, *Desulfovibrio desulfuricans* in the quartz–hematite separation was also demonstrated (Sabari Prakasan and Natarajan, 2010).

High amounts of mineral-specific extracellular proteins were secreted by quartz-grown *D. desulfuricans*. Secretion of extracellular polysaccharides was found to be the highest in the case of hematite-grown cells. Protein profiles of unadapted and mineral-adapted *D. desulfuricans* were then established. Quartz-adapted proteins exhibited the highest adsorption density on quartz. Hematite exhibited the highest affinity towards exopolysaccharides.

Flotation separation of hematite and quartz using *D. desulfuricans* was then established as shown in Table 1.12.

In the absence of bacterial interaction, no significant flotation of quartz and hematite was achieved. Higher surface hydrophobicity was exhibited by quartz due to adsorption of bacterially produced hydrophobic proteins. Flotation recovery of hematite decreased to 2% with hematite-grown cells due to its higher affinity towards polysaccharides, which rendered them increasingly hydrophilic. Selective separation of quartz from a binary mixture of quartz and hematite was also established. Efficient preferential flotation of quartz from hematitic could be achieved after interaction with quartz-adapted cells or metabolites of *D. desulfuricans*.

1.5.4 Biobeneficiation Using *P. polymyxa*

P. polymyxa is a Gram-positive, neutrophilic heterotroph occurring in iron ore deposits. The use of *P. polymyxa* and their metabolic products in iron

TABLE 1.12

Flotation Behaviour of Some Iron Ore Minerals before and after Interaction with *D. desulfuricans*

| | Percent Flotation | | | |
| | | | Mixture | |
	Hematite	Quartz	Hematite	Quartz
Without biotreatment	5	17	5	15
Hematite-adapted cells	2	39	5	37
Quartz-adapted cells and metabolites	9	75	9–12	76–85

Source: Adapted from Sabari Prakasan, M.R., Natarajan, K.A. *Colloids and Surfaces B: Biointerfaces* 78; 2010: 163–70.

TABLE 1.13

Flotation Behaviour of Iron Ore Minerals in the Presence and Absence
of *P. polymyxa*

	Control	Cells Alone	Protein	Polysaccharides
Quartz (silica)	4	60	98	4
Corundum (alumina)	5	2	7	1
Hematite	4	3	4	1

Source: Adapted from Deo, N., Natarajan, K.A. *Minerals Engineering* 10; 1997: 1339–354; Deo, N., Natarajan, K.A. *International Journal of Mineral Processing* 55; 1998: 41–60.

ore beneficiation was studied (Deo and Natarajan, 1997, 1998, 1999, 2001; Somasundaran et al., 2000; Natarajan, 2009).

Flotability of different iron ore minerals under different biopretreatment conditions is illustrated in Table 1.13.

Interaction with cells of *P. polymyxa* and their metabolic products affected the flotation and flocculation–dispersion of various minerals as shown in Tables 1.13 and 1.14. Flotation and dispersion of quartz were promoted while hematite and corundum were depressed or flocculated after interaction with bacterial cells and metabolic products. Bacterial proteins promoted surface hydrophobicity and flotation (dispersion) of quartz while interaction with exo-polysaccharides promoted flocculation and depression of hematite and corundum (Deo and Natarajan, 1997, 1998, 1999, 2001; Somasundaran et al., 2000; Natarajan, 2009).

Selective flocculation of 1:1 mixtures of hematite and alumina with silica (quartz) after interaction with cells of *P. polymyxa* indicated efficient separation of quartz from hematite and alumina. However, efficient separation of alumina (corundum) from hematite required prior adaptation of bacterial cells in the presence of corundum. Such preadapted cells secreted alumina-specific bioproteins (Deo and Natarajan, 1999). Corundum-adapted cells of *P. polymyxa* and their metabolic products were efficient in the separation of hematite from corundum through bioflocculation (Table 1.15).

Applicability of bioflotation and bioflocculation using *P. polymyxa* was also demonstrated using real iron ore samples. From an alumina-rich iron

TABLE 1.14

Settling of Minerals under Different Biopretreatment Conditions

Mineral	Percent Weight Settled at Neutral pH			
	Control	Cells	Proteins	Polysaccharides
Quartz (silica)	58	20	18	45
Corundum (alumina)	85	95	78	96
Hematite	90	98	85	95

Source: Adapted from Deo, N., Natarajan, K.A. *Minerals Engineering* 10; 1997: 1339–354; Deo, N., Natarajan, K.A. *International Journal of Mineral Processing* 55; 1998: 41–60.

TABLE 1.15

Selective Bioflocculation of Corundum-Hematite (1:1)
Mixture Using Corundum-Adapted Cells and
Bioproteins Separated from Adapted Cells

Deslimings	Percent Hematite Separation
1	50–55
3	67–70
5	83–90
6	95–98

Source: Adapted from Deo, N., Natarajan, K.A. *Minerals and Metallurgical Processing* 16(4); 1999: 29–34.

ore containing up to 15% aluminium oxides and silicates, a very significant reduction in alumina and silica could be brought about through bioflotation or bioflocculation using adapted cells of *P. polymyxa*. Similarly, a significant removal of silica could be obtained through bioflotation and bioflocculation from another silica-rich iron ore, containing up to 32% silica and 45% iron.

1.5.5 Bioremoval of Phosphorous and Clays from Iron Ores

The separation of phosphorus from hematite can be achieved by microbially mediated solubilisation in an environment-friendly and cost-effective manner. In the mining environment many microorganisms take part in the phosphorus cycle and solubilise mineral phosphates through the production of organic and inorganic acids. In nutrient-limited environments, organisms are capable of mobilising the phosphorus contained in minerals. Phosphorus mobilising bacteria *Burkholdaria caribensis* isolated from a Brazilian iron ore deposit could be used to liberate mineral phosphate from iron ore. The production of gluconic acid and exopolysaccharides solubilised the phosphorus contained in the ore. Selective bioleaching of phosphorus from high phosphorus iron ores can be achieved. Indigenous sulphur-oxidising bacteria from municipal wastewater could grow well in iron ore slurries containing phosphorous (Nautryal, 1999; Delvasto et al., 2009). High acidic conditions resulting from bacterial growth lead to efficient solubilisation of phosphorous. Fungal species such as *A. niger* could be used for phosphate solubilisation since the organisms produce chelating organic acids.

Selective flocculation of hematite–apatite mixtures using yeast cells (*Saccharomyces*) was attempted as a function of pH. At a pH in the range of 9–10, significant selective flocculation of hematite particles was observed, while apatite fines were dispersed (Natarajan and Padukone, 2011).

Kaolinite can be removed from hematite using selective bioflocculation using *B. subtilis*. Kaolinite-adapted cells and their metabolic products were found to be efficient in the separation of kaolinite from hematite. Kaolinite particles were dispersed while hematite was selectively flocculated (Poorni and Natarajan, 2013, 2014).

1.6 Biobeneficiation of Sulphide Minerals Using *Acidithiobacillus* Bacteria

The sulphur-bacteria cycle in mining environments constitutes iron-sulphur-sulphide oxidising acidophiles such as *A. ferrooxidans* and *A. thiooxidans* as well as SRB, which are sulphate-reducing neutrophilic heterotrophs. Biogenesis of various sulphide minerals such as pyrite, chalcopyrite, arsenopyrite, sphalerite and galena is due to participation and activity of bacteria in the sulphur cycle. Sedimentary mineral sulphides are of biogenic origin. Many sulphide minerals can be formed due to bacterial reduction of sulphates by SRB.

Biobeneficiation of sulphide-bearing ores such as those containing pyrite, chalcopyrite, arsenopyrite, sphalerite and galena can be achieved using *A. ferrooxidans, A. thiooxidans* as well as SRB. A few examples are illustrated below.

A. ferrooxidans has been shown to adhere selectively onto pyrite leading to significant changes in its surface chemical properties. Profuse attachment of *A. ferrooxidans* cells onto pyrite renders them hydrophilic, leading to significant depressions in subsequent flotations. Such a bio-modification can be beneficially used in desulphurisation of pyritic coals (Ohmura et al., 1993).

In a pyrite–chalcopyrite system as those occurring in copper ores, selective microbial depressions of pyrite would prove to be beneficial. *A. ferrooxidans* can be effectively used for selective pyrite removal from chalcopyrite through prior bacterial conditioning. For example, when a mixture of pyrite and chalcopyrite were conditioned with bacterial cells after the addition of small quantities of a xanthate collector, the selective depression of pyrite and significant chalcopyrite flotation could be obtained. Similarly, the selective flotation of arsenopyrite from pyrite could also be achieved through bacterial conditioning using *A. ferrooxidans*. Bacterial cells exhibited a significantly higher affinity towards pyrite compared to arsenopyrite. Even shorter periods of interaction with bacterial cells resulted in a significant depression of pyrite, while the flotability of arsenopyrite was promoted. Very good flotation separations between pyrite and arsenopyrite could be achieved through bacterial conditioning followed by the addition of a xanthate collector and copper activator (Chandraprabha et al., 2004a).

The settling behaviour of pyrite and arsenopyrite in the presence and absence of *A. ferrooxidans* is illustrated in Table 1.16. The settling rate (flocculation) of pyrite was promoted at even lower cell densities at acidic and neutral pHs. On the other hand, there was no significant change on the settling rate of arsenopyrite. The control of cell density is essential to achieve selective flocculation of pyrite from arsenopyrite, since higher cell densities would result in a decrease in selectivity (Chandraprabha et al., 2004b).

TABLE 1.16

Settling Behaviour of Pyrite and Arsenopyrite
in the Presence of *A. ferrooxidans*

Mineral	pH	Percent Settled	
		Cells	Control
Pyrite	2.5	96	32
	6.5	95	16
Arsenopyrite	2.5	40	37
	6.5	24	20

Source: Adapted from Chandraprabha, M.N., Natarajan, K.A., Somasundaran, P. *Journal of Colloid and Interface Science* 276; 2004b: 323–32.

Good separation of pyrite from a ternary mixture containing chalcopyrite and arsenopyrite could also be achieved through bacterial conditioning. In the presence of copper sulphate, selective pyrite depression from arsenopyrite and chalcopyrite could be achieved by suitably conditioning with bacterial cells and a xanthate collector. For example, 85% chalcopyrite and about 80% arsenopyrite could be recovered with significant pyrite depression.

The use of *A. ferrooxidans* in the flotation separation of galena from sphalerite was also demonstrated (Yelloji Rao et al., 1992). Control of cell density and interaction time is critical in achieving selective galena—sphalerite separation as illustrated in Table 1.17.

The enhancement of the flotability of sulphide minerals in the presence of *A. ferrooxidans* could be attributed to surface sulphur formation through biooxidation. Bacterial oxidation over an extended period of time leads to reoxidation of the sulphur to sulphoxy compounds and to sulphate; a gradual build-up of such oxidised sulphate layer on mineral surfaces would impede flotation. Similarly, higher surface coverage of bacterial cells could also lead to depression.

TABLE 1.17

Effect of Initial Cell Density on the Flotability of
Galena and Sphalerite

Cells/mL	Percent Flotation	
	Sphalerite	Galena
10^4	96	75
10^8	92	45
10^9	50	20

Source: Adapted from Yelloji Rao, M.K., Natarajan, K.A., Somasundaran, P. *Minerals and Metallurgical Processing* 9; 1992: 395–400.

Initial cell population as well as the duration of biotreatment are important factors influencing the flotability of sphalerite and galena. With an increase in cell population and duration of bacterial interaction, the flotability of sphalerite and galena was observed to decrease. Significantly higher numbers of cells were seen to be attached to galena unlike on sphalerite. Surface oxidation of galena lead to formation of oxidised sulphate layers that may hinder its flotation. Enhanced hydrophobicity of sphalerite is due to the presence of elemental sulphur due to bacterial oxidation.

Microbially induced flotation of sulphide minerals using *A. thiooxidans* has also been reported (Santhiya et al., 2000).

Electrophoretic mobilities of pyrite, chalcopyrite and arsenopyrite were observed to be affected after interacting with *A. thiooxidans*. For example, the isoelectric point of pyrite shifted from a pH value of about 3.2 to about 4.2, while that of chalcopyrite and arsenopyrite shifted from pH 2.4 to 3.2 after bacterial interaction. The highest surface adsorption of *A. thiooxidans* could be observed on pyrite, followed by chalcopyrite and arsenopyrite.

The differential flotation of pyrite–chalcopyrite mixtures after conditioning with xanthate followed by cells of *A. thiooxidans* resulted in 85% flotation recovery for chalcopyrite while only about 15% of pyrite could be recovered. In the presence of copper activation, chalcopyrite–arsenopyrite recoveries could be substantially increased to more than 90% with significant pyrite depression from a ternary mixture of the three sulphides.

Flotation behaviour of galena and sphalerite was also studied before and after interaction with *A. thiooxidans*. Flotation recovery of sphalerite was observed to be significantly enhanced while galena was depressed. Significant differences in surface adsorption of cells onto galena and sphalerite along with the nature of the surface reaction products are responsible for the observed flotation behaviour of galena and sphalerite. Bacterial interaction also promoted the dispersion of sphalerite particles while flocculating galena, especially at pH values near about 8–9.

$ZnSO_4$ formed on sphalerite surfaces due to bacterial interaction is soluble, unlike the case with lead sulphate on galena surfaces. Adsorption of *A. thiooxidans* was found to be significantly higher on galena along with formation of insoluble lead sulphate as a reaction product. Sphalerite surfaces exhibited lower cell adsorption and an increased presence of elemental sulphur.

1.6.1 Use of SRB in Mineral Processing

Unlike acidophilic, aerobic autotrophs, SRB such as *Desulfovibrio* spp. are anaerobic, neutrophilic heterotrophs reducing sulphate to sulphide (Solozhenkin and Lyubavina, 1985).

H_2S and other metal sulphides are generated during their growth. From a flotation view point, SRB could function as biological sulphidising agents. Many oxidised ores could be sulphidised through interaction with SRB. For example, flotation of cerusstie was improved after pre-treatment with SRB. Selective flotation of lead and zinc from sulphide ores can be achieved after conditioning with SRB. Lead recovery in concentrate was enhanced due to bacterial treatment by about 95% with almost complete depression of sphalerite after SRB interaction. Selective separation of chalcopyrite and molybdenite after biotreatment with SRB has been attempted. Besides, they can also be used as desorbents for adsorbed xanthate from floated mineral surfaces.

SRB generate large volumes of H_2S, which can convert metals and their salts to sulphides. Through sulphate reduction, SRB can produce biogenic sulphides such as silver, cadmium and other metal sulphides. Bacterial interaction can be so controlled as to generate either a sulphide product (promotion of flotation) or H_2S gas (depressant). Bioconversion of iron oxides to ferrous sulphide, malachite to covellite, silver carbonate/chloride to argentite and smithsonite to sphalerite would facilitate easy flotation recovery of the otherwise non-floatable oxide/carbonate minerals using xanthate collectors. SRB has been increasingly used in the remediation of acid mine/drainage and in biomaterials processing. Dissolved metal ions in acidic and neutral effluents can be precipitated as sulphides and recovered after treatment with SRB. Biogenic metal sulphides can be produced through interaction with SRB from different industrial minerals.

1.7 Biobeneficiation Mechanisms

Microorganisms bring about selective mineral dissolution through direct and indirect mechanisms. Direct mechanism involves microbial adhesion to the mineral surfaces leading to enzymatic attack. Bacterial cells can attach and colonise on minerals forming biofilms, which consist of various microbial consortia along with their metabolic and metal-reacted products. For example, microorganisms such as *Acidithiobacillus* sp., *Bacillus* sp., *P. polymyxa*, SRB, fungi and yeasts used in this study were observed to adhere on minerals such as kaolinite, alumina, silica, calcite, hematite apatite, pyrite, chalcopyrite, sphalerite and galena. Mechanisms of bacterial adhesion involve electrostatic, chemical and hydrophobic forces. Since mineral particles as well as bacterial cells are charged in aqueous solutions, electrostatic forces may come into play in initial bacterial adhesion. Subsequent to surface adhesion, attached organisms secrete various biopolymers that serve as adhesives facilitating irreversible attachment. Such biopolymers contain amino acids, exopolysaccharides and organic acids. Spore-forming microbial cells when exposed to mineral and metal ion containing environments form protective capsules around the cell walls. For example, cells of *P. polymyxa* on growth in the presence of calcite were found to form polysaccharide-enriched capsules (Somasundaran et al., 2000). However, such cells did not exhibit capsule formation when grown in the presence of silica. As discussed above, microorganisms when grown in the presence of different minerals secrete mineral-specific bioreagents such as proteins and exopolysaccharides. In the presence of hematite and calcite, *P. polymyxa* cells were found to secrete significant amounts of polysaccharides, while the presence of silica and kaolinite promoted the secretion of proteins. Similarly, fungi such as *A. niger* and bacteria such as *B. subtilis* and *P. polymyxa* secrete several organic acids such as citric, oxalic and gluconic acids promoting the dissolution of the mineral in the medium. Exopolysaccharides are known to form calcium-based complexes in the presence of calcite.

Indirect mechanism of mineral dissolution involves the role of microbial metabolites acting as mineral solvents. Both oxidative and reductive biodissolution can take place. A typical example of oxidative dissolution is solubilisation of sulphides such as pyrite in the presence of sulphur and iron oxidising bacteria such as *A. ferrooxidans* and *A. thiooxidans*. Microbial iron removal from pyritic coals using *Acidithiobacillus* bacteria falls into this category. Since the acidic metabolites of *A. thiooxidans* also contain reducing ions such as HS^-, thiosulphates and tetrathionates, the dissolution of minerals through reductive processes could also occur. The use of iron-reducing bacteria such as SRB, *Shewanella* and *Desulfuromonas*, on the other hand, bring about reduction of ferric to the bivalent state facilitating solubilisation of ferric oxides such as hematite in an acid medium or in the presence of chelating agents. Direct bioreduction of iron oxides exploiting the iron-reducing metabolisms of iron-reducing bacteria is a treatment option for removal of ferric oxides from kaolin clays and bauxite.

Other than selective biodissolution, interfacial changes brought about by microorganisms due to mineral adsorption can induce surface hydrophobicity or hydrophilicity leading to selective flotation or flocculation. Such microbially induced surface chemical changes occur at a faster rate (unlike biodissolution) facilitating rapid separation of different mineral constituents present in an ore matrix. All the bacterial and yeast species used in biobeneficiation brought about significant surface chemical changes on interacted minerals such as kaolinite, silica, alumina, calcite, apatite, hematite, pyrite, chalcopyrite, sphalerite and galena. Significant shifts in mineral IEP values and measured zeta potentials could be observed after microbial interaction. Microbial interactions enhanced surface hydrophobicity of silica and kaolinite while hematite and calcite were rendered more hydrophilic. Increased settling (flocculation) of hematite and calcite and dispersion of silica and kaolin clays could be observed after interacting with microbial cells and their metabolic products. Adsorption of bacterial proteins conferred enhanced surface hydrophobicity (as in the case of sphalerite, silica and kaolinite) and of exopolysaccharides rendered minerals such hematite, calcite and galena increasingly hydrophilic. During bioflocculation, mineral interface containing the biofilm is modified and inter-particle bridging through surface biopolymers takes place.

Acidophiles such as *A. ferrooxidans* and *A. thiooxidans* were observed to alter the surface chemistry of sulphide minerals such as pyrite, chalcopyrite, arsenopyrite, sphalerite and galena. Different adsorption behaviours of *Acidithiobacillus* cells led to selective flotation depression of pyrite and galena among the above sulphides.

1.8 Conclusions

Biomineralisation and biobeneficiation aspects of clays, bauxites, sulphide and iron ores are critically discussed. The role of native microorganisms in their genesis, mineralisation and mobilisation is clearly established. It is significant that native mining microorganisms isolated from the above ore deposits are capable of bringing about beneficiation-related surface chemical changes on kaolin, alumina, aluminium silicates, iron oxides, as well as sulphide minerals such as pyrite, chalcopyrite, sphalerite and galena. It becomes very clear that microbially induced beneficiation processes could become the most promising technologies for processing low-grade resources of complex multisulphides, iron ores, bauxite and industrial clay deposits. The most promising areas for potential commercialisation are iron removal from clays and bauxites, calcium and silica removal from bauxites, alumina, phosphorous and clay removal from iron ores as well as flotation of complex multisulphide ores. Biobeneficiation processes are especially suited to treat fines and slimes including processed tailings with a view to value addition and economical utilisation of waste resources. Ever-increasing demands for cost effective, energy efficient and environment-friendly process technologies for low grade and waste mineral resources will be a major incentive for adopting bio-beneficiation processes in place of existing physiochemical methods. Some attractive features of biobeneficiation processes are

a. Mineral-specific bioreagents are generated by different microorganisms when grown and adapted in the presence of minerals.
b. Harvesting of such mineral-specific biopolymers containing proteins and polysaccharides will pave the way for development of bioflocculants, biocollectors and depressants.

All the research studies on biobeneficiation have so far been confined to bench scale only and it becomes highly essential to carry out scaled-up and pilot-scale studies in close collaboration with concerned mining industries to establish techno-economic process viabilities of commercially relevant bioprocesses.

Acknowledgements

The author is thankful to the Department of Science and Technology, Government of India, New Delhi, for providing financial support to this research work. Thanks are also due to the National Academy of Sciences (India) for Platinum Jubilee Senior Scientist Fellowship.

References

Ajayi, O.A., Adefila, S.S. Comparative study of chemical and biological methods if beneficiation of Kankara kaolin. *International Journal of Science and Technology Research* 1; 2012: 13–18.

Arakaki, A.H., Nakazawa, M., Nemoto, T.M., Matsunaga, T. Formation of magnetite by bacteria and its application. *Journal of Royal Society Interface* 5; 2008: 977–99.

Bonneville, S., Cappellen, P.V., Behrends, T. Microbial reduction of iron (III) oxy-hydroxides; Effect of mineral solubility and availability. *Chemical Geology* 212; 2004: 255–68.

Chandraprabha, M.N., Natarajan, K.A. Microbially-induced mineral beneficiation. *Mineral Processing and Extractive Metallurgy Review* 31; 2010: 1–29.

Chandraprabha, M.N., Natarajan, K.A., Modak, J.M. Selective separation of pyrite and chalcopyrite by biomodulation. *Colloids and Surfaces B: Biointerfaces* 37(3–4); 2004a: 93–100.

Chandraprabha, M.N., Natarajan, K.A., Somasundaran, P. Selective separation of arsenopyrite from pyrite by biomodulation in the presence of *At. ferrooxidans*. *Journal of Colloid and Interface Science* 276; 2004b: 323–32.

Delvasto, P., Ballester, A., Munoz, J.A., Gonzalez, F., Blazquez, M.L., Igual, J.M., Valverde, A., Garcia-Balbua, C., Mobilization of phosphorous from iron ore by the bacterium Burkholderia Caribensis FeGLO3. *Minerals Engineering* 22; 2009: 1–9.

Deo, N., Natarajan, K.A. Interaction of *Bacillus polymyxa* with some oxide minerals with reference to mineral beneficiation and environmental control. *Minerals Engineering* 10; 1997: 1339–354.

Deo, N., Natarajan, K.A. Studies on interaction of *Paenibacillus polymyxa* with iron ore minerals in relation to beneficiation. *International Journal of Mineral Processing* 55; 1998: 41–60.

Deo, N., Natarajan, K.A. Role of corundum-adapted strains of *Bacillus polymyxa* in the separation of hematite and alumina. *Minerals and Metallurgical Processing* 16(4); 1999: 29–34.

Deo, N., Natarajan, K.A. Role of bacterial interaction and bioreagents in iron ore flotation. *International Journal of Mineral Processing* 62; 2001: 143–57.

Deo, N. Studies on biobeneficiation and bioremediation using *Bacillus polymyxa* with reference to iron ore and bauxite processing. PhD thesis, 1998, Indian Institute of Science, Bangalore, India.

Fortin, D., Langley, S. Formation and occurrence of biogenic iron-rich minerals. *Earth Science Reviews* 72; 2005: 1–19.

Gao, M., Lin, Y., Xu, X., Chan, Z. Bioleaching of iron from kaolin using Fe (III)—Reducing bacteria with various carbon, nitrogen sources. *Applied Clay Science* 48; 2010: 379–83.

Gilbert, P.U.P.A., Abrecht, M., Frazer, B.H. The organic-mineral interface in Biominerals. *Reviews in Mineralogy and Geochemistry* 59; 2005: 157–85.

Groudev, A.N. Biobeneficiation of mineral raw materials, in: S.K. Kawatra and K.A. Natarajan (editors) *Mineral Biotechnology* Littleton: SME, 2001: 37–54.

Groudev, S.N. Use of heterotrophic microorganisms in mineral biotechnology *Acta Biotechnologica* 7; 1987: 299–306.

Groudev, V.I., Groudeva, S.N. Bauxite dressing by means of *Bacillus circulans. Travaux ICSOBA* 13(18); 1983: 257–63.

Jones, B.E.H., Haynes, R.J., Philips, I.R. Addition of an organic amendment and/or residue mud to bauxite residue sand in order to improve its properties as a growth medium. *Journal of Environmental Management* 95; 2012: 29–38.

Karavaiko, G.I., Avakyan, Z.Z., Ogurtsova, L.V., Safonova, O.F. Microbiological processing of bauxite. In J. Salley, R.G.L. McGready and L. Wichlacz (Editors), *Biohydrometallurgy*, Ottawa: CANMET, 1989: 99–103.

Kostka, J.E., Dalton, D.D., Skelton, H., Dollhopf, S., Stucki, J.W. Growth of iron (III)-reducing bacteria on clay minerals as the sole electron acceptor and comparison of growth yields on a variety of oxidized iron forms. *Applied and Environmental Microbiology* 68; 2002: 6256–262.

Laskou, M., Eliopoulos, M.E. The role of microorganisms on the mineralogical and geochemical characteristics of the Parnassos—Ghiona bauxite deposits, Greece. *Journal of Geochemical Exploration* 93; 2007: 67–77.

Laskou, M., Eliopoulos, M.E. Biomineralization and potential biogeochemical processes in bauxite deposits: Genetic and ore quality significance. *Miner Petrol* 107 2013: 471–86.

Lee, E.Y., Cho, K.S., Ryu, H.W. Microbial refinement of kaolin by iron-reducing bacteria. *Applied Clay Science*, 22(1–2); 2002: 47–53.

Liu, Y., Gao, M., Dai, S., Peng, K., Jia, R. Characterization of magnetotactic bacteria and their magnetosomes isolated from Teishan iron ores, in Huber province of China. *Material Science and Engineering* 26; 2006: 597–601.

Mandal, S.K., Banerjee, P.C. Iron leaching from China clay with oxalic acid—Effect of different physicochemical parameters. *International Journal Mineral Process* 74; 2004: 263–70.

Modak, J.M., Vasan, S.S., Natarajan, K.A. Calcium removal from bauxite using *Paenibacillus polymyxa. Minerals and Metallurgical Processing* 16; 1999: 6–12.

Natarajan, K.A. Microbial aspects of environmentally benign iron ore beneficiation. In *Proceedings of Iron Ore Conference* 2009: July 27–29, AUSIMM, Perth, WA, 27.

Natarajan, K.A. Developments in biotechnology for environmentally benign iron ore beneficiation, *Transaction of Indian Institute of Metals* 66(5–6); 2013: 457–65.

Natarajan, K.A., Modak, J.M., Anand, P. Some microbiological aspects of bauxite mineralization and beneficiation. *Minerals and Metallurgical Processing* 14(2); 1997: 47–53.

Natarajan, K.A., Padukone, S.U. Microbially induced separation of quartz from calcite using *Saccharomyces cerevisiae. Colloids and Surfaces B: Biointerfaces* 88; 2011: 45–50.

Natarajan, K.A., Padukone, S.U. Microbially-induced separation of quartz from hematite using yeast cells and metabolites. *Minerals and Metallurgical Processing* 29(2); 2012: 81–87.

Natarajan, K.A. NPTEL Course, *Metals Biotechnology*, http://nptel.ac.in/courses.php

Nautryal, C. An efficient microbiological growth medium for screening phosphate solubilizing microorganisms. *FEMS Microbiology Letter* 170; 1999: 265–70.

Ohmura, N., Kitamura, K., Saiki, H. Mechanism of microbial flotation using *At. ferrooxidans* for pyrite suppression. *Biotechnology and Bioengineering* 14; 1993: 671–76.

Papassiopi, N., Vazevanidou, K., Paspaliaris, I. Effectiveness of iron reducing bacteria for the removal of iron from bauxite ores. *Mineral Engineering* 23; 2010: 25–31.

Phalguni, A., Modak, J.M., Natarajan, K.A. Biobeneficiation of bauxite using *Bacillus polymyxa*: Calcium and iron removal. *International Journal of Mineral Processing* 48; 1996: 51–60.

Poorni, S., Natarajan, K.A. Microbially induced selective flocculation of hematite from kaolinite. *International Journal of Mineral Processing* 125; 2013: 92–100.

Poorni, S., Natarajan, K.A. Flocculation behavior of hematite-kaolinite suspension in presence of extracellular bacterial proteins and polysaccharides. *Colloids and Surfaces B: Biointerfaces* 114; 2014: 186–92.

Roberts, J.A., Fowle, D.A., Hughes B.T., Kulczycki, E. Attachment behavior of *Shewanella putrefaciens* onto magnetite under aerobic and anaerobic conditions. *Geomicrobiology Journal* 23; 2006: 631–40.

Sabari Prakasan, M.R., Natarajan, K.A. Microbially induced separation of quartz from hematite using sulfate reducing bacteria. *Colloids and Surfaces B: Biointerfaces*, 78; 2010: 163–70.

Santhiya, D., Subramanian, S., Natarajan, K.A. Surface chemical studies on galena and sphalerite in the presence of *At. thiooxidans* with reference to mineral beneficiation. *Minerals Engineering* 13; 2000: 747–63.

Sarvamangala, H., Natarajan, K.A. Microbially induced flotation of alumina, silica/calcite from haematite. *International Journal Mineral Processing* 99; 2011: 70–77.

Sarvamangala, H., Natarajan K.A., Girisha, S.T. Biobeneficiation of iron ores. *International Journal of Mineral Processing* 1(2); 2012: 21–30.

Shelobolina, E.S., Pickering, S.M., Lovely, D.R. Fe-cycle bacteria from industrial clays mined in Georgia, USA. *Clay and Clay Minerals* 53; 2005: 580–6.

Solozhenkin, P.M., Lyubavina, L.L. Sulfate reducing bacteria and some microscopic fungi in ore preparation and hydrometallurgy, Chapter 6. In: *Modern Aspects of Microbiological Hydrometallurgy*. G.I. Karavaiko and S.N. Groudev (Editors), Moscow: Center for International Projects. 1985: 409–414.

Somasundaran, P., Deo, N., Natarajan, K.A. Utility of bioreagents in mineral processing. *Minerals and Metallurgical Processing* 17; 2000: 112–115.

Styriakova, I., Styriak, I. Iron removal from kaolins by bacterial leaching. *Ceramics Silikaty* 44; 2000: 135–41.

Vasan, S.S., Modak, J.M., Natarajan, K.A. Some recent advances in the bioprocessing of bauxite. *International Journal of Mineral Processing* 62; 2011: 173–86.

Williams, P.J., Cloete, T.E. Microbial community study of the iron ore concentrates of the Sishen iron mine, South Africa. *World Journal of Microbiology and Biotechnology* 24(11); 2008: 2531–538.

Yelloji Rao, M.K., Natarajan, K.A., Somasundaran, P. Effect of biotreatment with *At. ferrooxidans* on the flotability of sphalerite and galena. *Mineral and Metallurgical Processing* 9; 1992: 395–400.

Zhou, Y., Wang, R., Lu, X., Chen, T. Roles of adhered *P. polymyxa* in the dissolution and flotation of bauxite: A dialytic investigation. *Frontiers of Earth Science in China* 4(2); 2010: 167–73.

2

Biomining of Base Metals from Sulphide Minerals

Bhargav C. Patel, Devayani R. Tipre and Shailesh R. Dave

CONTENTS

Biomining has a long history, although the early miners did not know that microbes were involved in the mining process. The use of microbes in ore processing has some distinct advantages over the traditional pyrometallurgical and hydrometallurgical processing.

The world-wide high-grade ore reserves are falling every day at an appall-
ing rate as most of them are worked out because of high metal demand,
traditional techniques such as pyrometallurgy and chemical processing
are becoming more and more economically incompatible. Microorganisms
bear a clear advantage over it, as not only they offer an economically viable option but also a clean technology (Siddiqui et al., 2009). They do not
require the high amounts of energy used during roasting or smelting and
do not produce sulphur dioxide or other environmentally harmful gaseous
emissions (Rawlings, 2002). To some extent it holds the promise of reducing
the capital costs (Devasia and Natarajan, 2004). Additionally, microbiological solubilisation processes are also applicable to recover metal values from
industrial and mineral wastes, which can serve as secondary raw materials
(Siddiqui et al., 2009).

2.1 Biomining Processes for Base Metals

There are two main types of biomining processes for the commercial-scale
microbially assisted metal base metal recovery. These are irrigation-type
and stirred-tank-type processes (Figure 2.1). Irrigation processes involve
the percolation of leaching solutions through the crushed ore that have
been stacked in columns, heaps or dumps. There are also several examples
of the irrigation of an ore body *in situ*, that is, without bringing the ore to
the surface. Stirred-tank-type processes employ continuously operating,
highly aerated stirred-tank reactors. One feature of both types of processes
is that, unlike most other commercial fermentation processes, neither is
sterile, nor any attempt is made to maintain the sterility of the inoculum
(Rawlings, 2002).

2.1.1 Bioleach Heap Technology: The Most Commercialised Irrigation-Based Biomining Expertise

Irrigation-based processes can be categorised based on the type of resources
to be processed such as dump leaching, heap leaching and *in situ* leaching.
Heap leaching deals with the newly mined materials (intermediate-grade
oxides and secondary sulphides deposited in the form of a heap on an imper-
vious natural surface or a synthetically prepared pad leached with circulation, percolation and irrigation of the leaching medium) (Pradhan et al.,
2008). Primary sulphides such as chalcopyrite are also suitable for this type
of leaching. Bioleach heap technology has emerged as the predominant technology route for the recovery of metals from low-grade ores. In terms of revenue generated, it is the most significant industrial application of biomining
(Rawlings, 2002). The technology has been in use since the 1960s for the acid

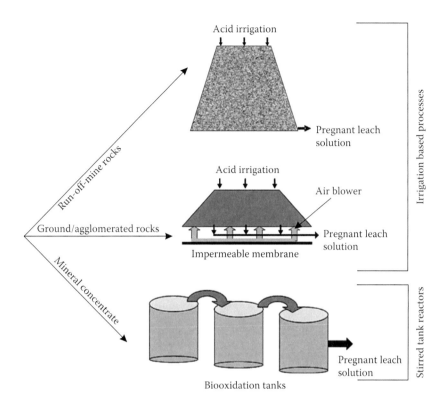

Acid irrigation

Pregnant leach solution

Irrigation based processes

Acid irrigation

Air blower

Pregnant leach solution

Impermeable membrane

Run-off-mine rocks

Ground/agglomerated rocks

Mineral concentrate

Stirred tank reactors

Pregnant leach solution

Biooxidation tanks

FIGURE 2.1
Major process options used in biomining. (Modified from Johnson, D. B. 2010. In: *Geomicrobiology: Molecular and Environmental Perspective*, 401–426. Netherlands: Springer.)

leaching of copper oxide minerals, and since the 1970s for the cyanide leaching of gold and silver.

The static bioleaching techniques are based on the principle of circulating acidic water and air through heaps of ore coarsely fragmented to activate the growth of microorganisms that amplify the oxidation of the sulphide minerals (Morin et al., 2006). This process involves stacking crushed ore into piles constructed on an impermeable layer fitted with a solution drainage system, or arranged on a slope to facilitate drainage. In many cases, the ore is agglomerated through tumbling with acid and/or irrigation solution prior to stacking.

2.1.2 Fundamental Principles of Bioleach Heap Technology

Although heap leaching appears to be a very simple process in concept, the sub-processes taking place within the ore bed are rather complex and their interactions not yet fully understood.

To unravel the processes underlying heap bioleaching, it is useful to distinguish between the phenomena taking place at different scales within the heap (Petersen and Dixon, 2007a). Beginning at the heap scale, we can distinguish a number of transport effects. More specifically, these include:

1. *Solution flow:* In unsaturated, coarsely granular packed beds, the solution generally flows along tortuous pathways but remains stagnant in pores and crevices between particles. This strongly influences heap performance in terms of reagent delivery and product removal from the reaction sites within the ore particles.

2. *Heat flow:* Heat of reaction, which is significant in sulphide leaching, is transported downwards through the heap as sensible heat with the flowing solution and upwards as latent heat with the flow of humid air. Depending on air and solution flow rates, heaps can assume certain temperature profiles, and a judicious manipulation of these variables may even allow a certain degree of control.

3. *Gas flow:* Although gas flow is usually well distributed in aerated heaps, ensuring ample supply of oxygen throughout, the supply of CO_2 may be limited under certain conditions. In non-aerated heaps, O_2 availability may also be limited, and gas distribution patterns are complex.

The next level, at the meso-scale, represents a cluster of particles within a heap bed. Here, two processes contribute to the overall rate of leaching:

1. *Diffusion transport:* Diffusion is the main mode of transport for dissolved constituents from and to the moving solution into pore spaces between particles, and into cracks and fissures within particles. The effect of pore diffusion on the overall kinetics is determined by the length of the diffusion path, which can be significant for systems with a poor solution distribution (Petersen and Dixon, 2007b).

2. *Microbial population dynamics:* This encompasses the complex interactions of a variety of microorganisms in the liquid phase and on the mineral surface. It includes the growth behaviour of each strain as a function of temperature and concentration of dissolved constituents (acid, Fe^{2+} and Fe^{3+} iron, O_2, CO_2 as the carbon source, etc.), and any synergies between these and the concomitant iron and sulphur oxidation reactions.

Finally, the smallest scale at which sub-processes of heap bioleaching need to be analysed is that of the individual mineral grain. Here, leaching is governed by the electrochemical interactions between the mineral grains and reagents in solution.

2.1.2.1 Microbiology of Bioleach Heap

A wide variety of microorganisms are found in natural leaching environments such as bioheap. The majority of known acidophilic microorganisms have been isolated from such natural environments. Understanding the microbiology of a bioheap is important for the advancement in commercial bioheap applications. It will also enable better control of conditions to improve upon the leaching rates, metal recoveries and cost of production. Limited comprehension is available on what actually occurs in a full-scale microbiologically operated bioheap, despite the commercial achievement in the copper ore bioheap leaching (Pradhan et al., 2008). Although oxidative dissolution of simple and complex sulphide ores and concentrates may be mediated by pure cultures of iron-oxidising acidophiles, as has often been described in laboratory studies, axenic cultures are never found in actual biomining operations. A consortia of microorganisms with synergistic (and sometimes complimentary) metabolic physiologies have been identified in all commercial-scale systems that have been examined (Johnson, 2010).

Microbial diversity is far greater in heaps and dumps, which are highly heterogeneous and uncontrolled environments, than in stirred tanks where conditions are far more homogeneous. Both operate essentially as 'inorganic' systems in that, inorganic nutrients are added to stimulate microbial activity, not organic carbon. This, together with the primary energy sources available being the sulphide minerals themselves, means that the dominant prokaryotes present are invariably chemo-autotrophic iron and sulphur oxidisers. However, organic carbon derived from living (as exudates) and dead (as lysates) primary producers can accumulate in leach liquors, and can support the growth of mixotrophic and heterotrophic acidophiles. Hence, it is possible, as noted by Johnson and Hallberg (2009), to divide microorganisms in biomining operations into three groups: (i) 'primary acidophiles', iron-oxidising prokaryotes that generate ferric iron and are responsible for initiating mineral dissolution; (ii) 'secondary acidophiles', sulphur-oxidising acidophiles that generate sulphuric acid from reduced sulphur produced during mineral dissolution and help to maintain the required pH conditions that are conducive for the biooxidation of sulphide minerals and (iii) 'tertiary acidophiles', heterotrophic and/or mixotrophic microorganisms that degrade soluble organic carbon wastes originating from the autotrophs, thereby detoxifying the environment of some highly sensitive primary and secondary prokaryotes into organic matter.

Okibe and Johnson (2004) evidently demonstrated the importance of microbial consortia and their interactions for optimising sulphide mineral dissolution in laboratory studies. As recently reviewed by Kondrat'eva et al. (2012), the microbial analysis of commercial-scale bioprocessing operations has shown, in all cases reported so far, that bacteria and/or Archaea that fulfill these primary, secondary and tertiary roles are present. Recently, Soto et al. (2013) have extensively studied and analysed the parameters influencing the microbial oxidation activity in the industrial bioleaching heap at the Escondida mine in Chile. In their study, the industrial run of mine (ROM) bioleaching heap of the Escondida mine is monitored monthly from pregnant leach solution (PLS) to assess the concentration of microorganisms, microbial activity and physicochemical parameters; this study has generated a huge amount of information. 'Knowledge Discovery in Databases' (KDD) was used to obtain a better description of the iron microbial activity and the dissolution rate of sulphide ores occurring in the leaching cycle. Thus, such data-mining tools have been useful to propose and confirm hypotheses on data obtained from industrial bioleaching systems. Industrial heap bioleach operations process very low-grade ores, often <1% of the mineral of interest. Understanding the dynamics of microbial interactions on low-grade ores, containing substantial proportions of gangue minerals, becomes essential for the effective exploitation of more recalcitrant and complex ore bodies and informs industrial operations. In this study, they have successfully studied the interaction of thermophilic microorganisms with metal sulphides in a simulated heap environment. Such an approach was also investigated by Tupikina et al. (2013) to determine the effect of acid stress on the persistence and growth of thermophilic microbial species after the mesophilic colonisation of low-grade ore in a heap-leach environment. Microbial analysis of selected PLS and ore samples provided useful information on the effect of pH on microbial colonisation. Their study indicated the low temperature during the initial phase of operation that negatively affected the persistence of the thermophilic microorganisms in the ore bed. Furthermore, their subsequent growth, on reaching thermophilic conditions, was negatively affected as a function of decreasing pH.

2.1.2.2 Current Applications and New Developments of Bioleach Heap Technology

A substantial number of heap-leaching metal-recovery processes have been in operation for many years (Rawlings et al., 2003). 'Thin layer' leaching, where crushed and acid-cured ore is stacked 2–3 m high and then rinsed, was first applied at the Lo Aguirre Copper Mine in Chile in 1980, and is regarded as the first instance of heap bioleaching. A further significant milestone in heap bioleaching was the introduction of forced aeration for the heap bioleaching of secondary copper sulphide ores at the Girilambone Copper Mine in Australia in 1993 (Gerick et al., 2009).

As estimated by Brierley (2008a), heap bioleaching of copper accounts for some 7% (about 10^6 t/year) of the total global annual production of approximately 1.7×10^7 t of copper. This does not include copper recovered using dump bioleaching processes. It is estimated that if dump bioleaching is included, some 20%–25% of the world's copper production is attributable to bioleaching. Examples of very large copper-leaching operations are those by Sociedad Contractual Minera El Abra and the Codelco Division Radimiro

Tomic in Chile producing 225,000 and 180,000 tons of Cu per annum, respectively (Pradhan et al., 2008). An excellent example of a current commercial bioleach application is the Quebrada Blanca operation in northern Chile (Brierley and Brierley, 2001) located on the Altiplano at an elevation of 4400 m under the cold temperatures and low oxygen partial pressure of high altitudes. The list of industrial heap bioleaching operations for secondary copper ores and mixed oxide/sulphide ores is given in Table 2.1 (Gerick et al., 2009). GeoBiotics, LLC has developed and patented several technologies for biooxidising or bioleaching of sulphide ores and concentrates in an engineered heap environment. The two principal technologies are the GEOCOAT™ and GEOLEACH™ processes. The process entails coating refractory sulphide gold concentrates onto a screened support rock or ore. Biooxidation pre-treatment takes place in a stacked heap configuration. The oxidised concentrate is removed from the support rock for gold extraction by conventional metallurgical processes. If the support rock is also a refractory ore, this can also be leached following biooxidation to recover additional metal values. The process can be applied to the biooxidation of refractory sulphide gold concentrates and to the bioleaching of copper, nickel, cobalt, zinc, and polymetallic base metal concentrates (Pradhan et al., 2008). The GEOCOAT process has also been tested for bioleaching copper from chalcopyrite concentrate using thermophilic microorganisms. The GEOLEACH technology is designed to maximise heat conservation through careful control of aeration and irrigation rates. Both the processes are simple, robust and ideally suited to operate in remote locations.

Many heap-leach processes have targeted the extraction of marginal ores that are not suitable for the production of concentrates or smelting. The development of heap-leaching technology has been largely engineering focussed rather than microbiology focussed. The main advances have therefore come from improved acidification methods, solution management and heat containment. Much of the progress in heap leaching can be attributed to research into the modelling of leach liquor distribution, oxygen diffusion and heat management (Bouffard and Dixon, 2001; Petersen and Dixon, 2002). The BIO SHALE Project has been running since 2004 in Finland for the extraction of nickel from black shale (Talvivaara deposits) using the bio-heap-leaching process (Watling, 2008). Recently in 2011, a high-temperature bioheap project was called off on which Nicico and Mintek were collaboratively working in Iran for the treatment of Sarcheshmeh copper ore (www.mintek.co.za, last accessed on 23 May, 2012). It was operating at the Sarcheshmeh copper complex in southern Iran producing some 170,000 tons of copper a year.

The HydroZinc™ process of Teck Cominco and the BioHeap™ process developed by Pacific Ore Technology for the heap bioleaching of zinc and nickel–copper ores, respectively, are reported to be in an advanced stage of development. Large-scale trials have demonstrated that high recoveries of nickel, copper and cobalt can be achieved using the BioHeap proprietary

TABLE 2.1

Industrial Heap Bioleaching Operations for Secondary Copper Ores and Mixed Oxide/Sulphide Ores

Industrial Heap Bioleach Plant and Location/Owner	Cathode Copper Production (t/a)	Years of Operation
Lo Aguirre, Chile/Sociedad Minera Pudahuel Ltda.	15,000	1980–1996 (deposit depletion)
Mount Gordon (formerly Gunpowder), Australia/ Western Metals Ltd.	33,000	1991–present
Mt. Leyshon, Australia/(formerly Normandy Poseidon)	750	1992–1995 (stockpile depleted)
Cerro Colorado, Chile/BHP Billiton	1,15,000	1993–present
Girilambone, Australia/Straits Resources Ltd. & Nord Pacific Ltd.	14,000	1993–2003 (ore depleted)
Ivan-Zar, Chile/Compañía Minera Milpro	10,000–12,000	1994–present
Punta del Cobre, Chile/Sociedad Punta del Cobre, S.A.	7000–8000	1994–present
Quebrada Blanca, Chile/Teck Cominco Ltd.	75,000	1994–present
Andacollo Cobre, Chile/Aur Resources, del Pacifico & ENAMI	21,000	1996–present
Dos Amigos, Chile/CEMIN	10,000	1996–present
Skouriotissa Copper Mine (Phoenix pit), Cyprus/ Hellenic Copper Mines	8000	1996–present
Zaldivar, Chile/Barrick Gold Corp.	1,50,000	1998–present
Lomas Bayas, Chile/XSTRATA plc	60,000	1998–present
Cerro Verde, Peru/Freeport McMoran & Buenaventura	54,200	1997–present
Lince II, Chile	27,000	1991–present
Monywa, Myanmar/Ivanhoe Mines Ltd, Myanmar No.1 Mining Enterprise	40,000	1998–present
Nifty Copper, Australia/Straits Resources Ltd.	16,000	1998–present
Equatorial Tonopah, Nevada/Equatorial Tonopah, Inc.	25,000 (projected)	2000–2001 failed
Morenci, Arizona/FreeportMcMoran	3,80,000	2001–present
Lisbon Valley, Utah/Constellation Copper Corporation	27,000 (projected)	2006–present
Jinchuan Copper, China/Zijin Mining Group Ltd.	10,000	2006–present
Spence, Chile/BHP Billiton	2,00,000	Commissioned in 2007
Whim Creek and Mons Cupri, Australia/Straits Resources	17,000	2006–present
Escondida, Chile	2,00,000	–
Toquepala, Peru	40,000	–
S&K Copper, Monywa, Myanmar	40,000	1999–present
Phoenix deposit, Cyprus	8000	1996–present

Source: Adapted from Brierley, J. A. 2008b. *Hydrometallurgy* 94 (1–4): 2–7; Gerick, M., J. W. Neale and P. J. van Staden. 2009. *J. S. Afr. I. Min. Metall.* 109: 567–585; Watling, H. R. 2006. *Hydrometallurgy* 84 (1–2): 81–108.

bacteria and patented processes. The process is also applicable to other sulphide ores such as zinc, polymetallic and refractory gold ores.

   In addition, a substantial amount of unpublished operational research has been carried out by companies such as Newmont Mining, Phelps Dodge, BHP Billiton, Mintek, POT-Titan and Rio Tinto. A novel sequential heap-leaching process has been identified as a possible alternative to the conventional concentrate–smelt–refine route for processing Platreef ore, a platinum group metals-containing ore with palladium predominance (Mwase et al., 2014).

2.1.2.3 Major Advantages and Limitations of Bioleach Heap Operations

Bioleach heap technology of metal sulphides is a stand-alone technology that is

- Robust and proven under different climatic conditions for oxides and secondary sulphides;
- Flexible—Heap engineering and management can accommodate site peculiarities in remote localities; suited to small deposits;
- Simple—A technology that can be communicated to non-scientific personnel;
- Low cost—Stacking, irrigation, aeration and solution collection are all basic infrastructure.

   However, this technology to date is still suffering from low metal extraction rates and low ultimate metal recoveries. The drawbacks listed below may outweigh the lower capital and operating costs of heap processes (Brierley, 2010):

- Good permeability, porosity, adequate wettability and reagent accessibility are problems. In addition, the presence of 'dead zones', may be due to the gangue mineral reactivity that further decreases the metal recovery (Palencia et al., 2002).
- Heap reactors are more difficult to aerate efficiently. Moreover, the rates of oxygen and carbon dioxide transfer that can be obtained are low, and extended periods of operation are required in order to achieve sufficient conversions (Pradhan et al., 2008).
- In heap leaching, it is difficult to maintain the solution pH within the range of 1.8–2.2. When the pH rises above 2.8, ferric iron precipitation occurs, this coats the mineral surfaces and reduces the rate of metal solubilisation.
- The effective provision of nutrients is more complicated and difficult to achieve. The addition of ammonium can result in the formation of jarosite precipitate, which removes ferric iron from the solution and

also coats mineral surfaces. Once the jarosite is formed, it precipitates on the mineral surfaces and decreases metal oxidation mediated by acidophilic microorganisms (Pradhan et al., 2008).

- Pyrrhotite, present in complex ores, releases substantial amounts of heat rather quickly and consumes acid-creating operating conditions that must be carefully managed in order to effectively utilise microbial leaching.
- Heaps are usually irrigated with raffinate and a gradual build up of inhibitory ions, such as sulphates and aluminium creates problems for microbes.
- Heap reactors are also more difficult to inoculate than are tank reactors. Different microbes exhibit different mineral adsorption isotherms, and this might cause uneven initial microbial species distribution within a heap.
- When a heap is constructed, fine material is agglomerated to the coarse particles using acid and a microbial inoculum can be added at this stage. A disadvantage of this option is that if the levels of acid used during agglomeration are too high, cell viability can be reduced.
- Even in carefully designed heap reactors, larger particle sizes, less-effective aeration and reduced process control make the process less efficient. Owing to this, the biomineralisation process is extended to months rather than days.

2.1.3 Stirred-Tank Reactors: The Most Controlled Biomining Operation

Bioleaching in stirred tanks for the selective recovery of precious and base metals from sulphide minerals is a fascinating front of biomining for at least two reasons (Morin, 2007). The first reason is the high potential of industrial-scale applications of this technology that has acted as a driving force for countless investigations required to make the economics of this process really attractive. The second reason is the growth selectivity and the steadiness of the microbial ecosystems in the bioreactors as compared to the diversity of the natural environment that has given the opportunity to study the interactions between microorganisms and minerals in privileged conditions.

Bioleaching in agitated tanks at an industrial scale is only relevant to the treatment of sulphide concentrates and is using the catalytical enhancement exerted by some microorganisms to oxidise sulphides and release the valuable metals they contain (Rawlings, 2002). The sulphide compounds to (bio) oxidise in the existing industrial plants are essentially pyrite and arsenopyrite in various proportions. Many successful investigations and demonstration operations have been run on other metal-bearing sulphides, such as sphalerite (ZnS), pentlandite ($[FeNi]_9S_8$]), covellite (CuS), chalcocite (Cu_2S) and chalcopyrite ($CuFeS_2$). Many reports are also published for bioleaching of uranium in stirred-tank reactors or column reactors (Abhilash et al., 2011;

Eisapour et al., 2013); along with stirred-tank operations, the rotating-drum reactors are also employed for leaching or catalyst production. A number of support materials such as polyurethane foam, activated carbon, chitosan beads and high-molecular-weight plastics were exploited for the growth of the biomining microorganisms (Patel et al., 2012b; Dave, 2008).

2.1.3.1 Fundamental Criteria of the Continuous-Flow Stirred-Tank Reactors for Biomining

From the chemical engineering point of view, bioleaching in agitated tanks is a continuous-flow steady-state process. The circuit of bioreactors configured in series or in parallel or in combination of the two is fed with the slurry of the sulphide concentrate, nutrients and air to bring the oxygen required to the sulphide oxidation. Microorganisms are injected once at the beginning of the start-up of an operation; a batch culture is maintained until a certain point as close as possible to the middle of the exponential phase of the bacterial growth when the feed in the fresh substrate can begin. A continuous flow of the substrate and nutrients through the tanks is then ensured to keep the optimum growth of microorganisms required for the fastest degradation of the sulphide minerals. The situation of equilibrium between the substrate feed rate and stable microbial growth is called chemostat (Morin, 2007). Mineral biooxidation is an exothermic process, and the bioreactors have to be cooled to remove excess heat (Rawlings, 2002). A discussion of very important criteria is as follows.

2.1.3.1.1 Microbiology of Stirred-Tank Operations

In general, various biomining microorganisms are found in stirred-tank reactors. However, the proportions of the microbes may vary depending on the mineral and the conditions under which the reactors are operated.

The environment in a stirred-tank mineral-oxidising bioreactor is highly homogeneous as it is operated at a set pH and temperature and controlled aeration. Some operations use single tanks, while others use a series of in-line tanks. Conditions, such as concentrations of soluble metals and metalloids, and often also pH, vary from tank to tank in a continuous-flow system as mineral oxidation becomes increasingly extensive, and this can have a significant impact on diversity and the number of indigenous microbial species (Okibe et al., 2003). The homogeneity within an individual tank in terms of pH, temperature, aeration and dissolved solids results in a limited ecological niche that often tends to be dominated by 2–4 species of acidophiles, although a smaller number of other microorganisms may be present.

There are only a few commercial processes that operate in the 45–50°C temperature range and therefore, studies on microorganisms that dominate these bioleaching consortia have been less well reported. There are even fewer reports on the types of microbes that occur in mineral treatment processes that operate at temperatures >70°C as opposed to processes at lower temperatures.

However, it is clear that these biomining consortia are dominated by Archaea rather than bacteria, with species of *Sulfolobus* and *Metallosphaera* being the most prominent. It is noteworthy that cultures with mixed populations result in higher pyrite-leaching efficiencies than cultures with individual pure bacterial components (Morin, 2007; Yang et al., 2014; Yu et al., 2014).

2.1.3.1.2 Carbon Supply

The microbial consortia currently used at an industrial scale are considered to be autotrophic, even if a number of species of heterotrophs are observed. They do not use other carbon sources other than carbon dioxide. The specific CO_2 supply is the subject of discussion as an operating parameter in the industrial units of bioleaching, though there is enough carbon dioxide in the air to feed the continuous growth of bacteria in tank bioleach processes through air sparging. At high bacterial growth rate, carbon uptake rate reaches levels such as CO_2 supply by the air injected is no more sufficient and the carbon transfer rate becomes a limiting step of the sulphide oxidation rate. However, the possibility that bacterial growth could be amplified or reduced by an excess of carbon dioxide may vary from one microorganism to another.

2.1.3.1.3 pH

The acidity level of the bioleaching medium results from the balance of protons between net-consuming reactions and the reactions of sulphuric acid production and iron hydrolysis. The optimal pH range is variable from one system to the other, and one microorganism to the other (Olson et al., 2003; Rawlings, 2002). In the case of refractory gold concentrates, it is observed that low pH reduces the risk of gold being encapsulated and less soluble during cyanidation treatment (Spencer, 2001). From the point of view of thermodynamics, the higher the pH, the faster the acid-producing reactions. On the other hand, a low pH value, close to 1.0, of course is harmful to the microorganism's metabolism and can be very selective for acid-tolerant species making the biological system fragile. For pH balance during the operation, a neutralising agent such as limestone may be added, but it enhances the formation of jarosite. A pH range of 1.4–1.7 is probably a good compromise between the risks mentioned above and the technical feasibility of controlling pH in huge stirred tanks.

2.1.3.1.4 Nutrients

The two essential components to provide for an optimal growth of the microorganisms are ammonium as the nitrogen source and phosphate as phosphorus. Two others may also be required, which are potassium and magnesium. However, they are often naturally available from the ore or concentrate. The required amounts of each nutrient are variable from one sulphide substrate to the other in the range of 1.0–10.0 kg/t of the sulphide concentrate and must be optimised in continuous leaching conditions where water recycling is also considered. Fertilisers can be used as sources of nitrogen and phosphorus: diammonium phosphate (DAP) and monoammonium phosphate (MAP) to make the process economical. For a long time, it has been considered that jarosite precipitated in proportion to the ammonium concentration in the slurry, which could be a cause of reduction of the bacterial activity by sulphide surface passivation.

2.1.3.1.5 *Temperature Control*

Three temperature ranges are considered for the current and future industrial-scale units of bioleaching in agitated tanks: the mesophiles typical range, which is 35–40°C, the moderate thermophiles, 45–55°C and the extreme thermophiles, 60–85°C. All the existing bioleach units in agitated tanks are at a low temperature, near 40°C, one used moderate thermophiles, near 50°C, and several projects are in progress to use extreme thermophiles (Rawlings et al., 2003) for the recovery of copper from chalcopyrite. As sulphide oxidation is an exothermic reaction, the heat removal is an important aspect for the design of the bioleach tanks. The heat generated by agitation is also an input to be accounted for in the design of the heat-removal system. The heat removal is generally ensured by transfer to water circulating in internal reactor coils that can be more than a kilometre long in one tank.

2.1.3.1.6 O_2 *Transfer*

Oxygen transfer is a key parameter for the design and the cost of a bioleaching installation. In order to use relatively simple and not excessively expensive equipment, in a way respecting the viability of the microorganisms and copying their natural conditions of growth, oxygen is supplied by the injection of air spurged and dispersed by mechanical means. There are two mixing systems currently in use at an industrial scale for processing refractory sulphide concentrates. The first system is a Rushton turbine, which is the traditional impeller used in processes requiring high gas dispersion rates. In all the other installations dealing with refractory gold concentrates, it is the Lightnin A315® impeller. In the demonstration BioCOP™ plant at Chuquicamata, the world's largest A315 impeller used in each tank has a diameter close to 5 m. Oxygen solubility in thermophilic systems is reduced to 1/3 of what could be expected at mesophilic conditions as suggested by BHP Billiton, and an automated dissolved oxygen strategy was used to control the oxygen consumption and prevent the microorganisms to suffer from their high sensitivity to this gas (Morin, 2007).

2.2 Recent Applications and New Developments of Stirred-Tank-Leaching Technology

Gencor has pioneered the commercialisation of biooxidation of refractory gold ores. The development of the BIOX™ process started in the late 1970s at

Gencor Process Research, Johannesburg, South Africa. The successful development of the technology led to the commissioning of a BIOX pilot plant in 1984, followed by the first commercial BIOX plant at the Fairview mine in Barberton, South Africa in 1986 (Rawlings, 2002; van Aswegen, 2007). Since then, at least six similar plants have been built in other countries, including Australia, Brazil, Ghana and Peru. The plant at Sansu, Ghana, was commissioned in 1994, expanded in 1995 and is probably the largest reactor process in the world. It consists of 24 tanks of 1,000,000 L each, processes 1000 tonnes of gold concentrate per day, and earns about half of Ghana's foreign exchange. The robustness, simplicity of operation, environmental friendliness and cost-effectiveness of the technology has been demonstrated at all of these operations. 'BACOX', the BacTech process for the pre-treatment of pyritic gold concentrates was commercialised in 1994 at Youanmi Mine, Western Australia (Watling, 2008). The plant utilised moderate thermophiles with optimum temperatures for growth between 45°C and 55°C, and was operated at 50°C.

The next stage of development came with the adaptation by Billiton of this technology for base metal sulphide treatment. The mesophile cultures of the BIOX® gold process could be successfully adapted to nickel sulphide concentrates and when coupled with a solvent extraction or ion exchange, followed by electrowinning, they were termed as the BioNIC® process. Success in nickel spurred the Billiton team to develop a process to treat concentrates arising from secondary copper minerals, such as chalcocite. This copper bioleach process was named BioCOP and a pilot-plant facility was established in 1997 with Codelco at its Chuquicamata operation in Chile (Clark et al., 2006). This, a thermophile tank leaching and downstream solvent extraction/electrowinning (SX/EW) process, encompasses both mesophile and thermophile technologies. The pilot unit on-site at Chuquicamata was converted into thermophile operations in May 2000 (Watling, 2008). A similar technology named BioZINC™ was also employed for zinc concentrates. A demonstration plant was commissioned on February 2002 at Hutti Gold Mines in the Karnataka state in India. The plant was designed for treating gold- and silver-bearing concentrates, but it can be used for the bioleaching of copper, zinc, nickel and other base metal concentrates. Another exciting experience is being carried out in Mexico. Peñoles S.A., in association with Mintek, has been able to produce several tonnes of copper cathodes in their demonstration plant in Monterrey. The plant is an integrated tank bioleaching, SE and EW facility capable of producing 500 kg of copper per day. The latest process operated at an industrial scale for bioleaching in tanks and currently in use at the cobalt bioleaching plant in Kasese, Uganda, is the BROGIM® system, designed by Robin Industries (now Milton Roy Mixing) (Bouquet and Morin, 2005).

2.2.1 Shortcomings of the Stirred-Tank Reactors

Although stirred-tank biooxidation of different sulphide minerals has been practiced for some 25 years and most problems have been successfully resolved over this period of time, the following current limitations must not be omitted (Morin, 2007).

It is a slow process and the solids concentration of the feed is limited, which means large pieces of equipment for high feed flow rates requiring sufficient ground surface and relatively high power consumption. Reactors are expensive to construct and operate; their use is restricted to high-value ores and concentrates (Rawlings, 2002). Chalcopyrite concentrate bioleaching with the extremely thermophilic Archaea requires more exotic materials of construction for the bioreactors to mitigate corrosion, which increases the capital cost (Brierley, 2008b).

2.3 Biphasic Leaching Operation: The Latest Alternative in Biomining

It is very clear from the difficulties faced (Sections 2.1.2.3 and 2.2.1) by the two major established biomining processes, that is, bioleach heap and stirred-tank reactors respectively in the treatment of sulphide minerals that the metal extraction kinetics is very low as compared to pyrometallurgical and hydrometallurgical approaches. Moreover, the time, site specificity, competitiveness and capital investment daunted the commercialisation of both the biomining approaches (Brierley, 2010).

2.3.1 Origin and Fundamentals

Ferric iron is the key oxidant in biohydrometallurgical processes where sulphide minerals such as pyrite (FeS_2), pyrrhotite ($[Fe_{(1-x)}S, [x = 0.2]]$), covellite (CuS) and sphalerite (ZnS) are oxidised. One of the main mechanisms of bacterial catalysis in the dissolution of sulphide minerals is based on the biological oxidation of ferrous iron with oxygen as the electron acceptor. Ferric iron thus produced chemically oxidised sulphide minerals and is reduced in this redox reaction to Fe^{2+}. Iron re-oxidation is essential in the bioleaching process because Fe^{3+} is an important electron shuttle and a chemical oxidant. Recirculation of Fe^{3+} and Fe^{2+} containing leach solutions back to the process is a common practice, but leads to the accumulation of high concentrations of dissolved iron (Nurmi et al., 2009). Ferrous iron can be chemically oxidised in acid solutions, but microbial oxidation occurs 10^5–10^6 times faster as compared to the chemical oxidation (Bosecker, 1997).

In relation to the importance of Fe^{3+} iron explained above, the biphasic leaching operation is based on 'non contact bioleaching'. This is a brilliant non-conventional option for the treatment of base metal concentrates, which have been extensively evaluated over the years (Carranza et al., 1997a, 2004; Fomchenko and Biryukov, 2009; Palencia et al., 2002; Patel et al., 2012a, 2014;

Romero et al., 1998, 2003). In the biphasic process, the bacterial oxidation of ferrous iron to ferric iron is performed in a separate vessel/reactor (first phase), which is physically separated from the leach reactor (second phase). The sulphide feed material in the leach reactor is contacted with Fe^{3+} iron solution generated by bacteria. From the reactor product, the liquid and solid phases are separated, with the liquid phase proceeding to metal recovery by SX-EW and returning to the first phase, which completes the liquor circulation loop between the leach reactor and the biooxidation vessel. This process is also known as IBES (indirect bioleaching with effect separation) (Carranza et al., 1997a) or BRISA (Biolixiviación Rápida Indirecta con Separación deAcciones: Fast indirect bioleaching with actions separation) (Carranza et al., 1997b).

The process is based on bioleaching by the indirect contact mechanism (Sand et al., 2001). According to this mechanism, metallic sulphides are chemically oxidised by ferric sulphate, leading to elemental sulphur and metals in the solution (Equations 2.1 through 2.3). The resulting ferrous (in Equations 2.1 through 2.3) is regenerated by iron-oxidising microorganisms (Equation 2.4)

$$CuFeS_2 \text{ (chalcopyrite)} + 2Fe(SO_4)_3 \rightarrow CuSO_4 + 5FeSO_4 + 2S^0 \qquad (2.1)$$

$$ZnS \text{ (sphalerite)} + 2Fe^{3+} \rightarrow Zn^{2+} + 0.125S_8 + 2Fe^{2+} \qquad (2.2)$$

$$Cu_2 \text{ (chalcocite)} + 4Fe^{3+} \rightarrow 2Cu^{2+} + 4Fe^{2+} + S^0 \qquad (2.3)$$

$$2Fe^{2+} \rightarrow 0.5O_2 + 2H^+ \xrightarrow{\text{Iron-oxidising acidophiles}} +2Fe^{2+} + H_2O \qquad (2.4)$$

A typical example of the biphasic leaching process is the BRISA process for the treatment of chalcopyrite concentrates. It consists of a stage of ferric leaching in which a silver salt is added as a catalyst. The solid residue contains elemental sulphur and all the silver added as a catalyst. The recovery of silver is accomplished after the removal of elemental sulphur. The two major reasons explained by the authors (Carranza et al., 1997a, 2004; Fomchenko and Biryukov, 2009; Palencia et al., 2002; Romero et al., 1998, 2003) to separate the chemical from the biological stage are: (1) the possibility to perform the chemical leaching at a high temperature in order to increase the kinetics as suffered in heap and stirred-tank bioleaching reactors; and (2) the inhibition of the bacterial growth by a heterogeneous ecosystem (heap leaching) and the harmful physico-chemical effects that exert on the bacteria when using a single-stirred-tank bioleaching reactor (Ballester et al., 2007).

2.3.2 Important Components of the Biphasic Leaching Technology

2.3.2.1 Microorganisms

From a known diversity of iron-oxidising microorganisms as discussed in Section 2.4, *Acidithiobacillus ferrooxidans*, *Leptospirillum ferrooxidans* and

Leptospirillum ferriphilum play a pivotal role in the oxidation of Fe^{2+} iron. For many years, *A. ferrooxidans* was considered the dominant iron-oxidising microorganism. However, based on kinetics reasoning (Boon et al., 1999a,b) and molecular ecology studies (Rawlings et al., 1999), it was reasoned that *L. ferrooxidans* might be the most important microorganism for ferrous iron oxidation. It is noteworthy that most of the ferric-generating bioreactor study used *A. ferrooxidans*, *L. ferrooxidans* and *L. ferriphilum*. The optimal pH for the growth of *A. ferrooxidans* is reported to be within the range of 1.8–2.5 (Rawlings et al., 1999). *L. ferrooxidans* is more acid resistant than *A. ferrooxidans* and can grow at lower pH values (<1.0). With regard to the temperature, *A. ferrooxidans* is considered to be more tolerant of low temperatures and less tolerant of high temperatures than *L. ferrooxidans*. There are several factors that play a role in the rate of oxidation of ferrous ions by these microorganisms. These factors include the ratio of ferrous/ferric iron concentration, cell and oxygen concentrations, pH, temperature, and the type of reactor (Daoud and Karamanev, 2006). The growth of *A. ferrooxidans* is inhibited above 36 mM ferric iron concentration, whereas *L. ferrooxidans* is resistant even to 500 mM ferric iron concentration (Curutchet et al., 1992). Moreover, as reported by Galleguillos et al. (2009), the metal resistance ability of the *L. ferriphilum* is far greater than *A. ferrooxidans* and *L. ferrooxidans*, which make it the candidate of choice for the bioreoxidation of Fe^{2+} iron from leachate containing different metals. The organism is quite useful in dual-stage bioleaching processes such as BRISA and IBES (Carranza et al., 1997a; Fomchenko and Biryukov, 2009; Palencia et al., 2002; Romero et al., 2003). Brierley and Brierley (2001) have also suggested the genus *Leptospirillum* along with thermophilic bacteria and Archaea as an important candidate for commercial applications. *Leptospirillum* spp., which can oxidise iron but not reduced forms of sulphur, has been found to be predominant in commercial biooxidation tanks (Coram and Rawlings, 2002). Currently, a novel biphasic leaching operation for the polymetallic bulk concentrate was demonstrated by Patel et al. (2014) which exploits a very-high-pulp density with the multimetal-resistant iron-oxidising *L. ferriphilum*-dominated bacterial consortium. The technology was fully optimised for both the stages, that is, ferric-generating bioreactor and chemical-leaching reactor. The flow sheet of the technology is illustrated in Figure 2.2.

2.3.2.2 Bioreactors

A critical module of an indirect biphasic system is the efficiency of the ferric-iron-regenerating bioreactor. Although scanty data on Fe^{3+}-generating bioreactors are available in literature, the various types of bioreactors including fluidised beds, packed beds, trickle beds, circulating beds, agitated reactors and rotating biological contactors, have been tested for their potential for high-rate Fe^{2+} iron oxidation by acidophilic microorganisms (Dave, 2008). The development of high-efficiency bioreactors for ferrous iron biooxidation

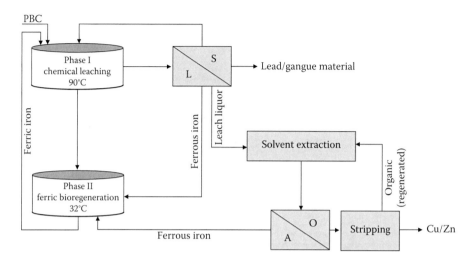

FIGURE 2.2
A novel biphasic leaching approach for the polymetallic bulk concentrate.

has opened the way to succeed in the potential commercial application of this innovative biomining technology, being especially attractive to bene-fit zinc and lead bulk or polymetallic concentrates containing other valu-able metals such as copper and silver (Frías et al., 2009). Most of the reactor designs studied exploited the attached bacterial growth on different types of support materials, as the microbial biofilm on the support material gives far better kinetics as compared to the planktonic growth. These studies show the successful immobilisation and high oxidation rates using packed-bed or fluidised-bed reactors with different support materials, such as low-grade sulphide minerals, calcium alginate, polyvinyl chloride (PVC), polyurethane foam, glass, activated carbon, exchange resin, nickel, siliceous stone, polyeth-ylene, extruded and expanded polystyrene and so on (Dave, 2008; Grishin and Tuovinen, 1988; Mazuelos et al., 2000; Patel et al., 2014).

2.3.2.3 Advantages of Biphasic Operation

The biphasic leaching operation is one of the most promising future tech-nologies that make use of autotrophic as well as heterotrophic acidophiles in controlled bioreactors (Ehrlich, 2001).

- It is the most economic option especially for the treatment of high-grade ores and the polymetallic concentrate as compared to other biomining technologies.
- It presents an important advantage when applied to the bioprocess-ing of zinc and zinc polymetallic concentrates such as high metals recovery, and fast kinetics of 4–5 h to get complete metals extraction,

sulphur oxidation to elemental sulphur (avoiding sulphuric acid generation, while the pyrite mineral remains untouched) and the use of conventional reactors (aeration is not required) and normal process equipments (Frías et al., 2009).

- Moreover, this procedure allows the separate optimisation of two stages: the biological one by the use of suitable packed-bed bioreactors and the chemical one by the use of stirred-tank reactors favouring the metal extractions by thermal activation and the use of catalysts such as silver.

- The separation of the processes has several important implications. If the processes are separated, the bacteria are not in direct contact with the elemental sulphur that produced byte sulphide-mineral reactions, and so, the oxidation product would be elemental sulphur and not sulphate. The chemical oxidation reactions do not require oxygen, and oxygen only needs to be supplied to the bacterial reactors. Separation of the processes would also allow optimisation of each of the processes separately: the leaching reactors could be operated at high temperatures, high Fe^{3+} iron concentration and very low pH; and the use of immobilised bacteria could be considered for the intensification of the bacterial reactions (Gerick et al., 2009).

- The connection of both stages (leaching and biooxidation) in a continuous operation in which the regenerated ferric ion is recycled to the leaching stage permits the use of high-pulp densities in leaching.

- The biomass produced in the separate bioreactor is much more as compared to single-stage bioleaching, because of the different kinetics: when bacteria use the metal sulphide directly as a source of energy, their doubling time ranges from days to weeks, whereas when they oxidise Fe^{2+} iron (indirect mechanism), that time ranges from 4 to 8 h.

- Ferric leaching as a separate phase offers several advantages over the complete oxidation of chalcopyrite. Copper recoveries in the high 90% range are achieved, especially when the chalcopyrite is finely ground. The sulphur moiety of chalcopyrite is oxidised to elemental sulphur, not sulphuric acid, which significantly reduces downstream neutralisation costs and the sulphur may be a saleable product.

- Operating costs are greatly minimised because less air is required to oxidise ferrous iron than to oxidise chalcopyrite.

- Silver ion inhibits the growth of iron-oxidising bacteria used for ferric iron regeneration in the biological stage of the BRISA process; however, the separation of chemical and bacterial actions enables the use of this catalytic system because silver is always in the ferric-leaching circuit and it will never reach the bioreactor.

This technology still suffers certain disadvantages as discussed below to be resolved in futuristic development.

- Heating the ferric solution before contact with the concentrate and cooling the ferrous-containing liquor before recirculation to the bacterial reactor. Thermophilic bacterial generation of the Fe^{3+} iron solution may be an answer. Also, the development of the equipment to improve the mass transfer between the sulphide concentrate and the Fe^{3+} solution (Brierley and Brierley, 2001).
- Jarosite formation leading to clogging of the bioreactors is a major issue to be resolved. However, the positive effect of jarosite in biomass retainment is evaluated by Kinnunen and Puhakka (2004a).
- Oxygen supply to satisfy the oxygen demand by the iron-oxidising microorganisms in high-rate bioreactors requires further development.

2.4 Future Aspects

The population dynamics in the different bioleaching experiments are being monitored (Zhang et al., 2014). So, various molecular techniques, such as metagenomics, metabolomics, microarray and advancement in microscopy techniques need to be used. Based on these techniques, newer and newer microbes need to be characterised from the mining habitats for further opportunities to exploit them in an optimum way.

Bioassisted processes have been continuously expanding their applications for the discovery of a new eco-friendly, economic and efficient technique for metal extraction from e-waste scrap as a valuable mineral resource. Therefore, developing a new technology for recovering precious metals is urgently demanded for improving the added value of waste polychlorinated biphenyl (PCB) recovery. Bioextraction of copper, nickel and zinc from PCBs is in progress in the author's laboratory using various reduced-iron and sulphur-compound-oxidising microorganisms.

References

Abhilash, K. D. Mehta, V. Kumar, B. D. Pandey and P. K. Tamrakar. 2011. Bioleaching—An alternate uranium ore processing technology for India. *Energy Procedia* 7: 158–162.
Ballester, A., M. L. Blázquez, F. González and J. A. Muñoz. 2007. Catalytic role of silver and other ions on the mechanism of chemical and biological leaching.

In: *Microbial Processes for Metal Sulphides*, eds., E. Donati, and W. Sand, 77–102. Dordrecht, The Netherlands: Springer.

Boon, M., H. J. Brasser, G. S. Hansford and J. J. Heijnen. 1999a. Comparison of the oxidation kinetics of different pyrites in the presence of *Thiobacillus ferrooxidans* or *Leptospirillum ferrooxidans*. *Hydrometallurgy* 53 (1): 57–72.

Boon, M., T. A. Meeder, C. Thne, C. Ras and J. J. Heijnen. 1999b. The ferrous iron oxidation kinetics of *Thiobacillus ferrooxidans* in continuous cultures. *Appl. Microbiol. Biotechnol.* 51 (6): 820–826.

Bosecker, K. 1997. Bioleaching: Metal solubilization by microorganisms. *FEMS Microbiol. Rev.* 20 (3–4): 591–604.

Bouffard, S. C., C. Sylvie and D.G. Dixon. 2001. Investigative study into the hydrodynamics of heap leaching processes. *Metall. Mater. Trans. B* 32 (5): 763–776.

Bouquet, F. and D. Morin. 2005. BROGIM®: A new three-phase mixing system testwork and scale-up. In: *Proceedings of the 16th International Biohydrometallurgy Symposium*. eds., S. T. L. Harrison, D. E. Rawlings and J. Petersen, 173–182. 25–29 September, Cape Town, South Africa.

Brierley, C. L. 2008a. How will biomining be applied in future? *Trans. Nonfer. Metals Soc.* 18 (6): 1302–1310.

Brierley, C. L. 2010. Biohydrometallurgical prospects. *Hydrometallurgy* 104 (3–4): 324–328.

Brierley, J. A. 2008b. A perspective on developments in biohydrometallurgy. *Hydrometallurgy* 94 (1–4): 2–7.

Brierley, J. A. and C. L. Brierley. 2001. Present and future commercial applications of biohydrometallurgy. *Hydrometallurgy* 59 (2–3): 233–239.

Carranza, F., I. Palencia and R. Romero. 1997a. Silver catalyzed IBES process: Application to a Spanish copper–zinc sulphide concentrate. *Hydrometallurgy* 44: 29–42.

Carranza, F., I. Palencia, R. Romero and N. Iglesias. 1997b. Application fields of the BRISA process. Influence of the ore mineralogy on the process flowsheet. In: *Proceedings of the International Biohydrometallurgy Symposium IBS97—BIOMINE 97 'Biotechnology Comes of Age'*, Australian Mineral Foundation, M2.1.1–M2.1.10. 4–6 August, Sydney, Australia.

Carranza, F., N. Iglesias, A. Mazuelos, I. Palencia and R. Romero. 2004. Treatment of copper concentrates containing chalcopyrite and non-ferrous sulphides by the BRISA process. *Hydrometallurgy* 71 (3–4): 413–420.

Clark, M. E., J. D. Batty, C. B. van Buuren, D. W. Dew and M. A. Eamon. 2006. Biotechnology in minerals processing: Technological breakthroughs creating value. *Hydrometallurgy* 83 (1–4): 3–9.

Coram, N. J. and D. E. Rawlings. 2002. Molecular relationship between two groups of the genus *Leptospirillum* and the finding that *Leptospirillum ferriphilum* sp. nov. dominates South African commercial biooxidation tanks that operate at 40 degrees C. *Appl. Environ. Microbiol.* 68 (2): 838–845.

Curutchet, G., C. Pogliani, E. Donati and P. Tedesco. 1992. Effect of iron (III) and its hydrolysis products (jarosites) on *Thiobacillus ferrooxidans* growth and on bacterial leaching. *Biotechnol. Lett.* 14 (4): 329–334.

Daoud, J. and D. Karamanev. 2006. Formation of jarosite during Fe^{2+} oxidation by *Acidithiobacillus ferrooxidans*. *Miner. Eng.* 19 (9): 960–967.

Dave, S. R. 2008. Selection of *Leptospirillum ferrooxidans* SRPCBL and development for enhanced ferric regeneration in stirred tank and airlift column reactor. *Bioresour. Technol.* 99 (16): 7803–7806.

Devasia, P. and K. A. Natarajan. 2004. Bacterial leaching biotechnology in the mining industry. *Resonance – J. Science Edu.* 9: 27–34.

Ehrlich, H. L. 2001. Past, present and future of biohydrometallurgy. *Hydrometallurgy* 59: 127–134.

Eisapour, M., A. Keshtkar, M. A. Moosavian and A. Rashidi. 2013. Bioleaching of uranium in batch stirred tank reactor: Process optimization using Boxâ€Behnken design. *Ann. Nucl. Energy* 54 (0): 245–250.

Fomchenko, N. V. and V. V. Biryukov. 2009. A two-stage technology for bacterial and chemical leaching of copper–zinc raw materials by Fe^{3+} ions with their subsequent regeneration by chemolithotrophic bacteria. *Prikl. Biokhim. Mikrobiol.* 45 (1): 64–69.

Frías, C., C. Sánchez, D. Martín, G. Díaz and S. Sanguilinda. 2009. Indirect bioleaching and ZINCEX™ process: A profitable marriage to process zinc polymetallic concentrates. *Adv. Mater. Res.* 71–73: 429–432.

Galleguillos, P. A., K. B. Hallberg and D. B. Johnson. 2009. Microbial diversity and genetic response to stress conditions of extremophilic bacteria isolated from the Escondida copper mine. *Adv. Mater. Res.* 71–73: 55–58.

Gerick, M., J. W. Neale and P. J. van Staden. 2009. A Mintek perspective of the past 25 years in minerals bioleaching. *J. S. Afr. I. Min. Metall.* 109: 567–585.

Grishin, S. I. and O. H. Tuovinen. 1988. Fast kinetics of Fe^{2+} oxidation in packed-bed. *Appl. Environ. Microbiol.* 54 (12): 3092–3100.

Johnson, D. B. 2010. The biogeochemistry of biomining. In: *Geomicrobiology: Molecular and Environmental Perspective*, eds., L. L. Burton, M. Mandl and A. Loy, 401–426. Netherlands: Springer.

Johnson, D. B. and K. B. Hallberg. 2009. Carbon, iron and sulfur metabolism in acidophilic micro-organisms. In: *Advances in Microbial Physiology*, ed., K. P. Robert, 54:201–255. UK: Academic Press.

Kinnunen, P. H. M. and J. A. Puhakka. 2004a. Chloride-promoted leaching of chalcopyrite concentrate by biologically-produced ferric sulfate. *J. Chem. Technol. Biotechnol.* 79 (8): 830–834.

Kondrat'eva, T., T. Pivovarova, I. Tsaplina, N. Fomchenko, A. Zhuravleva, M. Murav'ev, V. Melamud and A. Bulayev. 2012. Diversity of the communities of acidophilic chemolithotrophic microorganisms in natural and technogenic ecosystems. *Microbiol.* 81 (1): 1–24.

Mazuelos, A., F. Carranza, I. Palencia and R. Romero. 2000. High efficiency reactor for the biooxidation of ferrous iron. *Hydrometallurgy* 58: 269–275.

Morin, D. H. R. 2007. Bioleaching of sulfide minerals in continuous stirred tanks. In: *Microbial Processes for Metal Sulphides*, eds., E. Donati and W. Sand, 133–150. Dordrecht, the Netherlands: Springer.

Morin, D., A. Lips, T. Pinches, J. Huisman, C. Frias, A. Norberg and E. Forssberg. 2006. BioMinE: Integrated project for the development of biotechnology for metal-bearing materials in Europe. *Hydrometallurgy* 83 (1–4): 69–76.

Mwase, J. M., J. Petersen and J. J. Eksteen. 2014. A novel sequential heap leach process for treating crushed Platreef ore. *Hydrometallurgy* 141 (0): 97–104.

Nurmi, P., B. Ã-zkaya, A. H. Kaksonen, O. H. Tuovinen, M. L. Riekkola-Vanhanen and J. A. Puhakka. 2009. Process for biological oxidation and control of dissolved iron in bioleach liquors. *Process Biochem.* 44 (12): 1315–1322.

Okibe, N. and D. B. Johnson. 2004. Biooxidation of pyrite by defined mixed cultures of moderately thermophilic acidophiles in pH-controlled bioreactors: Significance of microbial interactions. *Biotechnol. Bioeng.* 87 (5): 574–583.

Okibe, N., M. Gericke, K. B. Hallberg and D. B. Johnson. 2003. Enumeration and characterization of acidophilic microorganisms isolated from a pilot plant stirred-tank bioleaching operation. *Appl. Environ. Microbiol.* 69: 1936–1943.

Olson, G. J., J. A. Brierley and C. L. Brierley. 2003. Bioleaching review part B: Progress in bioleaching: Applications of microbial processes by the minerals industries. *Appl. Microbiol. Biotechnol.* 63 (3): 249–257.

Palencia, I., R. Romero, A. Mazuelos and F. Carranza. 2002. Treatment of secondary copper sulphides (chalcocite and covellite) by the BRISA process. *Hydrometallurgy* 66 (1–3): 85–93.

Patel, B. C., D. R. Tipre and S. R. Dave. 2012a. Optimization of copper and zinc extractions from polymetallic bulk concentrate and ferric iron bioregeneration under metallic stress. *Hydrometallurgy* 117–118 (0): 18–23.

Patel, B. C., D. R. Tipre and S. R. Dave. 2012b. Development of *Leptospirillum ferriphilum* dominated consortium for ferric iron regeneration and metal bioleaching under extreme stresses. *Bioresour. Technol.* 118 (0): 483–489.

Patel, B. C., M. K. Sinha, D. R. Tipre, A. Pillai and S. R. Dave. 2014. A novel biphasic leaching approach for the recovery of Cu and Zn from polymetallic bulk concentrate. *Bioresour. Technol.* 157 (0): 310–315.

Petersen, J. and D. G. Dixon. 2002. Thermophilic heap leaching of a chalcopyrite concentrate. *Miner. Eng.* 15: 777–785.

Petersen, J. and D. G. Dixon. 2007a. Modeling and optimization of heap bioleach processes. In: *Biomining*, eds., D. E. Rawlings and D. B. Johnson, 153–176. Berlin: Springer-Verlag.

Petersen, J. and D. G. Dixon. 2007b. Principles, mechanisms and dynamics of chalcocite heap bioleaching. In: *Microbial Processing of Metal Sulfides*, eds., E. Donati and W. Sand, 193–218. The Netherlands: Springer.

Pradhan, N., K. C. Nathsarma, K. Srinivasa Rao, L. B. Sukla and B. K. Mishra. 2008. Heap bioleaching of chalcopyrite: A review. *Miner. Eng.* 21 (5): 355–365.

Rawlings, D. E. 2002. Heavy metal mining using microbes. *Annu. Rev. Microbiol.* 56: 65–91.

Rawlings, D. E., D. Dew and C. du Plessis. 2003. Biomineralization of metal-containing ores and concentrates. *Trends Biotechnol.* 21 (1): 38–44.

Rawlings, D. E., H. Tributsch and G. S. Hansford. 1999. Reasons why '*Leptospirillum*'-like species rather than *Thiobacillus ferrooxidans* are the dominant iron-oxidizing bacteria in many commercial processes for the biooxidation of pyrite and related ores. *Microbiol.* 145 (Pt 1): 5–13.

Romero, R., A. Mazuelos, I. Palencia and F. Carranza. 2003. Copper recovery from chalcopyrite concentrates by the BRISA process. *Hydrometallurgy* 70 (1–3): 205–215.

Romero, R., I. Palencia and F. Carranza. 1998. Silver catalyzed IBES process: Application to a Spanish copper–zinc sulphide concentrate: Part 3. Selection of the operational parameters for a continuous pilot plant. *Hydrometallurgy* 49 (1–2): 75–86.

Sand, W., T. Gehrke, P. G. Jozsa and A. Schippers. 2001. (Bio)chemistry of bacterial leaching—Direct vs. indirect bioleaching. *Hydrometallurgy* 59 (2–3): 159–175.

Siddiqui, M. H., A. Kumar, K. K. Kesari and J. M. Arif. 2009. Biomining—A useful approach toward metal extraction. *Am.-Eurasian J. Agron.* (2): 84–88.

Soto, P. E., P. A. Galleguillos, M. A. Serón, V. J. Zepeda, C. S. Demergasso and C. Pinilla. 2013. Parameters influencing the microbial oxidation activity in the industrial bioleaching heap at Escondida mine, Chile. *Hydrometallurgy* 133: 51–57.

Spencer, P. A. 2001. Influence of bacterial culture selection on the operation of a plant treating refractory gold ore. *Int. J. Miner. Process.* 62: 217–229.

Tupikina, O. V., S. H. Minnaar, R. P. van Hille, N. van Wyk, G. F. Rautenbach, D. Dew and S. T. L. Harrison. 2013. Determining the effect of acid stress on the persistence and growth of thermophilic microbial species after mesophilic colonisation of low grade ore in a heap leach environment. *Miner. Eng.* 53 (0): 152–159.

van Aswegen, P. C., J. van Niekerk and W. Olivier. 2007. The BIOX™ process for the treatment of refractory gold concentrates. In: *Biomining*, eds., D. E. Rawlings and D. B. Johnson, 1–34. Berlin: Springer-Verlag.

Watling, H. R. 2006. The bioleaching of sulphide minerals with emphasis on copper sulphides—A review. *Hydrometallurgy* 84 (1–2): 81–108.

Watling, H. R. 2008. The bioleaching of nickel–copper sulfides. *Hydrometallurgy* 91 (1–4): 70–88.

Yang, H., S. Feng, Y. Xin and W. Wang. 2014. Community dynamics of attached and free cells and the effects of attached cells on chalcopyrite bioleaching by *Acidithiobacillus* sp. *Bioresour. Technol.* 154 (0): 185–191.

Yu, R., L. Shi, G. Gu, D. Zhou, L. You, M. Chen, G. Qiu and W. Zeng. 2014. The shift of microbial community under the adjustment of initial and processing pH during bioleaching of chalcopyrite concentrate by moderate thermophiles. *Bioresour. Technol.* 162(0): 300–307.

Zhang, L., J. Wu, Y. Wang, L. Wan, F. Mao, W. Zhang, X. Chen and H. Zhou. 2014. Influence of bioaugmentation with *Ferroplasma thermophilum* on chalcopyrite bioleaching and microbial community structure. *Hydrometallurgy* 146 (0): 15–23.

3

Microbial Extraction of Uranium from Ores

Abhilash and B. D. Pandey

CONTENTS

3.1 Introduction

The continued depletion of high-grade ores and growing awareness of environmental degradation associated with the traditional methods have provided impetus to explore simple, efficient and less polluting biological methods in uranium mining, processing, and waste water treatments (Torma, 1983; Bosecker, 1990). Hydrometallurgical methods have some disadvantages such as poor recovery, involvement of high process and energy cost, and increases in the pollution load of water resources (Bruynesteyn, 1989; Dwivedy and Mathur, 1995). Uranium could also be recovered by microorganisms that catalyse the oxidation and reduction of uranium and associated metals also, and hence influence their mobility in the environment. Industrial-scale bioleaching of uranium is carried out by spraying stope walls with acid mine drainage and the in situ irrigation of fractured underground ore deposits. The recent upsurge of interest in this area is motivated by the fact that it is simple, effective, and potentially a relatively less-expensive process involving low-energy that is environmentally benign; this is due to the uranium solubilising and accumulating properties of certain microorganisms. Besides, its industrial application to ensure the supply of raw material for producing energy, microbial leaching has a definite potential for remediation of mining sites, treatment of wastes and detoxification of sewage sludge (Brierley and Brierley, 1999). The commercial application of bioleaching of uranium from low-grade ores has been practiced since the 1960s. The seven leading uranium-producing countries in descending order are Canada, Australia, Niger, the Russian Federation, Kazakhstan, Namibia and Uzbekistan. Currently, the two largest producers, namely, Canada and Australia alone account for over 50% of global uranium production. In regard to bioleaching, Canada produced about 70,000 lb of U_3O_8 in 1977 at Agnew Lake Mine, Ontario, Canada from its ore using *Acidithiobacillus ferrooxidans* (*A. ferrooxidans*) (McCready and Gould, 1990). Commercial scale experience has been limited to the operations at Denison's Elliot Lake, Ontario, Canada in the 1980s and the dump bioleaching at the Gibraltar Mine, British Columbia. The presence of microorganisms in leaching operations has been found to be beneficial in catalysing the uranium dissolution process (Brierley, 1997; Brierley and Brierley, 1999). Currently, this technology is applied on a commercial scale not only for the recovery of uranium but also for extraction of copper, nickel, gold and so forth through heap, dump and in situ leach techniques (Torma, 1983; Torma and Banhegyi, 1984; Mwaba, 1991; Elshafeea et al., 2014). The process flow sheet for the bioleaching of uranium is shown in Figure 3.1.

The microbial consortia responsible for removing uranium from its minerals are considered to be the complex mixtures of acidophilic, autotrophic and heterotrophic bacterial strains (de Siloniz et al., 1993; Johnson, 1998). In most ores, uranium occurs as a mixture of minerals containing uranium in

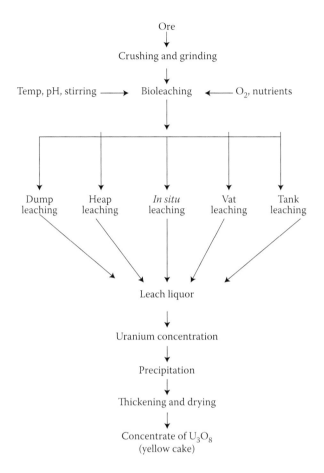

FIGURE 3.1
Flowchart for uranium bioprocessing.

either U(IV) or U(VI) state (Saraswat, 1988). Uranium is mostly soluble in its most oxidised state, that being U(VI) (Brierley, 1978; Lundergren and Silver, 1980). U(IV) can be oxidised to the soluble form by ferric ions as reported by Dutrizac and MacDonald (1974) and the oxidation occurs much more rapidly in the presence of *A. ferrooxidans*. A pH 1.0–2.5 is considered suitable for the bioleaching; however, for rapid uranium extraction kinetics, a high Fe(III)/Fe(II) ratio is necessary, which determines the redox potential (Eh) of the leach liquor (Abhilash et al., 2009). For the leaching to be effective, it must reach the potential (Eh) above 400 mV in accordance with the Eh–pH diagram (Figure 3.2). As can be seen from this diagram, the stability region of uraninite at the lower redox potential value (–80 mV), typically considered for leaching, goes up to pH 8.0 and extends up to pH 11.25 at still lower redox potential (–100 mV). The solubility of uraninite in the form of U(IV) is typically observed at the higher acidity, which is pH below 0.6 with the redox

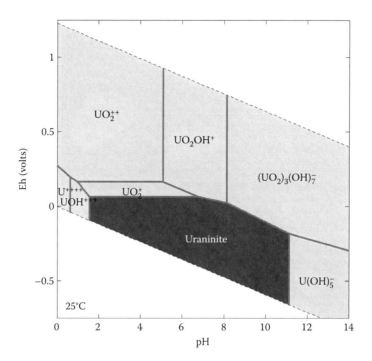

FIGURE 3.2
Pourbaix diagram for U–H_2O system at 25°C and C = 10^{-2}M.

potential of <300 mV. Though, the solubility of U(VI) species from uraninite are facilitated by higher redox potential (>300 mV) in the wider pH range (0–5.5). The bioleaching catalysed by *A. ferrooxidans* creating strong oxidising conditions at redox potential above 400 mV would thus favour uranium dissolution in the normal working pH range (1.5–2.5) of such bacteria.

3.2 Uranium Leaching Methodologies

3.2.1 Chemical Leaching

There are several methods of chemical leaching, but the most common one is the process involving the use of sulphuric acid. The choice of acid or alkali concentration and the temperature of leaching are based on techno-economic considerations. The leaching rate is faster in acid solution as compared to alkali (carbonate) solution, and is consistent with the simultaneous attack of sites with different affinities for uranium. Higher temperature and leachant concentration favour uranium recovery, but more refractory matrix materials also get decomposed, thereby increasing the amount of impurities in the solution (Youlton and Kinnaird, 2013). The most effective leach solutions contain anions capable of forming stable dissolved complexes with uranyl (UO_2^{2+}) ions. At the industrial level, the Fe(III) regeneration is carried out by adding strong oxidising agents that are capable of maintaining the redox potential between 400 and 550 mV. Groudeva and Groudeva (1990) used H_2SO_4, $H_2SO_4/Fe_2(SO_4)_3$, $H_2SO_4/$ oxone($KHSO_5$), HCl, KCl, HCl/Oxone ($KHSO_5$) and oxalic acid as the acidic medium and concentrated alkaline solutions of >0.01 M $Na_2CO_3/NaHCO_3$, $Na_2CO_3/NaHCO_3 + H_2O_2$, $Na_2CO_3/NaHCO_3^+$ bivalent ion, $NaHCO_3$-CO_3, $Na_4P_2O_7 \cdot 10H_2O$ to recover uranium from flotation tailings. The most commonly used oxidants are H_2O_2 (Brierley, 1979), H_2SO_5 (Muñoz et al., 1995a), $NaClO_3$ (Michel and Jouin, 1985), MnO_2, and pyrolusite (Shakir et al., 1992). The H_2O_2/NH_3 has been used quite often for the recovery of U_3O_8 (Brierley, 1979). Shakir et al. (1992) reported the use of H_2SO_4/HNO_3 and $Fe_2(SO_4)_3/$ H_2SO_4 (Ring, 1980) to leach uranium from phosphatic sandstone. Ketzinel et al. (1984) used Israeli Zefa-Ef'eh phosphorite with Na_2CO_3 to leach uranium from the raw phosphorites. Amme et al. (2005) used natural, commercially available groundwater spiked with solutions of $CaCl_2$, $NaHCO_3$ and Na_2SiO_3 with pellets of UO_2 under argon atmosphere (<2 ppm O_2). Wang et al. (2013) have used supercritical CO_2 fluid leaching of uranium from sandstone-type ores. In India, MnO_2 (pyrolusite ore) in H_2SO_4 has been used to leach uranium from Jaduguda, Bhatin and Narwapahar ores (Agate, 1983).

3.2.2 Bioleaching

Depletion of high-grade mineral resources necessitates mining of ores/minerals deeper underground. The percentage extractability of uranium from different ores through bioleaching is summarised in Table 3.1. A look at the table shows various degrees of leachability of uranium from different locations depending upon the type of minerals present, their association with other ore minerals and gangue, and their leachability under standard practiced conditions. Minerals such as betafite (samiresite), brannerite (Frank et al., 1920) and parsonite have lower dissolution characteristics (16%–35%) whereas most other minerals display moderate (40%–55%) to higher levels of uranium dissolution (60%–88%). The role of microorganisms in mining, ore processing and waste-water treatment is likely to become increasingly important. A comparison of acid, alkali and bio-leaching is shown in Table 3.2. The modest nutritional requirements of microorganisms are provided by the aeration of an iron- and/or sulphur-containing mineral suspension in water or by the irrigation of a heap. Small quantities of inorganic fertiliser can be added to ensure that nitrogen, phosphate, potassium and trace element limitation does not occur. A further advantageous characteristic of bioleaching processes is that they are usually not contaminated by unwanted microorganisms. Table 3.3 shows different microorganisms, which are capable of leaching uranium

TABLE 3.1

List of Various Uranium Minerals with Their Composition across Various Locations, and Respective Conventional Leachability

Mineral Name	Chemical Formula	% Uranium Extractability	Locality Found	References
Autunite	$Ca(UO_2)_2(PO_4)_2 \cdot 12H_2O$	62.4	Many worldwide	Rolfe and Oxon (1911), Casas et al. (1998), Ohnuki et al. (2005)
Becquerelite	$Ca\ U_6O_{19} \cdot 11H_2O$	72.5	Madagascar	Rakotoson et al. (1983)
Betafite (Samiresite)	$(Ca,Na,U)_2(Ti,Nb,Ta)_2O_6(OH)$	16	Madagascar, many worldwide	Rakotoson et al. (1983)
Brannerite	$(U, Ca, Ce)(Ti, Fe)\ O_6$	33.5	Many world wide	Frank et al. (1920), Ring (1980)
Carnotite	$K_2(UO_2)_2(VO_4)_2 \cdot 3H_2O$	52.8	South-western USA	John et al. (1999)
Chadwickite	$UO_2\ HAsO_3$	60.4	SE China	Maozhong et al. (2005a)
Coffinite	$U(SiO4)_{1-x}(OH)_{4x}$	72.6	NW China, Congo	Currier (2002), Maozhong et al. (2005b)
Cuprosklodowskite	$Cu[UO_2(SiO_2OH]_2 \cdot 6H_2O$	55.2	Congo	Frondel (1958), Currier (2002)
Curienite	$Pb_2U_5O_{17} \cdot 4H_2O$	61.1	Congo	Frondel (1958)
Davidite	$(Fe, Ce, U)_2(Ti, Fe, V, Cr)_5\ O_{12}$	3.2	Canada	Mostafa and Janeczek (1997)
Francevillite	$(Ba,Pb)\ UO_2\ 2\ V_2O_8 \cdot 5H_2O$	48.7	France	Peter et al. (1983)
Lanthinite	$UO_2 \cdot 5UO_3 \cdot 10H_2O$	78.3	Kasolo, Belgian Congo	Currier (2002)
Jachymovite	$(UO_2)_8SO_4(OH)_{14} \cdot 13H_2O$	69.8	Jachymov, Czech Republic	Frondel (1958)
Johannite	$Cu(UO_2)_2(SO_4)_2(OH)_2 \cdot 12H_2O$	48.9	Jachymov, Bohemia, Czech Republic	Frondel (1958)
Joliotite	$UO_2\ CO_3 \cdot 2H_2O$	65	Menzeenschaward, Schwarzwald, Germany	Frondel (1958)
Nenadkevite	$H\ K(UO_2\ SiO_4) \cdot 1/5H_2O$	55.4	Russia, Ukraine	Polikarpova (1957)

Mineral	Formula		Location	Reference
Parsonsite	$Pb_2UO_2(PO_4)_2 \cdot 2H_2O$	26.1	Zaire	Shoep (1923)
Pitchblende	UO_2	88.1	Many world wide	Ring (1980), Agate (1983), Ketzinel et al. (1984), Shakir et al. (1992), Chadwick (1997), Amme et al. (2005)
Renardite	$Pb(UO_2)_4 PO_4 (OH)_4 \cdot 7H_2O$	57	Zaire	Frondel (1958), Deffeyes and MacGregor (1980)
Rutherfordine	$UO_2 CO_3$	72	East Africa	Frondel (1958), Deffeyes and MacGregor (1980)
Sabugalite	$H Al(UO_2)_4 (PO_4)_4 \cdot 16H_2O$	53.6	Sabugal in Portugal, France	Deffeyes and MacGregor (1980)
Sklodowskite	$Mg(UO_2)_2 Si_2O_7 \cdot 6H_2O$	55.5	Congo	Frondel (1958), Currier (2002)
Torbernite	$Cu(UO_2)(PO_4)_2 8H_2O$	63.5	Czech Republic	Frondel (1958)
Tyuyamunite	$Ca (UO_2)_2 (VO_4)_2 \cdot 8H_2O$	51.9	Tyuya-Muyun hill, Tukestan	Deffeyes and MacGregor (1980)
Uramphite	$NH_4 UO_2 \cdot (PO_4) \cdot 3H_2O$	54.5	Kyrgyzstan	John and Andrew (1997)
Uraninite (bragerite)	UO_2	88.2	Many worldwide	Saraswat (1988), Rao et al. (1989), Roy and DhanaRaju (1997), Alexandre et al. (2005), Chi and Yun (2006)
Uranophane (Uranotile)	$Ca UO_2SiO_3 (OH)_2 \cdot 5H_2O$	40.6	Oberpfalz, Bavaria, Brazil	Leonardos et al. (1987), Williams et al. (1988)
Uranopilite	$(UO_2)_6 SO_4(OH)_2 \cdot 4H_2O$	63.9	Czech Republic, USA	Mattus and Torma (1980), John and Andrew (1997)
Zeunerite	$Cu (UO_2)_2(AsO_4)_2 \cdot 12H_2O$	44.8	Schneeberg, Saxony, Germany	John and Andrew (1997)
Zippeite	$(UO_2)_2 (SO_4) (OH)_2 \cdot 4H_2O$	61.9	Jackpile Mine, Laguna, New Mexico, USA	John and Andrew (1997)

TABLE 3.2

Comparison of Acid, Alkali and Bioleaching Methods

Acid Leaching	Alkali Leaching	Bioleaching
It is not simple and hence sophisticated training is needed.	It is also not simple and hence sophisticated training is needed.	It is simple and no sophisticated training is needed.
It is applied to large operations.	It is also applied to large operations.	It may be applied to small and large operations.
Acid leaching achieves a high uranium extraction, typically 70%–90% (Bhatti et al., 1998).	Extraction from alkaline leaching is low, typically 60%–70%.	Bioleaching using microorganisms achieves moderately high uranium extraction, typically 50%–98% (Fisher et al., 1966; Abhilash et al., 2009).
Acid leaching yields faster dissolution of uranium, requiring 40–70 pore volume (Taylor et al., 2004).	Slower kinetics of uranium dissolution; requires typically more pore volumes than acid leaching (Taylor et al., 2004).	Bioleaching yields very slower dissolution of uranium, requires more pore volumes (Taylor et al., 2004).
No elevated concentration of leaching agents.	No elevated concentration of leaching agents.	The formation of a microclimate around the ore particle with elevated concentration of leaching agents (Krebs et al., 1997).
High acid consumption of carbonate bearing ores (Gupta et al., 2004).	Potential to treat ores containing high levels of carbonates.	Potential to treat all minerals including silicates.
Mandatory use of corrosion for carbonate-bearing ores.	Common material and equipment can be used.	Common material and simple equipments can be used.
Addition of oxidant not always required because of presence of iron in recycled solution (Taylor et al., 2004).	Addition of oxidant always required and accurate amounts of alkali is necessary (Taylor et al., 2004).	Addition of oxidant not required as it is generated in situ. The excess acid generated must be neutralised (Taylor et al., 2004).
Accurate amounts of acid is necessary.		

Possibility of recovering by-products.	Leaching chemistry is very selective.	Leaching chemistry is highly selective.
Leaching efficiency is high.	Leaching efficiency is lower than acid leaching.	Leaching efficiency is generally high.
Additional processing on surface may be required to produce contaminant-free product.	Product dissolution from ion-exchange should produce product of required quality.	Additional processing is not required.
Acid leaching causes the emission of gaseous.	Alkali leaching causes the emission of gaseous pollutants.	No emission of gaseous pollutants (eco-friendly).
The air pollution by sulphur oxides cannot be eliminated but sulphur by-products can be recovered (Taylor et al., 2004).	The air pollution by sulphur oxides cannot be eliminated but sulphur by-products can be recovered (Taylor et al., 2004).	The air pollution by sulphur oxides can be eliminated but sulphur by-products cannot be recovered (Taylor et al., 2004).
Ore cannot be upgraded at mining site.	Ore cannot be upgraded at mining site.	Ore can be upgraded at mining site.
Leaching agents are to be transported.	Leaching agents are transported.	Leaching agents are produced *in situ* (no need for the transportation).
It is not dependent on the climate.	It is also not dependent on the climate	Dependence on the climate.
Capital costs are high.	Capital costs are low compared to acid leaching.	Capital costs are low.
High-energy demand.	High-energy demand.	Low-energy demand.

TABLE 3.3

Amenability of Different Microorganism for Leaching of Uranium from Its Ores at Various Conditions

Microorganism	Characteristics	Leaching Agent	Carbon Req.	Oxygen (Optimum)	pH (Optimum)	Temp	Leachability	References
Acidithiobacillus ferrooxidans	Oxidise: Fe^{2+}, S°, U^{4+}, Cu^+, Se^{2+}, thiosulphate, tetrathionate, S^+	Fe^{3+}, H_2SO_4	OC	A	1.2–6.0 (1.5–2.5)	5–40 (28–35)	70%–98%	Harrison and Norris (1986)
Consortia of Acidithiobacillus thiooxidans and Acidithiobacillus ferrooxidans	Oxidise: S°, thiosulphate, tetrathionate	H_2SO_4	OC	SA	0.5–6.0 (1.5–2.5)	10–40 (28–30)	65%–72%	Rawlings (2001)
Acidithiobacillus acidophilus	Oxidise: S°, organic compounds	H_2SO_4	OC	A	1.5–6.0 (3.0)	5–35 (25 – 30)	60%–75%	Harrison and Norris (1986)
Leptospirillum ferrooxidans	Oxidise: Fe^{2+}, pyrite	Fe^{3+}	OC	A	1.5–4.5 (2.5–3.0)	20–40 (35–45)	70%–85%	Coram and Rawlings (2002), Johnson (2001), Hippe (2000)
Pseudomonas fluorescens	Accumulation by siderophores	Fe^{3+}	H	A	5.2–8.5	22–25 (24)	54%	Kalinowski et al. (2004)
Aspergillus niger	Metabolite-mediated	Oxalate	H	SA	3.0–6.0 (3.5–4.5)	20–40 (35)	85%	Strasser et al. (1994)
Aspergillus terreus	leaching, with accumulation on	Oxalate	H	SA	3.0–6.3	25–30 (28)	75%	Hefnawy et al. (2002)
Penicillium spinulosum	surface	Fumarate	H	SA	4.5–7.5	25–30 (28)	82%	
Cladosporium oxysporum		Unknown	H	SA	3.0	30	71%	
Aspergillus flavus and Curvularia clavata		Unknown	H	SA	3.0	30	59% and 50%	Mishra et al. (2009)

Note: OC = Obligate chemolithotroph; H = heterotroph; A = aerobic; SA = strict aerobic.

from its ores and the optimum conditions for the leaching. The applicability of the various classes of microbes is correlated with their characteristics by virtue of which they utilise these raw materials. For instance, the *Acidithiobacillus* group of bacteria utilises the energy from Fe(III) in an acid medium in the presence of oxygen at an optimum pH of 1.5–2.5 in mesophilic range to leach 70%–98% metal content of the substrate. The optimum levels of nutrients, pH and temperature are very essential to maintain the exponential phase activity of these microbes. In the case of fungus, the primary action is initiated through the metabolite-mediated reactions.

The important microorganisms involved in dissolving uranium from minerals are those that are responsible for producing Fe(III) and H_2SO_4 required for the bioleaching reactions. These are the iron- and sulphur-oxidising chemolithrophic bacteria and archaea (Rossi, 1990). The microorganisms have a number of features in common that make them especially suitable for their role in mineral solubilisation. They grow autotrophically by fixing CO_2 from the atmosphere, which means that it is not necessary to feed them with an organic carbon source. These chemolithotrophs obtain their energy by using either Fe(II) or reduced inorganic sulphur compounds (some use both) as an electron donor and oxygen as the electron acceptor. From the point of view of bio-hydrometallurgy, these organisms use the difference in redox potential between the pair Fe(II)/Fe(III) or S^o/SO_4^{2-} and O_2/H_2O as their source of energy. As sulphuric acid is produced during the oxidation of inorganic sulphur and the microorganisms grow at low pH environments, most mineral dissolution processes operate at a pH between 1.5 and 3.0. This permits these microbes to make use of the iron cycle as both Fe(II) and Fe(III) are soluble at low pH. Besides Fe(II) serving as the electron donor, many of the sulphur-oxidising organisms are able to use Fe(III) in place of oxygen as an electron acceptor (Rossi, 1990; Waksman and Joffe, 1992). This ability is relevant in non-aerated heap leaching in which oxygen might not penetrate to the bottom of the heap. The modest nutritional requirements of these organisms are provided by the aeration of an iron- and/or sulphur-containing mineral suspension in water or the irrigation of a heap or dump. As may be expected, microorganisms that grow in mineral-rich environments are remarkably tolerant to a wide range of metal ions, though there are considerable variations within and between the species (Waksman and Joffe, 1992).

In the bioleaching processes that operate at 40°C or less, the most important microorganisms are believed to be a consortium of Gram-negative bacteria. These include the iron- and sulphur-oxidising *A. ferrooxidans*, the sulphur-oxidising *Acidithibacillus thiooxidans* (*A. thiooxidans*), *Acidithiobacillus caldus* (*A. caldus*), and the iron-oxidising *Leptospirillum ferrooxidans* and *L. ferriphilum* (Markosyan, 1972; Harrison and Norris, 1986; Norris et al., 1986; Barrett et al., 1993). Limited investigations have been carried out on with the microorganisms such as *A. caldus* and *L. ferrooxidans* that dominate at 45–50°C temperature (Sugio et al., 1985; Norris and Parrot, 1986; Goebel and

Stackebrandt, 1994; Hallberg and Lindstrom, 1994; Leduc and Ferroni, 1994; Kelly and Wood, 2000; Rawlings, 2001; Coram and Rawlings, 2002). At temperatures above 65°C, the consortia are dominated by archaea rather than bacteria, with species of *Sulfolobus* and *Metallosphaera* being most prominent (Rawlings et al., 1999; Hallberg and Johnson, 2001). Archaea belong to the genus *Acidianus* such as *Ad. ambivalensi* or *Ad. infernus* that are also capable of growing at high temperatures (90°C for *Ad. infernus*) on reduced sulphur and at low pH (Clark and Norris, 1996a,b). However, the use of these organisms to contribute to industrial uranium bioleaching is not well established compared to those growing at 50°C or less. There are a few reports on the evaluation of microbial species in the metal leaching environment, in which eukaryotic life appears to be restricted largely to certain fungi including yeasts and protozoa (de Siloniz et al., 2002).

The enthusiasm of the microbiologists working on the development of the new biomining techniques is matched by a need in the minerals industry to find alternatives to conventional methods of mining, ore processing and wastewater treatment. So far, *A. ferrooxidans* and *A. thiooxidans* were manipulated aiming at higher leachability of metals; however, the use of indigenous microorganisms derived from the source ore body offers many advantages (Lissette et al., 2006).

3.3 Modes and Mechanism of Uranium Bioleaching

There has been a long-standing debate as to whether the microbially assisted leaching of uranium ore follows a direct (contact) or indirect (noncontact) mechanism (Helmut, 2001). In the leaching of uranium, the bacteria do not directly attack the uranium mineral. But they generate Fe(III) by oxidising pyrite to soluble Fe(II) form. Fe(III) readily attacks minerals incorporating U(IV) converting them into U(VI) which is soluble in dilute sulphuric acid (Ivarson, 1980; Boon and Heijnen, 1998). The schematic representation of the mechanisms for bioleaching of uranium from its ore by *A. ferrooxidans* is shown in Figure 3.3a,b.

3.3.1 Direct Mechanism

In direct leaching, there is a physical contact between the microbial cell and the ore surface. The oxidation of S into SO_4^{2-} takes place through several enzyme catalysed reactions (Figure 3.3a). In the process of obtaining energy from the inorganic material the microbial cell causes electrons to be transferred from Fe or S to O_2 (Chander and Briceno, 1987; Sand et al., 1998; Boon et al., 1999). It should be noted that the inorganic ions never enter the bacterial cell; the electrons released by the oxidation reaction are transported through a protein system in the cell membrane and finally (in aerobic organisms)

(a)

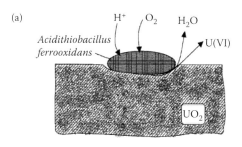

(b)

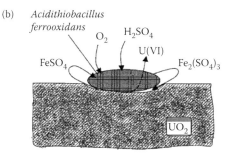

FIGURE 3.3

Schematic representation of bioleaching of uranium ore (a) direct leaching, (b) indirect leaching, using *Acidithiobacillus ferrooxidans*. (From Abhilash and Pandey, B. D., 2013c. *Mineral Processing and Extractive Metallurgy Review: An International Journal*, 34(2), 81–113.)

to oxygen atoms forming H_2O molecules. The transferred electrons give up energy that is coupled to the formation of adenosine tri-phosphate (ATP), the energy currency of the cell (Mignone and Donati, 2004).

Most of the iron containing uranium ores, or iron externally added in the form of salt, is oxidised to iron (III) sulphate according to the following reactions (Sand et al., 2001):

$$4FeS_2 + 14O_2 + 4H_2O \rightarrow 4FeSO_4 + 4H_2SO_4 \tag{3.1}$$

$$4FeSO_4 + O_2 + 2H_2SO_4 \rightarrow 2Fe_2(SO_4)_3 + 2H_2O \tag{3.2}$$

The above reactions can be summarised as

$$4FeS_2 + 15O_2 + 2H_2O \rightarrow 2Fe_2(SO_4)_3 + 2H_2SO_4 \tag{3.3}$$

Therefore, the direct bioleaching of uranium can be described according to the following reaction:

$$2UO_2 + O_2 + 2H_2SO_4 \rightarrow 2UO_2SO_4 + 2H_2O, \Delta G_{303} = -130.4\,kJ \cdot Mol^{-1} \tag{3.4}$$

There is some evidence to show that the microbial cells have some intimate contact with the ore surface (Pradhan et al., 2008; Abhilash and Pandey,

2011). However, the mechanism of attachment and the initiation of uranium solubilisation are not completely understood. The microorganisms attach to the specific sites of the ore surface paving the uranium dissolution due to electrochemical interactions (Sand et al., 1995).

3.3.2 Indirect Mechanism

Indirect leaching is usually referred to as microbial-assisted leaching with *A. ferrooxidans* in which the sulphide leaching proceeds by reaction with Fe(III). The microorganisms need not be in contact with the ore surface but they accelerate the reoxidation of Fe(II) in solution to generate Fe(III), which chemically oxidises the sulphur present in the ore (Figure 3.3b). The bio-oxidation is about 10^5–10^6 times faster than the chemical oxidation (Dziurla et al., 1998). The uranium solubilisation by an indirect mechanism can be described as

$$\text{Fe} \cdot \text{S}_2 + \text{H}_2\text{SO}_4 \rightarrow 2\text{FeSO}_4 + \text{H}_2\text{O} + 2\text{S}° \tag{3.5}$$

$$\text{FeSO}_4 + \text{H}_2\text{SO}_4 + \text{O}_2 \rightarrow \text{Fe}_2(\text{SO}_4)_3 + \text{H}_2\text{O} \tag{3.6}$$

The Fe(II) can be oxidised by microbes to Fe(III), which takes part in the oxidation process again. The sulphur formed is simultaneously oxidised depending on the species to H_2SO_4 which aids (oxidising agent) the dissolution of uranium as follows:

$$2\text{S}° + 3\text{O}_2 + 2\text{H}_2\text{O} \rightarrow 2\text{H}_2\text{SO}_4 \tag{3.7}$$

The insoluble uranium(IV) is oxidised to the water-soluble uranium(VI) sulphate as

$$\text{UO}_2 + \text{Fe}_2(\text{SO}_4)_3 \rightarrow \text{UO}_2\text{SO}_4 + 2\,\text{FeSO}_4 \tag{3.8}$$

The Fe(III) can be generated by the oxidation of pyrite (FeS_2), which is often associated with the uranium ore. In an acidic solution without the bacteria Fe(II) is stable and the leaching mediated by Fe(III) would be slow. *A. ferrooxidans* can accelerate such an oxidation reaction by a factor of more than a million (Brierley, 1982a). There are some evidences that U(IV) can be oxidised to U(VI) enzymatically while using some of the energy of this reaction for the assimilation of CO_2 (Abhilash et al., 2009). Despite the evidence of bacterial adhesion to pyrite and the conviction of many authors that this process is the first stage of direct bio-oxidation (Brierley, 1982b; Boon and Heijnen, 1993), doubts still remain concerning the effective contribution of the attached bacteria on pyrite of the uranium ore. Even its existence is sometimes discussed and recent studies seem to confirm the exclusive occurrence of indirect mechanism (Boon and Heijnen, 1993; Shrihari et al., 1995). However, some investigations show that the kinetics of pyrite leaching is improved when the bacterial sulphide contacts are favoured (Sand et al., 1995; Shrihari et al., 1995). In practice, the leaching of uranium is far more complex than the above analysis might suggest (Rashidi et al., 2012). There are numerous processes in addition to direct enzymatic oxidation and bacterial generation of Fe(III). Some chemical reactions between Fe(III) and metal-sulphide minerals result in the formation of secondary minerals and elemental sulphur, which can bind or inactivate the reactive surfaces. When sulphur is formed, *A. thiooxidans* plays an indispensable role in oxidising the sulphur to sulphuric acid, thus exposing uranium for further leaching.

3.3.3 Bioleaching by Heterotrophs

Since the minerals are not sterile and cannot be sterilised on a commercial scale, bioleaching using heterotrophs presents some process design challenges that are not realised in the bioleaching with autotrophic acidophiles. The acidophilic autotrophs grow in highly selective environments that tolerate hardly any competitor that can displace them, but it is not the case with the heterotrophs. The mechanism of uranium dissolution/accumulation by heterotrophs can be visualised (Mishra et al., 2009; Fomina et al., 2010) by considering the combination of a process involving acidolysis and complexolysis. The free-living and symbiotic fungi secrete carboxylic acids such as oxalic, citric and gluconic acids, which decrease the pH of the media and also act as metal chelators (Mishra et al., 2009) in transforming insoluble uranium to soluble uranium species. There are reports of XANES spectra consistently reporting uranyl coordination by carboxylate ligands, suggesting the involvement of fungal carboxylic acids in uranium solubilisation (Fomina et al., 2010). The fungal species like *Penicillium* sp., *Aspergillus* sp. (Burgstaller and Schinner, 1993) and some algae are known to accumulate uranium on their cell walls or in the cytosol (Calmoi and Cecal, 2007). The microbiological processes for the removal of metals from solution can be divided into three categories: the adsorption of metal ions onto the surface of a microorganism, the intracellular uptake of metals and the chemical transformation of metals by biological agents.

The recovery of uranium from pyritic and sulphidic/nonsulphidic ores using microorganisms such as *Aspergillus, Penicillium, Cladosporium* and *Curvularia* species has been used to upgrade the mineral raw material by the removal of impurities (Mishra et al., 2009). It is based on the generation/secretion of some organic acids such as oxalic, citric and gluconic acids from the heterotrophic microorganisms, particularly fungus, which act as chelating agents. *Aspergillus* species, which synthesises oxalic acid as a metabolite, could leach appreciable amounts of uranium from geological rocks (Strasser et al., 1994). The leach ability of uranium from the ferruginous siltstone (Hefnawy et al., 2002) using native fungal species such as *Aspergillus terreus* and *Penicillium spinulosum* was found to vary with the presence of SiO_2 and

Fe_2O_3 contents in the ore. The ores with high SiO_2 and low Fe_2O_3 resulted in respective uranium leaching of 88.2% and 62% with *A. terreus* at 1% concentration of ore, whereas with *P. spinulosum*, the recovery was 81.5% and 77.6% from the two ores. Greater quantities of uranium leached by these fungal species may be attributed to the production of carboxylic acids in the media that shift the pH to a lower acidity while forming soluble complexes. Quartz sand, kaolins and clays which lower the quality of minerals can be removed by these microorganisms. Best results were achieved when oxalic and citric acids were the main components in the leachant/solution.

3.4 Factors Influencing Bioleaching

The effectiveness of uranium dissolution depends upon the properties of the microorganisms, ore/mineral species including surface area of the minerals, particle size, water availability, temperature, pH, redox potential, oxygen and the carbon dioxide supply, supply of other nutrients (nitrogen compounds and phosphate) and toxic substances and formation of the secondary minerals (Muñoz et al. 1995a; Tuovinen, 1984). Bioleaching of uranium needs to be optimised with respect to the rate of bioleaching reactions and growth of the microorganism involved. To optimise the leaching, it is necessary to understand the nature of biotic and abiotic reactions of the system. However, the effectiveness largely depends on the efficiency of the microorganisms and the mineralogy of the ore to be leached. The maximum yields of uranium can be achieved only when the leaching conditions correspond to the optimum growth conditions of the microorganisms.

3.4.1 Mineralogy of the Ore

No two ores are identical and within each ore deposit the mineralogical composition and the concentration of uranium display heterogeneity. Bioleaching profiles of mineral samples from each mineralisation are unique because of the variation in inclusion and mineral composition, acid demand characteristics and concentration of uranium. Miller et al. (1963) suggested that refractory ores such as brannerite would be difficult to treat microbially. However, the bacterial leaching has been used with good results in the mines near Elliot Lake, Canada, where the brannerite ore abounds. The pyrite content of the mineral is suggested to be of fundamental importance, and hence the extraction of uranium is limited to the minerals with high sulphide content (Lowson, 1975). Reports elicit the beneficial effect of adding pyrite to the uranium containing ores during bioleaching (Miller et al., 1963; Bhurat et al., 1973). Hence, the most suitable quantity of pyrite depends on the mineral content and the characteristics of the pyrite to be added. For example, adding pyrite to the concentration of 5 kg/t to the Indian (Bhurat et al., 1973) and the Portuguese ores, and 3 kg/t to the Spanish ore (Muñoz et al., 1995a) was quite effective.

3.4.2 Microorganisms

A variety of bacteria as well as fungi are known to dissolve uranium from its ores. The process is affected by the bacterial population density, its metal tolerance (Phyllis et al., 1983) and its ability to adapt to the mineral surface and the environment. The rate and the extent of dissolution also depend on the microbial strain within the same species. Inoculum density has been found to affect the bio-leaching rate. Besides, mineralogy of the ore, particle size, pH of the medium, temperature and concentration of oxidant (Fe^{3+}) are known to influence the rate of uranium bioleaching. These parameters are reported to be as just as critical for chemical leaching of uranium in acidic conditions (Bhatti et al., 1998). The toxicity of heavy metals from uranium ore and waste may influence the rate of leaching, and hence the use of metal-tolerant species enhances the leaching process. So far, *A. ferrooxidans*, *A. thioxidans*, *Sulfolobus* and *Leptospirillium* sp. have been used industrially (Muñoz et al., 1995a,b,c).

3.4.3 Temperature

Temperature plays a decisive role in the uranium extraction since the behaviour of certain species depends on it. Seasonal variation in the temperature greatly affects the bioleaching processes. During summer, when the temperature and humidity are high, high rates of uranium dissolution will be observed compared to the lower temperature. The optimum temperature for the oxidation of Fe(II) and sulphide by microorganisms depends on the species (Table 3.2). The effect of temperature can be very well seen with the application of the thermophiles that can grow at higher temperature, so oxidation kinetics may be compared when mesophiles and moderate thermophiles are used. The bioleaching kinetics in the presence of thermophiles (*Sulfolobus* sp., *Acidianus* sp.) is higher than that of mesophiles (*A. ferrooxidans*) and moderate thermophiles of *Leptospirillum* genera (Clark and Norris, 1996a). The bioleaching dissolution reaction is exothermic therefore, the temperature increases during the reaction. The optimum temperature for ferrous iron oxidation was found to be in the range 28–32°C. *A. ferrooxidans*, *A.thiooxidans* and *L. ferrooxidans* are conventionally implemented in bioleaching processes at temperature ranging from 30°C to 45°C (Nemati and Webb, 1998). In recent years, there has been interest in the application of high-temperature processes (65–85°C) utilising thermophilic archaea such as *Sulfolobus* sp. (Norris et al., 1986; Clark and Norris, 1996a,b). Higher temperature leads to more uranium recovery, but the more refractory matrix materials also get decomposed and the impurity level increases in the leach solutions (Blake et al., 1992).

3.4.4 pH

An acidic environment must be maintained in order to keep Fe(III) and uranium in solution (Bhatti et al., 1998) and facilitate leaching. Acidity is controlled by the oxidation of iron, sulphur and metal sulphides, and by the hydrolysis of ferric iron. The hydrogen ions may also be detrimental due to the formation of precipitate of basic ferric sulphates that inactivate the surfaces of sulphide minerals and in some cases may even prevent the flow of the growth medium. The pH values in the range 1.0–2.5 are optimum for the oxidation of iron and S^{2-}; retarded microbial growth at pH below 1.0 occurs because of the considerable inhibitions of the cells to the higher acidity. However, one can adapt the cells either by increasing or decreasing the pH. *Acidithiobacillus* species can attack some suphide minerals and certain soluble sulphur compounds under neutral conditions. A lower activity limit of *A. ferrooxidans* lies around pH 1.0 and is inhibited above pH 5.5 (Rossi, 1990).

3.4.5 Nutrients and Energy Source

In the bio-leaching tests, sufficient nutrients such as nitrogen, magnesium and iron are necessarily required to support the growth of *Acidithiobacillus* species (McCready and Gould, 1990). McCready et al. (1986) have studied the effect of various nutrients on the growth of *A. ferrooxidans* at various conditions and observed that the phosphate ion is an essential nutrient for most cells. The starvation for this component leads to a stressing condition (bacterial growth rate, capacity to oxidise Fe(II) and to fix CO_2 will be reduced) which affects the bioleaching capacity (Seeger and Jerez, 1993). The energy source is the oxidation of Fe(II), elemental sulphur and other sulphur-containing uranium minerals which is mediated primarily by the genera such as *Acidithiobacillus, Leptospirillum, Sulfolobus, Sulfobacilli* and *Acidianus* at low pH values to release uranium (Norris et al., 1986). Ferrous iron is the easily oxidisable substrate for most of the bacteria.

3.4.6 Importance of Iron Oxidation

Small amounts of iron are sufficient to oxidise even large excesses of uranium. Biological oxidation of Fe(II) has been proven to be a significant step in dissolving uranium. Fe(II) is readily oxidised to Fe(III) which can serve as an electron donor (Cabral and Ignatiadis, 2001). The Fe(II)/Fe(III) redox couple has a very positive standard electrode potential (~770 mV at pH 2.0). The concentrations of Fe(III) in excess of 3 g/L have no significant effect and it has been established that a Fe(III) concentration of 1–2 g/L is normally sufficient for the effective dissolution of U(IV) with a minimum of 0.5 g/L Fe(II) (Tuovinen, 1984). As a result only oxygen is able to act as a natural electron acceptor in the presence of protons with water as the reaction product (O_2/H_2O of 820 mV at pH 7.0). In industries, Fe(III) generation has been carried out by adding strong oxidising agents while maintaining an Eh between 400 and 500 mV. The use of Fe as an electron donor will therefore occur only during aerobic respiration since Fe(II) spontaneously oxidises to Fe(III) unless the pH is low. Extreme acidophilic bacteria are able to use Fe(II) as an electron donor in a manner that is not possible for bacteria that grow at neutral pH due to minor difference in redox potential between the Fe(II)/Fe(III) and O_2/H_2O redox couples and only one mole of the electron is released per mole of iron oxidised. The vast amounts of Fe(II) need to be oxidised to produce relatively little cell mass (Kawatra et al., 1989). The large quantity of iron is not transported through the cell membrane but remain outside of the cell and each Fe(II) ion simply delivers its electron to a carrier situated in the cell envelope. Earlier, it was reported that *A. ferrooxidans* can increase the rate of Fe(II) oxidation by half a million to a million times, compared with the abiotic chemical oxidation of ferrous iron by dissolved oxygen (Ohnuki et al., 2005). Nemati and Webb (1997, 1998) have studied extensively the kinetics of iron oxidation for the bacterium *A. ferrooxidans*. Blake et al. (1992) have investigated components of iron oxidation by five different acidophilic microorganisms, three bacteria (*A. ferrooxidans*, unidentified bacterium M1, *L. ferrooxidans*), and two archaea (*S. metallicus* and *M. sedula*). In all five organisms, the components of the electron transport chains were very different and it was concluded that the ability to use Fe(II) as an electron donor had probably evolved independently several times, but the growth of thermophiles on Fe(II) was rather limited. Other soluble metal ions are frequently present in fairly high concentrations in highly acidic environments. Metal ions which exist in more than one oxidation state and which have redox potentials that are more negative than the O_2/H_2O redox couple, have the potential to serve as electron donors for acidophilic bacteria (DiSpirito and Tuovinen, 1982; Hutchins et al., 1986). Whenever Fe(III) is present, it is, however, difficult to unequivocally demonstrate the biological oxidation as opposed to the chemical oxidation by ferric iron.

3.4.7 Sulphur Reduction

The acid responsible for the very low pH environment in which extreme acidophiles are found is most often sulphuric acid, which is produced by the oxidation of inorganic sulphur compounds. For bio-oxidation to occur, the reduced inorganic sulphur serves as an electron donor with O_2 serving as the energetically most favourable electron acceptor. A variety of reduced inorganic sulphur compounds are released as a result of the chemical reaction of suphide minerals with H_2O, Fe(III) and O_2. The kinetics of sulphur oxidation is more difficult to study as in the case of the substrate, which is a solid and no concentration can be determined. The kinetics of sulphur oxidation is greatly depressed for bacteria grown with sulphur as a substrate (Espejo et al., 1988). Sugio et al. (1988a,b, 1992) found that Fe(III) plays an important role in the biochemical mechanisms of sulphur oxidation in *A. ferrooxidans* and other bacteria.

3.5 Bioleaching Techniques

3.5.1 Laboratory Investigations

Experimental approaches for bioleaching of uranium minerals have usually involved at least one of the following techniques: (1) submerged, (2) column and (3) stirred-tank reactors. Each technique has specific/unique characteristics with respect to experimental variable such as aeration, particle size, pulp density (PD) or the solid-to-liquid ratio.

3.5.1.1 Submerged Leaching

Submerged leaching is also known as suspension or shake flask leaching; a very interesting method using fine-grained ore (particle size <100 μm) that is suspended in the nutrient medium and kept in motion by shaking or stirring. Suspension leaching on a laboratory scale in agitated shake flasks is a convenient tool to investigate the bioleach ability of an ore and to reveal the optimal leaching conditions. It is a very effective method and has the advantage that it can easily be controlled and regulated. Considering high yields in uranium extraction by submerged leaching, the change from shake flask to the bioreactor is useful. It may be possible to choose a favourable temperature and to add phosphate, ammonia, carbon dioxide, sulphuric acid, iron, or other additives to accelerate the leaching process. Grinding the ore down to particle size below 100 μm increases the specific surface area and the leaching rate substantially, but these fines are difficult for percolation leaching and thus have to be treated in suspension; therefore, it is done in a flask with agitation/shaking and aeration (Muñoz et al., 1995b). The pulp may contain 10%–30% of solids in suspension (PD). However, it has major limitations of continuously changing conditions. Steady states cannot be reached and it is difficult to control experimental variables such as pH or dissolved oxygen. The lack of steady-state operation makes it difficult to examine the effect of experimental factors because they may be amplified or negated by continuously changing conditions (Gong-Xin et al., 2010). It is also an expensive tool and its application is restricted to special purposes, for instance to the leaching of concentrates. The influence of several variables on the bioleaching of uranium is reported with *A. ferrooxidans*, *A. thiooxidans*, *L. ferrooxidans* and their mixed culture using Spanish ore from the FE Mine, Spain (Muñoz et al., 1995b). Munoz et al. (1995b) reported 98% uranium recovery with *A. ferrooxidans* using Spanish uranium ore at 35°C and pH 2.0. About 49% uranium recovery was reported in 30 days from Turamdih uranium ore (0.024% U_3O_8) using 10% PD at 30°C and pH 2.0 in shake flask experiments (Pal et al., 2010). Bhatti et al. (1998) reported about 93% uranium leaching by chemical lixiviants (acid) using brannerite at 28°C and pH 1.5 with shaking at 180 rpm. Uranium recovery of 60% was achieved by Garcia Jr. (1993) using *A. ferrooxidans* as against 30% leaching in control experiments from

Brazilian low-grade ore (<1400 mm) at 30% PD, 30°C, pH 2.8 and shaking at 200 rpm in 60 days. Recently, Abhilash et al. (2006) reported 98% uranium recovery from Turamdih ore at 35°C, pH 1.7 and 20% PD in 40 days. The bioreactors permit aeration and complete mixing of suspended solids, and control of various parameters can provide most useful information on the factors that influence the kinetics of bioleaching of uranium minerals. Earlier studies focused on the improvement of the bioleaching rate through verification of optimum particle size and PD of the mineral as well as bioreactor configuration. However, in spite of several investigations, there seems to be several technical problems associated with the corrosiveness of the leach liquor, high attrition encountered with the ore particles, difficulties in feeding mineral suspensions, and ensuring sample homogeneity on a large scale (Eisapour et al., 2013). Therefore, the bioreactor technology for uranium bioleaching is yet to find acceptance in the industries.

3.5.1.2 Column/Percolation Bioleaching

Column leaching works on the percolation principle and utilises coarser ore material that, depending on the experimental scale, may be more imitative of the material in dump, heap or in situ bioleaching applications. These studies are of long duration, lasting several months to several years, thereby imposing various constraints that may be difficult to resolve. A characteristic feature of the column leaching is that it appears to result in the formation of different zones within the ore sample that have distinct chemical and physical gradients. These zones may exhibit differences in redox potential, iron precipitation and elemental sulphur formation. These zones may have some analogy with those in large-scale dump and heap bioleaching operations (Qiu et al., 2011). In industrial operations, such zones have not been characterised at mine sites and are likely to include temperature gradients also (Dexin et al., 2012). In the simplest case, the column consists of a glass tube, plastic-lined concrete or steel provided with a sieve plate in its bottom part and filled with ore particles. Muñoz et al. (1995c) reported leaching studies in columns based on their size into three types, namely, (i) small columns (7 cm diameter and 70 cm height), (ii) medium columns (14 cm diameter and 70 cm height) and (iii) large columns (25 cm diameter and 250 cm height). They have reported 60%–85% of uranium recovery from Spanish uranium ore in bioleaching of small, medium and large columns using *A. ferrooxidans*. Sometimes, airlift percolators (5 mm diameter and 30–60 cm height) can be considered as the smallest and the simplest of the class of percolation devices. The coarse size ore particles are not advisable due to irregular flow of leach liquor thereby leading to unreliable data. The ore charge of the percolator ranges from 150 to 300 g, and the average particle size up to 10 mm. The ore bed is irrigated or flooded with microbial cells in a nutrient medium. The leach liquor trickling through the column is pumped up by the compressed sterile air to the top of the column for recirculation. Simultaneously, the stream of air takes care of the aeration of the system. The percolators can be arranged in series and the flow of leach liquor is accomplished by means of peristaltic pumps. To monitor the course of leaching the effluent samples are collected at a certain interval and state of the process is determined on the basis of the pH measurements, micro-biological investigations and the chemical analysis of iron and uranium. The correct operation of percolation leaching needs the periodic addition of distilled water to compensate for the evaporation loss while restoring liquor volume. The percolation leaching has, however, a few disadvantages such as inefficiency and fairly slow rate of metal dissolution; a series of experiments lasting 100–300 days are not unusual due to an inadequate oxygen supply and the unfavourable surface ratio.

In a column leaching experiment (40 cm height and 7.5 cm diameter) packed with 500 g of Figueira-PR ore, Brazil, uranium extraction of approximately 50% was obtained at 30°C and pH 2.8 using *A. ferrooxidans* in 45 days (Garcia Jr., 1993). About 50% of uranium was recovered with *A. ferrooxidans* at 10–12°C and pH 1.2 at Agnew Lake Mine, Canada, from 2 tonnes of <8-inch ore packed in a 2-ft × 16-ft column (McCready and Gould, 1990). Abhilash et al. (2007) reported ~70% uranium bio-recovery on 80 kg and later validated the results on a 2-ton scale for Turamdih ore, India at pH 1.7 in 60 days' time with the average particles of 10–12 mm size.

3.5.2 Large Scale/Industrial Operations

Industrial bioleaching processes can be categorised into two types, namely: (i) irrigation-based principles (dump and heap bioleaching and in situ leaching), and (ii) vat- and stirred-tank bioleaching. The choice of method is influenced by the factors such as the size, shape, grade, depth and thickness of the ore deposits. The rate of recovery of uranium values from heap and dump leach operations has been simulated by mathematically modeled reactions of several minerals and reagents. The rates of all reactions are controlled by molecular diffusion within the particles of heap or dump leaching (Box and Prosser, 1986). It is still arguable that no fundamental differences exist among dump, heap and in situ bioleaching since all these processes are carried out on coarse and unsized fragmented ores. However, dump and heap bioleaching offers advantages such as limited adverse influence of mineral geology and maximum uranium recovery over in situ bioleaching processes.

3.5.2.1 Dump Leaching

The earliest engineering technology used dump leaching that involved gathering low-grade (otherwise waste) ore of large rock/boulder size into vast mounds or dumps (100–2000 m length, 25–400 m height and 100–300 m width) and irrigating it with dilute sulphuric acid to encourage the growth and activities of mineral-oxidising acidophiles, primarily iron-oxidising mesophiles (Rossi, 1990). The residence time of leach liquor ranges from a few hours for dump limited in height to three days for dumps of 80 m height and as much as 12 days for 150 m high dumps. In commercial dump-leaching operations a million tonnes of ore over-burden and waste rock containing small but valuable quantities of uranium are transported by truck or train from open-pit uranium mines to the dump site (Morais et al., 2008). To these impressive formations thousands of gallons of leach liquor is applied by flooding or sprinkling the top surface. Sprinkling introduces air, a vital component of both the chemical and the biological oxidation reactions in the leach liquor. The dumps are not inoculated with the leaching microorganism. They are ubiquitous and when conditions in the rock pile become suitable for their growth then they proliferate. The leach liquor percolates through the leach dump and the pregnant or uranium-laden solution is collected in catch basins or reservoirs at the foot of the dump (Figure 3.4). Uranium is recovered from the solution by solvent extraction or ion exchange. The barren or uranium-free solution is then recycled to the top of the dump. Because of the construction methods employed and the volume of the solid material treated, dump leaching is a crude operation. The placement of the dump in a natural valley can impede the flow of air to the interior of the pile (Figure 3.4). However, the large size of some of the rocks limits contact among the metal-suphide minerals, the oxidising solution and the bacteria.

The oxidation of mining heaps containing pyrite is modeled based on the coupling of the macroscopic transport and the microscopic particle reaction kinetics by incorporating oxygen and heat transport based on the two major mechanisms of convection and diffusion (Fisher, 1966). It was established that the interaction of macroscopic transport with the microscopic particle kinetics, water infiltration, particle size distribution and the irrigation rate have an impact on the oxidation rate of the heap containing pyrite (Davis and Ritchie, 1987; Pantelis and Ritchie, 1991, 1992). The bacterial stope leaching was initiated in 1977 at Agnew Lake Mine, Canada, but terminated in 1980 (McCready and Gould, 1990). The macroscopic transport is the main limiting factor in dump leaching. Using this technique, Rio Algom's Milliken Mine, Elliot Lake area, Canada, has been producing 250,000 lb of uranium per year (Pantelis and Ritchie, 1992).

3.5.2.2 Heap Leaching

Heap leaching offers somewhat more regulation of biological, chemical and engineering factors than dump leaching. It is used to extract uranium from suphide and oxide minerals of a slightly higher grade ore than those subjected to dump leaching. In heap leaching, the rocks are often crushed to avoid the solution contact problems encountered when leaching large boulders and the heaps are built up on impermeable pads to prevent loss of the leach solution into the underlying soil (Figure 3.4). The controlling dimension being the heap's height, which can range from 2 to 3 m to as

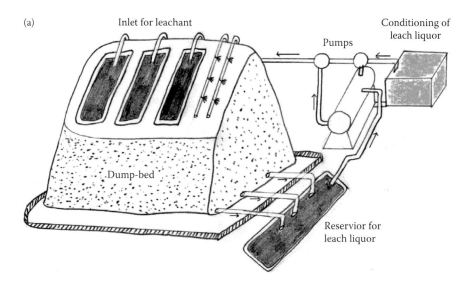

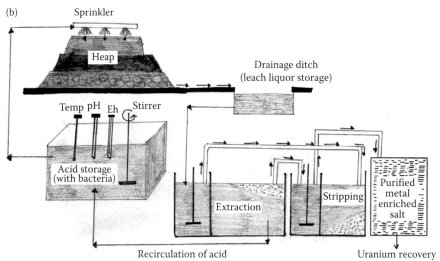

FIGURE 3.4
Various industrial operations of bioleaching, namely, (a) dump leaching; (b) heap leaching; (c) *in situ* leaching. (From Box, J. C. and Prosser, A. P., 1986. *Hydrometallurgy*, 16(1), 77–92; Taylor, G. et al., 2004. *CSIRO Land and Water Client Report August 2004.* Available on line at: www.epa. sa.gov.au/pdfs/isl_review.pdf)

much 7–8 m when the greater part of the ore lumps are of 60–150 mm or greater (McCready and Gould, 1990; Rossi, 1990). The volume of heap can range from a few thousands to 15,000 tonnes accordingly. Boutonnet and Henry (1981) reported 90% uranium recovery at Rio Tinto mines from the heaps made with a run-of-mine ore containing about 4% of <147 mm and

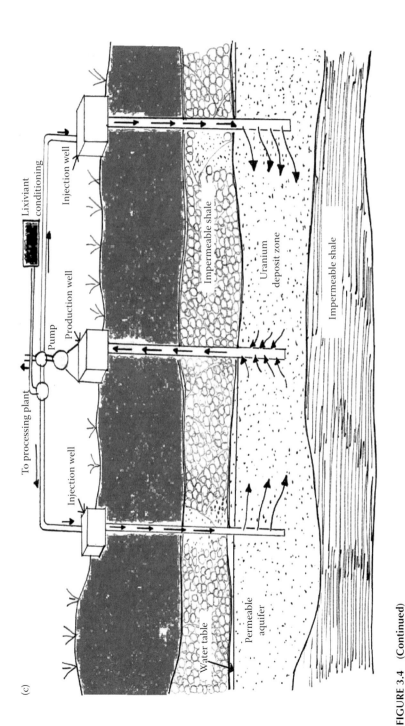

FIGURE 3.4 (Continued)
Various industrial operations of bioleaching, namely, (a) dump leaching; (b) heap leaching; (c) *in situ* leaching. (From Box, J. C. and Prosser, A. P., 1986. *Hydrometallurgy*, 16(1), 77–92; Taylor, G. et al., 2004. *CSIRO Land and Water Client Report August 2004*. Available on line at: www.epa.sa.gov.au/pdfs/isl_review.pdf)

25% of <19 mm particles prevailing particulate leaching. Aeration systems have been installed to increase the flow of air in the piles. The degree of compaction ranges from 80 to 120 proctor for a slope heap leaching. In the early 1960s, the uranium recovery process was made more efficient by the construction and irrigation of specially designed heaps (Schnell, 1997; Brierley, 1982b; Rawlings et al., 2003). However, commercial applications of the heap method are relatively few since preference has usually been given to the much simpler although less-effective dump leaching method for sub-marginal ores and to richer run-off-mines ore. About 57% uranium was recovered in 30 weeks from the stope containing 50-ton ore at the 900-ft depth with 2000 gallons of leach liquor in the temperature range 10–12°C (McCready and Gould, 1990). Mashbir (1964) reported the commercial application of heap leaching from low-grade ores and mine wastes in North America. The heap leaching (850 tonnes) of Figueira-PR ore, Brazil, showed uranium extraction of approximately 50% using *A. ferrooxidans* at pH 2.8 in 45 days (Garcia Jr., 1993). The total Spanish production of uranium (250 t U_3O_8/year) was obtained by heap bioleaching from ENUSA mines (MINA FE), near Ciudad Rodrigo, Spain (Cerda et al., 1993). Thus, heap leaching is applied to ores of low grade to bear the transportation and operating costs associated with the conventional beneficiation processes (Cerda et al., 1993).

3.5.2.3 In Situ Leaching

Many problems encountered with the conventional mining techniques such as exposure to radon gas and mine roof failures are avoided with in situ mining. The process has the advantage such that the ore need not be removed from the ground and is usually carried out on the haloes of the low-grade ore that are left behind after the high-grade ores have been removed. It is most economical when old mine workings can be used to access the low-grade deposits. It is also a promising technique for the recovery of uranium from the low-grade ores in inaccessible sites. This involves fracturing underground working using explosives, percolating with acidic leach liquors containing bacteria, pumping the pregnant liquor to the surface and extraction of solubilised uranium by the established technique. This technology has minimal impact on the environment, and is currently employed to extract residual minerals from abandoned mine working as mentioned above. To dissolve uranium from depleted mine working, the leaching liquor is applied directly to the walls and the roof of an intact stope or to the rubble of fractured working (Figure 3.4). Best known is the in situ leaching operations in the underground uranium mines in Elliot Lake district of Canada, including the Stanrock, Milliken and Denison mines. At that time, the annual production of uranium from the Stanrock Mine was about 50,000 kg U_3O_8 whereas 60,000 kg U_3O_8 was produced in the Milliken Mine, Ontario, Canada after improvement of the leaching conditions (McCready and Gould, 1990). Young uranium deposits (post gla-cial) are of apparent economic interest in view of their common occurrence and amenability to the in situ leaching and lack of radioactive components (Culbert and Leighton, 1988). Nguyen et al. (1983) studied the effect of leach solution quality, initial and transport conditions on the leaching from ura-ninite and pitchblende. In 1988, approximately 300 tonnes of uranium was recovered from Denison mine, Lake Elliot district, Canada (McCready and Gould, 1990; Cabral and Ignatiadis, 2001). However, with the reduction in demand for uranium in subsequent years, this mine stopped production and no microbially assisted uranium recovery is currently taking place. The recoveries of uranium (Mars, 1970; McCready and Gould, 1990; Cerda et al., 1993; Garcia Jr., 1993) in heap, dump and in situ bioleaching processes along with the amount of acid consumption are summarised in Table 3.4. About 1084 tonnes of uranium was produced in 2004 at Beverley mine, Australia, with *A. ferrooxidans*. There are several in situ mining operations that have experienced decreased production due to heavy accumulations of inorganic particles and bacteria on well screens, casings and submers-ible pumps.

3.5.2.4 Tank Leaching

With the development of the first commercial tank bioleaching plant at the Fairview Gold Mine in South Africa for gold processing, efforts were directed to treat other ore concentrates to recover valuable metals (Acevedo, 2000). Stirred-tank leaching may be most effective for the ore concentrates and more than 80% of total uranium can be extracted, if the technology becomes fully adoptable. It is more expensive to construct and to oper-ate tank leaching than dump, heap or in situ leaching processes, but the rate of uranium extraction is much higher and can be used successfully. Even though a relatively limited number of papers has appeared on the stirred reactors, most of them emphasise theoretical and technical difficul-ties, which have discouraged the use of the process on a commercial scale. The highly aerated and continuous-flow reactors placed in series are used to treat the ore. The reactor may be extremely large (up to 1350 m^3) and may allow for greater control (temperature, suphide mineral oxidation being an exothermic reaction) of the process. Finely milled ore concentrate is added to the first tank together with nutrients in the form of salts containing fer-tilisers. The ore suspension flows through series of highly aerated tanks that are pH and temperature controlled. The dissolution of uranium takes place in days when compared with weeks and months in heap or dump leaching operations. The tank reactors that operate at 40–50°C (i.e. meso-philes and thermo-tolerant acidophiles) have proved to be highly robust, and very little process adaptation is required for the treatment of different ore types. The major constraint using tank reactor is the pulp density (PD) that should be maintained in the suspension. Generally, the pulp density

TABLE 3.4

Industrial Operations on Bioleaching of Uranium

Mine Name	Microorganism	Type of Operation	Acid Consumption (kg/tonnes)	% U recovery/days (d) or year(s) (yr)	References
Brazil					
Figueria Mine	A. ferrooxidans	Heap	30	51.3/83d	Garcia Junior (1993)
Canada					
Rio Algom Mine	A. ferrooxidans	Heap	1–6	47/210d	McCready and
Dension Mine	A. ferrooxidans	Stope	1–4	50/200d	Gould (1990)
Angew Lake Mine	A. ferrooxidans	Stope	1–4	57/210d	
Australia					
Ranger Mine	A. ferrooxidans or natural	Heap and dump	1–3	78/150d	Kawatra and
Olympic Dam Mine	Mixed culture	In situ	1–3	70/yr	Natarajan (2001)
Beverley Mine	Mixed culture	In situ	1–3	70/yr	
Kazakhstan					
Stepnogorsk Mine	A. ferrooxidans	Heap	1–5	82/yr	Fyodorov (1999)
Southern ISL Mines	A. ferrooxidans	In situ	1–10	51/yr	
Spain					
ENUSA Mine	A. ferrooxidans and A. thiooxidans	Heap	1–2	75/150d	Muñoz et al. (1995a)
France					
Saint Pierre Mine	A. ferrooxidans	Heap	1–3	90/yr	Rossi (1990)
India					
Turamdih Mine	A. ferrooxidans	Column	1.26	70/60d	Abhilash et al. (2007)

is limited to that of 20%, as with any further increase the liquor becomes too thick for efficient gas transfer and the shear force induced by the stirrer or impeller causes damage to the microbial cells as in vat leaching. This limitation in PD, and higher capital and running costs as compared to the heap or dump leaching discourage their use in uranium dissolution. As per known literature, no country is producing uranium using this technology, whereas it is considered very useful for the recovery of copper and gold on a large scale.

3.6 Bioleaching of Uranium in the Indian Scenario

Uranium was first detected at Jaduguda (quartz–chlorite–biotite–schist type) in 1951. The first uranium processing plant was commissioned in 1967 at Jaduguda, Jharkhand. Different mining methods and leaching techniques were studied to find the most suitable alternative keeping the cost and the environmental impact as low as possible (Banerjee, 1999). The uraninite in Jaduguda, Bhatin and Narwapahar ores is entirely UO_2 (IV) and UO_3 (VI) types, and hence UO_2(IV) remains undissolved or dissolves slowly in the absence of iron in the traditional method (Gupta et al., 2004). The use of solid–liquid leaching technique as unit operation (H_2SO_4/MnO_2, $NaClO_3$, $Fe_2(SO_4)_3$ or pyrolusite leaching) is the most important process by which uranium ores are being processed in India. It is reported that at 1 kg/ton ferric sulphate consumption with 1 kg/ton acid, the same amount of uranium can be leached with 20 kg/ton of acid and 4 kg/ton pyrolusite (Agate, 1983). The sulphate leach liquor (pH 1.0– 2.0) typically carries 0.5–0.6 g/L uranium and other impurities like Fe, Al, V, Cu, Mn and so forth in varying concentrations depending on the nature of the mineral and leaching conditions employed.

For a country with limited energy resources and for long-term energy security, bioleaching exploiting microorganisms is an alternate, highly selective, eco-friendly and economically attractive option (Price et al., 2004). The procedures are not complicated and are easy to control, and extensive technical knowledge is not required (Acevedo et al., 1993). Moreover, the microorganisms used in these processes are able to grow in acidic environment with high metal content such as U, Th, Cu and Ni. Table 3.5 summarises the application of bioleaching of uranium in the Indian scenario in the light of commercial potentials with respect to the available deposits. *A. ferrooxidans* are reported to be the dominant species in the Jaduguda and Turamdih mine waters (Agate, 1983). Mathur et al. (2000) have reported 79% uranium recovery in 12 h using 9 g H_2SO_4/kg of Jaduguda ore adopting BACFOX (bacterial film oxidation) method containing *A. ferrooxidans*. Pal et al. (2010) reported 49% recovery on bench scale using *A. ferrooxidans* at 30°C, pH 2.0 and 10% PD from the

TABLE 3.5

Bioleaching of Uranium Ores in the Indian Scenario

Sl. No.	Type of Ore	Conditions	Remarks	References
1	Keruadungri ore	Bench ferric sulphate leaching	92.8% extraction	Mathur and Dwivedy (1994), Dwivedy et al. (1972)
2	Mosabani plant tailings with uranium oxide	Percolation bioleaching using ferric sulphate (0.5 t)	64% extraction	Muralikrishna and Bhurat (1993)
3	Uraninite ore of Chhinjra, Himachal Pradesh	Heap leaching (sulphuric acid as leachant and manganese dioxide as oxidant.)	84% extraction	Sankaran et al. (1988)
4	Quartz chlorite-sericite (Jaduguda mine)	Bench-scale biogenically generated ferric sulphate as an oxidative leachant using *A. ferrooxidans* by Bacfox process	95% extraction	Mathur and Dwivedy (1994)
5	Uraninite ore of Turamdih, Jharkhand	Bench scale leaching by mine isolate of bacteria (*A. ferrooxidans*)	98% extraction	Abhilash et al. (2007)
6		80 kg and 2.0T column leaching by mine isolate of bacteria (*A. ferrooxidans*)	68%–70% extraction	Abhilash et al. (2007)
7		Bench-scale leaching by mine isolate of fungus (*Cladosporium oxysporum*)	71% extraction	Mishra et al. (2009)
8		Bench-scale leaching by mine isolate of bacteria (*A. ferrooxidans*)	49% extraction	Pal et al. (2010)

Turamdih ore. Recently, 98% and 70% uranium biorecovery from Turamdih ore was reported from the author's laboratory (Abhilash et al., 2006, 2007, 2009, 2010, 2011a,b, 2012a,b) in shake flasks and column experiments, respectively, at pH 1.7. Agitation leaching adopting biogenically produced Fe(III) (BACFOX process) at ambient temperature may be considered viable, techno-economically feasible and an eco-friendly approach (Abhilash and Pandey, 2013a,b), for which a separate analysis is required and the same is not included here. In situ bioleaching plants can be built in the immediate vicinity of the ore deposit of very low-grade, saving transport costs. This technology should be of great interest for the other developing countries. In the near future, bioleaching in Pachuca with dump and heap using *A. ferrooxidans* may be considered at the industrial scale to accommodate the energy demand of the country.

3.7 Conclusions

To meet the growing demand of uranium for power generation and other applications, efficient use of the below cut-off grade ores and secondary resources by bioleaching is often considered because it is technically feasible, economically viable and an eco-friendly option. Development of the bioleaching technology advanced rapidly during 1960s and 1970s. The implementation of heap, dump and stope bioleaching processes has been found to be quite successful and utilised the world over. In particular, mines in Canada and Portugal were running on large-scale heap leaching processes for recovering uranium. In the 1980s, the commissioning of the first commercial tank bioleaching plant at Fairview Gold mine in South Africa paved the way for an efficient bioleaching option for gold, but the same has not been found favourable for the uranium industry and there needs to be concerted efforts to exploit the technology. Limited research activities on uranium bioleaching are being pursued. A few such programs are aimed at developing the heap bioleaching for certain deposits. Mineralogical features of each ore differ from one another and so is their amenability to bioleaching by microorganisms. The use of indigenous microorganisms isolated from the specific mine environment has been the most practical approach for establishing a bioleaching process for a particular deposit because of their natural tolerance to the chemical and environmental conditions. Among these, the acidophilic, mesophilic microbe, *A. ferrooxidans* in combination with *A. thiooxidans* has been mostly exploited. Currently, operating industrial biomining processes have used bacteria that grow optimally from ambient to 50°C and therefore, the role of *Leptospirillium* species – a moderate thermophile along with *A. ferrooxidans* has also been described. The thermophilic microorganisms have the potential to enable mineral bio-oxidation to be carried out at still higher temperatures of 70°C or more. The potential of *Sulfolobus* species to leach uranium from ores is being recognised because of the extraordinary ability of these organisms to attack resistant mineral structures while improving the kinetics. The development of high-temperature processes will extend the variety of minerals that can be commercially processed.

Further study is needed to assess the factors such as the effect of hydrostatic pressure on the leaching microorganisms and the loss of permeability resulting from their growth in industrial leaching operations. Bio-leaching for recovering uranium has a great potential for exploitation worldwide, including India, to treat poor ore bodies, secondary resources and even the rock adjacent to the currently acceptable mining area. By understanding the actual mechanism of bacterial oxidation (direct/indirect) of uranium mineral, it would be possible to tailor microbial species (single/consortia) to a particular environment for enhanced efficiency.

References

Abhilash, Singh, S., Mehta, K. D., Kumar, V., Pandey, B. D. and Pandey, V. M., 2006. Microbially catalysed process for uranium dissolution from its ore. In *Proceedings of the International Sem. Min. Process. Tech* (G. Bhaskar Raju, S. Prabhakar, S. Subba Rao, D. S. Rao and T. V. V. Kumar, Eds.) Chennai: Allied Publishers, pp. 717–723.

Abhilash, Mehta, K. D., Kumar, V. and Pandey, B. D., 2007. Bio-hydrometallurgical approach in processing of low grade Indian uranium ore in column reactor. *Proceedings of Biohydrometallurgy*, BIOHYDROMET 2007, Falmouth, UK: Minerals Engineering International, pp. 531–536.

Abhilash, Singh, S., Mehta, K. D., Kumar, V., Pandey, B. D. and Pandey, V. M., 2009. Dissolution of uranium from silicate-apatite ore by *Acidithiobacillus Ferrooxidans. Hydrometallurgy*, 95, 70–75.

Abhilash, Mehta, K. D., Kumar, V., Pandey, B. D. and Tamrakar, P. K., 2010. Column bioleaching of a low grade silicate ore of uranium. *Mineral Processing and Extractive Metallurgy Review*, 31, 224–235.

Abhilash and Pandey, B. D., 2011. Role of ferric ions in bioleaching of uranium from a low tenor Indian ore. *Canadian Metallurgical Quarterly*, 50(2), 102–112.

Abhilash, Mehta, K. D., Kumar, V., Pandey, B. D. and Tamrakar, P. K., 2011a. Bioleaching—An alternate uranium ore processing technology for India. *Energy Procedia*, 7, 158–162.

Abhilash, Pandey, B. D. and Ray, L., 2012. Bioleaching of apatite rich low grade Indian uranium ore. *Canadian Metallurgical Quaterly*, 51(4), 390–402.

Abhilash, Pandey, B. D. and Ray, L., 2011b. Column bioleaching of apatite rich low grade Indian uranium ore. In. *Proceedings of the 19th International Biohydrometallurgy Symposium (IBS-2011)*, (G. Qiu, T. Jiang, W. Qin, X. Liu, Y. Yang and H. Wang, Eds.) Changsha, China: CSU, China Press, Vol. 1, pp. 680–683.

Abhilash and Pandey, B. D., 2013a. Microbial processing of apatite rich low grade Indian uranium ore in bioreactor. *Bioresource Technology*, 128, 619–623.

Abhilash and Pandey, B. D., 2013b. Bioreactor leaching of uranium from a low grade Indian silicate ore. *Biochemical Engineering Journal*, 71, 111–117.

Abhilash and Pandey, B. D., 2013c. Microbially assisted leaching of uranium—A review. *Mineral Processing and Extractive Metallurgy Review: An International Journal*, 34(2), 81–113.

Acevedo, F., 2000. The use of reactors in biomining processes.' *Electronic Journal of Biotechnology*, 3(3), 1–11.

Acevedo, F., Gentina, J. C. and Bustos, S., 1993. Bioleaching of minerals—A valid alternative for developing countries. *Journal of Biotechnology*, 31, 115–123.

Agate, A. D., 1983. Bioleaching of Indian uranium ores, Progress in biohydrometallurgy. In *Proceedings of International Biohydrometallurgy Symposium*, Calgiri, pp. 325–330.

Amme, M., Wiss, T., Thiele, H., Boulet, P. and Lang, H., 2005. Uranium secondary phase formation during anoxic hydrothermal leaching processes of UO 2 nuclear fuel. *Journal of Nuclear Materials*, 341, 209–223.

Banerjee, D. C., 1999. Uranium exploration in the petrozoic basin in India—Present status and future strategy. *Developments of Uranium Resource, Production, Demand and Environment*, Vienna: IAEA Bulletin, 81–94.

Barrett, J., Hughes, M. N., Karavaiko, G. I. and Spencer, P. A., 1993. Metal extraction by bacterial oxidation of minerals. In *Inorganic Chemistry* (E. H. J. Burgess, Ed.) Chichester: Ellis Horwood, pp. 212–221.

Bhatti, T. M., Vuorinen, A., Lehtinen, M. and Tuovinen, O. H., 1998. Dissolution of uraninite in acid solutions. *Journal of Chemical Technology and Biotechnology*, 73(3), 259–263.

Bhurat, M. C., Dwivedy, K. K., Jarayam, K. M. L. and Dar, K. K., 1973. Some results of microbial leaching of uranium ore samples from Narwapahar, Bhatin and Keruandri, Singhbhum District, Bihar. *NML Technical Journal*, 15(4), 47–51.

Blake, R. C., Schute, E. A., Waskovsky, J. and Harrison, A. P. Jr., 1992. Respiratory components in acidophilic bacteria that respire on iron. *Geomicrobiology Journal*, 10, 173–192.

Boon, M., Brasser, H. J., Hansford, G. S. and Heijnen, J. J., 1999. Comparison of the oxidation kinetics of different pyrites in the presence of *Thiobacillus ferrooxidans* or *Leptospirillum ferrooxidans*. *Hydrometallurgy*, 53(1), 57–72.

Boon, M. and Heijnen, J. J., 1993. Mechanisms and rate-limiting steps in bioleaching of sphalerite, chalcopyrite and pyrite with *Thiobacillus ferrooxidans*. In *Biohydrometallurgical Technologies* (A. E. Torma, J. E. Wey and V. L. Lakshmanan, Eds.) Warrendale, PA: The Minerals, Metals and Materials Society, pp. 217–236.

Boon, M. and Heijnen, J. J., 1998. Chemical oxidation kinetics of pyrite in bioleaching processes. *Hydrometallurgy*, 48, 27–41.

Bosecker, K., 1990. Microbial leaching in environmental clean-up programmes. Biohydrometallurgy and environment towards the mining of 21st century. In *Proceedings of International Biohydrometallurgy Symposium* (Jackson Hole, Ed.), Wyoming, pp. 533–545.

Boutonnet, G. and Henry, J., 1981. How new plant designs make small U 3 O 8 deposit economic. *World Minister*, 34(5), 50–56.

Box, J. C. and Prosser, A. P., 1986. A general model for the reaction of several minerals and several reagents in heap and dump leaching. *Hydrometallurgy*, 16(1), 77–92.

Brierley, C. L., 1978. Bacterial leaching. *CRC Critical Reviews in Microbiology*, 6, 207–262.

Brierley, C. L., 1979. Effect of hydrogen peroxide on leach dump bacteria. *AIME Transactions*, 266, 1860–1863.

Brierley, C. L., 1982a. Microbial mining. *Scientific American*, 247, 42–50.

Brierley, C. L., 1982b. Ferric iron reduction by *Thiobacillus ferrooxidans* at extreme low pH values. *Scientific American*, 247(2), 42–51.

Brierley, C. L., 1997. Mining biotechnology: Research to commercial development and beyond. In *Biomining: Theory, Microbes and Industrial Processes* (D. E. Rawlings, Ed.) Berlin, Germany: Springer Verlag, pp. 3–17.

Brierley, J. A. and Brierley, C. L., 1999. Present and future commercial applications of biohydrometallurgy. In *Biohydrometallurgy and the Environment toward the Mining of the 21st Century* (R. Amils and A. Ballester, Eds.) Amsterdam, Netherlands: Elsevier, pp. 81–89.

Bruynesteyn, A., 1989. Mineral biotechnology. *Journal of Biotechnology*, 11(1), 1–10.

Burgstaller, W. and Schinner, F., 1993. Leaching of metals with fungi. *Journal of Biotechnology*, 27, 91–116.

Cabral, T. and Ignatiadis, I., 2001. Mechanistic study of the pyrite–solution interface during the oxidative bacterial dissolution of pyrite (FeS_2) by using electrochemical techniques. *International Journal of Mineral Process*, 62, 41–64.

Calmoi, R. and Cecal, A., 2007. Bioleaching of uranyl ions from uranium ores by some algae. *Environmental Engineering and Management Journal*, 6(1), 27–30.

Cerda, J., Gonzalez, S., Rios, J. M. and Quintana, T., 1993. Uranium concentrates bio-production in Spain: A case study. *FEMS Microbiology Reviews*, 11, 253–259.

Chadwick, J., 1997. McArthur River uranium. *Mining Magazine*, 19, 1–191.

Chander, S. and Briceno, A., 1987. Kinetics of pyrite oxidation. *Minerals and Metallurgical Processing*, 4, 171–173.

Chi, S. J. and Yun, U., 2006. Radioactive intensity and geochemistry of uranium-bearing black shale from the Ogcheon Fold Belt, South Korea. *Geophysical Research Abstracts*, 8, 03861–2.

Clark, D. A. and Norris, P. R., 1996a. Oxidation of mineral sulfides by thermophilic microorganisms. *Minerals Engineering*, 9(11), 1119–1125.

Clark, D. A. and Norris, P. R., 1996b. *Acidimicrobium ferrooxidans* gen. nov. sp. nov.: Mixed culture ferrous iron oxidation with *Sulfobacillus* species. *Microbiology*, 141, 785–790.

Coram, N. J. and Rawlings, D. E., 2002. Molecular relationship between two groups of the genus *Leptospirillum* and the findings of *Leptospirillum ferriphylum*. sp. dominates. *Applied and Environmental Microbiology*, 68, 838–845.

Culbert, R. R. and Leighton, D. G., 1988. Young uranium. *Ore Geology Reviews*, 3, 313–330.

Currier, R. H., 2002. Searching for minerals in the Democratic Republic of the Congo. *Mineralogical Record*, 33 (6), 473–487.

Davis, G. B. and Ritchie, A. I. M., 1987. A model of oxidation in pyritic mine wastes: Part 3: Import of particle size distribution. *Applied Mathematical Modelling*, 11, 417–422.

de Siloniz, M. L., Eva-Maria, P., Miguel-Angel, C., Domingo, M. and Jose, M. P., 2002. Environmental adaptation factors of two yeasts isolated from the leach ate of a uranium mineral heap. *FEMS Microbiology Letters*, 210, 233–237.

de Siloniz, M. I., Lorenzo, P., Murua, M. and Perera, J., 1993. Characterization of a new metal-mobilizing *Thiobacillus* isolate. *Archives of Microbiology*, 159, 237–243.

Deffeyes, K. S. and MacGregor, I. D., 1980. World uranium resources. *Scientific American*, 242(1), 50–60.

Dexin, D., Yulong, Liu., Guangyue, Li., Nan, Hu., Yongdong, W., 2012. Two stage column leaching of uranium from uraninite ore. *Advanced Science Letters*, 5(1), 96–100.

DiSpirito, A. A. and Tuovinen, O. H., 1982. Uranous ion oxidation and carbon dioxide fixation by *Thiobacillus ferrooxidans*. *Archives of Microbiology*, 133(1), 28–32.

Dutrizac, J. E. and MacDonald, R. J. C., 1974. Ferric iron as a leaching medium. *Minerals Science and Engineering*, 6, 59–62.

Dwivedy, K. K. and Mathur, A. K., 1995. Bioleaching-our experience. *Hydrometallurgy*, 38(1), 99–109.

Dwivedy, K. K., Bhurat, M. C. and Jayaram, K. M. V., 1972. Heap and bacterial leaching tests on sample of low grade uranium ore from Keruadungri, Singhbhum, Bihar. *NML Technical Journal*, 14, 72–75.

Dziurla, M. A., Achouak, W., Lam, B.-T., Heulin, T. and Berthelin, J., 1998. Enzyme-linked immunofiltration assay to estimate attachment of *Thiobacilli* to pyrite. *Applied and Environmental Microbiology*, 64, 2937–2942.

Eisapour, M., Keshtkar, A., Moosavian, M. A. and Rashidi, A., 2013. Bioleaching of uranium in batch stirred tank reactor: Process optimization using Box–Behnken design. *Annals of Nuclear Energy*, 54, 245–250.

Elshafeea, H. Y., Abow Slama, Etemad Ebraheem, Adam K. Sam, 2014. Precipitation and purification of uranium from rock phosphate. *Journal of Radioanalytical and Nuclear Chemistry*, 299(1), 815–818.

Espejo, R. T., Escobar, B., Jodlicki, E., Uribe, P. and Badilla-Ohlbaum, R., 1988. Oxidation of ferrous iron and elemental sulfur by *Thiobacillus ferrooxidans*. *Applied and Environmental Microbiology*, 54, 1694–1699.

Fisher, J. R., 1966. Bacterial leaching of Elliot Lake uranium ore, Canada. *Mining and Metallurgical Bulletin*, 59, 588–592.

Fomina, M., Charnock, J. M., Hillier, S., Alvarez, R., Livens, F. and Gadd, G. M., 2010. Role of fungi in the biogeochemical fate of depleted uranium. *Current Biology*, 18, R375–R377.

Frank, L., Hess, A. B. and Roger, C. W., 1920. Brannerite: A new uranium mineral. *Journal of the Franklin Institute*, 189(2), 225–237.

Frondel, C., 1958. Systematic minerology of uranium and thorium, *United States Geology Survey Bulletin*, 1064, United States Government Press, Washington, DC.

Fyodorov, G. V., 1999. Uranium deposits of the Inkay—Mynkuduk ore field, Kazakhstan. *Proceedings of the Developments in Uranium Resources, Production, Demand and the Environment*, Vienna, 15–18 June, 1999, pp. 95–112.

Garcia, Jr., O., 1993. Bacterial leaching of uranium ore from Figueira-PR, Brazil, at laboratory and pilot scale. *FEMS Microbiology Reviews*, 11, 237–242.

Goebel, B. M. and Stackebrandt, E., 1994. Culture and phylogenetic analysis of mixed microbial population found in natural and commercial bioleaching environments. *Applied and Environmental Microbiology*, 60, 1614–1621.

Gong-Xin, C., Guan-Chai, W. and Jin-Hui, L., 2010. Study on bioleaching of uranium ore in magnetic stirring reactor and gas stirring reactor. *Geochemica et Cosmochimica Acta*, 74(11), Supplement 1.

Groudeva, V. I. and Groudeva, S. N., 1990. Microbial leaching of uranium from flotation tailings in alkaline media. *Process Metallurgy*, 9, 319–326.

Gupta, R., Pandey, V. M., Pranesh, S. R. and Chakravarty, A. B., 2004. Study of an improved technique for precipitation of uranium from eluted solution. *Hydrometallurgy*, 71, 429–434.

Hallberg, K. B. and Johnson, D. B., 2001. Biodiversity of acidophilic prokaryotes. *Advances in Applied Microbiology*, 49, 37–84.

Hallberg, K. B. and Lindstrom, E. B., 1994. Characterisation of *Thiobacillus cladus*. sp. nov, a moderately acidophile. *Microbiology*, 140, 3451–3456.

Harrison, A. P. and Norris, P. R., 1986. Characteristics of *Thiobacillus ferrooxidans* and other iron oxidizing bacteria with emphasis on nucleic acid analyses. *Biotechnology and Applied Biochemistry*, 8, 249–257.

Hefnawy, M. A., El-Said, M., Hussein, M. and Amin, M. A., 2002, Fungal leaching of uranium from its geological ores in Alloga Area, West Central Sinai, Egypt. *Online Journal of Biological Sciences*, 2(5), 346–350.

Helmut, T., 2001, Direct versus indirect bioleaching. *Hydrometallurgy*, 59(2–3), 177–185.

Hippe, H., 2000, *Leptospirillum* gen. nov. (ex Markosyan 1972), nom. rev., including *Leptospirillum ferrooxidans* sp. nov. (ex Markosyan 1972) nom. rev. and *Leptospirillum thermo-ferrooxidans* sp. nov. (Golovacheva et al. 1992). *International Journal of Systematic and Evolutionary Microbiology*, 50, 501–503.

Hutchins, S. R., Davidson, M. S., Brierley, J. A. and Brierley, C. L., 1986, Microorganisms in reclamation of metals. *Annual Review of Microbiology*, 40, 311–336.

Ivarson, K. C., 1980, Enhancement of uranous-ion oxidation by *Thiobacillus ferrooxidans*. *Current Microbiology*, 3, 253–254.

John, L. J. and Andrew, C. R., 1997, New mineral names. *American Mineralogist*, 82, 207–210.

John, L. J., Nikolain, P. and Andrew, C. R., 1999, New mineral names. *American Mineralogist*, 84, 195–1198.

Johnson, D. B., 1998, Biodiversity and ecology of acidophilic microorganisms. *FEMS Microbiology Ecology*, 27, 307–317.

Kalinowski, B. E., Oskarsson, A., Albinsson, Y., Arlinger, J., Degaard-Jensen, A. O., Andlid, T. and Pedersen, K., 2004. Microbial leaching of uranium and other trace elements from shale mine tailings at Ranstad. *Geoderma*, 122, 177–194.

Kawatra, S. K., Eisele, T. C. and Bagley, S. T., 1989. Studies of pyrite dissolution in Pachuca tanks and depression of pyrite flotation by bacteria. In *Biotechnology in Minerals and Metal Processing* (B. J. Scheiner, F. M. Doyle and S. K. Kawatra, Eds.) Littleton, CO: Society of Mining Engineers, pp. 55–62.

Kawatra, S. K. and Natarajan, K. A., 2001. *Mineral Biotechnology: Microbial Aspects of Mineral Beneficiation, Metal Extraction, and Environmental Control*, Englewood, CO: Society of Mining, Metallurgy and Exploration (SME) Inc.

Kelly, D. P. and Wood, A. P., 2000. Reclassification of some species of *Thiobacillus* to newly designated genera *Acidithiobacillus* gen. nov., and *Thermobacillum* gen. nov. *International Journal of Systematic and Evolutionary Microbiology*, 50, 511–516.

Ketzinel, Z., Volkman, Y., Hazzid, M. and Azaria, M., 1984. On the possibility of high temperature selective leaching of uranium from raw phosphorites. *Hydrometallurgy*, 12, 29–132.

Krebs, W., Brombacher, C., Bosshard, P. P., Bachofen, R. and Brandl, H., 1997. Microbial recovery of metals from solids. *FEMS Microbiology Reviews*, 20(3–4), 605–617.

Leduc, L. G. and Ferroni, G. D., 1994. The chemolithotrophic bacterium *Thiobacillus ferrooxidans*. *FEMS Microbiology Reviews*, 14(2), 103–120.

Leonardos, O. H., Fernandes, S. M., Fyfe, W. S. and Powell, M., 1987. The microchemistry of uraniferous laterites from Brazil: A natural example of inorganic chromatography. *Chemical Geology*, 60(1–4), 111–119.

Lissette, V., Chi, A., Simon, B., Alvaro, O., Nicolas, G., Jeff, S., Donald, F. H. and Carlos, A. J., 2006. Genomics, metagenomics and proteomics in biomining microorganisms. *Biotechnology Advances*, 24(2), 197–211.

Lowson, R. T., 1975. Bacterial leaching of uranium ores: A review. *Australian Atomic Energy Commission (AAEC)*, Publ. E356, 24, 1–54.

Lundergren, D. G. and Silver, M., 1980. Ore leaching by bacteria. *Annual Review Microbiology*, 34, 263–283.

Maozhong, M., Changquan, F. and Mostafa, F., 2005a, Petrography and genetic history of coffinite and uraninite from the Liueryiqi granite-hosted uranium deposit, SE China. *Ore Geology Reviews*, 26(3–4), 187–197.

Maozhong, M., Huifang, X., Jia, C. and Mostafa, F., 2005b, Evidence of uranium biomineralization in sandstone-hosted roll-front uranium deposits, northwestern China. *Ore Geology Reviews*, 26(3–4), 198–206.

Markosyan, G. E., 1972. A new iron-oxidizing bacterium *Leptospirillum ferrooxidans*. *Biology Journal of Armenia*, 25, 26–29.

Mars, L. F., 1970. Underground leaching of uranium at the Pitch mine. *Mining Congress Journal* 36–41.

Mashbir, D. S., 1964. Heap leaching of low-grade uranium ore. *Mining Congress Journal* (50), 50–54.

Mathur, A. K. and Dwivedy, K. K., 1994. Microbial leaching of uranium from low grade ores: A review. *Annual Review Microbiology*, 34, 263–283.

Mathur, A. K., ViswaMohan, K., Mohanty, K. B., Murthy, V. K. and Seshadrinath, S. T., 2000. Uranium extraction using biogenic ferric sulphate—A case study on quartz-chlorite ore from Jaduguda, Singhbhum Thrust Belt, Bihar, India. *Minerals Engineering*, 13(5), 575–579.

McCready, R. G. and Gould, W. D., 1990. Bioleaching of uranium. In *Microbial Mineral Recovery* (H. L. Erlich and C. L. Brierley, Eds.) New York: McGraw-Hill, pp. 107–126.

McCready, R. G. L., Wadden, D. and Marchbank, A., 1986. Nutrient requirements for the in-place leaching of uranium by *Thiobacillus ferrooxidans*. *Hydrometallurgy*, 17, 61–71.

Michel, P. and Jouin, J. P., 1985. Le traitement des minerais d'uranium a Bessines, Ind. Miner. *Industrie Minérale Mines et Carrières Les Techniques*, 5, 207–210.

Mignone, C. F. and Donati, E. R., 2004. ATP requirements for growth and maintenance of iron-oxidizing bacteria. *Biochemical Engineering Journal*, 18(3), 211–216.

Miller, R. P., Napier, E., Wells, R. A., Audsley, A. and Dabom, G. R., 1963. Natural leaching of uranium ores: Discussions and contributions. *Transactions Institute of Mining and Metallurgical*, 72, 217–254.

Mishra, A., Pradhan, N., Kar, R. N., Sukla, L. B. and Mishra, B. K., 2009. Microbial recovery of uranium using native fungal strains. *Hydrometallurgy*, 95(1–2), 175–177.

Morais, C. A., Gomiero, L. A. and Fitho, W. S., 2008. Leaching of uranium ore from Caetite's Facilities, Bahia State, Brazil. In *6th Proceedings of International Symposium on Hydrometallurgy, Arizona* (C. A. Young, P. R. Taylor, C. G. Anderson and Y. Choi, Eds.), Englewood, Colorado: Society for Mining, Metallurgy, and Exploration, pp. 1119–1122.

Mostafa, F. J. and Janeczek, R. C., 1997. Ewing mineral chemistry and oxygen isotopic analyses of uraninite, pitchblende and uranium alteration minerals from the Cigar Lake deposit, Saskatchewan, Canada. *Applied Geochemistry*, 12(5), 549–565.

Muñoz, J. A., Ballester, A., Gonzalez, F. and Blazquez, M. L., 1995b. A study of the bio-leaching of a Spanish uranium ore. Part II: Orbital shaker experiments. *Hydrometallurgy*, 38(1), 59–78.

Muñoz, J. A., Blazquez, M. L., Ballester, A. and Gonzalez, F., 1995c. A study of the bioleaching of a Spanish uranium ore. Part III: Column experiments. *Hydrometallurgy*, 38(1), 79–97.

Muñoz, J. A., Gonzalez, F., Blazquez, M. L. and Ballester, A., 1995a. A study of the bioleaching of a Spanish uranium ore. Part I: A review of the bacterial leaching in the treatment of uranium ores. *Hydrometallurgy*, 38, 39–57.

Muralikrishna, N. and Bhurat, M. C., 1993. Recovery of uranium by bacterial leaching from copper floatation tailings, Mosabani, Singhbhum district, Bihar. In *Proceedings of the International Symposium of the Importance of Biotechnology*, Andhra University, Waltair, A. P., pp. 92–95.

Mwaba, C. C., 1991, Biohydrometallurgy: An extraction technology for the 1990s. *Mining Magazine*, 160–161. http://www.highbeam.com/doc/1G1-11363278. htmh

Nemati, M. and Webb, C., 1997. A kinetic model for biological oxidation of ferrous iron by *Thiobacillus ferrooxidans*. *Biotechnology Bioengineering*, 48, 478–486.

Nemati, M. and Webb, C., 1998. Inhibition effect of ferric iron on the kinetics of ferrous iron biooxidation. *Biotechnology Letters*, 20(9), 873–877.

Nguyen, V. V., Pinder, G. F., Gray, W. G. and Botha, J. F., 1983. Numerical simulation of uranium in situ mining. *Chemical Engineering Science*, 38, 1855–186.

Norris, P. R. and Parrot, L., 1986. High temperature, mineral concentrate dissolution with Sulfolobus. In *Fundamental and Applied Biohydrometallurgy*, (R. W. Lawrence, R. M. R. Branian and H. G. Ebner, Eds.) Amsterdam, Netherlands: Elsevier, pp. 355–365.

Norris, P. R., Parrott, L. and Marsh, R. M., 1986. Moderately thermophilic mineral oxidizing bacteria. *Biotechnology and Bioengineering Symposium*, 16, 253–262.

Ohnuki, T., Yoshida, T., Ozaki, T., Samadfam, M., Kuzai, N., Yubuta, K., Mitsugashira, T., Kasama, T. and Francis, A. J., 2005. Interactions of uranium with bacteria and kaolinite clay. *Chemical Geology*, 220, 237–243.

Pal, S., Pradhan, D., Das, T., Sukla, L. B. and Roy Chaudhury, G., 2010. Bioleaching of low-grade uranium ore using *Acidithiobacillus ferrooxidans*. *Indian Journal of Microbiology*, 50(1), 70–75.

Pantelis, G. and Ritchie, A. I. M., 1991. Macroscopic transport mechanisms as a rate-limiting factor in dump leaching of pyritic ores. *Applied Mathematical Modelling*, 15, 136–143.

Pantelis, G. and Ritchie, A. I. M., 1992. Rate-limiting factors in dump leaching of pyritic ores. *Applied Mathematical Modelling*, 16, 553–560.

Peter, J. D., Louis, J. C., Andrew, M. C. and Michael, F., 1983. New minerals names. *American Mineralogist*, 68, 849–852.

Phyllis, A. W., Martin, I., Dugan, P. R. and Tuovinen, O. H., 1983. Uranium resistance of *Thiobacillus ferrooxidans*. *European Journal of Applied Microbiology Biotechnology*, 18, 392–395.

Polikarpova, V. A., 1957. Nenadkevite—A new silicate of uranium. *Journal Nuclear Energy*, 4(2), 262–265.

Pradhan, N., Pradhan, S. K., Sukla, L. B. and Mishra, B. K., 2008. Micro Raman analysis and AFM imaging of *Acidithiobacillus ferrooxidans* bio-film grown on uranium ore. *Research in Microbiology*, 159, 557–561.

Price, R. R., Blaise, J. R. and Vance, R. E., 2004. Uranium production and demand. *NEA News*, 22, 1–2.

Qiu, G., Li, Q., Yu, R., Sun, Z., Liu, Y., Chen, M., Yin, H., Zhang, Y., Liang, Y., Xu, I., Sun, L. and Liu, X., 2011. Column bioleaching of uranium embedded in granite porphyry by a mesophilic acidophilic consortium. *Bioresource Technology*, 102(7), 4697–4702.

Rakotoson, G., Raoelina, A. L. and Pai, G., 1983. Measurement of the escape rate of radon in uranium minerals from Madagascar. *International Journal of Radiation and Isotopes*, 34(7), 1017–1018.

Rashidi, A., Safdari, J., Roosta-Azad, R. and Zokaei-Kadijani, S., 2012. Modeling of uranium bioleaching by *Acidithiobacillus ferrooxidans*. *Annals of Nuclear Energy*, 43, 13–18.

Rao, M. V., Nagabhushana, J. C. and Jeyagopal, A. V., 1989. Uranium mineralization in the middle proterozoic carbonate rock of the Cuddapah super group, southern peninsular India. Exploration and research for atomic minerals. AMD, India, 2, pp. 29–38.

Rawlings, D. E., 2001. The molecular genetics of *Thiobacillus ferrooxidans* and other mesophilic, acidophilic, chemolithotrophic, iron-or sulfur-oxidizing bacteria. *Hydrometallurgy*, 59(2–3), 187–201.

Rawlings, D. E., Dew, D. and du Plessis, C., 2003. Biomineralisation-metal containing ore and concentrates. *Trends in Biotechnology*, 21, 38–44.

Rawlings, D. E., Tributsch, H. and Hansford, G. S., 1999. Reasons why leptospirillum-like species rather than *Thiobacillus ferrooxidans* are the dominant iron-oxidizing bacterium in many commercial process for the biooxidation of pyrite and related ore. *Microbiology*, 145, 5–13.

Ring, R. J., 1980. Ferric sulphate leaching of some Australian uranium ores. *Hydrometallurgy*, 6(1–2), 89–101.

Rolfe, M. A. H. and Oxon, C. E., 1911. Autunite (hydrated uranium calcium phosphate). *The Lancet*, 177, 766–767.

Rossi, G. L., 1990. *Biohydrometallurgy*, New York: McGraw-Hill.

Roy, M. and Dhana Raju, R., 1997. Petrography and depositional environment of the U-mineralized phosphatic siliceous dolostone of Vempalle formation, Cuddapah basin. *Journal of Geological Society, India*, 50, 577–585.

Sand, W., Gerke, T., Hallmann, R. and Shippers, A., 1995. Sulfur chemistry, biofilm and the (in)direct attack mechanisms—A crucial evaluation of bacterial leaching. *Applied Microbiology and Biotechnology*, 43, 961–966.

Sand, W., Gehrke, T., Hallmann, R. and Schippers, A., 1998. Towards a novel bioleaching mechanism. *Mineral Processing and Extractive Metallurgy Review*, 19, 97–106.

Sand, W., Tilman, G., Peter-Georg, J. and Schippers, A., 2001. (Bio)-chemistry of bacterial leaching-direct vs. indirect bioleaching. *Hydrometallurgy*, 59(2–3), 159–175.

Sankaran, R. N., Sah, V. N., Singh, T. P. and Dwivedy, K. K., 1998. Heap leaching on uranium ores at Baginda, Himachal Pradesh. A case study of effective pollution control. *Exploration and Research of Atomic Minerals*, 1, 165–167.

Saraswat, A. C., 1988. Uranium mineralization in the Indian perspective and strategy. Exploration and research for atomic minerals. AMD, India, 1, pp. 1–11.

Schnell, H. A., 1997. Bioleaching of copper. In *Biomining: Theory, Microbes and Industrial Processes* (D. E. Rawlings, Eds.), Berlin: Springer-Verlag, pp. 21–43.

Seeger, M. and Jerez, C. A., 1993. Response of *Thiobacillus ferrooxidans* to phosphate limitation. *FEMS Microbiology Review*, 11(1–3), 37–41.

Shakir, K., Aziz, M. and Beheir, Sh. G., 1992. Studies on uranium recovery from uranium-bearing phosphatic sandstone by a combined heap leaching-liquid-gel extraction process. 1- Heap leaching. *Hydrometallurgy*, 31, 29–40.

Shoep, A., 1923. A new radioactive mineral, parsonite. *Journal of the Franklin Institute*, 195(4), 584.

Shrihari, J., Modak, M., Kumar, R. and Gandhi, K. S., 1995. Dissolution of particles of pyrite mineral by direct attachment of *Thiobacillus ferrooxidans*. *Hydrometallurgy*, 38, 175–178.

Strasser, H., Burgstaller, W. and Schinner, F., 1994. High yield production of oxalic acid for metal leaching processes by *Aspergillus niger*. *FEMS Microbiology Letters*, 119, 365–370.

Sugio, T., Domatsu, C., Munakata, O., Tano, T. and Imai, K., 1985. Role of ferric ion-reducing system in sulfur oxidation of *Thiobacillus ferrooxidans*. *Applied and Environmental Microbiology*, 49, 1401–1406.

Sugio, T., Katagiri, T., Moriyama, M., Inagaki, K. and Tano, T., 1988b, Existence of a new type of sulphite oxidase which utilizes ferric iron as an electron acceptor in *Thiobacillus ferrooxidans*. *Applied and Environmental Microbiology*, 54, 153–157.

Sugio, T., Wada, K., Mori, M., Inagaki, K. and Tano, T., 1988a, Synthesis of an iron oxidizing system during the growth of *Thiobacillus ferrooxidans* on sulfur basal salts medium. *Applied and Environmental Microbiology*, 54, 150–152.

Sugio, T., White, K. J., Shute, E., Choate, D. and Blake, R. C., 1992. Existence of a hydrogen sulfide ferric ion oxidoreductase in iron oxidizing bacteria. *Applied and Environmental Microbiology*, 58, 431–433.

Taylor, G., Farrington, V., Woods, P. and Molloy, R., 2004. Review of environmental impacts of the acid *in situ* leach uranium mining process. *CSIRO Land and Water Client Report August 2004.* Available on line at: www.epa.sa.gov.au/pdfs/isl_review.pdf

Torma, A. E., 1983. Biotechnology applied to mining of metals. *Biotechnology Advances*, 1(1), 73–80.

Torma, A. E. and Banhegyi, I. G., 1984. Biotechnology in hydrometallurgical processes. *Trends in Biotechnology*, 2, 13–15.

Tuovinen, O. H., 1984. Effect of pH, iron concentration and pulp densities on the solubilisation of uranium from ore materials chemicals and microbiological acid solutions: Regression equation and confidence band analysis. *Hydrometallurgy*, 12, 141–149.

Waksman, S. A. and Joffe, I. S., 1992. Microorganisms concerned with the oxidation of sulfur in soil. II *Thiobacillus thiooxidans*, a new sulfur oxidizing organisms isolated from the soil. *Journal of Bacteriology*, 7, 239–256.

Wang, W. G., Tan, K., Xie, E., Liu, J. and Cai, G., 2013. Supercritical CO_2 fluid leaching of uranium from sandstone type ores. *Advanced Materials Research*, 634–638, 3517–3521.

Williams, H., Colman-Sadd, S. P. and Swinden, H. S., 1988. Tectonic–stratigraphic subdivisions of central Newfoundland. *Geological Survey of Canada*, 88-1B, 91–98.

Youlton, B. J. and Kinnaird, J. A., 2013. Gangue–reagent interactions during acid leaching of uranium. *Minerals Engineering*, 52, 62–73.

4

Biohydrometallurgy: From the Zijinshan Experience to Future Prospects

Liu Xingyu, Ruan Renman, Wen Jiankang and Wang Dianzuo

CONTENTS

4.1 Introduction

The Zijinshan copper mine is the largest secondary copper sulphide mine in China. It has an ore reserve of more than 400 million tonnes with an average copper grade of 0.43% and geological reserve of metal copper of 1.72 million tonnes. A pilot bioheap leaching plant with a production capacity of 300 t/year of copper cathode was built at the Zijinshan copper mine at the end of 2000, and then the plant was scaled up to a capacity of 1000 t/year of copper cathode by June 2002 (Renman et al., 2006). In December 2005, a commercial underground mining—bioheapleaching—solvent extraction–electrowinning (SX–EW) plant was commissioned with a capacity of 10,000 t/year of copper cathode. From year 2006 to 2009, the total copper cathode production was 37633.6 t, the average operation cost for copper cathodes was 1.10 USD/lb and the total profit and tax exceeded 149 million USD (Huang, 2010). This plant is the first commercial application of bioheapleaching in China. The overview of the plant was shown in Figure 4.1. By using bacterial-assisted heapleaching, low production cost was achieved in Zijinshan. Furthermore, life-cycle assessment of the process showed that this technology has advantages of energy savings and less environmental impact when compared with the traditional flotation-flash smelter process (Ruan et al., 2010). In 2011, the plant was scaled up to a design capacity of 20,000 t/year of copper cathode. The new plant is capable of treating even a lower grade of the ROM (run of mine) that enables a lower copper cutoff grade that subsequently increases copper's geological reserve of the mine.

In this chapter, the description of the case study will be provided, as well as the research work conducted during the past 10 years related to bioheapleaching process of Zijinshan copper sulphides. The practice and research experience presented may have the benefit of the development of biohydrometallurgy in the following years.

FIGURE 4.1
Overview of the plant.

4.2 Industrial Practice of the High-Pyrite-Content Copper Sulphide Bioheapleaching Plant

During the past few decades, bioheapleaching has been well established all over the world. The operating leaching temperature was usually at ambient or moderate temperatures up to 50°C (Brierley and Brierley, 2001; Rawlings and Johnson, 2007). The total iron concentration was lower than 10 g/L (Demergasso et al., 2010; Ibaceta and Garrido, 2005). The Eh is usually around 800–900 mV (SHE) (Demergasso et al., 2010). And the operating pH was around 1.5–2.0 (Demergasso et al., 2010). Comparing with other bioheapleaching plants worldwide (Brierley and Brierley, 2001; Domic, 2007), bioheapleaching at Zijinshan was facing special engineering challenges such as acid and iron accumulation in the solution system from the high content of pyrite and high mean annual rainfall (1676.6 mm) concentrated from March to August. According to the distinct mineralogy of the ore, by using a proper engineering design and operation parameter, a distinct low-redox potential bioheapleaching system was established at the Zijinshan plant. The unique characteristics of the system include: (a) low operating pH (0.8–1.0), high concentration of total iron (exceeded 50 g/L) and high temperature (up to 60°C) in the pregnant leach solution (PLS); and (b) sulphur oxidisers were dominant in the heap. In this

TABLE 4.1

The Chemical Composition of ROM

Elements	Cu	TS	SO$_3$	Fe	As	CaO	MgO	Al$_2$O$_3$	SiO$_2$	K$_2$O	Pb	Na$_2$O
Composition (%)	0.4	5.28	4.18	3.59	0.021	0.032	0.14	12.22	67.19	1.58	0.02	0.11

Source: Adapted from Ruan, R. et al., 2011. *Hydrometallurgy*, 108(1): 130–135.

TABLE 4.2

The Mineral Composition of ROM by MLA

Minerals	Pyrite	Digenite	Chalcocite	Covellite	Enargite	Quartz	Alum	Dickite
Composition (%)	5.80	0.2	0.03	0.20	0.10	64.2	11.67	15.24

Source: Adapted from Ruan, R. et al., 2011. *Hydrometallurgy*, 108(1): 130–135.

system, copper was recovered in high efficiency, excessive iron in the solution phase was removed by the formation of jarosite in the heap and free acid in the raffinate was neutralised by limestone. Thus, iron was balanced at a low cost.

4.2.1 Mineralogy

Table 4.1 shows the chemical composition of ROM of the commercial plant, indicating a high sulphur content. Recently, MLA (mineral liberation analysis, JKtech) and XRD (x-ray diffraction) analysis showed that the copper sulphide in the ore is mainly digenite and covellite. The ROM only contains a small amount of chalcocite (Table 4.2), which is different from previous publications (Renman et al., 2006; Ruan et al., 2010; Xingyu et al., 2010a,b). A high content of pyrite (5.8 wt.%) within the ROM would be responsible for a large amount of acid and iron generated during the bioleaching process. Meanwhile, there is 11.67% of the chemical formula of alum inside the ROM, a sign of dissolved potassium ion for jarosite precipitation.

4.2.2 Mining, Comminution and Ore Stacking

Before the year 2008, underground mining produced ROM ore that was transported to the crushing system with 20 t of trolley locomotive. From the year 2009, 60% of the ROM was from an open pit. The ROM ore smaller than 1 m was crushed by two-stage crushing to a size of 80% passing 40 mm. The dump truck was used to load ores to the heap area and the bulldozer was used to assist the ore stacking. The average copper grade of the stacked ores was decreased from 0.42% in 2006 to 0.34% in 2009 with an average grade of 0.38%. The overall heap area was 0.2 million square metres. After levelling the bottom of the heap area, a prepared base was covered with 1 m of soil and sands. On top of the base, an impermeable liner composing of a 2-mm high-density polyethylene (HDPE) liner and plastic grid liner was used. Multi-lift permanent stacking was used. After the first lift was launched, raffinate was irrigated for

30–60 days, then, on top of the first lift, the second lift began to stack and so on. Currently, the heap consists of three lifts, each with a height of 8–10 m. The ROM was crushed to coarse particle sizes, which prevented fine particles from agglomerating to ensure permeability. Although the dump truck was used for ore stacking, good permeability inside the heap was achieved.

4.2.3 Bioheapleaching

In October 2005, the heap was launched. Acid mine water was adjusted to a pH of 1.7 by dilute sulphuric acid (2%) before being irrigated into the heap. An irrigation pipe system was arranged in 3×3 m grid and sprinkler leaching was used with a solution irrigation rate of around 12–16 $L/m^{2}*h$. The heap was not aerated. The bioleaching system has five solution ponds for PLS (51,000 m^3), irrigation (55,000 m^3), raffinate (63,000 m^3), flood collection (88,000 m^3) and standby (41,000 m^3), respectively. From March 2006 to September 2006, 131,964 m^3 of solutions with a high concentration of ferric from the pilot plant were continually applied to the commercial leach circuit. After the ferric-rich solution from the pilot plant was applied, the copper production increased significantly (Table 4.3), indicating the fact that a high ferric concentration facilitates copper dissolution.

Since the ore in the Zijinshan copper bioheapleaching process contains a high content of pyrite, acid and iron were produced from pyrite dissolution. Furthermore, the oxidation of reduced inorganic sulphur compound (RISC) generated heat which increased the temperature inside the heap. As a result, copper sulphide and pyrite dissolution was promoted. By using multi-lift permanent stacking and non-aeration, the heat loss was reduced and the temperature of PLS was maintained around 45–60°C for the whole year (Figure 4.2); thus, we estimated that the temperature inside the heap may exceed 70°C Figure 4.3 showed the variation in pH of PLS and iron concentration. It is clear that due to the high content of pyrite, the acidity and iron concentration were significantly increased during the operation. In December 2007, the pH

TABLE 4.3

The Typical Chemical Composition of PLS in Start-Up Stage[a]

Elements \ Date	Jan 2006	Feb 2006	Mar 2006	May 2006	Jul 2006	Sep 2006	Oct 2006
Cu^{2+} (g/L)	2.16	1.54	1.96	5.61	2.35	2.37	3.87
Fe^{2+} (g/L)	0.26	0.38	0.52	1.49	3.16	3.76	4.93
TFe (g/L)	2.12	2.80	3.43	8.45	10.66	19.12	29.10
pH	1.53	1.54	1.55	1.28	1.12	1.06	1.02
Output of Cu cathode (t/a)	/	69.0	234.6	554.5	779.0	622.7	619.3

Source: Adapted from Ruan, R. et al., 2011. *Hydrometallurgy*, 108(1): 130–135.
[a] Data were collected from the monthly production report of Zijinshan copper mine.

(a)   (b)

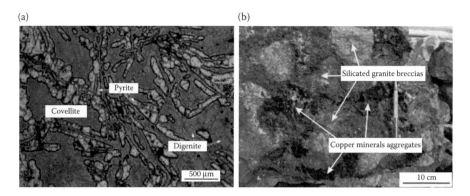

FIGURE 4.2
(See colour insert.) Optical microscope images showing coarse grains of sulphide aggregates (a) and copper sulphide aggregates associated with the matrix of minerals in the ores (b). (From Ruan, R. et al., 2011. *Hydrometallurgy*, 108(1): 130–135. With permission.)

of PLS decreased to 0.85, and at the same time, the acidity reached 30 g/L and the total iron concentration reached 50 g/L. The unusual low pH has a negative effect on the subsequent SE process. Thus, in 2008, a system was built for neutralisation of the free acid within the raffinate, and the capacity was 200 m³/h. In this system, limestone was used and the pH of raffinate was elevated to 1.2. Since October 2008, the pH and the Eh of PLS is maintained around 0.9–1.0 and 700–740 mV (SHE), respectively (Table 4.4). Owing to the high temperature and high iron concentration, jarosite was formed inside the heap (Figure 4.4). Scanning electron microscope (SEM) image showed that the formed jarosite crystal is crystallised with good integrality (Figure 4.5) which is unable to block the mineral dissolution. As excessive iron could form

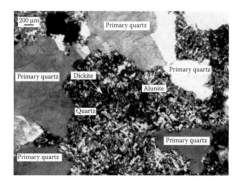

FIGURE 4.3
(See colour insert.) Optical microscope image showing the altered granite composed by quartz dickite and alunite in the ore body. (From Ruan, R. et al., 2011. *Hydrometallurgy*, 108(1): 130–135. With permission.)

TABLE 4.4

The Typical Chemical Composition of PLS in Recent 2 Years[a]

Year	pH	Eh	Cu^{2+}(g/L)	Fe^{2+}(g/L)	TFe (g/L)
2008	0.9–0.92	710–730	3.90	15.38	57.67
2009	0.90–0.94	710–740	4.07	12.71	54.89

Source: Adapted from Ruan, R. et al., 2011. *Hydrometallurgy*, 108(1): 130–135.
[a] Data were collected from the annual production report of Zijinshan copper mine.

(a)   (b)

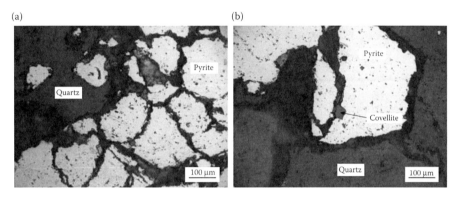

FIGURE 4.4

(See colour insert.) Optical microscope image showing pyrite in the bioleach residue of industrial heap. The voids between the pyrite grains are from leached copper sulphides (a). A large grain of copper sulphides was leached leaving unleached covellite (b). (From Ruan, R. et al., 2011. *Hydrometallurgy*, 108(1): 130–135. With permission.)

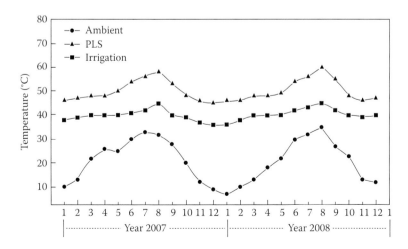

FIGURE 4.5

Temperature of ambient and solution flows. Symbols: •, ambient temperature; ▲, PLS and ■, irrigation solution. (From Ruan, R. et al., 2011. *Hydrometallurgy*, 108(1): 130–135. With permission.)

TABLE 4.5

Copper Extraction in 4 Years

Year	Cu in ROM (%)	Cu in Residue (%)[a]	Estimated Copper Extraction from Residue (%)[b]	Output of Cu Cathode (t)
2006	0.42	0.087	79.29	6781.25
2007	0.42	0.082	80.48	8007.54
2008	0.41	0.080	80.49	10004.57
2009	0.34	0.060	82.35	12840.24

Source: Adapted from Ruan, R. et al., 2011. *Hydrometallurgy*, 108(1): 130–135.

[a] At the end of every year, the dig machine was used for digging the leaching residues inside the heap, and 3–5 samples were picked inside the heap which was originally stacked roughly 1 year before. Each of the samples contained 25 kg of leaching residues. Cu in the residue was then analysed.

[b] Calculated from Cu in ROM and Cu in the residue of each year.

jarosite in the heap, the iron concentration in the solution phase was stabilised. In the steady stage, the leaching system was maintained at high acid and iron concentration, high salinity and low pH, which achieved high copper recovery (Table 4.5).

4.2.4 Solvent Extraction and Electrowinning

The SX unit was designed as a conventional 2E, 1 W and 1S circuit (two-stage extraction, one-stage wash and one-stage stripping). Zijin 988 (produced by Zijin regent plant) was selected as the copper extractant. 260# kerosene was used as the diluent. The extractant volume percent was around 15%–20%. EW is operating at a current density of 180 A/m^2 in 98 electrolytic cells. Copper and iron concentration in the tank house were 48 g/L and lower than 3 g/L, respectively. Each electrolytic cell contains 54 cathodes and 55 anodes with a current efficiency of around 85%–90%. Standard cathodes copper (Cu-CATH-2) is being stripped from stainless-steel blanks, at a cathode weight of around 40 kg each.

4.2.5 Water Treatment

The mean annual rainfall and evaporation in Zijinshan are 1676.6 and 1300 mm, respectively. As the rainfall is concentrated from March to August, the water balance in the leaching system is a challenge. There are three systems for water treatment that include one membrane system and two lime neutralisation systems. The acid mine water produced by the mining activity at the Zijinshan plant contains Cu^{2+} of 273 mg/L, total Fe (TFe) of 209 mg/L and pH of 2–3. Membrane technology was used to treat this acid mine water with a capacity of 3300–3600 m^3/day. The copper-rich water (2 g/L Cu^{2+}) retained by the membrane treatment plant was directly pumped into the SX system. In the rainfall season, a large amount of rainwater would enter the leaching system; thus, a flood collection and a standby pond were in use. Excessive water from the leaching system was reserved in the flood collection pond. This water was treated by lime neutralisation or SE depending on the copper concentration of the solution. Furthermore, excessive acid mine water in the rainfall season was treated by another lime neutralisation system. The solid waste residues from two lime neutralisation systems were properly stored in the tailings storehouse. To improve water balance, there is a plan to expand the volume of the flood collection ponds and increase the capacity of the membrane system next year. Thus, in the wet season, excessive water from the leaching system will be reserved and recycled, and all the acid mine water will be treated by the membrane system.

4.3 Pyrite Bioleaching Process: Effect of pH and Aeration to the Microbial Community

During the bioleaching process, the main mechanism of bacterial catalysis in the dissolution of sulphidic minerals is based on the bacterial oxidation of ferrous ion (Fe^{2+}), with oxygen as the electron acceptor (Ozkaya et al., 2007). Dissolution of mineral sulphides can be defined as either acid producing, which leads to a decrease in pH (e.g. pyrite dissolution) or acid consuming, which leads to an increase in pH (e.g. chalcocite dissolution) (Plumb et al., 2008b). Since a pH decrease is mainly due to pyrite dissolution, an understanding of the microbial dynamic change during pyrite bioleaching is necessary for limiting the leaching of the iron during copper bioleaching.

4.3.1 Bacterial Culture and Batch Experiments

The acidophilic iron-oxidising culture was taken from the bioleaching plant of the Zijinshan copper mine and maintained on a 9 K medium plus pyrite powder for 1 week. Two sets of 3 L stainless-steel tanks coated with Teflon were used for batch-leaching tests. Overhead stirrers were used to mix the ore powder and inoculated the microbial consortia at 200 rpm. Reactor A was aerated while intermittent aeration was used in reactor B (Table 4.6). The air flow rate was 15 L/h. The reactor solution was continually monitored with pH and oxidation reduction potential (ORP) probes (SUNTEX) which had good resistance to low pH (to pH 0). The pH and ORP probes were connected to a data recorder which recorded data at intervals of 1 h. 200 mL of the culture with OD_{600} value of 0.6 was sub-inoculated into reactor A, which already contained 1300-mL double-distilled H_2O, 1500 mL 2X 9 K base media and 100 g pyrite powder for the batch experiment. At 96 h, 200 mL culture was taken from reactor A and cultivated in the shake flask by using 9 K media until the OD_{600} reached 0.6. This culture was sub-inoculated

TABLE 4.6

Aeration Schedule of the Two Reactors

	Operation Time (h)				
Reactors	0–96	96–120	120–144	144–264	264–432
A	+	+	+	+	+
B	−	+	−	+	−

Source: Adapted from Xingyu, L. et al., 2009. *Hydrometallurgy,*
          95(3): 267–272.
+ aeration on; − aeration off.

into reactor B, which was filled with the same medium as reactor A. T Fe concentration in the liquid phase was detected using the colourimetric method described previously by Karamanev et al. (2002) for every 24 h during the 432-h leaching experiment.

4.3.2 Monitoring of Bacterial Population Change

In order to understand the dynamics of the microbial populations in the reactors, we determined time profiles, using the percentage of clones within the library as a surrogate of relative abundance. Figure 4.6 showed the temporal distribution of the bacterial assemblages. The percentage of other members ranging from 0% to 10% was not shown. The 16S rDNA inventory revealed that the major bacterial population contained four species: *Leptospirillum ferriphilum, Sulfobacillus thermotolerans, Acidithiobacillus caldus* and *Acidithiobacillus albertensis*. Within reactor A, the population change of *L. ferriphilum* had a reverse correlation with *S. thermotolerans*, indicating that these two species of bacteria were competing with each other. During leaching in reactor A, when the pH decreased from 1.3 to 1.0, the proportion of *L. ferriphilum* decreased from 45.83% to 11.11% while *S. thermotolerans* increased from 25.53% to 71.11%, indicating that, in this pH range, *S. thermotolerans* grew faster than *L. ferriphilum*. But when the pH decreased from 1.0 to 0.9, *S. thermotolerans* decreased very fast from 71.11% to undetectable, while the population of *L. ferriphilum* increased to 81.25%. This was consistent with the previous research (Kinnunen and Puhakka, 2005) that at less than pH 1, *L. ferriphilum* can still achieve high Fe oxidation rate. Below pH 0.9, the situation reversed that the population of *S. thermotolerans* again increased to 26.00% while the proportion of *L. ferriphilum* was decreased to 66% by the end of the test.

Under intermittent aeration, it is worth noting that when the air was fed into the reactor (192–288 h), the population of *L. ferriphilum* increased from 8% to 74.47% and *S. thermotolerans* decreased from 54% to 2.13%, which shows the same trends as in reactor A. Below pH 0.9, the population of *S. thermotolerans* increased to 42.60% while *L. ferriphilum* decreased to 42.59%, which was also consistent with reactor A. However, under a non-aerated condition

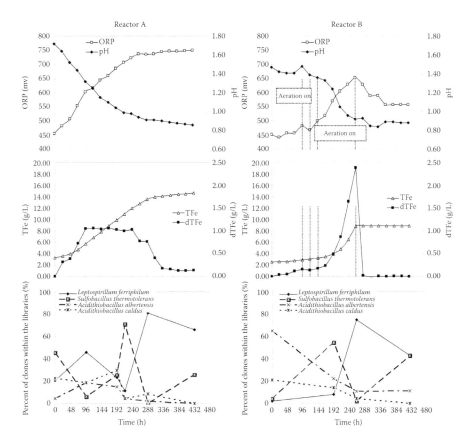

FIGURE 4.6

Profiles of pH, ORP, TFe leached out, total Fe leached per day (dTFe) and temporal distribution of the abundance of the 16S rRNA gene during the batch tests. (From Xingyu, L. et al., 2009. *Hydrometallurgy*, 95(3): 267–272. With permission.)

and pH below 0.9, the population of *S. thermotolerans* was higher than for the aerated reactor (reactor A), indicating that *S. thermotolerans* may have better growth under non-aerated conditions.

To compare the structures of the communities from the 10 samples in two tests at specific operational taxonomic unit (OTU) definitions, we calculated the community similarity index (θ) with OTUs defined at a distance level of 3%. The unweighted pair group method with arithmetic mean dendrogram comparing the pairwise θ values between the 10 samples from two reactors are shown in Figure 4.7. It seems that in reactor B, once the aeration started, the microbial community tends to develop these functional bacterial groups which utilise oxygen to oxidise iron and sulphur. Thus, samples taken from reactor B under the aeration condition tend to cluster with samples taken from reactor A.

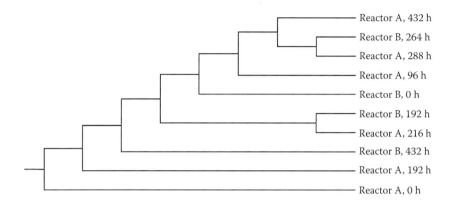

FIGURE 4.7
Unweighted pair group method with arithmetic mean dendrogram comparing the pair-wise θ values between the 10 clone libraries from two reactors. (From Xingyu, L. et al., 2009. *Hydrometallurgy*, 95(3): 267–272. With permission.)

4.3.3 Growth Inhibition of *Leptospirillum* at Different pH Range

In this study, we identify the major bacterial groups in all the samples by using 16S rRNA gene clone library. While the results showed general trends over time rather than a detailed quantitative estimation of the microbial population structure, they can still provide a rough image of the bacterial dynamic change during the pyrite bioleaching process. The most prominent organisms making up these communities are of the order *Leptospirillum, Sulfobacillus* and *Acidithiobacillus*. In both reactors, the pro-portion of *Acidithiobacillus*-like strains continuously decreased according to the decreasing pH and increasing redox, which was consistent with previous research (Demergasso et al., 2005). The tolerance of the genus *Leptospirillum* to low pH and the rise of proportion at high redox is a sig-nificant and much-reported finding (Battaglia-Brunet et al., 1998; Coram and Rawlings, 2002; Helle and Onken,1988; Hippe, 2000b; Sand et al., 1992b). *Leptospirillum*-like microorganisms have a greater tolerance to Fe^{3+} compared to *Acidithiobacillus ferrooxidans*, which has been suggested as one of the major reasons for the predominance of *Leptospirillum* sp. in indus-trial processes (Rawlings et al., 1999). Rawlings previously noted that the optimum pH for the growth of *L. ferriphilum* was between pH 1.4 and 1.8 (Coram and Rawlings, 2002). But in our research, for two pH ranges (pH 1.6–1.3, pH 1–0.9), *Leptospirillum* was dominant in the reactor. And it is worth noting that the decrease in *Leptospirillum* between pH 1.3 and 1.0 was not reported previously and needs further investigation. If the dTFe of the two reactors was compared, the highest dTFe value was achieved in reactor B, 264 h (2.38 g/L · d). This may be due to the lower pH (pH 0.93) and higher Fe^{3+} concentration which allowed *Leptospirillum* to domi-nate the system. *Acidithiobacillus* seems to be one of the main components at higher pH and lower conductivity, which is consistent with previous research (Bond et al., 2000).

Sulfobacillus-like sequences were related to strains and clones obtained from many acid environments such as Yellowstone National Park (Johnson et al., 2003), bioleaching thermophilic processes of pyrite, arsenopyrite and chalcopyrite (Dopson et al., 2004) and sediments of acid-mining lakes (GenBank description). The dominance of *Sulfobacillus*-related sequences has been found in a column bioreactor (d'Hugues et al., 2002), where the proportion of *Sulfobacillus*-related sequences usually increased at the end of the test, consistent with our experiments. It seems that *Sulfobacillus* can compete effectively with *Leptospirillum* at a lower pH (below 0.9). In this research, we also found the enhanced dominance of the genus *Sulfobacillus* over genus *Leptospirillum* under increasing oxygen limitation which was not reported previously.

4.4 Bioleaching of Chalcocite: Response of the Microbial Community to Environmental Stress and Leaching Kinetics

For understanding the pH affection to the bioleaching of secondary copper sulphide, in this research, chalcocite mineral bioleaching was conducted in two stirred reactors, the leaching pH started from pH 1.2 to pH 1.5 respectively and the microbial community at a different leaching stage in both reactors was studied using a cultivation-independent molecular approach.

4.4.1 Chalcocite Preparation and Leaching Test Description

The chalcocite mineral used in this study was hand selected from the Zijinshan copper mine (Fujian, China). Chemical analysis of the ore revealed a composition of (mass percent) 72.51% Cu, 4.64% Fe and 20.55% S. The size fraction of the concentrate used was 44–50 μm. The acidophilic mesophile culture was taken from the bioleaching plant of the Zijin copper mine and maintained on 9 K medium plus pyrite powder for 1 month. Three litres stainless-steel tanks coated with Teflon were used for batch-leaching tests. The effective volume of the reactors was 2.5 L. Both the reactors were aerated and the air flow rate was 15 L/h. Overhead stirrers with a rotation rate of 200 rpm were applied to both reactors. The initial leaching pH of reactor A was pH 1.2, while for reactor B, pH 1.5 was used. The reactor solution was continually monitored with pH and ORP probes.

4.4.2 Shrinking Particle and Shrinking Core Models

The model developed from previous research was used in this research (daSilva, 2004). According to the model, the following mixed-control equation was developed based on the shrinking particle and shrinking core models:

$$t - t_{tag} = (1/D)*\{[1 - 3*(1 - \lambda)^\wedge(2/3) + 2*(1 - \lambda)] -$$
$$[1 - 3*(1 - \lambda_{tag})^\wedge(2/3) + 2*(1 - \lambda_{tag})]\} + (1/K)* \qquad (4.1)$$
$$\{[1 - (1 - \lambda)^\wedge(1/3)] - [1 - (1 - \lambda_{tag})^\wedge(1/3)]\}$$

where t is the reaction time, λ is the metal extraction (%), D is diffusion coefficient, K is the observed rate constant and tag was the time point to begin modelling the extraction kinetics. At $t = t_{tag}$, $\lambda = \lambda_{tag}$. The ratio of K to D will reveal the relative contribution of diffusion and reaction effects, a K/D ratio greater than one would indicate that diffusion control dominates, while a K/D ratio of less than one would indicate that the reaction control dominates.

4.4.3 Model Fitting

The shrinking core model was used to determine whether the leaching tests were chemical control or diffusion control. Copper and iron dissolution data from the two reactors were fitted to the mixed-control rate equation (da Silva, 2004). Owing to the presence of lag period for copper extraction in both reactors, t_{tag} value of 24 and 96 h was used for reactor A and B, respectively. To carry out the model fitting, the parameters of K and D were varied, to get the best least-squares description of the experiment results. The model-fitting curves using this method are included along with the experiment results in Figure 4.8. Table 4.7 showed the K/D ratio of copper and iron dissolution from this research and iron dissolution of a typical pyrite bioleaching in the same reactor conducted in our previous research (Xingyu et al., 2009). High K/D ratio larger than 10^{20} for copper dissolution was obtained in both reactors, indicating that molecular diffusion control is significant for copper dissolution. However, for iron dissolution, all the leaching experiments exhibited a relative low K/D ratio compared with copper dissolution, indicating that diffusion control is less important during pyrite bioleaching compared with chalcocite bioleaching. Furthermore, if copper dissolution in the two reactors was compared, the copper-leaching recovery ratio of reactor A is higher than reactor B and the K/D ratio of reactor A is nearly 100-fold more than reactor B, indicating that a low pH could facilitate chalcocite bioleaching and diffusion control is more important at such a relatively low pH for copper dissolution. When comparing leaching results of iron from this research and the previous research, the lower iron dissolution rate of 23.01% and high K/D ratio of 4.03 were achieved at pH 1.2, indicating that leaching pH of 1.2 may limit pyrite dissolution during copper bioleaching, which was consistent with our previous research (Xingyu et al., 2009).

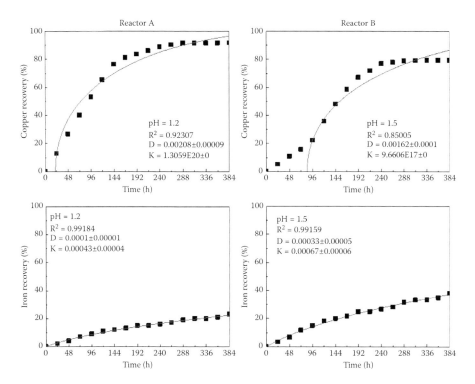

FIGURE 4.8
Comparison between leaching experiments in two reactors (closed squares) and model predictions (solid lines) obtained by fitting Equation 4.1. (From Xingyu, L. et al., 2010. *Hydrometallurgy*, 103(1): 1–6. With permission.)

4.4.4 Monitoring of Bacterial Population Change

Figure 4.9 showed the profiles of pH, ORP, copper concentration, TFe leached out, total cell counts and temporal distribution of the abundance of the 16S rRNA gene during the batch tests. In both reactors, the population change of the genus *Leptospirillum* had a nearly reverse dependence with the genus *Acidithiobacillus*, indicating that the iron oxidiser and sulphur oxidiser were

TABLE 4.7

Kinetic Parameters for Leaching Tests in Research and Our Previous Research

	Leaching Test	Initial pH	K/D	Final Recovery (%)
This research	Copper dissolution	1.2	6.49×10^{22}	91.77
	Iron dissolution	1.2	4.30	23.01
	Copper dissolution	1.5	5.96×10^{20}	79.35
	Iron dissolution	1.5	2.03	38.13
Liu et al. (2009)	Iron dissolution	1.72	3.64×10^{-18}	76.72

Source: Adapted from Xingyu, L. et al., 2010. *Hydrometallurgy*, 103(1): 1–6.

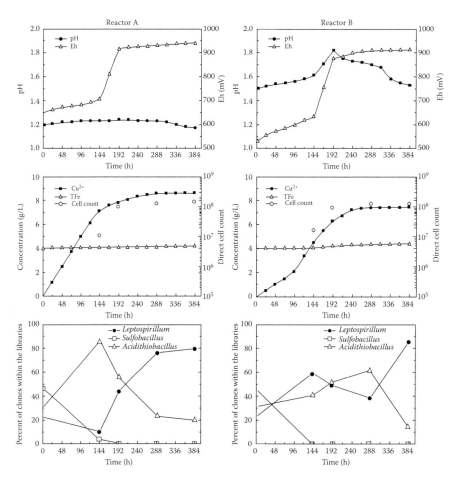

FIGURE 4.9
Profiles of pH, ORP, copper concentration, TFe leached out, total cell counts and temporal distribution of the abundance of the 16S rRNA gene during the batch tests. (From Xingyu, L. et al., 2010. *Hydrometallurgy*, 103(1): 1–6. With permission.)

competing with each other. In the initial leaching stage of reactor A, the decrease proportion of *Leptospirillum* from 22.2% to 10% and increase proportion of *Acidithiobacillus* from 31.1% to 86% was detected under a relatively higher redox potential (from 651 to 705 mV) and low pH (pH 1.2) compared with reactor B, indicating that the sulphur oxidiser was favoured at pH 1.2 compared with the iron oxidiser. But in the following leaching stage, the proportion of *Leptospirillum* increased significantly from 10% to 80% and the genus *Acidithiobacillus* decreased from 86% to 20% as the Eh was significantly increased from 705 to 940 mV, indicating that the accumulation of Fe^{3+} ions could affect the microbial community structure, thus making *Leptospirillum* dominate the system. A relatively high concentration of copper ion (more than 7.2 g/L) within the solution might be another reason.

For reactor B, in the initial leaching stage, the increase proportion of *Leptospirillum* and *Acidithiobacillus* under a relatively low redox potential (from 530 to 631 mV) and high pH (pH 1.5–1.6) was observed, each from 22.2% to 59.2%, 31.1% to 40.8%, respectively. The continuous increase proportion of *Acidithiobacillus* from 40.8% to 68.7% under a high redox potential (from 631 to 910 mV) was monitored in the following leaching stage, and the proportion of *Leptospirillum* was decreased from 59.2% to 38.3% at the same leaching period. Which suggested the sulphur oxidiser could effectively compete with an iron oxidiser at pH range of 1.5–1.6. In the final stage, the proportion of *Leptospirillum* was largely increased from 38.3% to 85.45% and the genus *Acidithiobacillus* was decreased from 61.7% to 14.55%. This was consistent with reactor A and may be caused by a high concentration of copper ion in the leaching solution.

Leaching experiment results that started from different pH showed that bacteria community succession in the two reactors differed with each other at the beginning of the leaching experiments. At 144 h, sulphur oxidisers (genus *Acidithiobacillus*) accounted for a high proportion of 86% in reactor A and high recovery of 76.38% for copper was achieved, while in reactor B, only 47.95% of copper was recovered, related to the dominance of iron oxidisers (genus *Leptospirillum* accounting 59.18% of the library). Furthermore, at this time point, iron recovery in reactors A and B was 12.02% and 19.89%, respectively. The optimum pH for the growth of *Leptospirillum* found in the previous research was between pH 1.4 and 1.8 (Coram and Rawlings, 2002), and Plumb et al. have found that the optimal iron oxidative rate of *Leptospirillum* occurred at pH 2.0 but the sulphur oxidiser *Acidithiobacillus thiooxidans* may have a better growth rate and sulphur oxidation rate at a pH range of 1.5–1.0 (Plumb et al., 2008a). This was consistent with our research that the effective oxidation of ferrous ion by *Leptospirillum* may need a slightly high pH while the sulphur oxidiser may favour a relatively low pH (1.2); thus, the dominance of sulphur oxidisers and low-leaching pH may facilitate chalcocite bioleaching and also prevent pyrite bioleaching. However, the dominance of *Leptospirillum* in both reactors at the final leaching stage (288–384 h) was observed. This was consistent with previous research (Demergasso et al., 2005; Xingyu et al., 2009), and may be due to *Leptospirillum*-like microorganisms that have a high growth rate at a high redox potential and also a greater tolerance to Fe^{3+} compared with other acidophiles (Rawlings et al., 1999).

Shrinking core models have been successfully applied to the bioleaching process in the last few years (daSilva, 2004; Lizama et al., 2005; Lizama and Suzuki, 1991). This study focusses on the relative importance of reaction and diffusion control during the mineral chalcocite bioleaching and pyrite bioleaching which started from different pH. From the modelling results, it seems that in both reactors, the copper dissolution is diffusion control and at a relatively low pH, the diffusion control is more important. However, a typical pyrite bioleaching carried out at relative high pH in the previous research was reaction controlled. In a typical sulphide ore bioleaching process, if it is diffusion control, the formation of the product layer is the major resistance upon the rate of metal extraction. The product layer often comes from sulphur formed by the bio-oxidation process (Lochmann and Pedlík, 1995; McGuire et al., 2001). Thus, for reducing the diffusion control of chalcocite bioleaching, abundant sulphur oxidisers are needed. Since pyrite bioleaching is less diffusion controlled, for limiting iron dissolution, the concentration of Fe^{3+}/Fe^{2+} coupled with the leaching solution should be maintained in a proper range.

4.5 Microbial Community Inside the Bioheapleaching System

The distribution and diversity of the bacterial community in the Zijinshan commercial non-aeration copper bioheapleaching system operated at pH 0.8 for 3 years were investigated. The 24-m high heap was cut off by a mechanical digger. On the trapezoidal cross-section of the heap, nine ore samples were taken from different vertical and horizontal locations and investigated by 16S rRNA gene clone library. Another three liquid samples from raffinate solution pond, spray solution pond and pregnant solution pond were also applied to 16S rRNA gene clone library analysis. These results may be used to advance the Zijinshan bioheapleaching operation.

4.5.1 Classification of the Clones

Figure 4.10a showed the ore-sampling sites, and the other three liquid samples were taken from the solution pond. For the 12 bacterial clone libraries, a total of 1166 16S rRNA gene clones were grouped into only 27 OTUs with a

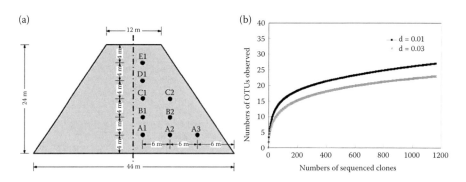

FIGURE 4.10
Ore-sampling sites (a) and rarefaction curves from DOTUR analysis (b). (From Liu, X.Y., Chen, B.W. and Wen, J.K., 2013. *Advanced Materials Research*, 825: 50–53. With permission.)

distance level of 3%. A range of 4–11 OTUs were detected and isolated from each sample. The greatest number of OTUs (9) was found in the sample of C2 and YS. Rarefaction curves, which plot the number of sequences screened versus the number of OTUs observed are shown in Figure 4.10. Results suggested that diversity at the genus/species level was fully detected in the 12 samples, and also supported by the high-estimated sample coverage of three clone libraries (Table 4.8).

The 1166 clones belong to four bacterial phyla: *Proteobacteria, Nitrospira, Firmicutes* and *Actinobacteria* (Figure 4.11). All generated phylogenetic trees resulted in stable branching. *Proteobacteria* were dominant in the clone libraries (52.66% of clones), and *Nitrospira* was also in the libraries in a relatively high amount (37.74% of clones). Other phyla such as *Firmicutes* and *Actinobacteria* were only presented in a small proportion, and each of them was detected for 6.52% and 3.09% of the clones, respectively. For all the identified bacterial sequences, 19 genera were represented. Among the 27 OTUs found in the 12 samples, six OTUs were related to members of the genus *Acidithiobacillus*, followed by *Leptospirillum* (three OTUs) and *Sulfobacillus* (two OTUs). Other detected OTUs belong to heterotrophic bacteria such as genera *Comamonas, Arthrobacter, Micrococcus* and so on. Among these genera, the genus *Acidithiobacillus* and *Leptospirillum* accounted for a high proportion; and each of them presented 494 and 440 clones, respectively.

TABLE 4.8

Diversity Indices of the Three Samples at Distance Level of 3%

Samples	Estimated Sample Coverage	Shannon Index[a]	Estimated OTUs[b]	Sequenced Clones	OTUs
A1	1	1.733 (1.588, 1.878)	8.0 (8.0, 8.0)	97	8
A2	1	1.328 (1.116, 1.540)	7.0 (7.0, 7.0)	96	7
A3	0.98	1.233 (0.988, 1.478)	9.5 (9.0, 17.3)	99	9
B1	0.989	1.407 (1.213, 1.602)	7.0 (7.0, 7.0)	91	7
B2	1	1.132 (0.951, 1.312)	5.0 (5.0, 5.0)	96	5
C1	0.96	1.251 (1.022, 1.481)	15.0 (10.0, 46.7)	100	9
C2	0.979	1.951 (1.770, 2.131)	12.0 (11.1, 25.1)	97	11
D1	0.99	1.703 (1.558, 1.847)	8.0 (8.0, 8.0)	96	8
E1	1	1.010 (0.866, 1.155)	4.0 (4.0, 4.0)	99	4
YP	1	1.113 (0.928, 1.299)	5.0 (5.0, 5.0)	99	5
YR	0.99	1.515 (1.363, 1.666)	7.0 (7.0, 7.0)	100	7
YS	0.958	1.516 (1.280, 1.751)	14.0 (11.4, 34.0)	96	11
All	0.995	2.321 (2.261, 2.381)	30.8 (27.6, 49.0)	1166	27

Source: Adapted from Liu, X.Y., Chen, B.W. and Wen, J.K., 2013. *Advanced Materials Research*, 825: 50–53.

[a] Maximum likelihood estimator.

[b] Bias-corrected form for the Chao1 richness.

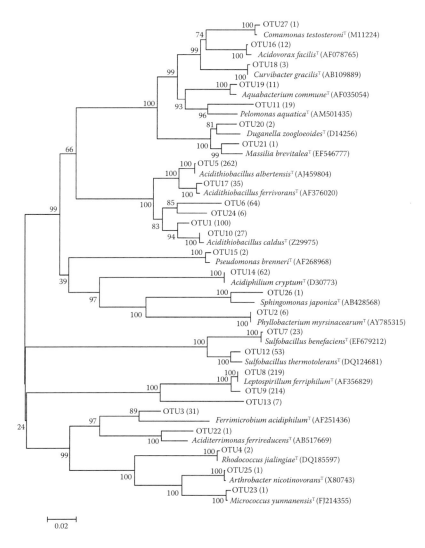

FIGURE 4.11
Phylogenetic affiliation of 16S rRNA gene sequences from 12 clone libraries belonging to the bacteria domain. Cloned 16S rRNA sequences from 12 clone libraries are arranged as: OTU name, and clone numbers. (From Liu, X.Y., Chen, B.W. and Wen, J.K., 2013. *Advanced Materials Research*, 825: 50–53. With permission.)

4.5.2 Copper and Iron Distribution in the Leached Ore Residue Samples

The distribution of copper and iron in the nine leached ore samples is shown in Table 4.9. The content of copper increased horizontally in low depth (2–3 years leaching time) from the inner to the outer site (A1, 0.50%; A3, 0.098%), and similar trends were detected for iron content. It is worth noting that

TABLE 4.9

The Distribution of Copper and Iron in the Nine Leached Ore Samples

Samples	Cu (%)	Fe (%)	S (%)
A1	0.050	2.06	4.15
A2	0.049	1.40	3.51
A3	0.098	2.65	4.27
B1	0.067	1.69	3.67
B2	0.074	2.03	3.89
C1	0.057	1.37	3.34
C2	0.087	1.76	3.36
D1	0.065	2.35	4.27
E1	0.091	2.11	3.73

Source: Adapted from Liu, X.Y., Chen, B.W. and Wen, J.K., 2013. *Advanced Materials Research*, 825: 50–53.

there is an interesting correlation between the horizontal distribution of the genus *Leptospirillum* and the iron/copper content in the ore samples when compared with Figure 4.12 at a lower depth. The reason for this correlation may be caused by the oxygen supply limitation inside the heap since this is a non-aeration heap.

4.5.3 Spatial Variation and Patterns of Bacterial Community Structure

The distribution of acidophilic microbes in the 12 samples is shown in Figure 4.12. The proportion of the genus *Leptospirillum* increased horizontally at a low depth (2–3 years leaching time) from the inner to the outer site (A1, 30.93%; A3, 67.68%) while the genus *Acidithiobacillus* decreased (A1, 41.24%; A3, 19.19%). Similar trends were detected in other depths. The proportion of the genus *Acidithiobacillus* decreased vertically from a higher to a lower

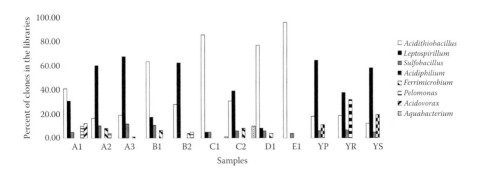

FIGURE 4.12
The distribution of acidophilic microorganisms in the 12 samples. (From Liu, X.Y., Chen, B.W. and Wen, J.K., 2013. *Advanced Materials Research*, 825: 50–53. With permission.)

depth (E1, 95.96%; A1, 41.24%) while the genus *Leptospirillum* increased from undetected (E1) to 30.93% (A1). A high proportion of heterotrophic bacteria related to the genus *Acidiphilium* were found in all the three liquid samples (accounting 11.11–32.00%). Other retrieved heterotrophic bacterial sequences mainly related to the genus *Acidovorax* (accounting 12.37% in A1 sample), genus *Pelomonas* (accounting 4.17–10.31% in several ore samples) and genus *Aquabacterium* (accounting 10.31% in C2 sample) were also identified.

Our research detected a relatively high amount of heterotrophic bacteria at the ore surface in the internal part of the heap and in the liquid samples, respectively. The retrieved heterotrophic bacterial sequences were mainly related to the genus *Acidiphilium*, genus *Acidovorax*, genus *Pelomonas* and genus *Aquabacterium*. Up to now, the contributions of heterotrophs to mineral dissolution have not received as much attention as the chemolithotrophs (Watling et al., 2014). The function of these bacteria differed a lot when compared with autotrophic acidophiles, and their roles inside the system are now under investigation. Factors which caused bacteria succession in this non-aerated heap still need further study.

The genus *Leptospirillum* was the most dominant group in the leaching solution (Figure 4.12). According to our previous research, growth inhibition of the genus *Leptospirillum* was revealed in the higher depth in the non-aeration heap and the genus *Acidithiobacillus*, especially sulphur oxidisers *A. albertensis*, *A .caldus* and *A. thiooxidans* were dominant at the ore surface (Xingyu et al., 2010b). In this research, we found the dominance of genus *Leptospirillum* in the outer part of the heap, while inside the heap, the bacteria community is still dominant by the sulphur oxidisers such as *A. albertensis* and *A. caldus*. Previous research noted that genus *Leptospirillum* cannot bear a relatively high temperature (Hippe, 2000a; Sand et al., 1992a); thus, compared with sulphur oxidisers, it cannot grow well inside the heap since the temperature inside the heap exceeds 60°C. Oxygen supply may be another reason for the growth inhibition of genus *Leptospirillum* in the inner part of the heap because aeration was not applied to this heap.

4.6 Leaching Kinetics of Copper Sulphides

Since the Zijinshan copper bioheapleaching plant works efficiently at low microbial activity (Xingyu et al., 2010b), it is important to understand why. This work describes dissolution kinetics of digenite and covellite in the acid ferric sulphate solution under a controlled redox potential and column bioleaching with raffinate from the plant at a different temperature and Fe^{3+} concentration. The rate-limiting factors for the leaching kinetics of the main copper minerals of Zijinshan copper ores and its practical implication for optimisation are discussed.

4.6.1 Dissolution Kinetics of Digenite and Covellite

4.6.1.1 Effect of Temperature

The reaction kinetics of digenite from Zijinshan leached by iron sulphate solution (36.4 g/L ferric sulphate and 0.67 g/L ferrous sulphate) was carried out in four different temperatures 30°C, 45°C, 60°C and 75°C at pH 1.0. The solution redox potentials were 720, 735, 745 and 755 mV (vs. SHE), respectively. The dissolution of digenite in subsequent graphs is defined as the percentage of copper leached in relation to the original concentration. The variation of dissolution at different temperatures is shown in Figure 4.13a. It showed rapid conversion of digenite to the secondary covellite in the first stage of copper dissolution and the rate reached 40% after only 2 min for all different temperatures. The digenite dissolution then slowed down at the second stage and was found to be temperature dependent. The dissolution reached 100% at 75°C as compared to 45% at 30°C. For chemical reaction control leaching, a plot of $1-(1-R)^{1/3}$ (R indicates the dissolution rate from Figure 4.13a) versus time gave straight lines for different temperatures. The rate constant k was obtained from the slope for the corresponding temperature. By plotting ln k versus 1/T derived from Arrhenius equation, activation energy is obtained from the slope. Figure 4.13b showed that the activation energy was 72.75 kJ/mol, and a similar range was reported in literature (Bolorunduro, 1999; Dutrizac and MacDonald, 1974a). Such high activation energy means that second stage dissolution of digenite is strongly related to temperature. Figure 4.14a shows covellite dissolution in ferric sulphate solution (Fe^{3+} concentration 39.7 g/L, Fe^{2+} 0.57 g/L at pH 1.0) at temperatures of 30°C, 45°C, 60°C and 75°C. The initial redox potentials were 720, 735, 745 and 755 mV (vs. SHE), respectively. This is similar to the results found in digenite: increases in

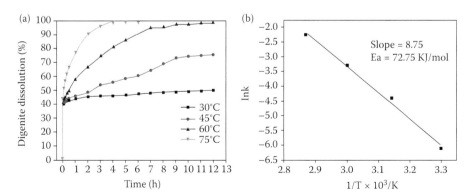

FIGURE 4.13
(See colour insert.) Effect of temperature on digenite dissolution: (a) Kinetic curve of the dissolution rate and (b) Arrhenius plot for the digenite second stage leaching. (From Ruan, R. et al., 2011. *Hydrometallurgy*, 108(1): 130–135. With permission.)

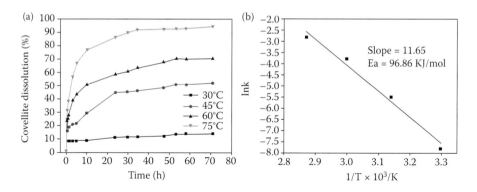

FIGURE 4.14
(See colour insert.) Effect of temperature on covellite dissolution: (a) Kinetic curve of the dissolution rate and (b) Arrhenius plot for the covellite leaching. (From Ruan, R. et al., 2011. *Hydrometallurgy*, 108(1): 130–135. With permission.)

temperature significantly increased the dissolution of covellite. The dissolution rate was much slower at 30°C compared with that at 75°C which was totally dissolved within 30 h. The Arrhenius plot based on the reaction rate is presented in Figure 4.14b and the best-fit straight line gave activation energy of 96.86 kJ/mol, which indicated that temperature was the key factor for covellite dissolution.

Compared with the second stage leaching of digenite (Figure 4.13a) and primary covellite (Figure 4.13a), the leaching rate of secondary covellite was much faster than that of primary covellite. Similar results were also reported in various other literatures (Cheng and Lawson, 1991; Miki et al., 2011; Walsh and Rimstidt, 1986).

4.6.1.2 Effect of Fe³⁺ Concentration

To investigate the effect of Fe^{3+} concentration, leaching of digenite was carried out in Fe^{3+} concentrations of 0.01, 0.03, 0.045, 0.06, 0.1, 0.65 and 1.0 M at pH 1.0 at 30°C, 45°C, 60°C and 75°C, respectively. The equivalent Fe^{2+} concentration was between 3×10^{-5} and 0.05 M, and was mainly Fe^{3+} in the leachate. The solution redox potential was stabilised at around 720, 735, 745 and 755 mV (vs. SHE) at different temperatures. At a fixed redox potential, there was a slight increase in TFe; however, due to the L/S ratio that was nearly 200; this small increase hardly had any effect on the dissolution. Results of dissolution at different ferric concentrations at a different temperature are shown in Figure 4.15. The slopes from the plot of log rate between 40% and 90% versus log Fe^{3+} concentration at different temperatures are summarised in Table 4.10. It showed that at Fe^{3+} concentrations below 0.1 M, the concentration has significant influence on the second stage leaching rate of digenite, whereas above this, the rate was independent of

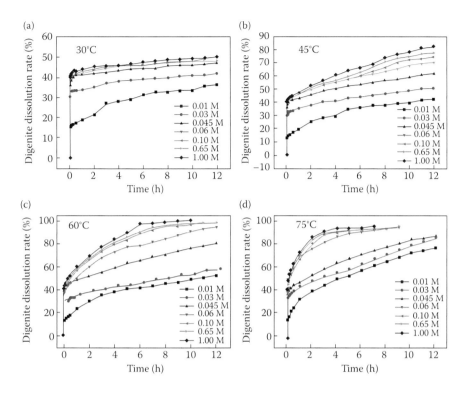

FIGURE 4.15
(See colour insert.) Effect of ferric concentration on digenite dissolution at (a) 30°C, (b) 45°C, (c) 60°C and (d) 75°C. (From Ruan, R. et al., 2011. *Hydrometallurgy*, 108(1): 130–135. With permission.)

the ferric concentration. Furthermore, it was found that the Fe^{3+} concentration had a more positive effect on the digenite-leaching rate at a higher temperature, evidenced by an order of 0.56 at 30°C and 0.82 at 75°C. Previous studies (Bolorunduro, 1999; Cheng and Lawson, 1991) also reported that the second stage of digenite dissolution increases with increasing ferric ion concentration in the special range.

TABLE 4.10

Reaction Order versus Fe^{3+} Concentration for the Second Stage Dissolution of Digenite at Different Temperature

Temperature (°C)	Fe (III) (M)	Slope	Fe (III) (M)	Slope
30	0.03–0.06	0.56	0.06–1.0	0.01
45	0.03–0.1	0.60	0.1–1.0	0.05
60	0.03–0.1	0.75	0.1–1.0	0.06
75	0.03–0.1	0.82	0.1–1.0	0.02

Source: Adapted from Ruan, R. et al., 2013. *Minerals Engineering*, 48: 36–43.

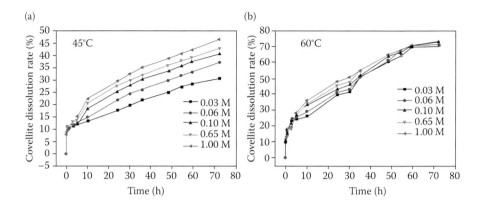

FIGURE 4.16
(See colour insert.) Effect of Fe³⁺ concentration on covellite dissolution at (a) 45°C and (b) 60°C. (From Ruan, R. et al., 2011. *Hydrometallurgy*, 108(1): 130–135. With permission.)

The covellite dissolution kinetics was carried out at pH 1.0 and the initial redox potential of 720 mV (vs. SHE). Ferric concentration of 0.03, 0.06, 0.1, 0.65 and 1.0 M with temperatures at 45°C and 60°C were examined. Results are shown in Figure 4.16. The slope of the log rate versus the log Fe³⁺ concentration plot gave a reaction order of 0.02 at 45°C and 0.05 at 60°C, respectively. It indicated that the Fe³⁺ concentration had a positive effect on covellite dissolution, but was not as significant as that on digenite. The dissolution kinetic results of covellite were consistent with previous studies (Dutrizac and MacDonald, 1974a; Mulak, 1971; Sullivan, 1930; Thomas and Ingraham, 1967).

4.6.1.3 Effect of Redox Potential

For the digenite-leaching experiment, redox potentials were set at 655, 705 and 755 mV (vs. SHE) by adding the prepared fer-326 ric/ferrous sulphate solution at a temperature of 60°C and pH of 1.0. For covellite leaching, the redox potentials were set at 655, 755 and 805 mV (vs. SHE) with the same temperature and pH. The total iron concentration was between 41 and 52 g/L and the Fe³⁺ concentration was about 39–40 g/L. The small range of the redox potential chosen for this experimentation was based on our industrial practice at Zijinshan (Ruan et al., 2011). Figure 4.17 shows digenite and covellite dissolution at different redox potentials. There was no significant increase in the leaching with increases in the redox potential. The slopes of the log rate versus the redox potential plot showed 0.21 and 0.045 for digenite and covellite, respectively.

In summary, both covellite dissolution and second stage dissolution of digenite were sensitive to temperature, but insensitive to the redox potential. At 75°C, leaching completed in about 3 h for digenite while it completed in about 30 h for covellite, which indicated that the dissolution of covellite was

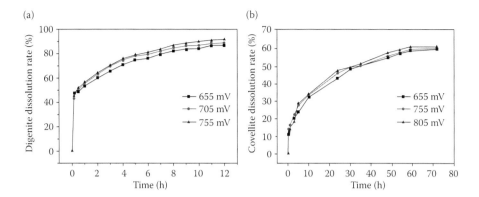

FIGURE 4.17
(See colour insert.) Effect of redox potential on (a) digenite dissolution and (b) covellite dissolution. (From Ruan, R. et al., 2011. *Hydrometallurgy*, 108(1): 130–135. With permission.)

one order of magnitude slower than that of digenite. Fe^{3+} concentrations had a positive effect on the second stage dis-343 solution of digenite and covellite dissolution, and the reaction order versus the Fe^{3+} concentration increased with an increase in temperature. The kinetic results indicated that elevated temperature and high ferric concentration accelerated the dissolution of digenite and covellite.

In Zijinshan copper ores, the main copper sulphides were digenite (47.2%), covellite (36.6%) and refractory copper mineral enargite (15.0%). To achieve a high rate of copper dissolution in heap leaching, temperature was the crucial factor. High concentrations of ferric ion would also benefit copper extraction.

4.6.2 Column Bioleaching of Zijinshan Copper Ores

Column bioleaching experiments (10-cm-diameter, 100-cm- high columns) were conducted to investigate the effects of temperature and ferric concentration on the bioleaching kinetics of Zijinshan copper ore.

4.6.2.1 Effect of Temperature

Column-leaching tests were conducted at temperatures 30°C and 60°C, with an initial TFe concentration of 50 g/L. The results are shown in Figure 4.18 and Table 4.11.

Figure 4.18b showed the inhibition of microorganism growth at 60°C from the low redox potential of about 700 mV (vs. SHE), whereas there was an active growth of microorganisms at 30°C with a redox potential of up to about 800 mV (vs. SHE). However, Figure 4.18a showed that the copper-leaching rate at 60°C was much faster than that at 30°C, and the copper extraction from the residue reached 90% at 60°C, but only 77% at 30°C after leaching for 60 days. Copper in the residues (Table 4.11) was 0.1% and 0.05% at 30°C

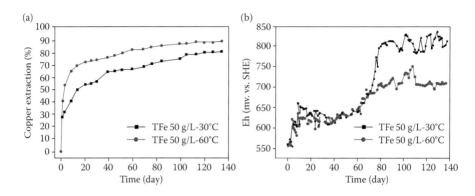

FIGURE 4.18
(See colour insert.) Column bioleaching at 30°C and 60°C. (a) Copper extraction from the solution and (b) changes of the redox potential. (From Ruan, R. et al., 2011. *Hydrometallurgy*, 108(1): 130–135. With permission.)

and 60°C, respectively after leaching for 135 days. The main copper minerals in Zijinshan copper ore were digenite, covellite and enargite. According to mineralogical investigation of the leached residue (data not shown), at 60°C, the dissolution of digenite, covellite and enargite was 100%, 88.7% and 58.9%, respectively; whereas at 30°C, the oxidation was 100%, 77.4% and 34.2%. The results showed that digenite was completely converted into covellite both at 30°C and 60°C in the bioleaching system, while an elevated temperature accelerated the covellite and enargite dissolution. Our previous study indicated that bacteria might accelerate digenite dissolution. At 30°, the copper extraction was 65% in ferric 380 sulphate, while it reached 95% with bacteria after leaching for 20 days (Ruan et al., 2010). Microorganisms also played a role in regenerating the leach solution. Lee et al. (2011) found that chalcocite showed good leachability by mesophilic bacteria at 22–24°C, while the covellite and enargite showed a refractory nature for bioleaching by mesophilic bacteria. These might explain why a high rate of digenite dissolution was achieved in column bioleaching of Zijinshan copper ore both at 30°C and 60°C.

TABLE 4.11

Copper Extraction from Residue at 30°C and 60°C

Leaching Time (days)	Cu in Residue (%)		Copper Extraction (%)	
	30°C	60°C	30°C	60°C
16	0.20	0.12	61.54	76.92
45	0.13	0.11	75.00	78.85
60	0.12	0.049	76.92	90.58
95	0.11	0.055	78.85	89.4
135	0.10	0.05	80.77	90.38

4.6.2.2 Effect of Fe³⁺ Concentration

Column bioleaching tests were carried out at initial total iron concentrations of 5 g/L (Fe^{3+} 3.15 g/L), 15 g/L (Fe^{3+} 11.51 g/L) and 50 g/L (Fe^{3+} 40 g/L) respectively, with temperature at 60°C. Figure 4.19 shows the copper extraction determined by the solution and changes of Fe^{3+} concentration.

Figure 4.19a shows that the copper dissolution rate was much faster at higher TFe concentrations (50 g/L) than that at low TFe concentrations (5 and 15 g/L) at an initial leaching time, while the copper extraction reached the same level after leaching for 120 days. During the course of leaching for all three TFe concentrations, pH was decreased from 1.5 to around 0.8 due to pyrite oxidation. Figure 4.19b shows an increase in ferric concentration due to pyrite oxidation and ferrous bio-oxidation. The results of column bioleaching were consistent with ferric- leaching kinetics that the dissolution rate of cop-403 per mineral was temperature dependent and a high Fe^{3+} concentration promoted digenite/covellite dissolution.

It is generally accepted that the bioleaching of sulphide minerals involves three major sub-processes, namely, the acid ferric leaching of the sulphide mineral, microbial oxidation of the sulphur moiety and the microbial oxidation of ferrous iron to the ferric form. In the bacterial leaching of sulphide minerals, ferric iron is the key oxidising agent and soluble iron species are the main determinants of redox potential (Ghorbani et al., 2011). Active iron-oxidising bacteria, such as *A. ferrooxidans* and *Leptospirillum ferrooxidans*, maintain high Fe^{3+}/Fe^{2+} ratios due to continued oxidation as part of their respiratory process (Ahonen and Tuovinen, 1995). Donati et al. (1997) confirmed that *A. ferrooxidans* was able to grow anaerobically on synthetic covellite with ferric ion as the electron acceptor; sulphur covered the sulphide surface as an electron donor and an efficient metal recovery was obtained under these conditions. Some studies found that, under anaerobic conditions

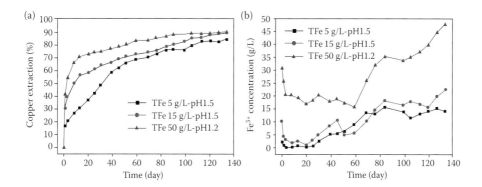

FIGURE 4.19
(See colour insert.) Effect of Fe^{3+} concentration on column bioleaching: (a) Copper extraction from the solution and (b) change of Fe^{3+} concentration during column leaching. (From Ruan, R. et al., 2011. *Hydrometallurgy*, 108(1): 130–135. With permission.)

(Pronk et al., 1992) or at extremely low pH values (Sand, 1989), *A. ferrooxidans* is able to reduce ferric to ferrous ion in the presence of sulphur or sulphide ores. Pronk et al. (1992) reported that this organism grows anaerobically using ferric ion as the electron acceptor and suggested that this could be the major biological process during oxidation of sulphide ores in leaching operations.

In most of the existing bioheapleaching practices, the plant design normally focussed on the bacterial requirements by using parameters such as ambient temperature, aeration, low heap height, low ferric concentration and high pH. However, Zijinshan copper bioheapleaching plant was operated with parameters far from the optimal condition for microorganisms growth (as about 104–432 cells/mL were detected in PLS), instead of elevated temperature (above 60°C inside the heap), high ferric concentration, low pH, high heap height and non-aeration. The operation system in Zijinshan is based on the ferric dissolution kinetics of copper minerals and ferric bio-regeneration outside the heaps.

4.6.3 The Rate-Limiting Factors for the Leaching Kinetics

Leaching kinetic experiments indicated that temperature is a crucial factor for abiotic dissolution of covellite, while ferric concentration and redox potential only had a slight positive effect. The second stage dissolution of digenite in acid ferric sulphate was relatively faster than that of covellite and sensitive to temperature and ferric concentration. Therefore, elevated temperatures play an important role in covellite and digenite dissolution. High ferric concentration is an additional benefit to the dissolution of digenite.

Column bioleaching of the Zijinshan copper ore indicated that copper extraction at 60°C was more efficient than that at 30°C and a high ferric concentration was an added benefit to leaching, despite low microorganism activity at an elevated temperature and high ferric concentration. The column test result was consistent with the mineralogy of the ore, dissolution kinetics of dominant copper minerals and industrial practice.

4.7 Current Status and Future Prospects

The commercial operation of the Zijinshan bioheapleaching plant has overcome the difficulties caused by a high content of pyrite in the feed and a high average annual rainfall. Technical measures were effectively implemented to achieve high copper recovery and low cost of acid and iron removal. A successful operation of the Zijinshan bioheapleaching system has demonstrated the adaptability of the system for broad applications according to mineralogy and the external environment.

Compared with many other copper mines, the copper grade of ROM at the Zijinshan copper mine is relative low (0.4%), with a high content of pyrite. Besides, the low content of acid-consuming gangues was presented in the ROM. As a result, acid and iron might be accumulated in the leaching solution system during the bioleaching process (Tables 4.3 and 4.4) and the leaching temperature may increase. With regard to the secondary copper sulphide (digenite and chalcocite), the leaching process in the acid and iron solution could be divided into two stages. In the first stage, copper was removed from secondary copper sulphide and the covellite-like product was produced. Previous research revealed that the first stage leaching of chalcocite is very fast compared with the second stage, and the second dissolution stage is slow (Dutrizac and MacDonald, 1974b), which makes the second stage the rate-determined step. Investigation for the second stage leaching kinetics of chalcocite showed that high iron concentration and high temperature could speed up the oxidation rate of the second leaching stage (Bolorunduro, 1999; Marcantonio, 1976). Thus, a relatively high copper recovery (more than 80 wt.%) was achieved in the Zijinshan bioheapleaching system as a result of high ferric concentration maintained in the leaching solution and high temperature inside the heap (Table 4.5).

In a bioheapleaching system, the dominance of iron oxidisers, especially the genus *Leptospirillum*, could effectively oxidise Fe^{2+} to Fe^{3+}. As a result, the ferric supply by microbial oxidation may be higher than the ferric demand for mineral oxidation and the redox potential may increase. This observation was verified by previous research that the dominance of *Leptospirillum* in the copper sulphide bioheapleaching system was correlated with fast ferrous oxidation and the increase of Eh (SHE) to 800 mV in the initial leaching stage (Demergasso et al., 2005, 2010). But for an Eh (SHE) that is lower, between 700 and 760 mV, pyrite dissolution was limited while secondary copper sulphide still has good leaching kinetics. However, pyrite dissolution could be accelerated when Eh (SHE) is higher than 800 mV as was observed by Wu (Wu et al., 2009). A high acidity, high iron concentration and high salinity were maintained in the Zijinshan bioheapleaching system by recycling the leaching solution system continuously. Multi-lift permanent stacking was used to keep the heap in high temperature. Under such an environment, the growth and ferrous oxidation of the genus *Leptospirillum* was inhibited, which was consistent with previous research outcomes (Özkaya et al., 2007; Penev and Karamanev, 2009; Petersen and Ojumu, 2007). Thus, the Eh of the Zijinshan leaching system was kept at a low level (710–740 mV), and the dissolution of pyrite was restricted.

Previous research noted that jarosite could be formed under conditions of high ferric concentration, low pH value and elevated temperature (Queneau, 1961). Such an environment was created in the Zijinshan heap due to the distinct mineralogy and technical measure listed above. As a result, jarosite was easily formed in the Zijinshan heap, and neutralisation was against free acid rather than iron. The expense for iron and acid removal was decreased.

A number of cost-cutting measures were implemented such as multi-lift permanent stacking, large particle size (80% passing 40 mm), ore stacking by the dump truck, non-aeration and non-agglomeration to achieve a low operating cost of 1.10 USD/lb, even though the average copper grade of ROM has continually decreased to as low as 0.38%.

Bioheapleaching practice in the Zijinshan plant showed that high concentration of ferric ion and elevated temperature could facilitate copper leaching with low bacteria activity. This is consistent with Marcantonio and Samuel's observation (Bolorunduro, 1999; Marcantonio, 1976). For secondary copper sulphide bioheapleaching that is operated at a low concentration of ferric ion and low temperature (Brierley, 2010; Demergasso et al., 2010), two technical measures might improve the kinetics of copper dissolution, namely, (a) increase of heap height or heap temperature; and (b) direct addition of $Fe_2(SO_4)_3$ into the leaching system or fast dissolution of pyrite during the start-up stage to increase ferric concentration of the solution.

To further optimise the iron balance in the Zijinshan plant, research projects are being conducted on the kinetics of pyrite dissolution and jarosite formation under conditions of high ferric concentration and elevated temperature as well as on optimum heap height and irrigation rate to control heap temperature better.

References

Ahonen, L. and Tuovinen, O.H., 1995. Bacterial leaching of complex sulfide ore samples in bench-scale column reactors. *Hydrometallurgy*, 37(1): 1–21.

Battaglia-Brunet, F. et al., 1998. The mutual effect of mixed *Thiobacilli* and *Leptospirilli* populations on pyrite bioleaching. *Minerals Engineering*, 11(2): 195–205.

Bolorunduro, S.A., 1999. Kinetics of leaching of chalcocite in acid ferric sulfate media: Chemical and bacterial leaching. University of British Columbia.

Bond, P.L., Druschel, G.K. and Banfield, J.F., 2000. Comparison of acid mine drainage microbial communities in physically and geochemically distinct ecosystems. *Applied and Environmental Microbiology*, 66(11): 4962–4971.

Brierley, C., 2010. Biohydrometallurgical prospects. *Hydrometallurgy*, 104(3): 324–328.

Brierley, J. and Brierley, C., 2001. Present and future commercial applications of biohydrometallurgy. *Hydrometallurgy*, 59(2): 233–239.

Cheng, C.Y. and Lawson, F., 1991. The kinetics of leaching chalcocite in acidic oxygenated sulphate–chloride solutions. *Hydrometallurgy*, 27(3): 249–268.

Coram, N.J. and Rawlings, D.E., 2002. Molecular relationship between two groups of the genus *Leptospirillum* and the finding that *Leptospirillum ferriphilum* sp. nov. dominates South African commercial biooxidation tanks that operate at 40 C. *Applied and Environmental Microbiology*, 68(2): 838–845.

d'Hugues, P., Foucher, S., Galle-Cavalloni, P. and Morin, D., 2002. Continuous bioleaching of chalcopyrite using a novel extremely thermophilic mixed culture. *International Journal of Mineral Processing*, 66(1): 107–119.

da Silva, G., 2004. Relative importance of diffusion and reaction control during the bacterial and ferric sulphate leaching of zinc sulphide. *Hydrometallurgy*, 73(3): 313–324.

Demergasso, C., Galleguillos, F., Soto, P., Serón, M. and Iturriaga, V., 2010. Microbial succession during a heap bioleaching cycle of low grade copper sulfides: Does this knowledge mean a real input for industrial process design and control? *Hydrometallurgy*, 104(3): 382–390.

Demergasso, C.S. et al., 2005. Molecular characterization of microbial populations in a low-grade copper ore bioleaching test heap. *Hydrometallurgy*, 80(4): 241–253.

Domic, E.M., 2007. A review of the development and current status of copper bioleaching operations in Chile: 25 years of successful commercial implementation. *Biomining*. Springer, New York, pp. 81–95.

Donati, E., Pogliani, C. and Boiardi, J., 1997. Anaerobic leaching of covellite by *Thiobacillus ferrooxidans*. *Applied Microbiology and Biotechnology*, 47(6): 636–639.

Dopson, M., Baker-Austin, C., Hind, A., Bowman, J.P. and Bond, P.L., 2004. Characterization of *Ferroplasma* isolates and *Ferroplasma acidarmanus* sp. nov., extreme acidophiles from acid mine drainage and industrial bioleaching environments. *Applied and Environmental Microbiology*, 70(4): 2079–2088.

Dutrizac, J. and MacDonald, R., 1974a. The kinetics of dissolution of covellite in acidified ferric sulphate solutions. *Canadian Metallurgical Quarterly*, 13(3): 423–433.

Dutrizac, J.E. and MacDonald, R., 1974b. *Ferric Ion as a Leaching Medium*. Department of Energy, Mines, and Resources, Ottawa.

Ghorbani, Y., Becker, M., Mainza, A., Franzidis, J.-P. and Petersen, J., 2011. Large particle effects in chemical/biochemical heap leach processes—A review. *Minerals Engineering*, 24(11): 1172–1184.

Helle, U. and Onken, U., 1988. Continuous microbial leaching of a pyritic concentrate by *Leptospirillum*-like bacteria. *Applied Microbiology and Biotechnology*, 28(6): 553–558.

Hippe, H., 2000a. *Leptospirillum* gen. nov. (ex Markosyan 1972), nom. rev., including *Leptospirillum ferrooxidans* sp. nov. (ex Markosyan 1972), nom. rev. and *Leptospirillum thermoferrooxidans* sp. nov. (Golovacheva et al. 1992). *International Journal of Systematic and Evolutionary Microbiology*, 50(2): 501–503.

Hippe, H., 2000b. *Leptospirillum* gen. nov.(ex Markosyan 1972), nom. rev., including *Leptospirillum ferrooxidans* sp. nov.(ex Markosyan 1972), nom. rev. and *Leptospirillum thermoferrooxidans* sp. nov.(Golovacheva et al. 1992). *International Journal of Systematic and Evolutionary Microbiology*, 50(2): 501–503.

Huang, X., 2010. Special Audit for Zijin Mining Group Co. Ltd., Chengxing Public Accounting Firm (Fujian), Fuzhou.

Ibaceta, D. and Garrido, J., 2005. SX development at Dos Amigos in Northern Chile. *ALTA Conference*, Perth, WA, Australia, pp. 17.

Johnson, D.B., Okibe, N. and Roberto, F.F., 2003. Novel thermo-acidophilic bacteria isolated from geothermal sites in Yellowstone National Park: Physiological and phylogenetic characteristics. *Archives of Microbiology*, 180(1): 60–68.

Karamanev, D., Nikolov, L. and Mamatarkova, V., 2002. Rapid simultaneous quantitative determination of ferric and ferrous ions in drainage waters and similar solutions. *Minerals Engineering*, 15(5): 341–346.

Kinnunen, P.H.-M. and Puhakka, J.A., 2005. High-rate iron oxidation at below pH 1 and at elevated iron and copper concentrations by a *Leptospirillum ferriphilum* dominated biofilm. *Process Biochemistry*, 40(11): 3536–3541.

Lee, J., Acar, S., Doerr, D.L. and Brierley, J.A., 2011. Comparative bioleaching and mineralogy of composited sulfide ores containing enargite, covellite and chalcocite by mesophilic and thermophilic microorganisms. *Hydrometallurgy*, 105(3): 213–221.

Liu, X.Y., Chen, B.W. and Wen, J.K., 2013. Insights into bacterial community structure of a commercial copper bioheapleaching plant: Growth of heterotrophic bacteria in the system. *Advanced Materials Research*, 825: 50–53.

Lizama, H., Harlamovs, J., McKay, D. and Dai, Z., 2005. Heap leaching kinetics are proportional to the irrigation rate divided by heap height. *Minerals Engineering*, 18(6): 623–630.

Lizama, H.M. and Suzuki, I., 1991. Kinetics of sulfur and pyrite oxidation by *Thiobacillus thiooxidans*. Competitive inhibition by increasing concentrations of cells. *Canadian Journal of Microbiology*, 37(3): 182–187.

Lochmann, J. and Pedlik, M., 1995. Kinetic anomalies of dissolution of sphalerite in ferric sulfate solution. *Hydrometallurgy*, 37(1): 89–96.

Marcantonio, P., 1976. Kinetics of dissolution of chalcocite in ferric sulfate solutions. Ph D thesis. University of Utah. United States.

McGuire, M.M., Edwards, K.J., Banfield, J.F. and Hamers, R.J., 2001. Kinetics, surface chemistry, and structural evolution of microbially mediated sulfide mineral dissolution. *Geochimica et Cosmochimica Acta*, 65(8): 1243–1258.

Miki, H., Nicol, M. and Velásquez-Yévenes, L., 2011. The kinetics of dissolution of synthetic covellite, chalcocite and digenite in dilute chloride solutions at ambient temperatures. *Hydrometallurgy*, 105(3): 321–327.

Mulak, W., 1971. Kinetics of dissolving polydispersed covellite in acidic solutions of ferric sulphate. *Roczniki Chemii*, 45(9): 1417–1424.

Özkaya, B., Nurmi, P., Sahinkaya, E., Kaksonen, A.H. and Puhakka, J.A., 2007. Temperature effects on the iron oxidation kinetics of a *Leptospirillum ferriphilum* dominated culture at pH below one. *Advanced Materials Research*, 20: 465–468.

Ozkaya, B., Sahinkaya, E., Nurmi, P., Kaksonen, A.H. and Puhakka, J.A., 2007. Iron oxidation and precipitation in a simulated heap leaching solution in a *Leptospirillum ferriphilum* dominated biofilm reactor. *Hydrometallurgy*, 88(1): 67–74.

Penev, K. and Karamanev, D., 2009. Kinetics of ferrous iron oxidation by *Leptospirillum ferriphilum* at moderate to high total iron concentrations. *Advanced Materials Research*, 71: 255–258.

Petersen, J. and Ojumu, T.V., 2007. The effect of total iron concentration and iron speciation on the rate of ferrous iron oxidation kinetics of *Leptospirillum ferriphilum* in continuous tank systems. *Advanced Materials Research*, 20: 447–451.

Plumb, J., Muddle, R. and Franzmann, P., 2008a. Effect of pH on rates of iron and sulfur oxidation by bioleaching organisms. *Minerals Engineering*, 21(1): 76–82.

Plumb, J.J., Muddle, R. and Franzmann, P.D., 2008b. Effect of pH on rates of iron and sulfur oxidation by bioleaching organisms. *Minerals Engineering*, 21(1): 76–82.

Pronk, J., De Bruyn, J., Bos, P. and Kuenen, J., 1992. Anaerobic growth of *Thiobacillus ferrooxidans*. *Applied and Environmental Microbiology*, 58(7): 2227–2230.

Queneau, P.B., 1961. *Extractive Metallurgy of Copper, Nickel, and Cobalt*, 1. Interscience Publishers, New York, London.

Rawlings, D., Tributsch, H. and Hansford, G., 1999. Reasons why '*Leptospirillum*'-like species rather than *Thiobacillus ferrooxidans* are the dominant iron-oxidizing bacteria in many commercial processes for the biooxidation of pyrite and related ores. *Microbiology-Reading*, 145: 5–13.

Rawlings, D.E. and Johnson, D.B., 2007. The microbiology of biomining: Development and optimization of mineral-oxidizing microbial consortia. *Microbiology*, 153(2): 315–324.

Renman, R., Jiankang, W. and Jinghe, C., 2006. Bacterial heap-leaching: Practice in Zijinshan copper mine. *Hydrometallurgy*, 83(1): 77–82.

Ruan, R. et al., 2011. Industrial practice of a distinct bioleaching system operated at low pH, high ferric concentration, elevated temperature and low redox potential for secondary copper sulfide. *Hydrometallurgy*, 108(1): 130–135.

Ruan, R. et al., 2010. Comparison on the leaching kinetics of chalcocite and pyrite with or without bacteria. *Rare Metals*, 29(6): 552–556.

Ruan, R. et al., 2013. Why Zijinshan copper bioheapleaching plant works efficiently at low microbial activity—Study on leaching kinetics of copper sulfides and its implications. *Minerals Engineering*, 48:36–43.

Sand, W., 1989. Ferric iron reduction by *Thiobacillus ferrooxidans* at extremely low pH-values. *Biogeochemistry*, 7(3): 195–201.

Sand, W., Rohde, K., Sobotke, B. and Zenneck, C., 1992a. Evaluation of *Leptospirillum ferrooxidans* for leaching. *Applied Environment Microbiology*, 58(1): 85–92.

Sand, W., Rohde, K., Sobotke, B. and Zenneck, C., 1992b. Evaluation of *Leptospirillum ferrooxidans* for leaching. *Applied and Environmental Microbiology*, 58(1): 85–92.

Sullivan, J.D., 1930. Chemistry of leaching covellite, 487. U.S. Govt. print. off.

Thomas, G. and Ingraham, T., 1967. Kinetics of dissolution of synthetic covellite in aqueous acidic ferric sulphate solutions. *Canadian Metallurgical Quarterly*, 6(2): 153–165.

Walsh, C.A. and Rimstidt, J.D., 1986. Rates of reaction of covellite and blaubleibender covellite with ferric iron at pH 2.0. *Canadian Mineralogist*, 24: 35–44.

Watling, H.R. et al., 2014. Bioleaching of a low-grade copper ore, linking leach chemistry and microbiology. *Minerals Engineering*, 56(0): 35–44.

Wu, B., Wen, J.K., Zhou, G.Y. and Ruan, R.M., 2009. Control of the redox potential by oxygen limitation inhibits bioleaching of pyrite. *Advanced Materials Research*, 71: 401–404.

Xingyu, L., Rongbo, S., Bowei, C., Biao, W. and Jiankang, W., 2009. Bacterial community structure change during pyrite bioleaching process: Effect of pH and aeration. *Hydrometallurgy*, 95(3): 267–272.

Xingyu, L. et al., 2010a. Bioleaching of chalcocite started at different pH: Response of the microbial community to environmental stress and leaching kinetics. *Hydrometallurgy*, 103(1): 1–6.

Xingyu, L., Bowei, C., Jiankang, W. and Renman, R., 2010b. *Leptospirillum* forms a minor portion of the population in Zijinshan commercial non-aeration copper bioleaching heap identified by 16S rRNA clone libraries and real-time PCR. *Hydrometallurgy*, 104(3–4): 399–403.

Xingyu, L., Rongbo, S., Bowei, C., Biao, W. and Jiankang, W., 2009. Bacterial community structure change during pyrite bioleaching process: Effect of pH and aeration. *Hydrometallurgy*, 95(3): 267–272.

5

Spectroscopic Study on the Bioleaching of Enargite Using Thermophile

Keiko Sasaki

CONTENTS

5.1 Introduction

Enargite (Cu_3AsS_4) is a representative copper–arsenic sulphide mineral. It occurs locally in some ore deposits, especially epithermal high sulphidation deposits (Arribas, 1995) and porphyry copper systems (Lattanzi et al., 2008). Although enargite is also valuable as a copper resource, the arsenic in the mineral makes it environmentally toxic through emissions associated with smelting (Dutré and Vandecasteele, 1995).

The oxidation rate of enargite is much slower than that of other Cu sulphides such as tennantite ($Cu_{12}AsS_{13}$) and chalcopyrite ($CuFeS_2$), especially at an acidic pH (Figure 5.1, Sasaki et al., 2010a). Because of the slow chemical oxidation rate of enargite, the process of bioleaching would be one potential approach for recovering copper from enargite. There have been a number of studies on the bioleaching of enargite by mesophilic bacteria (Escobar et al., 1997; Acevedo et al., 1998; Watling, 2006; Corkhill et al., 2008). Although the leaching mechanisms and performance were thoroughly discussed, the passivation mechanism and mineralogical alteration of enargite have not yet been elucidated. Previously, the formation of amorphous ferric arsenate was investigated, which can temporarily scavenge dissolved As during the bioleaching of enargite by As-adapted *Acidithiobacillus ferrooxidans* (Sasaki et al., 2010b). The recovery of Cu from this process was quite low (<3% within

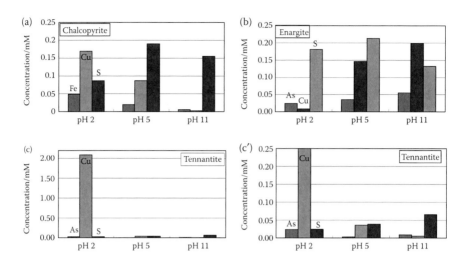

FIGURE 5.1
(**See colour insert**.) Dissolved Cu, S, Fe and As concentrations from (a) chalcopyrite, (b) enargite and (c) tennantite after 1 h oxidation by 0.013% H_2O_2 at pH 2, 5 and 11 under O_2 bubbling. Note the scale of the vertical axis in (c) and (c'). Mineral samples were ground to 74–105 μm. 0.5 g of each mineral sample was added into 50 mL of 0.013% H_2O_2 and 0.01 M KNO_3 at pH 2, 5 and 11 adjusted with 1 M HNO_3 and 1 M KOH. (Modified from Sasaki, K. et al., 2010a. *Hydrometal.*, 100, 144–151.)

130 days, Figure 5.2b). Bioleaching tests using thermophilic bacteria, such as *Sulfolobus metallicus*, demonstrated better performance when compared to mesophiles (Escobar et al., 2000; Muñoz et al., 2006). In the presence of As and Fe, highly crystalline scorodite was produced via microbial metabolism of the thermophilic iron-oxidising archaebacterium, *Acidianus sulfidivorans* (Gonzalez-Contreras et al., 2010). Several studies have indicated that *A. brierleyi* has inherent properties that expedite the extraction of various metals, such as Cu, Mo, Ni, Fe and Zn, from sulphide minerals (Konishi et al., 1998). In this chapter, the bioleaching of enargite by *A. brierleyi* at 70°C to obtain a high recovery of Cu from enargite and elucidate the passivation mechanisms involved is reviewed, especially the removal of As from the aqueous phase.

5.2 Bioleaching of Enargite Using Thermophile

Based on the indirect oxidation mechanism with thermophile, bioleaching of enargite was conducted in the presence of 50 mM Fe^{2+}. *A. brierleyi* had oxidised the dissolved Fe^{2+} almost completely within 10 days (Figure 5.2a). A decrease in the total dissolved Fe was also observed and 91.7% of the initial

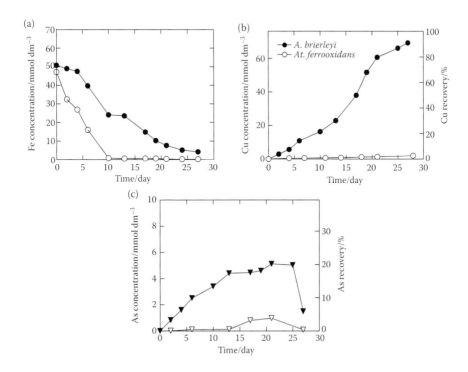

FIGURE 5.2
Changes in the concentrations of (a) Fe (•) and Fe(II) (○), (b) Cu (•) and (c) As (▼) and As(III) (▽) in bioleaching of enargite with *A. brierleyi* at 70°C. Note the symbol ○ in (b) indicates with *A. ferrooxidans*. (Modified from Takatsugi, K., Sasaki, K., Hirajima, T., 2011. *Hydrometal.*, 109, 90–96.)

Fe concentration was estimated to have precipitated into the solid phase. Using *A. brierleyi*, 90.9% of the Cu was recovered from enargite into the aqueous phase after 27 days. Regarding the stoichiometry of the dissolved Cu and As, preferential leaching of Cu compared to As was clearly observed (Figures 5.2b and c). This suggests that some of the dissolved As had precipitated, mainly as Fe–As compounds.

Additionally, a decrease in As of 3.5 mmol/L was observed after 25 days that was independent of Fe. This coincided with the decrease in the leaching rate of Cu from the enargite while most of the dissolved Fe had already precipitated. The Fe^{2+}-oxidising activity of the microorganisms is the most important factor governing the efficiency of Cu bioleaching from enargite. It is reasonable to expect that, once the oxidation of enargite had declined, the regeneration of Fe^{3+} also ceased and the solution chemistry, including the redox potential and ionic activities of the secondary minerals, changed. Then, other As compounds could be formed that do not incorporate any Fe species.

The chemical valence of As in the enargite was confirmed to be as As(III) by As K-edge XANES (Sasaki et al., 2010a). However, the dissolved As was specified to be mostly H_3AsO_3 by high-performance liquid chromatography–inductively coupled plasma–mass spectrometry (HPLC–ICP–MS) (Figure 5.2c). This suggests that dissolved H_3AsO_3 from the enargite was immediately oxidised to $H_2AsO_4^-$ (Smedley and Kinniburgh, 2002). Okibe et al. (2014) have recently found that the extent of As(III) oxidation by *A. brierleyi* became greater at elevated culture As(III) concentrations, especially in the presence of the yeast extract, and that microbial growth on the yeast extract was also implied to facilitate As(III) oxidation by Fe(III) on the cell surface. Although the mechanism of oxidation of H_3AsO_3 in the leaching solution is not yet clear, a similar phenomenon was also observed in the bioleaching of pyrite (FeS_2) by thermophilic bacteria at high Eh potential (>420 mV vs. Ag/AgCl) (Wiertz et al., 2006). In that case, the proposed mechanism was electrochemical reactions at the surface of the pyrite. Thermodynamically, H_3AsO_3 can be oxidised by Fe^{3+} at low pH due to the difference in redox potentials of the Fe^{3+}/Fe^{2+} (0.771 V at 25°C) and $H_3As^VO_4/H_3As^{III}O_3$ (0.560 V at 25°C) couples. However, in most cases, the reaction is quite slow because no oxidation of H_3AsO_3 is observed. Accordingly, it is conceivable that enargite can catalyse the oxidation of H_3AsO_3 in this system. The proposed mechanism follows electrochemical reactions. The following half-reactions may occur on the surface of the enargite during bioleaching.

$$Cu_3As^{III}S_4 + 19H_2O = H_3As^{III}O_3 + 3Cu^{2+} + 4SO_4^{2-} + 35H^+ + 33e^- \text{(anodic)} \quad (5.1)$$

$$H_3As^{III}O_3 + H_2O = H_3As^VO_4 + 2H^+ + 2e^- \text{ (anodic)} \quad (5.2)$$

$$Fe^{3+} + e^- = Fe^{2+} \text{ (cathodic)} \quad (5.3)$$

Once H_3AsO_3 has been oxidised to $H_2AsO_4^-$ in the presence of Fe^{3+}, ferric arsenate and/or crystallised scorodite ($FeAsO_4 \cdot 2H_2O$) could be formed to immobilise As in the solid phase.

$$H_3AsO_4 + Fe^{3+} + 2H_2O = FeAsO_4 \cdot 2H_2O + 3H^+ \quad (5.4)$$

5.3 Characterisation of Secondary Minerals in Bioleaching of Enargite with Thermophile

After 27 days of leaching, the solid residue was characterised by XRD (Figure 5.3). The XRD pattern for solid residues after 27 days indicated that scorodite and potassium jarosite ($KFe_3(SO_4)_2 \cdot (OH)_6$) were formed on the residual enargite as secondary minerals, suggesting that most of the dissolved Fe was consumed to form these minerals in the batch test (Figure 5.2a). Figure 5.4 shows the XP-spectra in the As 3d region for the original enargite and the solid residue after 27 days of leaching. The predominant As species in the original enargite is As(III) and part of the surface was

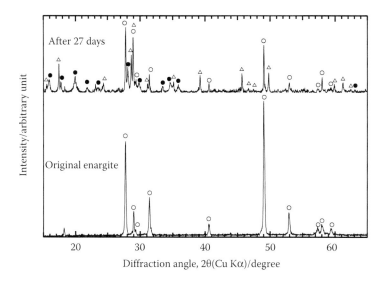

FIGURE 5.3

XRD patterns of enargite and the solid residue after 27 days of leaching of the enargite by *A. brierleyi* at 70°C. The symbols ○, △ and • are assigned to enargite (JCPDS 35-0755), potassium jarosite (JCPDS 22-0827) and scorodite (JCPDS 37-0468), respectively. (Modified from Takatsugi, K., Sasaki, K., Hirajima, T., 2011. *Hydrometal.*, 109, 90–96.)

oxidised to As(V). However, in the solid residue after leaching, the peaks assigned to As(III) completely disappeared and the dominant component at $E_B[\text{As } 3d_{5/2}] = 45.0$ eV is assigned to As(V). Thus, the secondary compound that precipitated in the latter stage of the bioleaching in the absence of Fe species was also arsenate.

Scorodite is by far the most common secondary As mineral formed under acidic conditions of an Fe(III)–As(V) system associated with arsenic-bearing primary minerals such as arsenopyrite and enargite (Flemming et al., 2005; Lattanzi et al., 2008). As scorodite is thermodynamically stable (K_{sp} range of $10^{-21.7}$–$10^{-25.86}$, Drahota and Fillippi, 2009) and suitable for arsenic fixation and storage, the synthesis of crystalline scorodite has been developed as one of the chemical-stabilising methodologies for controlling arsenic in waste streams in the smelting processes (Fujita et al., 2008). Arsenate oxyanion is ready to associate with several cations, such as NH_4^+, Mg^{2+} and Ca^{2+} from the ingredients of the medium and Al^{3+}, Zn^{2+} and Cu^{2+} from impurities/components in the enargite specimen, to form precipitates. The formation of the following arsenates was considered to be secondary Fe-free, As-bearing minerals. Although Cu^{2+} is the most predominant cation, the saturation index of $Cu_3(AsO_4)_2$ was still negative after 25 days. Gräfe et al. (2008) have reported that hydrated clinoclase ($Cu_3(AsO_4)_2 \cdot 2H_2O$) or a euchroite-like precipitate ($Cu_2AsO_4(OH)$) could be formed on the surface of goethite and natrojarosite by synergistic interaction between the metallic cation and the oxyanion

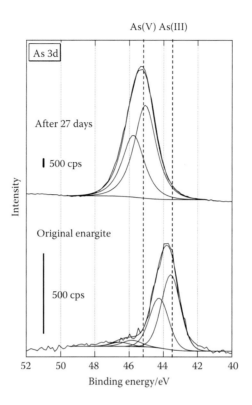

FIGURE 5.4
XP-spectra of the As 3d region for enargite and the residue after 27 days of leaching. Vertical bars indicate 500 cps. (Modified from Takatsugi, K., Sasaki, K., Hirajima, T., 2011. *Hydrometal.*, 109, 90–96.)

similar to co-sorption, even in a seemingly undersaturated solution. This suggests that precipitates such as $Cu_3(AsO_4)_2$ are formed, contributing to the immobilisation of the dissolved $H_2AsO_4^-$ after 25 days of leaching due to a seed-forming effect from the densely adsorbed Cu^{2+} and $H_2AsO_4^-$ on the surface of the solid residue.

According to the SEM-EDS mapping results for Cu, As, S, Fe, K and O, the mineralogical distribution was drawn (Figure 5.5). It indicates that enargite was firstly coated with elemental sulphur and/or sulphur-rich metal-deficient layers (Sasaki et al., 1995), and then other secondary minerals (such as potassium jarosite and scorodite) precipitated onto the surface of the residue. Although *A. brierleyi* is capable of oxidising sulphide and elemental sulphur, insoluble Fe(III) secondary minerals, such as scorodite and potassium jarosite, immediately precipitated onto the surface of the sulphur-rich layer preventing the contact between the microbial cells as well as Fe^{3+} ions with the elemental sulphur and enargite. As a result, enargite was partially passivated by multi-layers consisting of sulphur, scorodite and potassium jarosite.

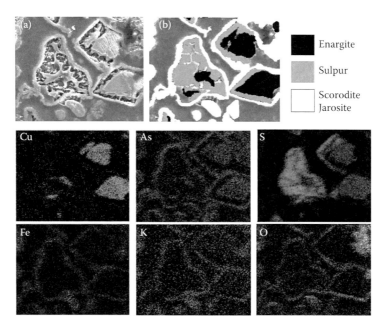

FIGURE 5.5
SEM image (a) and the expected mineralogical distribution (b) according to EDS elemental mapping of Cu, As, S, Fe, K and O for a cross-section of the residue after 27 days of leaching with *A. brierleyi* at 70°C. The scale bar indicates 10 µm in the SEM image (a). (Modified from Takatsugi, K., Sasaki, K., Hirajima, T., 2011. *Hydrometal.*, 109, 90–96.)

TEM–EDS elemental mapping results show a clear trend of an overlaid distribution of K–Fe–S on the right side and Cu–As–Fe on the left side of the narrow ditch in the square region of the DF image (Figure 5.6). The former elemental distribution suggests the formation of potassium jarosite, and the latter suggests mixed compounds of Fe–As minerals (scorodite) and a Cu–As phase such as $Cu_3(AsO_4)_2$. Observation of the narrow ditch in the nano-domain also suggests the presence of an organic layer with a thickness of approximately 100 nm between the two phases. This is expected to be a dense zone of extracellular polymeric substances (EPS) produced by *A. brierleyi*, in which carboxylate and phosphate and other functional groups accumulate cationic species, such as K^+, Fe^{3+} and Cu^{2+}, as components of secondary minerals.

5.4 Passivation of Arsenic on Solid Phase

Based on this comprehensive interpretation, the predominant passivation mechanisms were predicted as shown in Figure 5.7. Incongruent dissolution

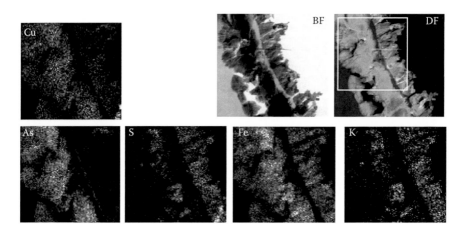

FIGURE 5.6
TEM and STEM images of the residue after 27 days of leaching with *A. brierleyi* at 70°C. EDS elemental mappings of Cu, As, S, Fe and K are from the square region in the DF image. Scale bars indicate 1 μm in the BF and DF images. (Modified from Takatsugi, K., Sasaki, K., Hirajima, T., 2011. *Hydrometal.*, 109, 90–96.)

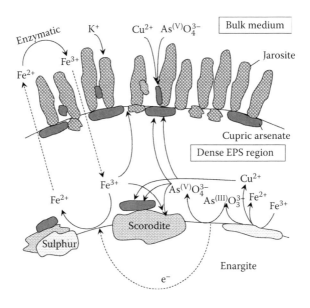

FIGURE 5.7
(**See colour insert.**) Schematic model of the bioleaching of enargite followed by the sequential formation of secondary minerals in the vicinity of the enargite surface. (Modified from Takatsugi, K., Sasaki, K., Hirajima, T., 2011. *Hydrometal.*, 109, 90–96.)

of the enargite led to the formation and accumulation of a secondary sulphur layer on the surface of the enargite. The formation of scorodite readily occurred on and/or near the surface of the enargite because $H_2AsO_4^-$ was produced by electron transfer between dissolved H_3AsO_3 and active catalytic sites on the surface of the enargite. At 70°C, the scorodite and/or ferric arsenate formed are well crystallised compared to what was found at room temperature, leading to greater stability. In contrast, an electrostatic interaction between metal cations such as Fe^{3+} and EPS densely distributed on the mineral triggered the nucleation of potassium jarosite in the vicinity of the EPS. Then, crystal growth of potassium jarosite progressed towards the bulk medium, which is rich in K^+, Fe^{3+} and sulphate. The formation of jarosite species not only causes suppression of the bioleaching of sulphide under acidic conditions, but is also accompanied by the co-sorption of cupric and arsenate ions on potassium jarosite, leading to the formation of cupric arsenate compounds. The formation of $Cu_3(AsO_4)_2$ also contributes to the immobilisation of some of the dissolved As species after the formation of scorodite was complete. Cupric arsenate is formed externally to the scorodite, but inside and/or near to the potassium jarosite. The observed thickness of approximately 100 nm for the EPS may be an underestimate due to shrinkage of the polymer during the preparation for TEM observation. It is insufficient physical space for a growth domain of *A. brierleyi*. The ion oxidation by *A. brierleyi* occurs mostly in the bulk medium and Fe^{3+} ions diffuse through the EPS region and then react with enargite to release Fe^{2+} and/or complex with arsenate to form scorodite.

In the case of bioleaching of enargite, it is necessary to consider minimising the As release in addition to maximising the Cu recovery. The predominant mechanism of As immobilisation in the bioleaching of enargite with the thermophilic iron-oxidising archaea, *A. brierleyi*, is the formation of scorodite ($FeAsO_4 \cdot 2H_2O$) (Takatsugi et al., 2011). In this process, the Fe^{3+} ions serve at least two functions: (i) as oxidants of enargite, and (ii) as a source of secondary Fe(III) precipitates, including jarosite and scorodite. Since both Fe and As are components of scorodite, the initial Fe^{2+} concentration and pulp density of enargite in the bioleaching system are important controlling factors of the bioleaching and secondary mineral formation. Two variables, the initial Fe^{2+} concentration and pulp density, are focussed in experiments to improve Cu recovery and immobilise As released from enargite.

Greater initial Fe^{2+} concentrations produced faster Fe^{2+} oxidation rate (Figures 5.8a through d); however, 2.7 g/L Fe^{2+} seems to be the optimal condition, where Cu recovery is maximised and As recovery is minimised (Figures 5.8e and f). After 76 days of leaching with an $[Fe^{2+}]_{ini}$ of 0.9 g/L, the formation of scorodite ($FeAsO_4 \cdot 2H_2O$) was observed in XRD for solids sampled from inoculated flasks (Figure 5.9b). At $[Fe^{2+}]_{ini} = 3.6$ g/L, XRD peaks assigned to K-jarosite were clearly observed (Figure 5.9e). This is in good agreement with the chemical composition of the aqueous phase (Figure 5.8). The x-ray patterns after 27 days of leaching with $[Fe^{2+}]_{ini}$ of 1.8 g/L did not show detectable

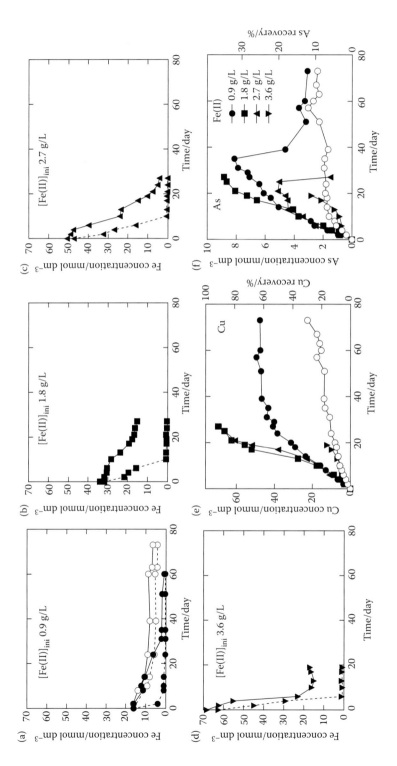

FIGURE 5.8
Effect of the initial Fe²⁺ on changes in the dissolved concentrations of (a) through (d) Fe²⁺ and the total dissolved Fe, (e) Cu and (f) As during the bioleaching of enargite by *A. brierleyi* at 70°C. Solid and dotted lines indicate the total dissolved Fe and Fe²⁺ in (a) through (d), respectively. Solid and open symbols indicate inoculated and sterile samples, respectively. (Modified from Sasaki, K., Takatsugi, K., Hirajima, T., 2011. *Hydrometal.*, 109, 153–160.)

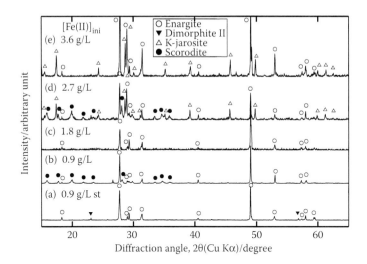

FIGURE 5.9
XRD patterns of solid residues after the bioleaching of enargite by *A. brierleyi* at 70°C. Reference minerals: enargite (JCPDS 35-0755), dimorphite II (JCPDS26-126), scorodite (JCPDS 37-0468) and K-jarosite (JCPDS 22-0827). The solid residues were analysed after (a) 76 days, (b) 76 days, (c) 27 days, (d) 27 days and (e) 19 days of bioleaching. (Modified from Sasaki, K., Takatsugi, K., Hirajima, T., 2011. *Hydrometal.*, 109, 153–160.)

secondary minerals, whereas K-jarosite and scorodite were formed when $[Fe^{2+}]_{ini}$ was 2.7 g/L (Figures 5.9c and d). These findings indicated that secondary minerals, including As-bearing precipitates, were either very few, poorly crystallised particles or amorphous, when $[Fe^{2+}]_{ini}$ was 1.8 g/L.

The As K-edge XANES spectra showed the chemical state of As in the solid residues (Figure 5.10). A peak assigned to enargite at 11,867.2 keV was clearly recognised in the solid residue when $[Fe^{2+}]_{ini}$ was 3.6 g/L in the inoculated and 0.9 g/L in the sterile systems. This was consistent with the low Cu recovery in both cases (Figure 5.8e). In other solid residues, the predominant peaks were assigned to As(V) at 11,871 keV. By fitting the observed XANES spectra using enargite and scorodite as standards, the molar fraction of As was obtained (Figure 5.11), assuming that no other As(V) precipitates were formed. The molar fraction of As in the enargite was at a minimum when $[Fe^{2+}]_{ini}$ was in the range of 1.8–2.7 g/L. A molar fraction of 83% As in the scorodite was achieved when $[Fe^{2+}]_{ini}$ was 2.7 g/L, coinciding with the maximum Cu recovery (Figure 5.8e). From these results, it seems that the optimal $[Fe^{2+}]_{ini}$ was 2.7 g/L for both Cu recovery and As immobilisation in the bioleaching of enargite with *A. brierleyi* at 70°C. XPS analysis demonstrated that As(V) species are predominant in all cases (data not shown). Thus, $H_3As^{(III)}O_3$ dissolved from enargite was oxidised to $H_2As^{(V)}O_4^-$ in the solution, resulting in the precipitation of stable scorodite and/or its precursor. The precipitates covered the surface of the residues even when the pulp density was 0.5 g/L.

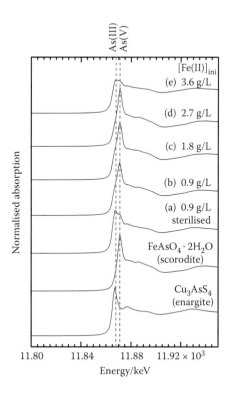

FIGURE 5.10
As K-edge XANES spectra for the solid residues after (a) 76 days, (b) 76 days, (c) 27 days, (d) 27 days and (e) 19 days of bioleaching with different $[Fe^{2+}]_{ini}$. The standards for scorodite and enargite are also included. (Modified from Sasaki, K., Takatsugi, K., Hirajima, T., 2011. *Hydrometal.*, 109, 153–160.)

The pulp density directly influenced the formation of As-bearing secondary minerals because the only source of As in this system was enargite. Figure 5.12a through c shows the changes of Fe and Fe(II) during the bioleaching of enargite with *A. brierleyi* when $[Fe^{2+}]_{ini}$ was 2.7 g/L and the pulp density was 0.5%, 1.0% and 2.0%. While the precipitation rate of ferric compounds was much faster in 1% of pulp density than in 2%, as shown in Figures 5.12b and c, the recovery of Cu was greater in the former than the latter. This suggests that the oxidative dissolution of enargite by Fe^{3+} might be suppressed due to a lack in Fe^{3+} provided by the microbial reaction. However, a lowering of the redox potential by complexation of Fe^{3+} with arsenate, which is a precursor to scorodite, is also possibly responsible for passivation in high pulp density.

Changes in the concentrations of Cu and As over time are shown in Figure 5.12d and e. Cu recoveries of 83% after 40 days, 91% after 27 days and 81% after 40 days were obtained with pulp densities of 0.5%, 1.0% and 2.0%, respectively (Figure 5.12d). The relative recovery of Cu was at a maximum when the pulp density was 1.0%. If $[Fe^{2+}]_{ini}$ is optimised with a pulp density of 0.5%, the

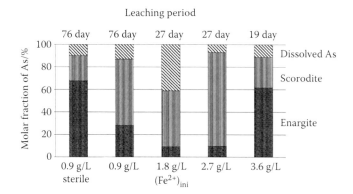

FIGURE 5.11

Molar fractions of As at the final point of leaching based on fitting the results of XANES spectra of the solid residues using the standard spectra for enargite and scorodite. (Modified from Sasaki, K., Takatsugi, K., Hirajima, T., 2011. *Hydrometal.*, 109, 153–160.)

optimal $[Fe^{2+}]_{ini}$ is predicted to be <2.7 g/L to minimise Fe(III) precipitates. For 2% pulp density, dissolved Fe^{3+} both oxidises enargite and forms Fe(III) precipitates on the surface of the enargite, and the corresponding optimal $[Fe^{2+}]_{ini}$ is difficult to quantify without additional tests. At 25 days, the release of As was slightly higher with 1.0% pulp density as compared to 0.5% (Figure 5.12e). Thus, a 0.5% pulp density was too low to form As-bearing secondary minerals. When the pulp density was 1.0% or 2.0%, a relatively rapid decrease in As concentration was observed after 22 days, independent of the Fe concentration. This was attributed to the formation of cupric arsenate.

The XRD patterns for the solid residues are shown in Figure 5.13. Peaks assigned to K-jarosite were observed in the solid residue after 40 days of leaching at 0.5% pulp density (Figure 5.13a). However, only scorodite was identified as a secondary mineral in the residue after 40 days of leaching at 2.0% pulp density (Figure 5.13c). Thus, the initial amount of enargite affected the secondary mineral phases in the solid residues. At 0.5% pulp density, dissolved Fe^{3+} was consumed to form K-jarosite. At 2% pulp density, scorodite rather than jarosite was preferentially formed because the solubility product (K_{sp}) of scorodite ($10^{-25.86} \sim 10^{-21.7}$ at 22°C~25°C, Drahota and Fillippi, 2009) is many orders of magnitude smaller than that of jarosite (10^{-11} at 25°C, Baron and Palmer, 1996). As shown in Figure 5.12d, Cu recovery was 81% at 2.0% pulp density. To determine the most appropriate combination of $[Fe^{2+}]_{ini}$ and pulp density for the most effective bioleaching of Cu and immobilisation of As within the confines of this fundamental study, further experiments would be necessary for the fine tuning of 1% pulp density and $[Fe^{2+}]_{ini}$ of 2.7 g/L. However, this approach must be examined at higher pulp densities and column-leaching studies, which are realistically closer to pilot- and demonstration-scale test designs.

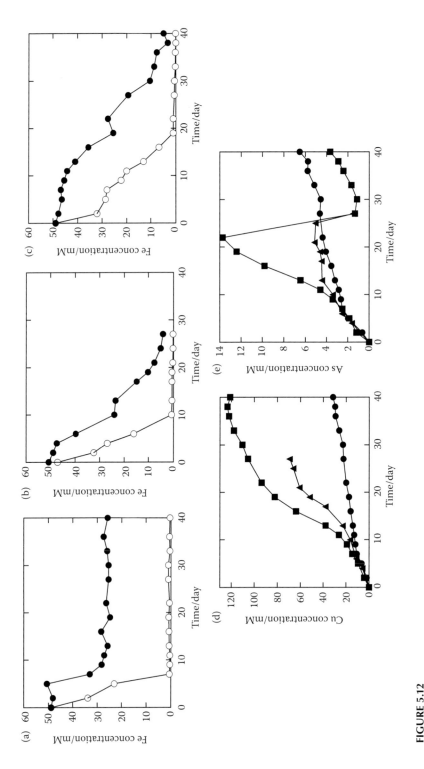

FIGURE 5.12
Effect of pulp density on the dissolved concentrations of (a) through (c) Fe, (d) Cu and (e) As. Pulp density: (a) 0.5%, (b) 1.0% and (c) 2.0%. Open and sold symbols are total Fe and Fe(II) concentrations in (a) through (c). Symbols in (d) and (e): ●, 0.5%; ▲, 1.0% and ■, 2.0%.

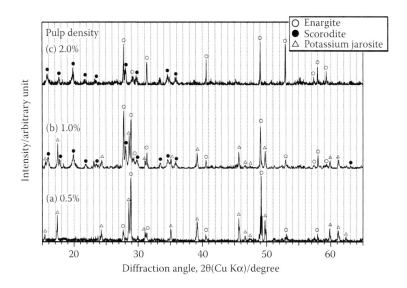

FIGURE 5.13
Effect of pulp density on XRD patterns for solid residues after bioleaching of enargite. (a) 0.5%;
(b) 1.0% and (c) 2.0%.

5.5 Conclusions

Spectroscopic investigation was introduced for bioleaching of enargite by the thermophilic iron-oxidising archaea, *A. brierleyi*. The preferential dissolution of Cu compared to As was mainly achieved by the formation of crystalline scorodite ($FeAsO_4 \cdot 2H_2O$). Although the chemical valence of As in enargite was confirmed to be As(III) by x-ray absorption fine structure (XAFS) and XPS, most of the dissolved As was in the compound $H_3As^VO_4$. During the bioleaching process, dissolved Fe^{3+} was consumed through the formation of precipitates, such as scorodite and jarosite depending on Fe^{2+} concentrations, as well as the oxidation of enargite. Even after the formation of scorodite had stopped due to the consumption of Fe^{3+}, As was further passivated by co-sorption of Cu^{2+} and $H_2As^VO_4$ on jarosite leading to the formation of copper arsenate. SEM-EDX and TEM investigation revealed that subsequent secondary mineral precipitation resulted in the formation of multi-precipitation layers consisting of scorodite, potassium jarosite, elemental sulphur and copper arsenate, after most of the Cu had been recovered from the enargite by microbially induced leaching. From the point of view of chemical engineering, the initial Fe^{2+} concentration and pulp density of enargite are the governing factors for Cu recovery in the bioleaching of enargite. When the initial Fe^{2+} concentration was increased, jarosite precipitated, leading to passivation of the enargite surface. When the initial Fe^{2+} concentration was lowered, the

Cu recovery was incomplete due to insufficient oxidation. Pulp density also clearly affected the makeup of secondary minerals in solid residues after the bioleaching of enargite. There is a tendency for dissolved Fe^{3+} to readily precipitate as potassium jarosite at relatively smaller pulp densities and as scorodite at relatively larger pulp densities. Under optimised conditions, the maximum Cu recovery of 90.9% and maximum As immobilisation of 94.1% were obtained.

References

Acevedo, F., Gentina, J. C., Garcia, N., 1998. CO_2 supply in the biooxidation of an enargite–pyrite gold concentrate. *Biotechnol. Lett.*, 20(3), 257–259.

Arribas Jr., A., 1995. Characteristics of high sulfidation epithermal deposits and their relation to magmatic fluid. *Mineral Assoc. Can. Short Course*, 23, 419–454.

Baron, D., Palmer, C. D., 1996. Solubility of jarosite at 4–35°C. *Geochim. Cosmochim. Acta*, 60, 185–195.

Corkhill, C. L., Wincott, P. L., Lloyd, J. R., Vaughan, D. J., 2008. The oxidative dissolution of arsenopyrite (FeAsS) and enargite (Cu_3AsS_4) by *Leptospirillum ferrooxidans*. *Geochim. Cosmochim. Acta*, 72, 5616–5633.

Drahota, P., Fillippi, M., 2009. Secondary arsenic minerals in the environment: A review. *Environ. Int.*, 35, 1243–1255.

Dutré, V., Vandecasteele, C., 1995. Solidification/stabilization of hazardous arsenic containing waste from a copper refining process. *J. Hazard. Mater.*, 40, 55–68.

Escobar, B., Huenupi, E., Godoy, I., Wiertz, J. V., 2000. Arsenic precipitation in the bioleaching enargite by *Sulfolobus* BC at 70°C. *Biotechnol. Lett.*, 22, 205–209.

Escobar, B., Huenupi, E., Wiertz, J. V., 1997. Chemical and biological leaching of enargite. *Biotechnol. Lett.*, 19(8), 719–722.

Flemming, R. L., Salzsauler, K. A., Sidenko, N. V., Sherriff, B. L., Sidenko, N. V., 2005. Identification of scorodite in fine-grained, high-sulfide, arsenopyrite mine-waste using micro x-ray diffraction (μ XRD). *Can. Mineral.*, 43(4), 1243–1254.

Fujita, T., Taguchi, R., Abumiya, M., Matsumoto, M., Shibata, E., Nakamura, T., 2008. Novel atmospheric scorodite synthesis by oxidation of ferrous sulfate solution. Part I. *Hydrometal.*, 90, 92–102.

Gräfe, M., Beattie, D. A., Smith, E., Skinner, W. M., Singh, B., 2008. Copper and arsenate co-sorption at the mineral–water interfaces of goethite and jarosite. *J. Colloid Interface Sci.*, 322, 399–413.

Gonzalez-Contreras, P., Weijma, J., Weijden, R., Van Der Buisman, C. J. N., 2010. Biogenic scorodite crystallization by *Acidianus sulfidivorans* for arsenic removal. *Environ. Sci. Technol.*, 44(2), 675–680.

Konishi, Y., Nishimura, H., Asai, S., 1998. Bioleaching of sphalerite by acidophilic thermophile *Acidianus brierleyi*. *Hydrometal.*, 47, 339–352.

Lattanzi, P., Pelo, S. D., Musu, E., Atzei, D., Elsener, B., Rossi, A., 2008. Enargite oxidation: A review. *Earth Sci. Rev.*, 86, 62–88.

Muñoz, J. A., Blázquez, M. L., González, F., Ballester, A., Acevedo, F., Gentina, J. C., González, P., 2006. Electrochemical study of enargite bioleaching by mesophilic and thermophilic microorganisms. *Hydrometal.*, 84, 175–186.

Okibe, N., Koga, M., Morishita, S., Tanaka, M., Heguri, S., Asano, S., Sasaki, K., Hirajima, T., 2014. Microbial formation of crystalline scorodite for treatment of As(III)-bearing copper refinery process solution using *Acidianus brierleyi*. *Hydrometal.*, 143, 34–41.

Sasaki, K., Tsunekawa, M., Ohtsuka, T., Konno, H., 1995. Confirmation of a previous sulfur-rich layer on pyrite after oxidative dissolution by Fe(III) ions around pH 2. *Geochim. Cosmochim. Acta*, 59, 3155–3158.

Sasaki, K., Takatsugi, T., Ishikura, K., Hirajima, T., 2010a. Spectroscopic study on oxidative dissolution of chalcopyrite, enargite and tennantite at different pH values. *Hydrometal.*, 100, 144–151.

Sasaki, K., Takatsugi, K., Kaneko, K., Kozai, N., Ohonuki, T., Tuovinen, O. H., Hirajima, T., 2010b. Characterization of secondary arsenic-bearing precipitates formed in the bioleaching of enargite by *Acidithiobacillus ferrooxidans*. *Hydrometal.*, 104, 424–431.

Sasaki, K., Takatsugi, K., Hirajima, T., 2011. Effects of initial Fe^{2+} concentration and pulp density on the bioleaching of Cu from enargite by *Acidianus brierleyi*. *Hydrometal.*, 109, 153–160.

Smedley, P. L., Kinniburgh, D. G., 2002. A review of the source, behaviour and distribution of arsenic in natural waters. *Appl. Geochem.*, 17, 517–568.

Takatsugi, K., Sasaki, K., Hirajima, T., 2011. Mechanism of the bioleaching of copper from enargite by thermophile iron-oxidizing archaea with the concomitant precipitation of arsenic. *Hydrometal.*, 109, 90–96.

Watling, H. R., 2006. The bioleaching of sulphide minerals with emphasis on copper sulphides—A review. *Hydrometal.*, 84, 81–108.

Wiertz, J. V., Mateo, M., Escobar, B., 2006. Mechanism of pyrite catalysis of As(III) oxidation in bioleaching solutions at 30°C and 70°C. *Hydromatal.*, 83, 35–39.

6

Microbial Biodesulphurisation of Coal

Anirban Ghosh, Sujata, Abhilash and B. D. Pandey

CONTENTS

6.1 Introduction

Coal has the distinction of being one of the most abundant and cheapest fossil fuel energy sources in the world and is therefore, increasingly used in industries, and for other commercial and domestic purposes (Barooah and Baruah, 1996). Because of these factors coal has been the key energy source for the industrial revolution and is expected to remain so for at least the next two to three decades (Aseefa et al., 2013). Since the time of the industrial revolution there has been an ever-increasing demand of coal (Acharya et al., 2005). This is because the requirement of energy in terms of heat and electricity generation for the industries has primarily depended on the process of combustion/burning of the coal. Thus, one of the components of enhanced industrial production is attributed to the increased consumption of energy and, for its generation, coal has undoubtedly been a major contributor. Coal-fired power plants, steel industries, refineries, smelters, paper and pulp mills and food processing plants are always in need of coal.

The stacks of the coal burning power plants are the major emitters of greenhouse gases such as carbon dioxide and sulphur dioxide. The sulphur dioxide, in particular, released into the atmosphere from the thermal power plants remains one of the most prominent and intractable issues of environmental concern since SO_2 is considered the major component of acid rain (Calkins 1994; Wang et al., 1996; Kadioglu et al., 1998). Therefore, while using coal, the industries need to keep a check on the SO_X emission due to coal combustion because industrial development at the cost of damage to the environment cannot be accepted by a civilised society for sustainable growth. To run industrial progress in parallel with environmental harmony, many governments have put into effect stringent air pollution control laws for the industries to follow. To abide by these laws, there has been an increasing demand for low ash and low sulphur coal throughout the world.

The presence of sulphur in the coal, as mentioned above, gives a negative impact on the environment causing air pollution when burnt (Calkins 1994). It is also known that all forms of sulphur present in coal contribute to SO_2 generation during the process of combustion. The sulphur in the coal gets oxidised into various oxides of sulphur (Bailey et al., 2002), which in gaseous forms are released primarily as SO_2 and in a much smaller quantity as SO_3, including gaseous sulphates. The sulphur oxides react with water molecules to form acid rain, which deteriorates the environment and infrastructure (Ryan and Ledda, 1998; Fecko et al., 2006). Acid rain is known to have an adverse effect on buildings and destroys forests and contaminates water bodies, such as ponds and lakes (Lakshmanraj et al., 2012). Moreover, SO_2 emission is a precursor of sulphated aerosols, which are considered as one of the main agents to form solid particulates that affect human health (Ryan and Ledda, 1998). Thus, in spite of being in abundance, full/efficient utilisation of coal is hindered due to the presence of sulphur in it (Kayser et al., 1993). Since 1971, oil and coal remain the most important primary energy sources, with coal increasing its share significantly since 2000 (Zhongxuan et al., 2002).

After the reduction of sulphur in the coal, the processed/clean coal could be an eco-friendly energy resource. Therefore, many coal-cleaning technologies are being adapted in big industries to reduce the sulphur in coal in order to reduce SO_2 emission. Nowadays, more emphasis is being given on eco-friendly coal-cleaning technologies, out of which bio-desulphurisation is a much sought after technique being investigated the world over.

6.2 Coal Reserves and Production

This section highlights the global coal reserves and production scenarios including the reserves and production of coal in India. It also describes the composition of coal and the relative distribution of various coal types. Location of the Indian coal reserves and its composition are also mentioned in some detail.

6.2.1 Global Coal Scenario

Globally, coal resources have been estimated at over 861 billion tons. While India accounts for 286 billion tons of coal resources (as of March 2011), other countries with a major chunk of resources are the United States, China, Australia, Indonesia, South Africa and Mozambique. Coal meets around 30.3% of the global primary energy needs and generates 42% of the world's electricity. In 2011, coal was one of the fastest growing forms of energy after renewable sources and its share in the global primary energy consumption increased to 30.3%—the highest since 1969. Since 2000, the global consumption of coal has grown faster than any other fuel. Currently, the five largest coal users are China, the United States, India, Russia and Germany, they account for 77% of the total global use. Anthracite and bituminous coals are the highest ranking coal and their abundance is estimated at ~404,762 million tons across the world. The quality of sub-bituminous and lignite coal is lower than that of anthracite and bituminous coal, but their known reserves are higher. Sub-bituminous and lignite coal globally account for ~456,176 million tons of the coal resources. Tables 6.1 and 6.2 show the proven coal reserves and the world coal production, respectively (BP Statistical Review of World Energy, 2013).

6.2.1.1 Coal Production

As per the demand in different sectors, coal production is increasing every year. Worldwide coal production in 2012 increased by 2% over 2011. In 2011, the coal production was recorded at 3759.1 million tons and in 2012 it increased to 3845.3 million tons (BP Statistical Review of World Energy, 2013). The Asia-Pacific region is the world's major coal producer.

6.2.2 Coal Reserves in India

India has the fifth largest coal reserve in the world. Of the total reserves, 88% are non-coking, 0.5% are tertiary and the rest is coking coal. Indian coals are designated as high ash and low sulphur coal. It contains ~45% ash which reduces the coal heat value (Www.pwc.com/india). According to the Geological Survey of India (GSI) and others [Central Mine Planning & Design Institute (CMPDI), Singareni Collieries Company Limited (SCCL) and Mineral Exploration Corporation Limited (MECL)] as of April 2012, 293.497 billion

TABLE 6.1

Proved Coal Reserves at the End of 2012

Country	Anthracite and Bituminous (million tons)	Sub-Bituminous and Lignite (million tons)	Total (million tons)
US	108,501	128,794	237,295
Canada	3474	3108	6582
Mexico	860	351	1211
Total North America	**112,835**	**132,253**	**245,088**
Brazil	–	4559	4559
Colombia	6366	680	6746
Venezuela	479	–	479
Others South and Central America	45	679	724
Total South and Central America	**6890**	**5618**	**12,508**
Bulgaria	2	2364	2366
Czech Republic	192	908	1100
Germany	99	40,600	40,699
Greece	–	3020	3020
Hungary	13	1647	1660
Kazakhstan	21,500	12,100	33,600
Poland	4338	1371	5709
Romania	10	281	291
Russian Federation	49,088	107,922	157,010
Spain	200	330	530
Turkey	529	1814	2343
Ukraine	15,351	18,522	33,873
UK	228	–	228
Other Europe and Eurasia	1440	20,735	22,175
Total Europe and Eurasia	**92,990**	**211,614**	**304,604**
South Africa	30,156	–	30,156
Zimbabwe	502	–	502
Other Africa	860	174	1034
Middle East	1203	–	1203
Total Middle East and Africa	**32,721**	**174**	**32,895**
Australia	37,100	39,300	76,400
China	62,200	52,300	114,500
India	56,100	4500	60,600
Indonesia	1520	4009	5529
Japan	340	10	350
New Zealand	33	538	571
North Korea	300	300	600
Pakistan	–	2070	2070
South Korea	–	126	126
Thailand	–	1239	1239

Continued

TABLE 6.1   (Continued)

Proved Coal Reserves at the End of 2012

Country	Anthracite and Bituminous (million tons)	Sub-Bituminous and Lignite (million tons)	Total (million tons)
Vietnam	150	–	150
Other Asia Pacific	1583	2125	3708
Total Asia Pacific	159,326	106,517	265,843
Total World	**404,762**	**456,176**	**860,938**

tons of coal up to a maximum depth of 1200 m has been estimated. Lignite reserves in India are currently estimated at around 41,962.79 million tons. Table 6.3 shows the state-wise geological resources of coal in India. Generally, coal is predominantly located in Jharkhand (80,356.20 MT), Chhattisgarh (50,896.15 MT), Odisha (71,447.41 MT), West Bengal (30,615.72 MT), Madhya Pradesh (24,376.26 MT), Andhra Pradesh (22,154.86 MT) and Maharashtra (10,882.09 MT) (Annual Report, Ministry of Coal, 2012–2013). Beside these, coal reserves are also located in other states in India such as Assam, Meghalaya and Arunachal Pradesh, though in smaller amounts.

6.2.2.1 Coal Production in India

Overall, 557.66 million tons of coal was produced in the year 2012–2013, out of which Coal India Limited and its subsidiaries produced 452.19 million tons during the period as against 435.84 million tons in the year 2011–2012 showing a growth of 3.7%. Singareni Collieries Company Limited supplies most of the coal to the southern part of India. Small quantities of coal are also produced by TISCO, IISCO, DVC and other companies in India (www.coal.nic.in).

6.2.2.2 Categorisation of Coal in India

The coal resources of India are available in the older Gondowana formation of peninsular India and the younger Tertiary formation (Table 6.4) of the northeastern region (Annual Report, Ministry of Coal, 2012–2013). The major amount of coal is available in Gondowana formation (292,004 MT) and the rest (1493 MT) is available in tertiary formation.

6.2.2.3 Grades of Coal in India

The grading of the coal depends upon the generation of useful heat value and the ash content present. Generally, coals are dividing into either a coking or non-coking category (Tables 6.5 and 6.6). Non-coking coal generates low useful heat value and the coking coals often have low ash contents (www. coal.nic.in). Depending upon the ash content present, coking coal is categorised as steel grade and washery grade. The steel grade of coal contains

TABLE 6.2

Coal Production Worldwide (million tons)

Country	2002	2003	2004	2005	2006	2007	2008	2009	2010	2011	2012
US	570.1	553.6	572.4	580.2	595.1	587.7	596.7	540.8	551.2	556.1	515.9
Canada	34.3	31.8	33.9	35.3	34.8	35.7	35.6	33.1	35.4	35.5	35.2
Mexico	5.3	4.6	4.7	5.2	5.5	6.0	5.5	5.0	5.8	7.0	6.6
Total North America	609.6	590.1	611.0	620.7	635.4	629.4	637.8	578.9	592.4	598.5	557.7
Brazil	1.9	1.8	2.0	2.4	2.2	2.3	2.5	1.9	2.0	2.1	2.2
Columbia	25.7	32.5	34.9	38.4	42.6	45.4	47.8	47.3	48.3	55.8	58.0
Venezuela	5.9	5.1	5.9	5.3	5.7	4.5	3.6	2.4	2.0	1.7	1.2
Other South and Central America	33.9	39.9	43.0	46.3	51.2	52.5	54.3	52.2	52.7	59.9	61.8
Bulgaria	4.3	4.5	4.4	4.1	4.2	4.7	4.8	4.5	4.9	6.1	5.4
Czech Republic	24.3	24.2	23.5	23.5	23.8	23.6	22.8	21.0	20.8	21.6	20.7
France	1.1	1.3	0.4	0.2	0.2	0.2	0.1	–	0.1	0.1	0.1
Germany	55.0	54.1	54.7	53.2	50.3	51.5	47.7	44.4	43.7	44.6	45.7
Greece	9.1	9.0	9.6	9.4	8.6	9.0	8.8	8.6	7.8	7.9	7.9
Hungary	2.7	2.8	2.4	2.0	2.1	2.0	1.9	1.9	1.9	2.0	1.9
Kazakhstan	37.8	43.3	44.4	44.2	49.1	50.0	56.8	51.5	54.0	56.2	58.8
Poland	71.3	71.4	70.5	68.7	67.0	62.3	60.5	56.4	55.5	56.6	58.8
Romania	6.6	7.0	6.7	6.6	6.5	6.7	6.7	6.4	5.8	6.7	6.4
Russian Federation	117.3	127.1	131.7	139.2	145.1	148.0	153.4	142.1	151.1	158.0	168.1
Spain	7.2	6.8	6.7	6.4	6.1	5.7	4.1	3.8	3.4	2.5	2.4
Turkey	11.5	10.4	10.1	12.6	13.7	16.0	16.8	17.1	15.8	16.3	15.4
Ukraine	42.8	41.6	42.2	41.0	41.7	39.9	41.3	38.4	39.9	44.0	45.9
UK	18.2	17.2	15.3	12.5	11.3	10.3	11.0	10.9	11.2	11.3	10.2
Other Europe and Eurasia	17.9	19.0	18.5	17.7	18.5	20.6	21.0	20.3	20.1	21.5	21.2

Total Europe and Eurasia	427.2	439.8	441.0	441.2	448.2	450.7	457.8	427.2	436.0	455.5	469.0
Total Middle East	0.6	0.7	0.8	1.0	1.0	1.0	1.0	0.7	0.6	0.7	0.7
South Africa	124.1	134.1	137.2	137.7	138.0	139.6	142.4	141.2	145.0	141.8	146.6
Zimbabwe	2.5	1.8	2.4	2.2	1.4	1.3	1.0	1.1	1.7	1.7	1.7
Other Africa	1.3	1.6	1.3	1.2	1.3	1.0	1.0	0.9	1.1	1.1	1.1
Total Africa	128.0	137.5	140.9	141.1	140.6	141.9	144.4	143.2	147.8	144.5	149.3
Australia	184.3	189.4	196.8	205.7	210.8	217.1	224.1	232.1	236.0	230.8	241.1
China	775.2	917.4	1061.3	1174.8	1264.3	1345.8	1401.0	1486.5	1617.5	1758.0	1825.0
India	138.5	144.4	155.7	162.1	170.2	181.0	195.6	210.8	217.5	215.7	228.8
Indonesia	63.5	70.3	81.4	93.9	119.2	133.4	147.8	157.6	169.2	217.3	237.4
Japan	0.8	0.7	0.7	0.6	0.7	0.8	0.7	0.7	0.5	0.7	0.7
New Zealand	2.8	3.2	3.3	3.3	3.6	3.0	3.0	2.8	3.3	3.1	3.1
Pakistan	1.6	1.5	1.5	1.6	1.7	1.6	1.8	1.6	1.5	1.4	1.2
South Korea	1.5	1.5	1.4	1.3	1.3	1.3	1.2	1.1	0.9	0.9	0.9
Thailand	5.7	5.3	5.6	5.8	5.3	5.1	5.0	5.0	5.1	6.0	5.1
Vietnam	9.2	10.8	14.7	18.3	21.8	22.4	23.0	25.2	24.6	24.9	23.5
Other Asia Pacific	19.6	20.3	22.1	24.9	25.3	24.0	25.8	28.8	36.9	41.1	40.0
Total Asia Pacific	1202.7	1364.9	1544.5	1692.2	1824.2	1935.6	2028.9	2152.1	2313.2	2499.9	2606.8
Total world	2401.9	2572.7	2781.3	2942.4	3100.7	3211.1	3324.2	3354.3	3542.7	3759.1	3845.3

TABLE 6.3

State-Wise Geological Resource of Coal in India

State	Geological Resources of Coal (in million tons)			
	Proved	Indicated	Inferred	Total
Andhra Pradesh	9566.61	9553.91	3034.34	22,154.86
Arunachal Pradesh	31.23	40.11	18.89	90.23
Assam	464.78	45.51	3.02	513.51
Bihar	0.00	0.00	160.00	160.0
Chhattisgarh	13,987.85	33,448.25	3410.05	50,896.15
Jharkhand	40,163.22	33,609.29	6583.69	80,356.20
Madhya Pradesh	9308.70	12,290.65	2776.91	24,376.26
Maharashtra	5667.48	3104.40	2110.21	10,882.09
Meghalaya	89.04	16.51	470.93	576.48
Nagaland	8.76	0.00	306.65	315.41
Odisha	25,547.66	36,465.97	9433.78	71,447.41
Sikkim	0.00	58.25	42.98	101.23
Uttar Pradesh	884.04	177.76	0.00	1061.80
West Bengal	12,425.44	13,358.24	4832.04	30,615.72
Total	118,144.82	142,168.85	33,183.48	293,497.15

TABLE 6.4

Category of Coal in India

Formation	Proved (million tons)	Indicated (million tons)	Inferred (million tons)	Total (million tons)
Gondowana	117,551	142,070	32,384	292,004
Tertiary	594	99	799	1493
Total	118,145	142,169	33,183	293,497

TABLE 6.5

Grades of Coking Coal

Grades	Ash Content
Steel grade- 1	Not exceeding 15%
Steel grade-2	Exceeding 15% but not exceeding 18%
Washery grade- I	Exceeding 18% but not exceeding 21%
Washery grade-II	Exceeding 21% but not exceeding 24%
Washery grade-III	Exceeding 24% but not exceeding 28%
Washery grade-IV	Exceeding 28% but not exceeding 35%

TABLE 6.6

Grades of Non-Coking Coal

Grade	Useful Heat Value-UHV (kcal/kg) UHV = 8900 – 138(A + M)	Corresponding Ash% + Moisture% (At 60% RH and 40°C)	Gross Calorific Value (kcal/kg) at 5% Moisture Value
A	Exceeding 6200	Not exceeding 19.5	Exceeding 6454
B	Exceeding 5600 but not 6200	19.6 to 23.8	Exceeding 6049 but not 6454
C	Exceeding 4940 but not 5600	23.9 to 28.6	Exceeding 5597 but not 6049
D	Exceeding 4200 but not 5600	28.7 to 34.0	Exceeding 5089 but not 5597
E	Exceeding 3360 but not 5600	34.1 to 40.0	Exceeding 4324 but not 5089
F	Exceeding 2400 but not 5600	40.1 to 47.0	Exceeding 3865 but not 4324
G	Exceeding 1300 but not 5600	47.1 to 55	Exceeding 3113 but not 3865

lower amounts of ash (~18%) in comparison to the washery grade coal (>18% but <35%). Based on the generation of useful heat value of non-coking coal, it is sub-divided into different grades (A–F). As such lower generation of useful heat value would not be acceptable for industrial purposes.

6.3 High Sulphur Coal

The quality of coal greatly depends upon the presence of the sulphur content, which has a bearing on its utilisation in industry and elsewhere. As mentioned above, the high sulphur coal is not suitable for most industrial purposes because of environmental concerns associated with the emission of sulphur dioxide. The Energy Information Administration (EIA) has prescribed a criteria based on the amounts of sulphur per million Btu of the coal to categorise the high sulphur coal. Based on the criteria (coal with 1.68 pounds of sulphur or more per million Btu) a large tonnage (167,269 million tons) is reported in different regions of the United States as high sulphur coal (US Coal reserves: EIA, 1993). Beside, Telkwa deposits in British Columbia, Tabas coal in Iran, Spanish coals and that of many other countries have deposits of high sulphur coal. Indian coals generally possess low sulphur content except the northeast region, basically the coals from Assam and Meghalaya, which contain high amounts of sulphur. Australian coals are known to contain exceptionally low amounts of sulphur (Olson and Brinckman, 1986). More details on the sulphur content of coal from different location across the globe will be discussed later in this chapter.

6.3.1 Chemistry of Sulphur in Coal

Samples of coal vary considerably in their composition and characteristics. Carbon content may range from 60% to 95%, with variable amounts of hydrogen, oxygen, sulphur and nitrogen in the coal in various forms. Sulphur is a common constituent in all the coals (Olson and Brinckman; 1986). Sulphur occurs in coal as pyrite (FeS_2), elemental sulphur, inorganic sulphates, minor amounts of inorganic sulphides other than pyrite, and organic sulphur. Organic sulphur is a general classification that includes all forms of sulphur organically bound to the actual coal matrix (Kadioglu et al., 1998, Ryan and Ledda, 1998).

The most important organic sulphur groups in coals are the following:

1. Different heteroaromatic compounds, such as thiophene and diben-zothiophene are generally present in different coals. These compounds are very stable and do not decompose easily by thermal treatment even up to very high temperatures.

2. Aryl sulphides are usually very stable because the sulphur is adjacent to an aromatic ring and takes part in the ring resonance.

3. Aliphatic sulphides are relatively unstable and tend to decompose to unsaturated compounds and H_2S.

4. Cyclic sulphides in which the sulphur is a part of non-aromatic ring are more stable than aliphatic, but are less stable than the aromatic sulphides.

5. Thiols (aryl and aliphatic thiols) and disulphides are unstable and tend to decompose easily to H_2S and to unsaturated compounds (Kadioglu et al., 1998).

6.3.1.1 Sulphur in Coal: A Worldwide Problem

Coal from different mines contains variable amounts of sulphur depending upon the geological location. Coal bearing low sulphur has high industrial prospects, while the presence of high sulphur restricts its utilisation and is often regulated for use. EIA surveyed across the United States and produced some vital information regarding the sulphur content in coals of different regions within the country. The EIA also divided the sulphur content in coal into three qualitative ratings—low sulphur (≤0.60 pounds of sulphur per million Btu), medium sulphur (0.61–1.67 pounds/million Btu) and high sulphur coal (≥1.68 pounds/million Btu). Three regions have clearly been identified with the high sulphur coal by the EIA in the United States out of the total (proven and prospective) reserves of 475,597.7 million short tons. These are (i) Appalachia regions such as Alabama, Georgia, Eastern Kentucky, Maryland, North Carolina, Ohio, Pennsylvania, Tennessee, Virginia and West Virginia, (ii) Interior regions such as Arkansas, Illinois, Indiana, Iowa, Kansas, Western Kentucky, Louisiana, Missouri, Oklahoma and Texas, and (iii) West regions such as Alaska, Arizona, Colorado, Idaho, Montana, New Mexico, North Dakota, Oregon, South Dakota, Utah, Washington and Wyoming (US Coal reserves: EIA; 1993). Telkwa deposits in British Columbia contain 1%–7% of sulphur in different coal seams (Ryan and Ledda; 1998). The Spanish coals from two mines, El Bierzo and Teruel have about 1%–3% of sulphur (Gomez et al., 1999). Tabas coal from Iran contains about 3% total sulphur, most of which is inorganic sulphur (Ehsani, 2006).

6.3.1.2 Sulphur Content in Indian Coal

There are major deposits of coal in India that are low in ash, but have high sulphur content. It is also found that 71% of Indian coals have high amounts of ash with varying amounts of sulphur (Barooah and Baruah, 1996). As mentioned above, the Indian coals from North east (NE) region possess low ash but high sulphur content varying from 3% to 8% in general with 75%–90% of it in the form of organic sulphur (Barooah and Baruah, 1996; Ali et al., 2009). Its potential for industrial utilisation is therefore, being hindered because of its high sulphur content. Northeast (NE) coal has the capacity to satisfy the need of low ash and low sulphur coals only after its high sulphur concentration is lowered.

Oxygen and sulphur on dry mineral matter basis as found in NE coals almost equals the oxygen content of the same rank of coals from other regions; that is, part of the oxygen is substituted by sulphur and so forth (Ali et al., 2009). The pyritic sulphur present in NE coals is highly disseminated in the organic matrix of the coal (Ali et al., 2009). Sulphur in these coals generally occurs in the range 2.7%–7.8%. Assam coal is known to have all three forms of sulphur—sulphate, pyritic and organic sulphur (Barooah and Baruah, 1996). Various types of organic sulphur functional groups, namely, thiol, disulphide, thioether, thioketone and thiophene are found in the high sulphur NE coal depending on the environmental conditions of coal formation (Barooah and Baruah, 1996; Ali et al., 2009). For example, the coal of Boragolai colliery from Makum coalfield contains 4.95% of sulphur. Coal mines in Meghalaya like Bapung and Khliehriat contains high amounts of sulphur from the desired level (Nayak, 2013). There are many other coalfields in Assam that contain high amount of sulphur (Barooah and Baruah, 1996). Tables 6.7 and 6.8 summarise the sulphur contents of Assam and Meghalaya coal.

TABLE 6.7

Sulphur Content in Coal of Assam (North-Eastern Region), India

Source	Sulphur%	Source	Sulphur%
Makum Coal		**Dilli-Jeypore Coal**	
Boragolai coal	4.95%	Dilli coal	1.18%–2.69%
Tipong coal	1.13%–1.94%	Jeypore coal	1.83%–3.82%

TABLE 6.8

Sulphur Content in Meghalaya Coal (Jaintia Hills), India

Source	Total Sulphur	Pyritic Sulphur	Sulphate Sulphur	Organic Sulphur
Bapung	7.31	0.97	0.80	5.54
Khliehriat	6.48	0.86	0.61	5.01
Sutnga	5.92	0.69	0.53	4.70
Mussiang Lamare	6.21	0.60	0.66	4.95

6.3.1.3 Need for Desulphurisation

In spite of coal being the main input material for the energy generation worldwide, it remains un-utilised for energy generation because of the presence of high amounts of sulphur. Different desulphurisation processes are being developed for the cleaning of coal and details are presented and discussed below.

6.4 Desulphurisation Processes

A clean coal technology is primarily developed by using various techniques to reduce the adverse environmental effects caused in the process of coal combustion. The desulphurisation methods are classified into three broad groups depending upon the type of application (Blazquez et al., 1993).

1. Physical desulphurisation
2. Chemical desulphurisation
3. Biological desulphurisation

An introduction to these methods in brief is given along with their merits and demerits. Bio-desulphurisation processes are the main focus and presented in detail.

6.4.1 Physical Methods of Desulphurisation

The physical method of sulphur removal from coal is based on its physical characteristics such as specific gravity, magnetic properties, surface and charge characteristics and size of the particles. Physical desulphurisation processes are the ones that remove pyritic sulphur from coals without chemical modification or destruction of the coal or impurities. There are many physical processes available to desulphurise coal (Leonard et al., 1981; Lewowski, 1993; Saydut et al., 2007; Lovas et al., 2011; Koca et al., 2012).

Flotation, gravimetric, magnetic and power ultrasound processes are the most common physical methods used. In the United States only 14 million tons of clean coal was produced in 1973 by the flotation process. Some demineralisation can also be achieved by simple physical processes based on the differences in the physical properties of the minerals and carbonaceous parts of the coal. Physical processes are cost-effective methods, but are less effective in the separation of finely dispersed pyrite and other minerals bound to the coal structure. Different types of physical processes followed for coal cleaning are described in Table 6.9.

6.4.2 Chemical Methods of Desulphurisation

Chemical coal desulphurisation methods utilise some chemical processes to separate coal from its impurities. A variety of chemical coal beneficiation processes are under development and some of these processes are also capable of removing organic sulphur from coal, which is beyond the removal ability of any physical desulphurisation process. Leaching is one of the most common methods employed for coal desulphurisation. Coal is mixed with acid or alkali and its sulphur is extracted while being heated or stirred. The type of reagent depends upon the type of sulphur in coal. Various chemical processes are presently being developed employing agents such as hydrochloric acid, nitric acid, ferric chloride, potassium hydroxide, methanol, tetralin, sodium hypochlorite, hydrofluoric acid, and sodium hydroxide for the extraction of sulphur from coal (Baruah et al., 1986a,b; Lolja, 1999; Ratnakandilok et al., 2001; Mukherjee and Borthakur, 2003; Li and Cho, 2005; Ehsani, 2006; Alam et al., 2009). Some chemical methods are illustrated in Table 6.10. It is found that the leaching behaviour of the mineral matter in coal towards aqueous HCl and HF solutions vary markedly. For instance, HCl can dissolve simple compounds such as phosphates and carbonates, whereas it cannot completely dissolve the clays in the coal. HF reacts with almost every mineral in the mineral matter except pyrite and most of the reaction products are water soluble.

6.4.3 Bio-Desulphurisation and Microorganisms

Desulphurisation of coal by physical methods have disadvantages since the removal of pyritic sulphur is possible at low level and elimination of organic sulphur is not possible. Employing the chemical methods for desulphurisation produces better results over the physical methods. But the chemical methods generally require high energy and chemical consumption. Furthermore, the development of technology for desulphurisation of coal by chemical means is very complex and is often not viable economically. Because of the disadvantages of physical and chemical methods, an increased thrust is currently being given on the bioprocessing. Several routes of microbial beneficiation of coal are possible.

TABLE 6.9

Types of Physical Processes for Coal De-Ashing/De-Sulphurisation

Multi gravity separation	Dense medium process is most prevalent out of the gravity separation techniques. It is efficient in ash removal for relatively coarse coal, but is not so useful in the fine coal cleaning and coal desulphurisation. The reasons are the requirement of feed size >0.5 mm and finer dissemination of sulphur minerals in coal matrix which can be liberated only by grinding to a finer size. The specific gravity of the pyrite and coal macerals are ~4.9 and 1.7–1.3, respectively. In current commercial practice, pyrites and other minerals are separated from the coal macerals on the basis of specific gravity difference in cyclones (Koca et al., 2012). *Demerits:* • Fine grinding of coal to liberate pyrite crystals. • Need of continuous supply of power.
Electrostatic separation	Electrostatic separation utilises the difference in conductivity or dielectric properties of coal and minerals to maintain or dissipate an induced charge under dynamic conditions. Coal is generally less conducting than the ash-forming minerals, but with pyrite it becomes conducting and hence suitable for the dry processing. This technique could circumvent the problems associated with water removal following fine coal cleaning (Lewowski, 1993). *Demerits:* Requires continuous power supply.
Oil agglomeration	In this process particles can be removed from the liquid suspension by selective wetting and agglomeration with specific oil. The process relies on the difference in the surface properties of coal and minerals in slurry. The oil dispersed into droplets on agitation with the suspension of coal particles make it hydrophobic in nature. The partially oil-coated particles stick together and form relatively large flocs or agglomerates which can be recovered on a screen. *Problem:* pyrite is readily wetted by fuel oil and agglomerated due to its weakly hydrophobic surface property as compared to most other minerals which are hydrophilic and do not become oil coated (Leonard et al., 1981). *Demerits:* Consumption of large amounts of oil and waste disposal.
Froth flotation	Flotation is the most widely used technique, based on the differences in surface-wetting characteristics. In coal flotation, the hydrophobicity difference between coal and other minerals is the driving force to separate pyrite and ash-forming minerals (Saydut et al., 2007).
Microwave treatment	Removes pyrite as pyrite absorbs microwave radiation. It is effective if magnetite is added and then the coal is processed with microwave, and finally by magnetic separation (Lovas et al., 2011).

TABLE 6.10

Chemical Methods of Desulphurisation

Coal Origin	Process	Conditions	Sulphur Removal	References
Mezino coal, Tabas, Iran	Leaching with nitric acid and hydrochloric acid.	Temp.—90°C. Acid conc.—30%, 1000 rpm	75.4%	Alam et al. (2009)
Tabas coal, Iran	1. Ferric chloride leaching for sulphur removal. 2. Ferric sulphate as a reagent to remove most of the pyrite sulphur.	Time—1 h, 15% ferric chloride Time—1 h, 38% ferric sulphate	71.2% of pyrite and 53.4% of total S. 45.8% of pyrite and 34.3% of total S.	Ehsani (2006)
Boragolai coal and Ledo coal, Assam, India	Desulphurisation with KOH solution.	1. 2% of KOH solution. Temp.—95°C. 2. 16% of KOH solution. Temp.—95°C.	94% of pyritic sulphur. 11%–15% of organic sulphur.	Mukherjee and Borthakur (2003)
Memaliaj coal, Stratum VI	Alkaline removal by using KOH solution.	1 M KOH Time—2 h.	50% of total sulphur	Lolja et al. (1999)
Mae Moh coal, Thailand	Desulphurisation with methanol water and methanol KOH.	1. 5% methanol, time—90 min, temp.—150°C. 2. 10% methanol. time—90 min, temp.—150°C	1. 69% pyritic and 39% organic sulphur. 2. 66% pyritic and 41% organic sulphur.	Ratnakandilok et al. (2001)
Assam coal	Hydrogenation process by using a solvent.	Tetralin used as a solvent.	27%–31% organic sulphur.	Baruah et al. (1986a,b)
Pittsburgh no. 8 coal	Treated with sodium hypochlorite.	Temp.—50°C and 90°C.	Pyritic sulphur 70% at 50°C and 78% at 90°C.	Li et al (2005)

TABLE 6.11

Role of Bacteria and Fungi in Bioprocessing of Coal

Microorganisms Type	Significant Role
Bacteria	These microbes are found regularly in the environment, and their very small size allows them to bind and act on the surface of the material. The properties of bacteria which mostly differ in their usage of energy and nutrients for the growth make them suitable for coal desulphurisation.
Fungi	Breaking down of coal structure into liquid and gaseous fluid by fungi is an interesting feature which can be judiciously used in coal desulphurisation.

Source: Adapted from Brock, T.D., Smith, D.W., Madigan, M.T. *Biology of Microorganisms*, 4th edition, Prentice-Hall, Englewood Cliffs, NJ, 1984.

Microorganisms may have the ability to attack either carbonaceous matter in the coal or the interspersed inorganic materials. De-polymerisation of coal and the breakage of various key structures could provide the basis for the liquefaction processes. Generally, bacteria play a major role in the bio-desulphurisation process (Table 6.11). Besides, some very common fungi that can also remove the sulphur from coal have been reported (Brock et al., 1984).

6.4.3.1 Bio-Removal of Pyritic Sulphur

The varying results of bio-desulphurisation of sulphur compounds and different by-products released in the coal processing have led to an investigation on the mechanism followed by microorganisms. Two different mechanisms have been proposed for pyrite removal by bacteria: indirect mechanism and the direct contact mechanism. According to the indirect mechanism, ferric ions are the primary oxidant, oxidising the metal sulphides while being reduced in turn to the ferrous state. The bacteria then enter the reaction by oxidising ferrous ions to the ferric state, thereby regenerating the primary oxidant. The direct contact mechanism is independent of the action of ferric ions, requiring only intimate physical contact between the bacteria and the sulphide mineral under aerobic conditions (Tributsch, 1999). Different authors have proposed the direct and indirect mechanisms of bacterial pyrite removal. Equations 6.1 and 6.2 describe the indirect mechanism and Equation 6.3 represents the direct mechanism of pyrite removal from coal.

$$FeS_2 + 8H_2O + 14Fe^{3+} \rightarrow 15Fe^{2+} + 2SO_4^{2-} + 16H^+ \tag{6.1}$$

$$14Fe^{2+} + 3.5O_2 + 14H^+ \rightarrow 14Fe^{3+} + 7H_2O \tag{6.2}$$

$$FeS_2 + H_2O + 3.5O_2 \rightarrow Fe^{2+} + 2SO_4^{2-} + 2H^+ \tag{6.3}$$

A range of microorganisms have been found that are capable of carrying out desulphurisation, each having its own significance based on the region from where it is isolated or the environmental conditions and the nature/origin of coal the world over. The most widely used microorganism in the process of desulphurisation till now is the meso-acidophilic bacteria, namely, *Acidithiobacillus ferrooxidans*. Thus, most literature on biodesulphurisation of coal accounts for the pyritic sulphur removal using *A. ferrooxidans*.

Initially, Zarubina et al. (1959) and Silverman et al. (1963) reported the experimental results of coal desulphurisation on lab scale, in which they carried out pyrite extraction using bacteria *A. ferrooxidans* type. All the experiments with this bacterium were carried out at an acidic pH. Zarubina et al. also reported the removal of 23%–27% of sulphur from the coal samples in 30 days. Eghbali and Ehsani (2010) used the culture (PTCC 1646 and PTCC 1647) of *A. ferrooxidans* to de-sulphurise the Tabas coal of Iran. The microorganisms were cultured in 9 K media at pH 2. Here, the pH provided was much more acidic compared to the normal pH found suitable for the environment. Ohmura et al. (1993) used *A. ferrooxidans* strain ATCC 23270 and some isolates (T-1, 9 and 11) to suppress pyrite floatability. Based on this result, it was suggested to use microbial flotation with *A. ferrooxidans* for de-sulphurisation of coal. This work illustrates that the selective and quick adhesion is the main function of the bacterium to achieve the coal desulphurisation.

Sulfolobus acidocaldarius, the thermophilic bacteria can remove 96% of inorganic sulphur and 50% of total sulphur from coal sample at 70°C (Kargi and Robinson, 1982). Olsson et al. compared the ability of desulphurisation of coal by thermophilic archaea such as, *Acidianus Brierleyi* (DSM 1651), *S. acidocaldarius* (DSM 639) and *Sulfolobus solfetaricus* (DSM 1616) with that of mesophilic bacterium *A. ferrooxidans* (DSM583) (Olsson et al., 1994). Two of the investigated microorganisms, *A. ferrooxidans* and *Acidianus brierleyi*, were found to be capable of oxidising pure pyrite as well as the organic sulphur in the coal. The kinetic analysis has shown higher rate constant for oxidation of pure pyrite (desulphurisation rate) by *A. brierleyi* as compared to *A. ferrooxidans*. Cismasiu (2010) investigated the desulphurisation of coal using cultures *of A. ferrooxidans* on pit coal from Lupeni mine and lignite from Turceni and Halanga mines. The experiments were conducted with 10 mL inoculum aged 7 days in 90 mL of specific medium (Leathen's) incubated at 28°C for 28 days at 150 rpm and S/L ratio of 5–10 g/100 mL. The diminished weight of different coals was found due to the action of *A. ferrooxidans* culture, which solubilised pyrite as soluble sulphate. Merrettig et al. (1989) described that *Leptospirillum* sp. was found to remove 85% of pyritic sulphur from coal in 40 days at pH 1.5 and 100 g/L pulp density in air-lift fermenter, though some formation of elemental sulphur was also observed during the process of bio-oxidation. The elemental sulphur formed could be further oxidised by the inoculation of *Acidithiobacillus thiooxidans*.

The isolated culture of *A. ferrooxidans* was mostly applied for bio-desulphurisation of different types of Indian coals (Mannivannan et al., 1994; Malik et al., 2000, 2001; Acharya et al., 2001, 2004). Acharya et al. reviewed the literature on bio-treatment of coal with the emphasis on the use of *A. ferrooxidans* to study the mechanism and kinetics of bio-depyritisation of coal (Acharya et al., 2001, 2004). It is suggested to achieve the bio-desulphurisation particularly that of pyritic sulphur in combination with physical separation, flotation/bio-flotation, micro-wave treatment and so forth. Mannivannan et al. used *A. ferrooxidans* isolated from the mine sludges for sulphur removal from the coal (Mannivannan et al., 1994). The rate of desulphurisation of coal was found to depend strongly on the physical and chemical parameters, pH, media composition and iron content in the material. Malik et al. also used *A. ferrooxidans* for desulphurisation of high sulphur Assam coal (Malik et al., 2000, 2001). Bio-desulphurisation of the coal with *A. ferrooxidans* was examined under four modes of operation namely conventional batch, constant volume pulse feeding (CVPF), increasing volume pulse feeding (IVPF) and leachate recycle (Malik et al., 2001). The effects of different pulse feeding strategies and leachate (product) recycle on biological performance were studied and compared with a conventional batch process. The sulphur removal rates for each of the four processes were calculated to be 0.04 g/day (batch), 0.09 g/day (CVPF), 0.19 g/day (IVPF) and 0.05 g/day (leachate recycle). The percentage of sulphur removal on the 30th day in batch, CVPF, IVPF and leachate recycle processes was found to be 72%, 93%, 97% and 90%, respectively. The increasing volume pulse feeding was the best operational strategy for bio-desulphurisation at enhanced rate for longer duration.

By using the indigenous fungal culture (*Aspergillus* sp.) isolated from the coal, de-sulphurisation of coals from Northeastern coalfields (Assam), India, was attempted. Removal of as high as 70%–80% of total sulphur from the coal was achieved (Acharya et al., 2005). Table 6.12 summarises the details of conditions and the extent of bio desulphurisation of coal using different bacteria (Beyer et al., 1985, 1987; Chandra and Mishra, 1988; Clark et al., 1993, Huifang and Yaqin, 1993; Pandey et al., 2005; Eghbali and Ehsani, 2010; Rajak et al., 2010; Caicedo et al., 2012; Aditiawati et al., 2013).

6.4.3.2 Bio-Desulphurisation of Organic Sulphur

As compared to the removal of pyritic sulphur from coal, which is relatively easy to achieve, removal of organic sulphur is difficult and little success has been achieved so far in this area. There have been very limited de-sulphurisation studies on the actual coal samples. The published and patented literatures are described in this section. Gomez et al. (1999) reported that a microorganism metabolically related to *Xanthomonas maltophila* is able to remove 69% of organic sulphur and 68% of inorganic sulphur from Spanish coal in M1 media at neutral pH. Thermophilic, facultative autotrophic, sulphur oxidising

TABLE 6.12

Coal Desulphurisation by Different Microorganisms

Sl. No.	Coal Source	Initial Sulphur (%)	Microorganisms	Condition	Media Used	Desulphurisation (%)	Reference
1	Pittsburgh no. 8 coal	Pyritic sulphur-2.16	*Metallosphaera sedula*	Temp.—70°C, RPM—150, PD—10%, pH—2.0	Mineral salt media	67% of pyritic sulphur	Olson et al. (1993)
2	German bituminous coal	Pyritic sulphur-3.6	*Acidithiobacillus ferrooxidans* strain BF 219	Temp.—30°C, PD—2%, PS—<50 mm, using airlift reactor at pH—1.9	Leathen's medium	97% of pyritic sulphur	Beyer et al. (1987)
3	Jeypore, Assam, India	Total sulphur-6.04	*Acidithiobacillus ferrooxidans* NCIB-8455	Temp.—28°C, RPM—200, PS—<65 µm, pH—2.0	Modified 9 K medium	45.3% of total sulphur	Chandra and Mishra (1988)
4	La Guacamaya, Colombia	Pyritic sulphur-1.03	Mixed culture of *Acidithiobacillus ferrooxidans* and *Acidithiobacillus thiooxidans*	Temp.—30°C, RPM—180, PS—<0.5 mm, PD—20%, pH—1.8	C1 medium with ferrous sulphate-1 g/L	59.22% of pyritic sulphur	Caicedo et al. (2012)
5	Nan Tong coal, China	Pyritic sulphur-0.62	*Acidithiobacillus ferrooxidans* strain T-4	Temp.—30°C, RPM—150, PD—20%, PS—<75 µm, pH—1.7	Leathen's medium	95% of pyritic sulphur	Huifang and Yaqin (1993)

Continued

TABLE 6.12 (Continued)

Coal Desulphurisation by Different Microorganisms

Sl. No.	Coal Source	Initial Sulphur (%)	Microorganisms	Condition	Media Used	Desulphurisation (%)	Reference
6	Rajur coal, Wardha, India	Pyritic sulphur–2.8	Mixed culture of *Acidithiobacillus ferrooxidans* and *Thiobacillus thiooxidans*	Temp.–32°C, PS–45–65 μm, PD–10%, pH–2.4	Ammonium chloride (0.4 g/L), dipottasium hydrogen phosphate (0.4 g/L) and magnesium chloride (0.4 g/L)	78%–81% of pyritic sulphur	Pandey et al. (2005)
7	Tabas coal, Iran	Initial sulphur–2.843	*Acidithiobacillus ferrooxidans*	Temp.–30°C, PS–0.5 mm, PD–10%, pH–2.0	9 K Medium	45% of pyritic sulphur	Reza et al. (2010)
8	Makum coal, Assam, India	Initial sulphur–2.843	*Acidithiobacillus ferrooxidans*	Temp.–32°C, PS–75 μm, PD–3%, 15 days, pH–2.5	9 K Medium	8.62% of total sulphur in bioreactor	Razak et al. (2010)
9	Muara Tigo Besar Utara, south Sumatra		Mixed culture of *Enterobacter, Lelcersia* and *Bacillus* sp.	RPM–120, PD–15%, pH–3–4, 14 days	Mineral salt sulphur free medium, Gunem et al. (2006)	82.36% of total sulphur	Aditiawati et al. (2013)
10	Ruhrgebiet coal, Germany	Pyrite–3.6	*Acidithiobacillus ferrooxidans* BF 219	Temp.–30°C, PS–<0.5 mm, PD–20%, pH–1.9, 28 days	Leathen's medium	94.7% of pyritic sulphur in modified air-lift bioreactor	Beyer et al. (1985)

S. acidocaldarius can remove 44% of initial organic sulphur from free sub-bituminous coal sample at 70°C (Kargi and Robinson, 1986).

Kilbane (1996) described in a patent the use of *Rhodococcus rhodochrous* (ATCC No. 53968) and/or *Bacillus sphaericus* (ATCC No. 53969) to selectively degrade the organic C–S bond in organic sulphur component of coal. The invention relates to the use of bacterial enzymes generated during the pre-adaptation of cells on various organo-sulphur compounds. The percentage of sulphur removal varied from 38% to 80% depending on the time period, organo-sulphur compound used and enzyme activity. The work was extended further in another US Patent by using both the strains as consortia wherein nearly 91% sulphur (organic) removal was reported in 8 weeks using coarser particle size of the coal (Kilbane 1992). Using a consortium of microorganisms, namely, *Hansenula sydowiorum, Hansenula ciferrii, Hansenula lynferdii* and *Cryptococcus albidus*, Stevens and Burgess patented a process wherein 46% reduction in the total sulphur content of the coal was achieved, this being mostly the reduction in the organic-S content at 20–25°C and pH 6–8 (Stevens et al., 1989).

Bio-desulphurisation of organic sulphur was carried out in most of the literature by using model compounds, which are frequently recognised as the organo-sulphur compounds in coal. Various organisms such as *Rhodococcus erythropolis* IGTS8 (ATCC 53968), *Shewanella putrefaciens, Brevibacterium, Rhodococcus* sp. IGTS8, thermophilic *Paenibacillus* sp. A11-2 and *B. subtilis* WUS2B, *Mycobacterium* sp., *Gordonia* sp., *Microbacterium, Lysinibacillus sphaericus* and so forth, were also explored for bio-desulphurisation of coals. Most of these investigations stressed upon the relevance of bio-desulphurisation of coal with the microbes while testing their potency to degrade the organic S-compounds expected in coals.

Out of these microbes *Rhodococcus* sp. (*R. erythropolis* and *R. rhodochrous*) has been extensively used to degrade organosulphur compounds. Kirimura et al. (2002), isolated a new strain of *Rhodococcus* sp. (WU-K2R) from the soil which was capable of utilising napthothiophene sulphone, benzothiophene, 3-methyl-benzothiophene or 5-methyl benzothiophene as the source of sulphur, but could not utilise dibenzothiophene, DBT-sulphone or 4,6-dimethyl-DBT. The microorganism could preferentially desulphurise asymmetric heterocyclic sulphur compounds such as napthothiophene and benzothiophene through the sulphur-specific degradation pathway. *R. erythropolis* strain (R1) and transformed *Rhodococcus* strain (mut 23) was found to degrade 100% of the original dibenzothiophene in 72 h (Etemadifar et al., 2008). Purdy et al. (1993), isolated two new bacterial strains, UM9 and UM3, identified as *Rhodococcus* sp., in which dibenzothiophene or DBT-sulphone served as the only bioavailable source of sulphur. Strain UM9 produced the desulphurised product, 2-hydroxybiphenyl with no other identifiable desulphurised products or release of sulphate or sulphites. Under optimised conditions of pH and temperature, UM9 exhibited up to 35% greater bio-desulphurisation of DBT-sulphone than that of UM3, and both the isolates desulphurised several other organic-sulphur compounds as well. The mechanism of bio-desulphurisation by UM3/UM9 is not fully understood, but appears to be different from those reported for other bacteria (Purdy et al., 1993).

Koyabahsi et al. isolated 35 bacterial strains, out of which the KA2-5-1 only, a variant of *R. erythropolis*, showed its ability to retain high desulphurisation activity. KA2-5-1 desulphurised a variety of alkyl dibenzothiophenes through the specific cleavage of their C–S bonds. In addition, unexpectedly, KA2-5-1 also attacked C–S-bond of alkyl benzothiophenes. The result raises the possibility that the same enzymatic step may be involved in desulphurisation of alkylated forms of both dibenzothiophene and benzothiophene in KA2-5-1 cells (Kobayashi et al., 2000). Another *Rhodococcus* sp. (strain FMF) isolated from an oil field, showed the ability to degrade dibenzothiophene (Haghighat et al., 2003). Dibenzothiophene or its oxide was used as the source for bacterium growth, and the accumulation of hydroxyl biphenyl as the end product was monitored. The isolate showed a desulphurised product in both the organic-S containing source media.

Regarding the activity of bacterial strain, *Shewanella putrefaciens*, it can de-sulphurise dibenzothiophene similar to that of *R. erythropolis* (IGTS8). *S. putrefaciens* converted dibenzothiophene into 2-hydroxybiphenyl in basal salt medium, which was detected by Gibbs assay and HPLC (Farahnaz et al., 2007). Furuya et al. (2001) isolated a bacterial strain WU-F1, which was found to exhibit dibenzothiophene-utilising activity over a wide temperature range (25–50°C) similar to that of mesophilic *Rhodococcus* sp. (IGTS8).

Another bacterial variant extensively used belongs to *Gordonia* genera. Kim et al. investigated the growth and desulphurisation characteristics of *Gordonia* sp. (CYKS1) for dibenzothiophene (Kim et al., 2004). The desulphurisation and the growth of the microbe are inhibited at 0.15 mM or higher concentration of 2-hydroxy biphenyl, a metabolic product of dibenzothiophene. Rhee et al. (1998) isolated a strain *Gordona* CYKS1, was able to convert dibenzothiophene to 2-hydroxybiphenyl and also utilised organic sulphur compounds other than dibenzothiophene as a sulphur source. A variant of *Gordonia alkanivorans*, 1B isolated from hydrocarbon-contaminated soil could desulphurise dibenzopthiophene, benzothiophene, DBT sulphone and alkylated thiophenic compounds (Alves et al., 2005). In particular, this strain (1B) was efficient for removing sulphur selectively from dibenzothiophene without breaking the C–C bond. At equimolar concentrations of dibenzothiophene and benzothiophene, the latter was the preferred compound as the S-source by this bacterium. Only when the benzothiophene concentration reached a very low level, the dibenzothiophene was utilised as the S-source. This clearly indicates that the bacteria will easily degrade benzothiophene compared to dibenzothiophene. Yet another strain of *Gordonia* sp. (F.5.25.8) was found to degrade dibenzothiophene while forming the same end product (Santos et al., 2006). Interestingly, the strain was able to utilise dibenzothiophene and carbazole as the source of S and N, simultaneously. High tolerance to DNA damage and its refractory nature to the induced mutagenesis, make it a potential species for bio-catalytic desulphurisation of fossil fuels.

Among other bacterial/microbial species *Brevibacterium* sp. was isolated and used with dibenzothiophene as a source of sulphur, wherein sulphite was released in a stoichiometrical amount and was further oxidised to sulphate (Afferden et al., 1990). Three metabolites of dibenzothiophene degradation were isolated and identified as dibenzothiophene-5-oxide, dibenzothiophene-5-dioxide and benzoate indicating a variation in the pathway of utilisation of dibenzothiophene by this species.

A thermophilic benzothiophene-degrading *Mycobacterium* sp. (SWU-4) was isolated from petroleum-contaminated soil samples from two provinces in Thailand, Chonburi and Rayong (Watanapokasin et al., 2002). The strain was grown in CMM medium containing thiophene, bromo (α) thiophene or 3-methylthiophene replacing benzothiophene as the source of carbon at 50°C. A mixed culture of the thermophilic bacteria isolated from crude oil was examined for anaerobic biodegradation of dibenzothiophene (Rath et al., 2012). The mixed culture degraded 98% of dibenzothiophene (0.5 mg/mL) at 65°C over 15 days both in the presence and absence of methyl viologen.

A newly isolated *Bacillus* sp. (strain KS1) could consume 68.75% of dibenzothiophene over a period of 7 days to produce 2-hydroxy biphenyl (Bahuguna et al., 2011). A strain DMT-7 (*Lysinibacillus sphaericus)* selectively desulphurised dibenzothiophene (Li et al., 2005). In addition, *L. sphaericus* has also shown the ability to utilise substrates such as benzothiophene, 3,4-benzo DBT, 4,6-dimethyl DBT, and 4,6-dibutyl DBT. *Microbacterium* sp. ZD-M2 obtained from the sludge showed potential to completely desulphurise model organic-S compounds (Bahrami et al., 2001).

Zahara et al. (2006), isolated yeast *Trichosporon* sp. from phenol-contaminated waste with phenol and dibenzothiophene as the only source of C and S, respectively. The yeast grown in dibenzothiophene + basal salt medium was able to reduce almost 100% of the dibenzothiophene in 14 days.

6.4.3.3 *Mechanisms of Organic Sulphur Removal*

The most common compound used as a model for organic sulphur compounds in coal is dibenzothiophene (DBT). Therefore, DBT is the compound of interest for which the mechanism of bio-desulphurisation needs to be investigated.

4S pathway for bio-desulphurisation: Kilbane (1989) proposed a pathway for oxidation of organic sulphur, which is essentially a sulphur-specific pathway regarded as the '4S' pathway as shown in Figure 6.1. According to this pathway, the organisms selectively oxidise the sulphur atom in DBT, without cleavage of C–C bond thereby maintaining the calorific value (Kirimura et al., 2001). The 4S pathway implies the conversion of dibenzothiophene to

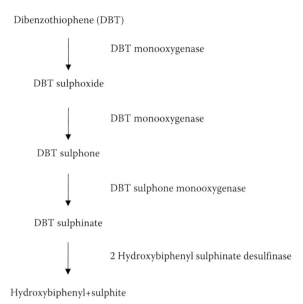

FIGURE 6.1
Sulphur specific 4S pathway proposed by Kilbane. (Adapted from Kilbane, J.J. *Trends of Biotechnology* 7, 1989: 97–101.)

2-hydroxybiphenyl. In this process, four different molecules are formed during the sulphur metabolism, and thus it is known as the 4S pathway.

The 4S pathway is a multienzymatic biodegradation process (Figure 6.1) and four genes are responsible for conversion of organic sulphur (Gupta et al., 2005). These are

1. DBT-monooxygenase is responsible for the first two oxidations of DBT to sulphoxide and then to sulphone (dszC genes),
2. DBT sulphone monooxygenase oxidising DBT-sulphone to 2-hydroxybiphenyl-2-sulphinic acid (dszA gene),
3. 2-Hydroxybiphenyl sulphinate desulphinase that converts HBP-sulphinate to 2-hydroxybiphenyl and sulphite (dszB genes), and
4. NADH-flavin mononucleotide oxidoreductase supplying FMNH2 needed for the first three oxidations (dszD genes).

R. rhodochrous (IGTS8) is able to desulphurise sulphur from DBT by the 4S pathway, which was first proposed by Gallagher et al. (1993). Besides, there are many bacteria such as *Gordona* CYKS1 (Rhee et al., 1998), *Paenibacillus* sp. (Konishi et al., 1997), *Xanthomonas* sp. (Constanti et al., 1994), *Mycobacterium* sp. (Li et al., 2003) that are also reported to remove sulphur from benzothiophene via the 4S pathway.

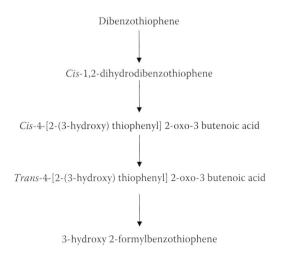

Dibenzothiophene

↓

Cis-1,2-dihydrodibenzothiophene

↓

Cis-4-[2-(3-hydroxy) thiophenyl] 2-oxo-3 butenoic acid

↓

Trans-4-[2-(3-hydroxy) thiophenyl] 2-oxo-3 butenoic acid

↓

3-hydroxy 2-formylbenzothiophene

FIGURE 6.2
Kodama pathway for desulphurisation of dibenzothiophene DBT. (Adapted from Kodama, K. et al. *Agricultural and Biological Chemistry* 37, 1973: 45–50.)

Kodama pathway for desulphurisation: Another pathway used for bio-desulphurisation is the Kodama pathway, which is most commonly used for DBT degradation. In this pathway the C–C bond of the DBT molecule is attacked (Figure 6.2). Oxidative cleavage of the C–C bond is the basis of the Kodama pathway (Kodama et al., 1973). It comprises of three major steps: hydroxylation, ring cleavage and hydrolysis.

Several microorganisms are reported to follow this pathway for desul-phurisation. *Pseudomonas jijani, Pseudomonas abikonesis* (Kodama et al., 1970), *Beijerincika* sp. (Laborde and Gibson, 1977) and fungi like *Cunninghamella elegans* (Crawford and Gupta, 1990) and *Pleutoris ostreatus* (Bezalel et al., 1996) are able to oxidise dibenzothiophene to DBT sulphoxide and DBT sulphone as the end product of the metabolism. As stated earlier, there are many microorganisms that are able to desulphurise different kinds of organo-sul-phur compounds. The microorganisms grow in a suitable temperature and produce different kinds of enzymes that help in the desulphurisation of the model organic sulphur compounds. Table 6.13 enlists the various microor-ganisms involved in desulphurisation of organic sulphur (Omori et al., 1992; Wang et al., 1996; Nekodzuka et al., 1997; Gilbert et al., 1998; Onaka et al., 2001; Maghsoudi et al., 2001; Matsui et al., 2002; Schilling et al., 2002; Rath et al., 2012). As can be seen, various mesophilic and thermophilic strains such as *Mycobacterium, Gordonia, Bacillus, Gordona, Corynebacterium, Paenibacillus, R. rhodochrous, R. erythropolis* and others, may utilise mainly the organo-sulphur compounds like dibenzothiophene, benzothiophene, napthothio-phene and alkylated dibenzothiophene.

TABLE 6.13

Microbes in Desulphurisation of Model Sulphur Compound

Microorganisms	Temperature Effect	Model Organic Sulphur	References
Mycobacterium phlei WU-F1	Thermophilic	Dibenzothiophene	Furaya et al. (2001)
Gordona sp. strain 213E (NCIMB 40816)	Mesophillic	Benzothiophene	Gilbert et al. (1998)
Bacillus subtilis WU-S2B	Thermophilic	Dibenzothiophene	Kirimura et al. (2001)
Rhodococcus sp. strain WU-K2R	Mesophilic	Benzothiophene and napthothiophene	Kirimura et al. (2002)
Rhodococcus sp. strain P32C1	Mesophilic	Dibenzothiophene	Maghsoudi et al. (2001)
Gordonia sp. CYKS1	Mesophilic	Dibenzothiophene	Rhee et al. (1998)
Rhodococcus sp. strain T09	Mesophilic	Dibenzothiophene	Matsui et al. (2002)
Mycobacterium sp. strain G3.	Mesophilic	Dibenzothiophene	Nekodzuka et al. (1997)
Corynebacterium sp. strain SY1.	Mesophilic	Dibenzothiophene	Omori et al. (1992)
Paenibacillus sp. strain A11-2	Thermophilic	Alkylated dibenzothiophene	Onaka et al. (2001)
Bacillus sp. KS1	Mesophilic	Dibenzothiophene	Rath et al. (2012)
Rhodococcus rhodochrous IGTS8 (ATCC-53968)	Mesophilic	Dibenzothiophene	Schilling et al. (2002)
Rhodococcus erythropolis	Mesophilic	Dibenzothiophene	Wang et al. (1996)

6.5 Future Research Directions

The chapter provides information on the general concept, mechanism and the advances made in the area of coal bio-desulphurisation. The acidophilic microorganisms can easily remove the inorganic sulphur (pyritic) from coal, whereas the removal of organic sulphur from coal is a challenging task. A number of isolates/microorganisms are reported, which show the potency in the presence of synthetic organic sulphur. These isolates were tested for desulphurising ability against different synthetic organic sulphur compounds such as dibenzothiophene, benzothiophene, thiophene, phenyl disulphide and so on. The microorganisms able to desulphurise the organo-sulphur compounds could be used for the removal of sulphur from coal. When microbes are suitably adapted to the desired conditions, in all likelihood it would increase its efficiency for sulphur removal from coal. Therefore, investigation on the kinetics of organo-sulphur removal by adapted microorganisms and optimisation of the organo-sulphur bioprocess are needed. Besides this, genetic manipulation of the potential microbial strains could possibly enhance their ability to desulphurise the organic sulphur present in such coals.

References

Acharya, C., Kar, R.N., Sukla, L.B. Bacterial removal of sulphur from three different coals. *Fuel* 80, 2001: 2207–216.

Acharya, C., Sukla, L.B., Misra, V.N. Biodepyritisation of coal. *Journal of Chemical Technology and Biotechnology* 79, 2004: 1–12.

Acharya, C., Sukla, L.B., Misra, V.N. Biological elimination of sulphur from high sulphur coal by *Aspergillus*-like fungi. *Fuel* 84, 2005: 1597–600.

Aditiawati, P., Akhmaloka, A., Sugilubin, D.I., Pikoli, M.R. Biodesulfurization of sub-bituminous coal by mixed culture of bacteria isolated from coal mine soil of south Sumatera. *Biotechnology* 12(1), 2013: 46–53.

Afferden, M.V., Schacht, S., Klein, J., Triiper, H.G. Degradation of dibenzothiophene by *Brevibacterium* sp. DO. *Archives of Microbiology* 153, 1990: 324–28.

Alam, H.G., Moghaddam, A.Z., Omidkhah, M.R. The influence of process parameters on desulphurization of Mezino coal by HNO_3/HCL leaching. *Fuel Processing Technology* 90, 2009: 1–7.

Ali, A., Srivastava, N.K., Srivastava, S.K., Goswami, R.N., Yadav, R.S., Hazra, S.K. Upgradation of high sulphur NE region Indian coals by pyrolysis in presence of hydrogen. *The Open Fuels and Energy Science Journal* 2, 2009: 40–46.

Alves, L., Salgueiro, R., Rodrigues, C., Mesquita, E., Matos, J., Girio, F.A. Desulfurization of dibenzothiophene, benzothiophene and other thiophene analogs by a newly isolated bacterium *Gordonia alkanivorans* strain 1B. *Applied Biochemistry and Biotechnology* 120(3), 2005: 199–208.

Annual report 2012–2013. Ministry of Coal. http://coal.nic.in.

Aseefa, Y., Mc Culley, H., Murray, M., Royales, S. Beyond the numbers. *U.S. Bureau of Labor Statistics* 2(3), 2013: 1–7.

Bahrami, A., Shojaosadati, S.A., Mohebali, G. Biodegradation of dibenzothiophene by thermophilic bacteria. *Biotechnology Letters* 23, 2001: 899–901.

Bahuguna, A., Lily, M.K., Manjula, A., Singh, R.N., Dangwal, K. Desulfurization of dibenzothiophene (DBT) by a novel strain *Lysinibacillus sphaericus DMT-7* isolated from diesel contaminated soil. *Journal of Environmental Sciences (China)* 23(6), 2011: 975–82.

Bailey, R.A., Clark, H.M., Ferris, J.M., Krause, S., Strong, R.L. *Chemistry of the Environment*, 2nd edition. US: Academic Press. 2002: 698–99.

Barooah, P.K., Baruah, M.K. Sulphur in Assam coal. *Fuel Processing Technology* 46, 1996: 83–97.

Baruah, B.P., Ghosh, J.L., Saikia, P.C., Bordoloi, C.S., Mazumdar, B., Baruah, J.N. Indian Patent application No. 516DEL, Filed 860612, 1986a.

Baruah, B.P., Ghosh, J.L., Saikia, P.C., Sain, B., Bordoloi, C.S., Mazumdar, B., Baruah, J.N. Indian Patent application No. 517DEL, Filed 860612, 1986b.

Beyer, M., Ebner, H.G., Assenmacher, H., Frigge, J. Elemental sulfur in microbiologically desulfurized coal. *Fuel* 66, 1987: 551–55.

Beyer, M., Ebner, H.G., Klein, J. Bacterial desulfurization of German hard coal. *Proceedings of the Sixth International Symposium on Biohydrometallugy, Process Metallurgy* 4, 1985: 151–64.

Bezalel, L., Hadar, Y., Fu, P.P., Freeman, J.P., Cerniglia, C.E. Initial oxidation products in the metabolism of pyrene, anthracene, fluorine, and dibenzothiophene by the white rot fungus *Pleurotus ostreatus*. *Applied and Environmental Microbiology* 62, 1996: 2554–559.

Blazquez, M.L., Ballester, A., Gonzalez, F., Mier, J.L. Coal bio desulphurization: A review. *Biorecovery* 2, 1993: 155–77.

BP Statistical Review of World Energy June 2013. http://www.bp.com/statistical review.

Brock, T.D., Smith, D.W., Madigan, M.T. *Biology of Microorganisms*, 4th edition, Prentice-Hall, Englewood Cliffs, NJ, 1984.

Caicedo, G., Prada, M., Pelaez, H., Moreno, C., Marquez, M. Evaluation of a coal bio desulfurization process (semi-continuous mode) on the pilot plant level. *Dyna* 79, 2012: 114–18.

Calkins, W.H. The chemical forms of sulphur in coals: A review. *Fuel* 73(4), 1994: 475–84.

Chandra, D., Mishra, A.K. Desulfurization of coal by bacterial means. *Resources, Conservation and Recycling* 1, 1988: 293–308.

Cismasiu, C.M. The acidophilic chemolithotropic bacteria involved in the desulphurization process of lignite and pit coal from Halunga, Mintia and Petrila mines. *Romanian Biotechnological Letters* 15(5), 2010: 5602–610.

Clark, T.R., Baldi, F., Olson, G.J. Coal depyritization by the thermophilic Archaeon *Metallosphaera sedula*. *Applied and Environmental Microbiology* 59(8), 1993: 2375–379.

Constanti, M., Giralt, J., Bordons, A. Desulfurization of dibenzothiophene by bacteria. *World Journal of Microbiology and Biotechnology* 10, 1994: 510–16.

Crawford, D.L., Gupta, R.K. Oxidation of dibenzothiophene by *Cunninghamella elegans*. *Current Microbiology* 21, 1990: 229–32.

Eghbali, F., Ehsani, M.R. Biodesulfurization of Tabas coal in pilot plant scale. *Iranian Journal of Chemistry and Chemical Engineering* 29(4), 2010: 75–78.

Ehsani, M.R. Desulphurization of Tabas coal using chemical reagents. *Iranian Journal of Chemistry and Chemical Engineering* 25, 2006: 59–66.

Etemadifar, Z., Emtiazi, G., Christofi, N. Enhanced desulfurization activity in protoplast transformed *Rhodococcus erythropolis*. *American-Eurasian Journal of Agriculturul and Environmental Science* 3(5), 2008: 795–801.

Farahnaz, A., Prayuenyong, P., Ibtisam, T. Biodesulfurization of dibenzothiophene by *Shewanella putrefaciens* NCIMB 8768. *Journal of Biological Physics and Chemistry* 7(2), 2007: 75–78.

Fecko, P., Sitavancova, Z., Cvesper, L., Cablik, V. Bacterial desulfurization of coal from mine CSA most. *Journal of Mining and Metallurgy* 42B, 2006: 13–23.

Furuya, T., Kirimura, K., Kino, K., Usami, S. Thermophilic biodesulfurization of dibenzothiophene and its derivatives by *Mycobacterium phlei* WU-F1. *FEMS Microbiology Letters* 204, 2001: 129–33.

Gallagher, J.R., Olson, E.S., Stanley, D.C. Microbial desulfurization of dibenzothiophene: A sulfur-specific pathway. *FEMS Microbiology Letters* 107(1), 1993: 31–35.

Gilbert, S.C., Morton, J., Buchanan, S., Oldfield, C., Mc Roberts, A. Isolation of a unique benzothiophene desulfurizing bacterium, *Gordona* sp. strain 213E (NCIMB

40816) and characterization of the desulfurization pathway. *Microbiology* 144(9), 1998: 2545–553.

Gomez, F., Amils, R., Marin, I. Bioremoval of organic and inorganic sulphur from coal samples. *Applied Microbiology and Biotechnology* 52, 1999: 118–21.

Gupta, N., Roychoudhury, P.K., Deb, J.K. Biotechnology of desulfurization of diesel: Prospects and challenges. *Applied Microbiology and Biotechnology* 66, 2005: 356–66.

Haghighat, F.K., Eftekhar, F., Mazaheri, M. Isolation of a dibenzothiophene desulfurizing bacterium from soil of Tabriz Oil Refinery. *Iranian Journal of Biotechnology* 1(2), 2003: 121–24. http://www.coal.nic.in/welcome.html.

Huifang, Z., Yaqin, L. Microbial desulfurization of coal. *Journal of Environmental Sciences* 5(1), 1993: 83–89.

Kadioglu, Y.Y., Karaca, S., Bayrakceken, S., Gulaboglu, M.S. The removal of organic sulphur from two Turkish lignites by chlorinolysis. *Turkish Journal of Chemistry* 22, 1998: 129–36.

Kargi, F., Robinson, J.M. Removal of organic sulfur from bituminous coal: Use of thermophilic organism *Sulfolobus acidocaldarius*. *Fuel* 65(3), 1986: 397–99.

Kargi, F., Robinson, J.M. Biological removal of pyritic sulfur from coal by the thermophilic organism *Sulfolobus acidocaldarius*. *Biotechnology and Bioengineering* 27(1), 1985: 41–49.

Kargi, F., Robinson, J.M. Removal of sulfur compound from coal by the thermophilic organism *Sulfolobus acidocaldorius*. *Applied and Environmental Microbiology* 44(4), 1982: 878–83.

Kayser, K.J., Bielaga-Jones, B.A., Jackowski, K., Odusan, O., Kilbane, J.J. Utilization of organosulphur compounds by axenic and mixed cultures of *Rhodococcus rhodochrous* IGTS8. *Journal of General Microbiology* 139, 1993: 3123–29.

Kilbane II, J.J. Enzymes from *Rhodococcus* strain ATCC no. 53968, *Bacillus sphaericus* strain ATCC no. 53969 and mixtures thereof for cleavage of organic C-S bonds of carbonaceous material. U.S. Patent 5,132,219, 1992.

Kilbane, J. Enzyme from *Rhodococcus rhodochrous* ATCC 53968, *Bacillus sphaericus* ATCC 53969 or a mutant thereof for cleavage of organic C-S bonds. U.S. Patent 5,516,677, 1996.

Kilbane, J.J. Desulfurization of coal: The microbial dissolution. *Trends of Biotechnology* 7, 1989: 97–101.

Kim, Y.J., Chang, J.H., Cho, K.S., Ryu, H.W., Chang, Y.K. A physiological study on growth and dibenzothiophene (DBT) desulfurization characteristics of *Gordonia* sp. CYKS1. *Korean Journal of Chemical Engineering* 21(2), 2004: 436–41.

Kirimura, K., Furuya, T., Sato, R., Ishii, Y., Kino, K., Usami, S. Biodesulfurization of naphthothiophene and benzothiophene through selective cleavage of carbon sulfur bonds by *Rhodococcus* sp. strain WU-K2R. *Applied and Environmental Microbiology* 68(8), 2002: 3867–872.

Kirimura, K., Furuya, T., Nishi, Y., Ishii, Y., Kino, K., Usami, S.H. Biodesulfurization of dibenzothiophene and its derivatives through the selective cleavage of carbonsulfur bonds by a moderately thermophilic bacterium *Bacillus subtilis* WU-S2B. *Biosciences and Bioengineering* 91(3), 2001: 262–66.

Kobayashi, M., Onaka, T., Ishii, Y. et al. Desulfurization of alkylated forms of both dibenzothiophene and benzothiophene by a single bacterial strain. *FEMS Microbiol Letters* 187(2), 2000: 123–26.

Koca, H., Koca, S., Aksoy, D.O. The effects of drum speed, shake amplitude combination on cleaning of lignite fines by multi gravity separator. *XII International Mineral Processing Symposium*, October 10–12, 2012, Bodrum, Turkey.

Kodama, K., Nakatini, S., Umehara, K., Shimizu, K., Minoda, Y., Yamada, K. Microbial conversion of petro-sulfur compounds. Part III. Isolation and identification of products from dibenzothiophene. *Agricultural and Biological Chemistry* 34, 1970: 1320–324.

Kodama, K., Umehara, K., Shimizu, K., Nakatani, S., Minoda, Y., Yamada, K. Identification of microbial products from dibenzothiophene and its proposed oxidation pathway. *Agricultural and Biological Chemistry* 37, 1973: 45–50.

Konishi J., Ishii, Y., Onaka, T., Okumura, K., Suzuki, M. Thermophilic carbon-sulfur-bond targeted bio desulfurization. *Applied and Environmental Microbiology* 63, 1997: 3164–169.

Laborde, A.L., Gibson, D.T. Metabolism of dibenzothiophene by a *Beijerinckia* species. *Applied and Environmental Microbiology* 34, 1977: 783–90.

Lakshmanraj, L., Ganesh, V., Chandiran, M. Desulfurization of diesel by *Pseudomonas* species isolated from Indian refinery sites and optimization of process parameters by factorial design. *Hitek Journal of Biological Science and Bioengineering* 1(1), 2012: 1–21.

Leonard, W.G., Greer, R.T., Markuszewski, R., Wheelock, T.D. Coal desulphurization and deashing by oil agglomeration. *Separation Science and Technology* 16(10), 1981: 1589–609.

Lewowski, T. Electrostatic desulphurization of Polish steam coals. *International Journal of Coal Preparation and Utilization* 13(1–2), 1993: 97–105.

Li, F.L., Xu, P., Ma, C.Q., Luo, L.L., Wang, X.S. Deep desulfurization of hydrodesulfurization-treated diesel oil by a facultative thermophilic bacterium *Mycobacterium* sp. X7B. *FEMS Microbiology Letters* 142, 2003: 65–70.

Li, W., Cho, E.H. Coal desulphurization with sodium hypochlorite. *Energy and Fuels* 19, 2005: 499–507.

Li, W., Zhang, Y., Wang, M.D., Shi, Y. Biodesulfurization of dibenzothiophene and other organic sulfur compounds by a newly isolated *Microbacterium* strain ZD-M2. *FEMS Microbiology Letter* 247, 2005: 45–50.

Lolja, S.A. A model for alkaline removal of sulphur from a low rank coal. *Fuel Processing Technology* 60, 1999: 185–95.

Lovas, M., Znamenackova, I., Zubrik, A., Kovacova, M., Dolinska, S. The application of microwave energy in mineral processing—A review. *Acta Montanistica* 16(2), 2011: 137–48.

Maghsoudi, S., Vossoughi, M., Kheirolomoom, A., Tanaka, E., Katoh, S. Biodesulfurization of hydrocarbons and diesel fuels by *Rhodococcus* sp. strain P32C1. *Biochemical Engineering Journal* 8, 2001: 151–56.

Malik, A., Dastidar, M.G., Roychoudhury, P.K. Biodesulfurization of coal: Rate enhancement by sulfur grown cells. *Biotechnology Letters* 22, 2000: 273–76.

Malik, A., Dastidar, M.G., Roychoudhury, P.K. Biodesulfurization of coal: Effect of pulse feeding and leachate recycle. *Enzyme and Microbial Technology* 28(1), 2001: 49–56.

Mannivannan, T., Sandhya, S., Pandey, R.A. Microbial desulfurization of coal by chemoautotrophic *Thiobacillus ferrooxidans*—An iron mine isolate. *Journal of Environmental Science and Health. Part A: Environmental Science and Engineering and Toxicology* 29(10), 1994: 2045–61.

Matsui, T., Noda, K., Tanaka, Y., Maruhashi, K., Kurane, R. Recombinant *Rhodococcus* sp. strain T09 can desulfurize DBT in the presence of inorganic sulfate. *Current Microbiology* 45, 2002: 240–44.

Merrettig, U., Wlotzka, P., Onken, U. The removal of pyritic sulphur from coal by *Leptospirillum*-like bacteria. *Applied Microbiology and Biotechnology* 31(5–6), 1989: 626–28.

Mukherjee, S., Borthakur, P. Effect of leaching high sulphur subbituminous coal by KOH and acid on removal of mineral matter and sulphur. *Fuel* 82, 2003: 783–88.

Nayak, B. Mineral matter and the nature of pyrite in some high-sulfur tertiary coal of Meghalaya, North-East India. *Journal of Geological Society of India* 81, 2013: 203–14.

Nekodzuka, S., Kambe, T.N., Nomura, N., Lu, J., Nakahara, T. Specific desulfurization of dibenzothiophene by *Mycobacterium* sp. strain G3. *Biocatalysis and Biotransformation* 15(1), 1997: 17–27.

Ohmura, N., Kitamura, K., Saiki, H. Mechanism of microbial floatation using *Thiobacillus ferrooxidans* for pyrite suppression. *Biotechnology and Bioengineering* 41(6), 1993: 671–76.

Olson, G.J., Brinckman, F.E. Bioprocessing of coal. *Fuel* 65(12), 1986: 1638–46.

Olsson, G., Pott, B.M., Larsson, L., Hoist, O., Karlsson, H.T. Microbial desulphurization of coal by *Thiobacillus ferrooxidans* and thermophilic archaea. *Fuel Processing Technology* 40(2–3), 1994: 277–82.

Omori, T., Monna, L., Saiki, Y., Kodama, T. Desulfurization of dibenzothiophene by *Corynebacterium* sp. strain SY1. *Applied and Environmental Microbiology* 58(3), 1992: 3911–915.

Onaka, T., Konishi, J., Ishii, Y., Maruhashi, K. Desulfurization characteristics of thermophilic *Penibacillus* sp. strain A11-2 against asymmetrically alkylated dibenzothiophenes. *Journal of Bioscience Bioengineering* 92(2), 2001: 193–96.

Pandey, R.A., Raman, V.K., Bodkhe, S.Y., Handa, B.K., Bal, A.S. Microbial desulphurization of coal containing pyritic sulphur in a continuously operated bench scale coal slurry reactor. *Fuel* 84, 2005: 81–87.

Purdy, R.F., Lepo, J.E., Ward, B. Biodesulfurization of organic sulfur compounds. *Current Microbiology* 27(4), 1993: 219–22.

Rajak, K.K., Yadav, B.K., Mandal, R.B., Saxena, V.K., Mandre, N.R. Bio-desulfurization of Makum (Assam) coal. *Proceedings of the XI International Seminar on Mineral Processing Technology* II, 2010: 976–80.

Rath, K., Mishra, B., Vuppu, S. Bio degrading ability of organo-sulphur compound of a newly isolated microbe *Bacillus* sp. KS1 from the oil contaminated soil. *Archives of Applied Science Research* 4(1), 2012: 465–71.

Ratnakandilok, S., Ngamprasertisita, S., Prasassarakich, P. Coal desulphurization with methanol/water and methanol KOH. *Fuel* 80, 2001: 1937–942.

Rhee, S.K., Chang, J.H., Chang, Y.K., Chang, H.N. Desulfurization of dibenzothiophene and diesel oils by a newly isolated *Gordona* strain CYKS1. *Applied and Environmental Microbiology* 64(6), 1998: 2327–331.

Ryan, B., Ledda, A. A review on sulphur in coal: With specific reference to the Telkwa deposit, North-Western British Columbia. *Geological Fieldwork* 1, 1998: 29–50.

Santos, S.C.C., Alviano, D.S., Alviano, C.S. et al. Characterization of *Gordonia* sp. strain F.5.25.8 capable of dibenzothiophene desulfurization and carbazole utilization. *Applied Microbiology and Biotechnology* 71, 2006: 355–62.

Saydut, A., Tonbul, Y., Baysal, A., Duz, M.Z., Hamamci, C. Froth flotation pretreatment for enhancing desulphurization of coal with sodium hydroxide. *Journal of Scientific and Industrial Research* 66, 2007: 72–74.

Schilling, B.M., Alvarez, L.M., Daniel, I.C., Cooney, C.L. Continuous desulfurization of dibenzothiophene with *Rhodococcus rhodochrous* IGTS8 (ATCC 53968). *Biotechnology Progress* 18, 2002: 1207–213.

Silverman, M.P., Rogoff, M.H., Wender, I. Removal of pyritic sulfur from coal by bacterial action. *Fuel* 42, 1963: 113–24.

Stevens, Jr., Stanley, E. Burgess, W.D. Microbial desulfurization of coal. U.S. Patent 4,851,350, 1989.

The Indian coal sector: Challenges and Future outlook, Indian chamber of Commerce. Www.pwc.com/india.

Tributsch, H. Direct versus indirect bioleaching. *Proceedings of the International Biohydrometallurgy Symposium*, 1999: 51–60.

US coal reserves: An update by heat and sulfur content. *Energy Information Administration* 1993: 1–150.

Wang, P., Humphrey, E., Kraweic, S. Kinetic analyses of desulfurization of dibenzothiophene by *Rhodococcus erythropolis* in continuous cultures. *Applied and Environmental Microbiology* 62(8), 1996: 3066–68.

Watanapokasin, Y., Nuchfoang, S., Nilwarangkoon, S., Sarangbin, S., Kakizono, T. Isolation and characterization of thermophillic benzothiophene degrading *Mycobacterium* sp. *Applied Biochemistry and Biotechnology* 98–100(1–9), 2002: 301–309.

Zahara, E., Giti, E., Sharareh, P. Removal of dibenzothiophene, biphenyl and phenol from waste by *Trichosporon* sp. *Scientific Research and Essay* 1(3), 2006: 72–76.

Zarubina, A.M., Lyalikova, N.N., Shmuk, E.I. Investigation of microbial oxidation of coal pyrite. *Izvest. Akad. Nauk, SSSR, Otdel, Tekh. Nauk Met i.Toplivo.* 1, 1959: 117–19.

Zhongxuan, G., Huizhou, L., Minfang, L., Shan, L., Jianmin, X., Jiayong, C. Isolation and identification of nondestructive desulfurization bacterium. *Science in China* 45(5), 2002: 521–31.

7

Application of Industrial Waste in Biohydrometallurgy: A Review on Its Use as Neutralising Agent and Potential Source for Metal Recovery

Ata Akcil and Chandra Sekhar Gahan

CONTENTS

7.1 Introduction

Biomining is the extraction of metal values from sulphidic ores and mineral concentrates using microorganisms. Microorganisms are well known for their active role in the formation and decomposition of minerals in the Earth's crust since the beginning of life on Earth. The utilisation of naturally available microorganisms for mineralisation of mineral deposits is an age-old process used in the Roman times during the first century BC, and probably by the Phoenicians before that. At its inception, microbial-mediated methods were used to leach copper without any knowledge of the microorganisms involved in the process. The discovery of the microbial world unravelled the hidden mysteries lying behind microbial processes involved in day-to-day human endeavours, out of which microbe-mediated mineral dissolution was well studied and developed with time. In recent years, remarkable achievements have been made in developing biomining to cater to the interest of the mineral industry to match the global demand for metals in the twenty-first century. Depletion of high-grade mineral deposits makes the traditional pyro-metallurgical process uneconomical for metal recovery. The search for alternative metal-recovery processes to achieve an economical advantage over conventional methods motivated the use of the biohydrometallurgical process, which in turn has accelerated the willingness of the metal industries to use low-grade minerals (Rawlings et al., 2003). Biomining is mostly carried out either by continuous stirred-tank reactors or heap reactors. Continuous stirred-tank reactors are used for both bioleaching and biooxidation processes collectively termed as biomining. Stirred-tank biooxidation processes are mostly applied on high-grade concentrates for the recovery of precious metals such as gold and silver, whereas the stirred-tank bioleaching process is used for the recovery of base metals such as cobalt, zinc, copper, and nickel from their respective sulphides, and uranium from its oxides. Continuous stirred-tank reactors are advantageous and widely used due to the following reasons (Rawlings and Johnson, 2007):

- The continuous flow mode of operation facilitates continual selection of those microorganisms that can grow more efficiently in the tanks, where the more efficient microorganisms will be subjected to less wash out leading to a dominating microbial population in the tank reactor.

- Rapid dissolution of the minerals due to the dominance of the most efficient mineral-degrading microorganisms utilising the iron and sulphur present in the mineral as the energy source. Therefore, there will be a continuous selection of microorganisms which will either catalyse the mineral dissolution or create the conditions favourable for rapid dissolution of the minerals.

- Process sterility is not required, as the objective of this process is to degrade the minerals stating less importance on the type of microorganisms involved in it. Therefore, more importance lies on an efficient dissolution process and the microorganisms that carry out the dissolution process efficiently are typically the most desirable ones.

Continuous stirred-tank biooxidation of refractory gold concentrates and in one case on a cobaltic pyrite concentrate is currently used in more than 10 full-scale operations using two different technologies with three more plants coming up in the near future (Brierley and Briggs, 2002; Rawlings et al., 2003; Olson et al., 2004; van Aswegen et al., 2007). Most of the plants have been commissioned by BIOX® and BachTech Bacox over the last 20 years as listed in Table 7.1. Canadian-based BacTech Mining Company's Bacox process is used for the treatment of refractory gold concentrates (Rawlings et al., 2003). Three plants using the Bacox process are in operation, with the most recent plant at Laizhou, in the Shandong province of China, owned by Tarzan Gold Co. Ltd (China Metals, Reports Weekly, Interfax China Ltd., 2004; Miller et al., 2004). Minbac Bactech bioleaching technology has been developed jointly by BATEMAN and MINTEK in Australia and Uganda. The BacTech Company signed an agreement on June 2008, to acquire Yamana Gold in two refractory gold deposits in Papua New Guinea. BacTech Mining Corporation has achieved significantly improved metal recoveries from the test work carried out on the tailing materials from the Castle Mine tailings deposit located in Gowganda near Cobalt, Ontario. This metallurgical work is a precursor to BachTech's plan to build a bioleaching plant near Cobalt, Ontario, to neutralise the arsenic-laden tailings prevalent in this area, and at the same time also to recover significant quantities of Co, Ni and Ag present in the tailings (BachTech Press Release, 2009). BHP Billiton Ltd operates pilot- and demonstration-scale processes for the recovery of base metals from metal sulphides of nickel, copper and zinc by stirred-tank bioleaching (Dresher, 2004). Bioleaching of zinc sulphides has been widely investigated on the laboratory scale by various researchers (Chaudhury and Das, 1987; Bang et al., 1995; Garcia et al., 1995; Sandström and Petersson, 1997; Pani et al., 2003; Rodríguez et al., 2003; Deveci et al., 2004; Shi et al., 2005). The possibilities to process low-grade complex zinc sulphide ores through bioleaching have received much attention and have been tested in pilot scale (Sandström and Petersson, 1997; Sandström et al., 1997). MIM Holdings Pty Ltd. holds a patent for a fully integrated process that combines bioleaching of zinc sulphides with solvent extraction and electrowinning of zinc metal (Steemson et al., 1994). New developments in stirred-tank processes have come with high-temperature mineral oxidation, which has been set up in collaboration between BHP Billiton and Codelco in Chile (Rawlings et al., 2003).

Biooxidation of refractory gold concentrates in continuous stirred-tank reactors and bioleaching of copper and nickel via heap reactors are

TABLE 7.1

Plants Using Continuous Stirred-Tank Biooxidation for Pre-Treatment of Refractory Gold Concentrates

Industrial Plant and Location and Owner	Concentrate Treatment Capacity (tons)	Operating Years	Current Status/Performance/Reasons for Closure
Fairview, Barberton, South Africa/Pan African Resources	62	1986–present	Gold field's BIOX
Sao Bento, Brazil/AngloGold Ashanti	150	1991–2008	Gold field's BIOX. A single-stage reactor was used to pre-treat the concentrate for pressure oxidation (under care and maintenance)
Harbour Lights, Western Australia	40	1991–1994	Gold field's BIOX. Ore deposit depleted (decommissioned)
Wiluna, Western Australia/Apex Minerals	158	1993–present	Gold field's BIOX
Ashanti, Obuasi, Ghana/AngloGold Ashanti	960	1994–present	Gold field's BIOX
Youanmi, Western Australia	120	1994–1998	BacTech Bacox
Tamboraque, San Mateo, Peru/Gold Hawk Resources	60	1998–2003	Gold field's BIOX
Beaconsfield, Tasmania, Australia/Beaconsfield Gold	~70	2000–present	BacTech Bacox
Laizhou, Shandong province, China/Eldorado Gold	~100	2001–present	BacTech Bacox
Suzdal, Kazakhstan/Centroserve	196	2005–present	Gold field's BIOX
Fosterville, Victoria, Australia/North gate Minerals	211	2005–present	Gold field's BIOX
Bogoso, Ghana/Golden Star Resources	750	2006–present	BacTech Bacox
Jinfeng, China/Eldorado Gold	790	2006–present	Gold field's BIOX
Kokpatas, Uzbekistan/Navoi Mining and Metallurgy	1069	2008–present	Gold field's BIOX
Coricancha, Peru	60	1998–20008	Gold field's BIOX temporarily stopped (under care and maintenance)

Source: Adapted from Van Niekerk, J. *Advanced Materials Research* 71–73, 2009: 465–468; Brierley, C.L. *Hydrometallurgy* 104, 2010: 324–328.

some of the established and commercialised technologies in present-day use. Bioprocessing of ores and concentrates provides economical, environmental and technical advantages over the conventionally used roasting and pressure oxidation (Liu et al., 1993; Lindström et al., 2003; Rawlings et al., 2003; van Aswegen et al., 2007). The increasing demand for gold motivates the mineral exploration from economical deposits and cheaper processing for their efficient extraction. Different chemical and physical extraction methods have been established for the recovery of gold from different types and grades of ores and concentrates. Generally, high-grade oxidic ores are pulverised and processed via leaching, while refractory ores containing carbon are roasted at 500°C to form oxidic ores by the removal of carbon due to combustion and sulphur as sulphur dioxide gas. However, the sulphidic refractory gold ores without carbon are oxidised by autoclaving to liberate the gold from sulphide minerals, and then is sent to the leaching circuit where gold is leached out using cyanide (Reith et al., 2007). In the global scenario today, pyro-metallurgical processes for the pre-treatment of refractory gold concentrates via roasting have been replaced with continuous stirred-tank reactors as a pre-treatment for the successful removal of iron and arsenic through biooxidation. Heap bioleaching is a rapidly emerging technology for the extraction of base metals from sulphide minerals. Significant attention has been focussed on the development of bioheap leaching in recent years (Brierley and Brierley, 2001). Heap bioleaching is mostly practiced on low-grade copper ores with 1%–3% copper and mainly on secondary copper sulphide minerals such as covellite (CuS) and chalcocite (Cu_2S). In heap leaching, the crushed secondary sulphidic ores are agglomerated with sulphuric acid, followed by stacking onto leach pads which are aerated from the base of the heap. Then the ore is allowed to cure for 1–6 weeks and further leached with acidic leach liquor for 400–600 days. A copper recovery of 75%–95% is obtained within this period of time. As the construction of heap reactors is cheap and easy to operate, it is the preferred treatment of low-grade ores (Readett, 2001). The commercial application of bioheap leaching designed to exploit microbial activity, was pioneered in 1980 for copper leaching. The Lo Aguirre mine in Chile processed about 16,000 tonnes of ore/day between 1980 and 1996 using bioleaching (Bustos et al., 1993). Numerous copper heap bioleaching operations have been commissioned since then (Brierley and Brierley, 2001). Overall, Chile produces about 400,000 tonnes of cathode copper by the bioleaching process, representing 5% of the total copper production (Informe al Presidente de la República, Comisión Nacional para el desarrollo de la Biotecnología, Gobierno de Chile, 2003). The Talvivaara Mining Company Plc. started an on-site pilot heap in June 2005 and the bioheap leaching commenced in August 2005. Talvivaara has planned to start full production in 2010. The estimated production of nickel is approximately 33,000 tonnes and has the potential to provide 2.3% of the world's current annual production of primary nickel by 2010. The first shipment of commercial-grade nickel sulphide started in February 2009 (www.talvivaara.com).

7.2 Microorganisms in Biomining

The sulphide mineral-oxidising microorganisms are acidophilic prokaryotes as their optimal growth varies between pH 2 and 4. They are autotrophic in nature as they use inorganic carbon (CO_2) as a carbon source. They are strictly chemolithotrophic, that is, they derive energy for growth from oxidation of reduced sulphur compounds, metal sulphides. Some species also derive energy through the oxidation of ferrous iron while others derive energy by the oxidation of hydrogen. They are classified into three groups such as mesophiles (20–40°C), moderate thermophiles (40–60°C) and thermophiles (60–80°C), based on the temperature requirements for optimal growth. The mesophiles, moderate thermophiles and thermophiles actively involved in bioleaching are given in the Table 7.2.

In a stirred-tank reactor, the acidophilic species vary from one type of concentrate to another; for details see Table 7.3. The most widely reported mesophilic iron oxidiser dominating in continuous stirred-tank reactors is *Leptospirillum ferrooxidans* but in some cases, *Leptospirillum ferriphilum* dominates (Dew et al., 1997; Norris, 2007; Sundkvist et al., 2008). The cause for the dominance of *L. ferriphilum* over *L. ferrooxidans* is due to its faster iron oxidation rate and tolerance to slightly higher temperature (Plumb et al., 2007).

7.3 Bioleaching Mechanisms

Microbial processes facilitating mineral biooxidation and bioleaching are defined in terms of the contact mechanism, the non-contact mechanism and the cooperative mechanism. In the contact mechanism (Figure 7.1a), the bacterial cells attach with the aid of extracellular polymeric substance (EPS) layers to the mineral surfaces, resulting in dissolution of the sulphide minerals

TABLE 7.2

Different Types of Microorganisms Actively Involved in Bioleaching

Mesophiles	Moderate Thermophiles	Thermophiles
Acidithiobacillus ferrooxidans	*Acidimicrobium ferroxidans*	*Sulfolobus metallicus*
Acidithiobacillus thiooxidans	*A. caldus*	*Sulphobacillus* sp.
Acidithiobacillus caldus	*Sulphobacillus thermosulphooxidans*	*Metallosphaera sedula*
L. ferrooxidans		
Leptospirillum ferrodiazotrophum		
Leptospirillum thermoferrooxidans		
L. ferriphilum		

TABLE 7.3

Acidophiles in Mineral Sulphide Concentrate Processing at Different Temperatures

Mineral Concentrate	Temperature (°C)	Major Types in Populations		
Pyrite/ arsenopyrite[a]	40	*L. ferrooxidans* (48%–57%)	*A. thiooxidans/A. caldus* (26%–34%)	*A. ferrooxidans* (10%–17%)
Mixed sulphides[b]	45	*A. caldus* (65%)	*L. ferrooxidans* (29%)	*Sulfobacillus* sp. (6%)
Nickel concentrate[c]	49	*A. caldus* (63%)	*Acidimicrobium* sp. (32%)	*Sulfobacillus* sp. (6%)
	55	*Sulfobacillus* sp. (93%)	*A. caldus* (5%)	*Acidimicrobium* sp. (2%)
Chalcopyrite[d]	75–78	'*Sulfolobus*' sp. (59%)	*Metallosphaera* sp. (1) (34%)	*Metallosphaera* sp. (2) (5%)

Source: Adapted from Norris, P.R. *Acidophile Diversity in Mineral Sulfide Oxidation.* In: D.E. Rawlings, D.B. Johnson (eds.), Biomining, Springer-Verlag, Berlin, Heidelberg, New York, 2007:199–212.

[a] Fairview and São Bento industrial plants (Dew et al., 1997).
[b] Mintek pilot plant (Okibe et al., 2003).
[c] Warwick University laboratory scale (Cleaver and Norris, unpublished data).
[d] Warwick University laboratory scale, HIOX culture (Norris, unpublished data).
  (The estimated proportions of species refer to continuous cultures and to primary reactors where several reactors were in series).

at the interface by an electrochemical process. In the non-contact mechanism (Figure 7.1b), the ferric iron, produced through biooxidation of ferrous iron comes in contact with the mineral surfaces, oxidises the sulphide mineral and releases ferrous iron back into the cycle. While, in the cooperative mechanism (Figure 7.1c), planktonic iron and sulphur oxidisers oxidise the colloidal sulphur, other sulphur intermediates and ferrous iron in the leaching solution, release protons and ferric iron which is further used in non-contact leaching (Rohwerder et al., 2003). Dissolution of the metal sulphides is controlled by two different reaction pathways, that is, the thiosulphate pathway and the polysulphide pathway (Figure 7.2).

7.3.1 Thiosulphate Pathway

The thiosulphate pathway is only applicable to the acid-insoluble metal sulphides such as pyrite (FeS_2), molybdenite (MoS_2) and tungstenite (WS_2). The thiosulphate pathway (Figure 7.2A) reaction mechanism followed in the bioleaching of pyrite is given below:

$$FeS_2 + 6Fe^{3+} + 3H_2O \rightarrow S_2O_3^{2-} + 7Fe^{2+} + 6H^+ \tag{7.1}$$

$$S_2O_3^{2-} + 8Fe^{3+} + 5H_2O \rightarrow 2SO_4^{2-} + 8Fe^{2+} + 10H^+ \tag{7.2}$$

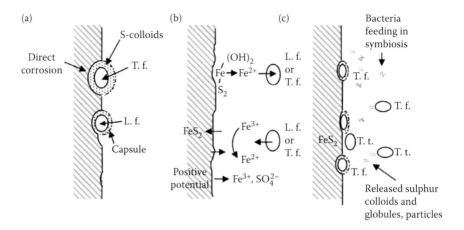

FIGURE 7.1
Patterns of direct and indirect interaction of the bacteria with pyrite (a) contact leaching; (b) non-contact leaching and (c) cooperative leaching. (Reprinted from Rawlings, D.E., Tributsch, H. and Hansford, G.S. *Microbiology* 145, 1999: 5–13.)

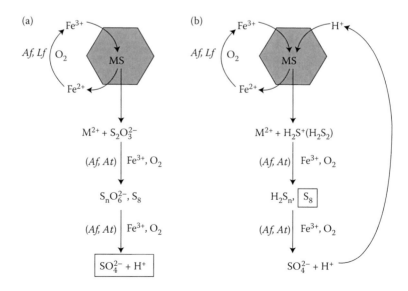

FIGURE 7.2
Schematic comparison of thiosulphate (a) and polysulphide (b) mechanisms in bioleaching of metal sulphides (Schippers and Sand, 1999). (Reprinted from Rohwerder, T. et al. *Applied Microbiology and Biotechnology* 63, 2003: 239–248.)

The above two equations sum up to give the following overall equation:

$$FeS_2 + 14Fe^{3+} + 8H_2O \rightarrow 2SO_4^{2-} + 15Fe^{2+} + 16H^+ \quad (7.3)$$

The main role of the microorganisms in this mechanism is to catalyse the regeneration of the consumed ferric ions by means of aeration as given below in Equation 7.4.

$$14Fe^{2+} + 3.5O_2 + 14H^+ \rightarrow 14Fe^{3+} + 7H_2O \quad (7.4)$$

The overall reaction based on the primary oxidant oxygen is given below:

$$FeS_2 + 3.5O_2 + H_2O \rightarrow Fe^{2+} + 2SO_4^{2-} + 2H^+ \quad (7.5)$$

7.3.2 Polysulphide Pathway

The polysulphide pathway is applicable for acid-soluble metal sulphides such as galena (PbS), sphalerite (ZnS), arsenopyrite (FeAsS) and chalcopyrite (CuFeS$_2$). The polysulphide pathway (Figure 7.2B) reaction mechanism of zinc sulphide bioleaching is stated below:

$$8ZnS + 14Fe^{3+} + 2H^+ \rightarrow 8Zn^{2+} + 14Fe^{2+} + H_2S_8 \quad (7.6)$$

$$H_2S_8 + 2Fe^{3+} \rightarrow S_8 + 2Fe^{2+} + 2H^+ \quad (7.7)$$

The microorganism's role in this mechanism is twofold:

- To catalyse the regeneration of the ferric ions consumed for the chemical oxidation of the intermediary hydrogen sulphide into elemental sulphur via the formation of polysulphides.
- To catalyse the generation of sulphuric acid in order to maintain the supply of protons required in the first reaction step for the dissolution of the mineral.

The further reaction steps are given below:

$$S_8 + 12O_2 + 8H_2O \rightarrow 8SO_4^{2-} + 16H^+ \quad (7.8)$$

$$16Fe^{2+} + 4O_2 + 16H^+ \rightarrow 16Fe^{3+} + 8H_2O \quad (7.9)$$

However, the overall reaction based on the primary oxidant oxygen is pH neutral as shown below:

$$ZnS + 2O_2 \rightarrow Zn^{2+} + SO_4^{2-} \quad (7.10)$$

It is evident from the above mechanism that a high microbial oxidation rate of ferrous to ferric iron is important for an efficient bioleaching process of sulphide minerals.

7.4 General Bioleaching Process

Continuously stirred-tank reactors are highly aerated reactors where pulp continuously flows through a series of reactors with a good control of pH, temperature and agitation, thereby creating a homogeneous environment for mineral biooxidation. The ores and concentrates used for the stirred-tank reactors are finely ground before they are used in the biooxidation process. The pulp density in the continuous stirred-tank reactors is limited to ~20% solids. A pulp density higher than 20% solids causes inefficient gas transfer and microbial cell damage by the high shear force caused by the impellers. The limitation in a pulp density to 20% solids and the relatively high cost for stirred-tank reactors have confined the process for use only with high-grade minerals (van Aswegen et al., 1991; Rawlings et al., 2003). The microorganisms used in bioleaching processes are chemolithotrophic and acidophilic having an optimum activity at a pH around 1.5; therefore, depending on the reactor configuration, the addition of neutralising agents is required to maintain the desired pH. Neutralisation of the acid produced during bioleaching of sulphide minerals is generally practiced using limestone (Arrascue and van Niekerk, 2006). In a bioleaching process of base metal recovery, neutralisation is required at different stages as stated in Figure 7.3.

Primary neutralisation to pH ~ 1.5 using limestone during the bioleaching process, secondary neutralisation to pH 3–4 using lime/limestone for precipitation of iron and arsenic and finally to pH 7–8 for effluent neutralisation by lime (Figure 7.3). Controlling pH at a proper level is important to the operation efficiency in bioleaching processes and generally, a pH range of 1.0–2.0 is maintained. Operating a bioleaching process at a pH above 1.85 may cause excessive iron precipitation as jarosite, while operation at a pH below 1.0 may result in foam formation, as observed at the BIOX process at Fairview and Wiluna (Dew, 1995; Chetty et al., 2000). After completion of bioleaching, the gold-containing residue is treated for gold recovery through cyanidation, leaving behind leach liquor with high levels of ferric iron (Fe^{3+}) and arsenate (AsO_4^{3-}). A study on the possibility to use oxidic by-products, such as steel slags, ashes and dusts as a neutralising agent in the bioleaching process at pH 1.5 and precipitation of Fe/As at pH 3 compared to slaked lime have proved to be successful and promising, due to the alkaline nature of these materials (Gahan, 2008; Gahan et al., 2008, 2009a; Cunha et al., 2008a). Neutralisation of the ferric iron (Fe^{3+}) and arsenate (AsO_4^{3-}) from the leachate at a pH of 3–4 with limestone or slaked lime

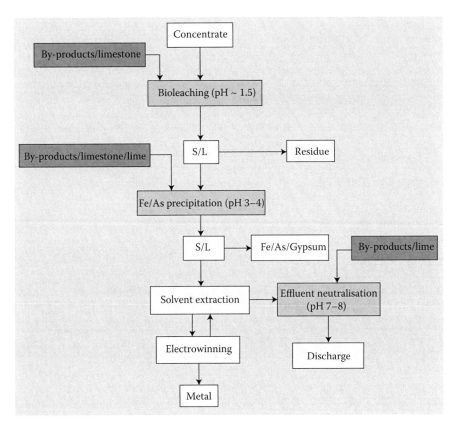

FIGURE 7.3
(**See colour insert.**) Flow sheet of a process for base metal production describing stages of neutralisation.

precipitates arsenic as a ferric arsenate ($FeAsO_4$) (Stephenson and Kelson, 1997). The ferric arsenate obtained is stable and environmentally acceptable according to the U.S. EPA (Environment Protection Agency) toxicity characteristic leaching procedure (TCLP) testing procedure (Broadhurst, 1994; Cadena and Kirk, 1995). Bioleached residues obtained from biooxidation of refractory gold concentrates have been reported to consume large amounts of cyanide during the subsequent cyanidation step. This is due to the formation of elemental sulphur or other reduced inorganic sulphur compounds, which react with cyanide to form thiocyanate (SCN^-) (Hackl and Jones, 1997; Shrader and Su, 1997; Lawson, 1997). In a study where a sequential two-step biooxidation of a refractory gold concentrate was done, moderate thermophiles were used in the first stage followed by a second stage with extreme thermophiles. It was found that the arsenic toxicity was lowered with respect to the extreme thermophiles while the NaCN consumption, due to SCN^- formation, was significantly decreased (Lindström et al., 2002, 2003). Later in 2004, van Aswegen and van Niekerk also reported similar studies as those conducted by Lindström et al. (2003) and stated the successful biooxidation of a refractory gold concentrate by a combination of mesophilic and thermophilic microorganisms and also achieved lower NaCN consumption and higher gold recovery (Lindström et al., 2003; van Aswegen and Niekerk, 2004).

7.5 Cost of Neutralising Agents in the Bioleaching Process

The cost for neutralisation is normally the second largest operation cost in BIOX plants and the limestone cost is directly proportional to the distance between the deposit and the operation plant (van Aswegen and Marais, 1999). Therefore, it is important to look for substitutes, such as dolomite, ankerite or calcrete (a low-grade limestone) deposits located close to the plant, in order to save operation costs. The Wiluna mine in Western Australia utilises locally mined cheap calcrete as a neutralising agent, which contributes to the economic viability of the BIOX process. The total cost involved in calcrete mining and transporting is 5 Australian dollars per tonne. Savings due to the use of calcrete helps in adjusting for Wiluna's high-power cost (van Aswegen and Marais, 1999; Marais (2008), the geologist guide to the BIOX process). Studies conducted to determine the neutralising capacity of the industrial oxidic by-products by chemical leaching at pH 1.5, the optimum pH level for bioleaching microorganisms, found the neutralising potential to be good enough to be used in bioleaching (Cunha et al., 2008b). Studies conducted on the use of oxidic industrial by-products as a substitute-neutralising agent to lime/limestone in the biooxidation of pyrite have shown positive results as stated by Gahan et al. (2008, 2009a). When a suitable alternative neutralising agent is to be chosen, some important criteria need to be fulfilled. First, the agent's neutralising capacity; second, it should be non-toxic to the microorganisms; and third, the overall net cost for delivery and handling of the agent, which is a function of freight cost and so on, but may also include an alternative cost for disposal.

7.6 Utilisation of Industrial Oxidic By-Products

7.6.1 Steel Slag

European steel industries produce large amounts of steel slag every year. The total amount of steel slag generated in 2004 was about 15 million tonnes, in which 62% was basic oxygen furnace (BOF) slag, 29% was electric arc furnace (EAF) slag and 9% was secondary metallurgical slag. Concerning the

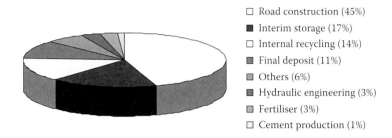

□ Road construction (45%)
■ Interim storage (17%)
□ Internal recycling (14%)
▣ Final deposit (11%)
▢ Others (6%)
■ Hydraulic engineering (3%)
▣ Fertiliser (3%)
□ Cement production (1%)

FIGURE 7.4
(**See colour insert.**) Use of steel slag produced in Europe. (Adapted from The European Slag Association (EUROSLAG). Legal status of slags. Position paper. Available in http://www.euroslag.org/media/Position_paper_Jan_2006.pdf, March 2006.)

use of these slags, 45% is used for road construction, 17% for interim storage, 14% for internal recycling, 11% for the final deposit, 6% for other purposes, 3% for a fertiliser, 3% for hydraulic engineering and 1% for cement production (Figure 7.4) (EUROSLAG, 2006).

The utilisation of steel slag in Sweden is different than what is practiced in mainland Europe. The main part of the steel slag produced in Sweden is sent to the final deposit, while some part is used for internal recycling and road construction, and a small amount is sent for interim storage and cement production (Figure 7.5).

Trials have been conducted for alternative applications on the 49% steel slag sent to the final deposit (Figure 7.5) to save the cost of landfill. The use of steel slag as a neutralising agent is expected to be viable due to its high alkalinity, ready availability and cost-effectivity in comparison to limestone. Comparative cost studies conducted on limestone with different neutralising agents (Hedin and Watzlaf, 1994) state that limestone was one-third the cost of slaked lime. As steel slag is much cheaper than limestone, its use as a neutralising agent could therefore be a benefit for the process cost-efficiency. Replacement of lime for steel slag in acid mine drainage (AMD) treatment

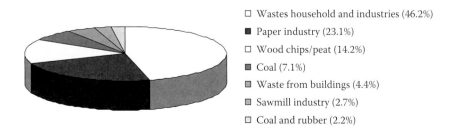

□ Wastes household and industries (46.2%)
■ Paper industry (23.1%)
□ Wood chips/peat (14.2%)
▣ Coal (7.1%)
▢ Waste from buildings (4.4%)
▣ Sawmill industry (2.7%)
□ Coal and rubber (2.2%)

FIGURE 7.5
(**See colour insert.**) Estimation of ash production per annum in Sweden 2003, estimation by C. Ribbing (2007).

was an innovative approach where its high alkalinity and neutralising capacity was utilised (Ziemkiewicz, 1998). The calcium–alumina–silicate complexes present in steel slag cause the pH to rise to high levels, thus precipitating metal ions and hindering the bacterial growth. However, if steel slags are to be used as neutralising agents in bioleaching, it is required that they should not contain elements which are toxic for the bacteria. As an example, fluoride that is common in certain types of slag is known to be toxic for microorganisms. It has been suggested that the reason for fluoride toxicity is due to the transport of fluoride through biological cell membranes, which occurs mainly through passive non-ionic diffusion of the free protonated form of fluoride, and hydrofluoric acid (HF), especially at pH < 6 (Gutknecht and Walter, 1981). HF is a weak acid with a pK_a-value of 2.9–3.4, depending on temperature and ionic strength (Pettit and Powell, 2001). Bioleaching is normally carried out at a pH range of 1–2. Consequently, free uncomplexed fluoride primarily exists as HF in the bioleaching solution. In a study by Sundkvist et al., it was shown that the fluoride toxicity greatly depends on pH and redox potential in the solution. In the presence of aluminium, strong aluminium fluoride complexes are formed, making the solution non-toxic by reducing the free HF concentration (Sundkvist et al., 2005; Brierley and Kuhn, 2009).

7.6.2 Combustion Ashes

Sweden produces a large quantity of non-coal ashes every year. The amount of ashes produced from different sources were 15%–25% from municipal waste, ~5% from peat, 10%–50% from sludge of the paper industry, 2%–4% from bark and 0.3%–0.5% from pure wood (Ribbing, 2007). In 2003, the estimated total amount of ash produced was 1,125,000 tonnes per annum, of which 715,000 tonnes was bottom ash and 410,000 tonnes was fly ash. The fly ash and bottom ash produced in Sweden varied from each other, depending on their fuel source and type of boilers used. The majority of the fly ash and bottom ash produced from different boilers came from the combustion of wastes of household and industries, paper industry and wood chips/peat (Figure 7.6) (Ribbing, 2007). All the non-coal ashes in Sweden have a high pH due to their high lime content. The use of ashes as a liner construction material in landfill is an option for their utilisation, but in Sweden, many landfills will be closed in the next 10–15 years (Tham and Ifwer, 2006; Ribbing, 2007). Therefore, alternative uses of ashes should be investigated. Studies conducted on the use of three types of coal combustion ashes generated in a power plant in Illinois, USA, suggested that ashes could be used as a neutralising agent in agriculture, waste treatment, fertilisers, wallboards, concrete and cement production, ceramics, zeolites, road construction and the manufacture of amber glass (Demir et al., 2001). AMD mitigation can be another alternative use of fly ash. Studies conducted by Hallberg et al. stated that the AMD generated in Falun, Sweden, could be prevented by covering the sulphide mine tailings with a mixture of fly ash and biosludge (Hallberg et al., 2005). Most of the non-coal

□ Final deposit (49%)
■ Internal recycling (30%)
□ Road construction (12%)
▨ Interim storage (8%)
▪ Cement production (1%)

FIGURE 7.6
(**See colour insert.**) Use of steel slag produced in Sweden. (Redrawn from Engström, F. Mineralogical influence of different cooling conditions on leaching behaviour of steel making slags. Licentiate thesis, Luleå University of Technology, ISSN: 1402-1757/ISRN LTU-LIC—07/58—SE/NR 2007:58.)

ashes produced in Sweden, as estimated in 2005, are utilised as construction material in landfill and construction of parking places and other surfaces, while the rest is used for various other purposes as described in Figure 7.7.

7.6.3 Dust

EAF dust recovered from the gas-cleaning system of scrap-based steel production is an industrial oxidic by-product with high zinc content. EAF dust is only about 1.5% of the total output from a typical steel mill, but can create major environmental problems, and therefore needs to be handled carefully. This dust contains zinc, calcium, iron and silicate with contaminants of heavy metals such as lead, cadmium, chromium and others. Since 1984, due to the presence of small quantities of heavy metals (mainly lead) in the EAF dust, it has been regulated as a hazardous waste under the US EPA's solid waste Resource Conservation and Recovery Act (RCRA, 1986). In the global scenario, a part of the EAF dust produced is shipped to hazardous waste landfills, while the other part is sent to industries for recycling. All steel industry and EAF users pay millions of dollars for the removal of unwanted elements, treatment and disposal of the EAF dust. In addition to posing a potential liability to the steel industries, the land-filled dust also contains significant and valuable quantities of recoverable zinc. In Sweden,

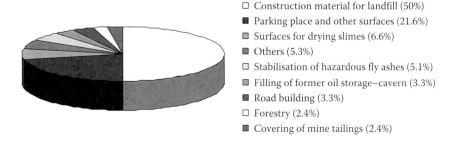

□ Construction material for landfill (50%)
■ Parking place and other surfaces (21.6%)
▨ Surfaces for drying slimes (6.6%)
▪ Others (5.3%)
□ Stabilisation of hazardous fly ashes (5.1%)
▨ Filling of former oil storage–cavern (3.3%)
▪ Road building (3.3%)
□ Forestry (2.4%)
▪ Covering of mine tailings (2.4%)

FIGURE 7.7
(**See colour insert.**) Uses of ashes in Sweden 2003, estimation by C. Ribbing (2007).

a major part of the EAF dust is recycled in a fuming process, which converts zinc ferrite into zinc oxide that is later sent to the zinc smelter for the recovery of zinc. Studies on hydrometallurgical processing for the recovery of zinc from EAF dust have been widely carried out by various researchers (Cruells et al., 1992; Leclerc et al., 2002; Bruckard et al., 2005; Havlík et al., 2006).

7.6.4 Lime Sludge

Lime sludge, a by-product generated from the paper and pulp industry, is reused for the production of lime (calcium oxide) by calcination at temperatures ranging from 1000°C to 1300°C and marketed as quicklime and hydrated lime (Sweet, 1986). A part of the lime sludge, called mesalime, is bled out from the process and has been found to be an excellent alternative neutralising agent during pyrite biooxidation (Gahan et al., 2008). Therefore, an alternative use of lime sludge generated from the paper industry can save the cost incurred for landfill.

7.7 Neutralisation Potential of Industrial Oxidic By-Products

Chemical leaching of steel slag, incineration ashes, dusts and lime sludge using sulphuric acid was investigated to determine their neutralising capacity for replacing normally used lime or limestone in bioleaching operations to maintain constant pH. These by-products included three ashes from combustion for energy production, five slag samples from ore- and scrap-based steelmaking, an EAF dust and mesa lime from a paper and pulp industry with reference to slaked lime ($Ca[OH]_2$). The neutralising potential of the by-products was evaluated by leaching them in sulphuric acid and comparing the amount of acid consumed compared to slaked lime. Most of the by-products showed a good neutralisation potential. The reaction kinetics based on neutralisation was lower for some slag products due to its slow dissolution of the silicates. Zinc recoveries from the zinc-containing EAF dust were high and have an additional benefit upon using them in neutralisation.

7.8 Use of Industrial Oxidic By-Products for Precipitation of Fe/As

A well-developed method in biohydrometallurgy, in the refractory gold concentrate after biooxidation of arsenopyrite and pyrite, there are intermediary steps where ferric arsenate is precipitated at a pH value of 3.0 prior to valuable metal extraction. This is usually achieved through hydroxide precipitation of ferric iron by the addition of lime or limestone. Cunha et al. (2008a,b) investigated the possibility to use oxidic industrial by-products for lime or limestone for ferric iron precipitation at pH 3.0. The neutralisation potential for 10 selected oxidic by-products such as slags, ashes and dusts was compared with slaked lime. Experiments were performed by decreasing pH to 3 by additions of H_2SO_4 to the slurry of the respective by-product at an S/L ratio of 1:10 at a temperature of 25°C and were continued till no changes in pH were observed during 10 days. Characterisation of the by-products revealed high concentrations of oxides such as lime, calcite and metal oxides as well as different forms of silicates in the materials, which all dissolved at pH 3. The neutralising potential was found to be high for most of the by-products with ladle slag being the highest. Slags generally had higher neutralisation potential and long-term effects while the ashes had high initial reactivity which is important for the continuous neutralisation in stirred tanks with limited retention time. Two of the by-products, that is, bioash and mesalime contained a reasonable amount of calcite, which is a very good neutralising agent without any toxicity. Replacement of the conventional lime and limestone with oxidic by-products for neutralisation of acidic leaching solutions was found to be successful in the investigations made and this would save cost of the neutralising agent.

7.9 Utilising Steel Slags as Neutralising Agents in Biooxidation of Pyrite

The comparative assessment of five different steel slags as neutralising agents during the biooxidation of pyrite was assessed with reference to a commercial-grade slaked lime. As the pyrite biooxidation is an acid-producing process, there was a requirement of controlling pH at 1.5 by periodic additions of the neutralising agent. The different steel slags used were argon oxygen decarbonisation (AOD) slag, BOF slag, EAF slag, composition adjustment by sealed argon bubbling–oxygen blowing (CAS–OB) slag and ladle slag, representing slags produced in both integrated steel plants and scrap-based steel plants. This study aimed to investigate the neutralising capacity and eventual toxic effects of the steel slags on the microbial activity. The results obtained from this study showed equally good or better efficiency compared to the reference material slaked lime. The pyrite biooxidation was found to be 75%–80% for all the experiments where steel slags were used as neutralising agents. Some of the slags used contained potentially toxic elements for the bacteria, such as fluoride, chromium and vanadium, but no negative effect of these elements could be observed on the microbial activity

(Gahan et al., 2009a,b). However, slags originating from stainless-steel production are non-eco-friendly. The neutralisation potential was determined by the amount of by-products required for neutralisation during bioleaching and was found to be very much promising for further scale-up of the study.

7.10 Ashes, Dust and Lime Sludge as Neutralising Agents in Biooxidation of Pyrite

Studies conducted on the utilisation of combustion ashes, EAF dust and lime sludge as neutralising agents with reference to slaked lime showed positive results for a few of the by-products, while others showed an eventually toxic effect for the microorganisms. The ashes used were bioash, waste ash and coal and tyres ash, representing ashes generated from the combustion of biomass, a mixture of wood chips and municipal waste, and a mixture of coal and tyres. The dust used was an EAF dust produced in a scrap-based steel plant, while the sludge used was mesalime produced in a paper and pulp plant. The bioleaching efficiency was similar for all the neutralising agents except waste ash compared to slaked lime. The extent of pyrite oxidation was in the range of 69%–75% for all neutralising agents, except waste ash, which had a pyrite oxidation of 59% (Gahan et al., 2008). The waste ash contained a large number of potentially toxic elements and the chloride concentration of 11% probably had a negative effect as observed on the lower redox potential and pyrite oxidation. The EAF dust has a good potential to be used as a neutralising agent in bioleaching processes for zinc recovery from zinc sulphides, due to the high content of zinc; however, it was recommended that the chlorides present should be removed prior to their use. However, waste ash and coal and tyres ash had lower neutralising capacities while others had a good neutralisation potential. It was suggested that the replacement of lime or limestone with ash, dust or lime sludge can render considerable cost savings to the bioleaching operation.

7.11 Mesalime and EAF Dust as Neutralising Agents in Continuous Biooxidation

Experiments carried out using the refractory gold concentrate in a continuous bioreactor at a retention time of 57 h on a one-stage reactor using mesalime and EAF dust as neutralising agents during biooxidation proved to be equally good compared to the reference slaked lime. The neutralising capacity of EAF dust was lower, while the mesalime was similar to the reference slaked lime. Arsenopyrite oxidation ranged from 85% to 90%, while pyrite oxidation was 63%–74% (Gahan et al., 2010). The gold recoveries were 85–90% for both by-products. The elemental sulphur content in the bioleach residue from the experiment with EAF dust encapsulates the partial content of gold which led to a slight decrease in the gold recovery. Cyanide consumption was relatively high and ranged from 8.1 to 9.2 kg/ton feed after 24 h of cyanidation. The by-products used can be feasible options as the neutralising agent operational units.

7.12 Ladle Slag and EAF Slag as Neutralising Agents in Continuous Biooxidation

EAF slag and ladle slag were used as neutralising agents in the continuous biooxidation of the refractory gold concentrate on a single-stage reactor at a retention time of 56 h (Gahan et al., 2011). The neutralisation capacity was determined as the amount needed, per ton of feed concentrate maintains the desired pH of 1.5 at a steady state. In this study, the reference slaked lime experiment had the highest neutralisation capacity with a requirement of 110 kg/ton feed followed by 152 kg/ton feed for ladle slag and 267 kg/ton feed for EAF slag. Arsenopyrite and pyrite oxidation was good for all the experiments with ladle and EAF slag compared to the reference slaked lime. Gold recoveries ranged between 86% and 89%, which was also reasonably good. Cyanide consumption per ton feed concentrate was 2 and 3 times for ladle slag and EAF slag respectively, compared to the reference slaked lime. The increased cyanide consumption could have been due to the elemental sulphur content which was reasonably high in the bioresidue. The elemental sulphur formed had different reaction kinetics due to thiocyanate formation. The cyanide losses by thiocyanate formation were 16%, 32% and 40% for EAF slag, slaked lime and ladle slag, respectively. The result obtained from ladle slag experiment could be a possible replacement for limestone if it is mixed in suitable proportions to meet the carbon dioxide demand.

7.13 Conclusion

The application of industrial oxidic by-products as neutralising agents in biohydrometallurgy had provided a very good insight into the research and development of the process engineering with respect to operation cost minimisation. A series of research carried out on the feasibilities to use industrial oxidic by-products resulted with a promising conclusion, which suggests that few steel slags and lime sludge together with EAF dust could be potential neutralising agents for the biohydrometallurgical operation both in biooxidation and bioleaching. However, some care should be taken while selecting the by-products to ensure that no toxic elements stay back in the system or go to the effluent pond. Many researchers have also shown that the utilisation of by-products could be a well-managed recycling technology if proper care is taken on the fate of the toxic chemicals in the system, which should be safe enough even after they are processed and disposed to the environment.

References

Arrascue, M.E.L. and van Niekerk, J. Biooxidation of arsenopyrite concentrate using BIOX® process: Industrial experience in Tamboraque, Peru. *Hydrometallurgy* 83, 2006: 90–96.

BachTech Press Release, 2009. http://www.marketwired.com/pressrelease/BacTech-Reports-Improved-Results-on-Tailings-Metallurgical-Test-Work-TSX-V-BM-941556.htm

Bang, S.S., Deshpande, S.S. and Han, K.N. The oxidation of galena using *Thiobacillus ferrooxidans*. *Hydrometallurgy* 37, 1995: 181–192.

Brierley, C.L. and Briggs, A.P. Selection and sizing of biooxidation equipment and circuits. In: A.L. Mular, D.N. Halbe, D.J. Barret (eds.), *Mineral Processing Plant Design, Practice and Control,* Society of Mining Engineers, Littleton, Colorado, 2002: 1540–1568.

Brierley, J. and Brierley, C. Present and future commercial applications of biohydrometallurgy. *Hydrometallurgy* 59, 2001: 233–239.

Brierley, J.A. and Kuhn, M.C. From laboratory to application heap bioleach or not. *Advanced Materials Research* 71–73, 2009: 311–317.

Brierley, C.L., Biohydrometallurgical prospects. *Hydrometallurgy* 104, 2010: 324–328.

Broadhurst, J.L. Neutralisation of arsenic bearing BIOX® liquors. *Minerals Engineering* 7, 1994: 1029–1038.

Bruckard, W.J., Davey, K.J., Rodopoulos, T. et al. Water leaching and magnetic separation for decreasing the chloride level and upgrading the zinc content of EAF steelmaking baghouse dusts. *International Journal of Mineral Processing* 75, 2005: 1–20.

Bustos, S., Castro, S. and Montealegre, R. The Sociedad Minera Pudahuel bacterial thin-layer leaching process at Lo Aguirre. *FEMS Microbiology Reviews* 11, 1993: 231–236.

Cadena, F. and Kirk, T.L. Arsenate precipitation using ferric iron in acidic conditions. New Mexico Water Resource Research Institute Technical Completion Report No. 293, New Mexico State University, Las Cruces, New Mexico, 1995.

Chaudhury, G.R. and Das, R.P. Bacterial leaching complex sulfides of copper, lead and zinc. *International Journal of Mineral Processing* 21, 1987: 57–64.

Chetty, K.R., Marais, H.J. and Kruger, M.J. The importance of pH control in the biooxidation process. In: *Proceedings of the Colloquium Bacterial Oxidation for the Recovery of Metals*, Johannesburg, 2000: 1–12.

China's first bacterial oxidation plant. Interfax Information Services, Interfax China, China Metals, Reports Weekly, Interfax China Limited, China Business News, 2004:1–7, Available from http://www.michelago.com/press/docs/200704_interfax.pdf

Cruells, M., Roca, A. and Núñez, C. Electric arc furnace flue dusts: Characterization and leaching with sulphuric acid. *Hydrometallurgy* 31, 1992: 213–231.

Cunha, M.L., Gahan, C.S., Menad, N. et al. Possibilities to use oxidic by-products for precipitation of Fe/As from leaching solutions for subsequent base metal recovery. *Minerals Engineering* 21, 2008a: 38–47.

Cunha, M.L., Gahan, C.S., Menad, N. et al. Leaching behaviour of industrial oxidic by-products: Possibilities to use as neutralisation agent in bioleaching. *Materials Science Forum* 587–588, 2008b: 748–752.

Demir, I., Hughes, R.E. and DeMaris, P.J. Formation and use of coal combustion residues from three types of power plants burning Illinois coals. *Fuel* 80, 2001: 1659–1673.

Deveci, H., Akcil, A. and Alp, I. Bioleaching of complex zinc sulphides using mesophilic and thermophilic bacteria: Comparative importance of pH and iron. *Hydrometallurgy* 73, 2004: 293–303.

Dew, D.W. Comparison of performance for continuous bio-oxidation of refractory gold ore flotation concentrates. In: T. Vargas, C.A. Jerez, J.V. Wiertz, H. Toledo (eds.), *Proceedings of International Biohydrometallurgy Symposium* IBS-95. Vina del Mar, Chile, 19–22 November, ISBN 956-19-0209-5, vol. 1, 1995: 239–251.

Dew, D.W., Lawson, E.N. and Broadhurst, J.L. *The BIOX@ Process for Biooxidation of Gold Bearing Concentrates*. In: D.E. Rawlings (ed.), Biomining, Springer, Berlin, Heidelberg, New York, 1997: 45–80.

Dresher, W.H. Producing copper nature's way: Bioleaching. *CWD: Innovations* May, 2004: 10.

Engström, F. Mineralogical influence of different cooling conditions on leaching behaviour of steel making slags. Licentiate thesis, Luleå University of Technology, ISSN: 1402-1757/ISRN LTU-LIC—07/58—SE/NR 2007: 58.

Gahan, C.S. Comparative study on different industrial oxidic by-products as neutralising agent in bioleaching. Licentiate thesis, Luleå University of Technology, ISSN: 1402-1757/ISRN LTU-LIC—08/19—SE/NR 2008:19.

Gahan, C.S., Sundkvist, J.E. and Sandström, Å. Use of mesalime and electric arc furnace (EAF) dust as neutralising agents in biooxidation and their effects on gold recovery in subsequent cyanidation. *Minerals Engineering* 23, 2010: 731–738.

Gahan, C.S., Cunha, M.L. and Sandström, Å. Study on the possibilities to use ashes, EAF dust and lime sludge as neutralizing agent in biooxidation. *The Open Mineral Processing Journal* 1, 2008: 26–36.

Gahan, C.S., Cunha, M.L. and Sandström, Å. Comparative study on different steel slags as neutralising agent in bioleaching. *Hydrometallurgy* 95, 2009a: 190–197.

Gahan, C.S., Sundkvist, J.E., Engström, F. and Sandström, Å. Utilisation of steel slags as neutralising agents in biooxidation of a refractory gold concentrate and their influence on the subsequent cyanidation. *Resources, Conservation and Recycling* 55, 2011: 541–547.

Gahan, C.S., Sundkvist, J.E. and Sandström, Å. A study on the toxic effects of chloride on the biooxidation efficiency of pyrite. *Journal of Hazardous Materials* 172, 2009b: 1273–1281.

Garcia Jr., O., Bigham, J.M. and Tuovinen, O.H. Sphalerite oxidation by *Thiobacillus ferrooxidans* and *Thiobacillus thiooxidans*. *Canadian Journal of Microbiology* 41, 1995: 578–584.

Gutknecht, J. and Walter, A. Hydrofluoric and nitric acid transport through lipid bilayer membranes. *Biochimica Biophysica Acta (BBA)—Biomembranes* 644, 1981: 153–156.

Hackl, R.P. and Jones, L. Bacterial sulfur oxidation pathways and their effect on the cyanidation characteristics of biooxidized refractory gold concentrates. In: *Biomine 97, International Biohydrometallurgy Symposium IBS-97*, Biotechnology Comes of Age, Australian Mineral Foundation, ISBN-0-908039-66-2, Sydney, Australia, 1997: M14.2.1–M14.2.10.

Hallberg, R.O., Granhagen, J.R. and Liljemark, A. A fly ash/biosludge dry cover for the mitigation of AMD at the Falun mine. *Chemie der Erde* 65, 2005: 43–63.

Havlík, T., Vidor e Souza, B., Bernardes, A.M. et al. Hydrometallurgical processing of carbon steel EAF dust. *Journal of Hazardous Materials* 135, 2006: 311–318.

Hedin, R.S. and Watzlaf, G.R. Passive treatment of acid mine drainage with limestone. *Journal of Environment Quality* 23, 1994: 1338–1345.

Lawson, E.N. The composition of mixed populations of leaching bacteria active in gold and nickel recovery from sulphide ores. In: *Biomine 97, International Biohydrometallurgy Symposium IBS-97*, Biotechnology Comes of Age, Australian Mineral Foundation, ISBN-0-908039-66-2, Sydney, Australia, 1997: QP4.1–QP4.10.

Leclerc, N., Meux, E. and Lecuire, J.M. Hydrometallurgical recovery of zinc and lead from electric arc furnace dust using mononitrilotriacetate anion and hexahydrated ferric chloride. *Journal of Hazardous Materials* 91, 2002: 257–270.

Lindström, E.B., Sandström, Å. and Sundkvist, J.E. A sequential two-step process using moderately and extremely thermophilic cultures for biooxidation of refractory gold concentrates. *Hydrometallurgy* 71, 2003: 21–30.

Lindström, E.B., Sandström, Å. and Sundkvist, J.E. Two-stage bioleaching of sulphidic material containing arsenic. United States Patent, Patent No. U.S. 6,461,577 B1, 2002.

Liu, X., Petersson, S., and Sandström, Å. Evaluation of process variables in benchscale biooxidation of the Olympias concentrate. *FEMS Microbiology Reviews* 11, 1993: 20–214.

Marais, H. The geologist guide to the BIOX® process. Available from http://www.bioxgf.co.za/content/publications/pdfs/THE%20GEOLOGISTS%-20GUIDE%20TO%20THE%20BIOX%20PROCESS.pdf, 2008.

Miller, P., Jiao, F. and Wang, J. The bacterial oxidation (Bacox) plant at Liazhou, Shandong province, China—The first three years of operation. *Bac-Min Conference*, 8–10 November, Bendigo, Australia, 2004: 1–9.

Norris, P.R. *Acidophile Diversity in Mineral Sulfide Oxidation*. In: D.E. Rawlings, D.B. Johnson (eds.), Biomining, Springer-Verlag, Berlin, Heidelberg, New York, 2007: 199–212.

Okibe, N., Gericke, M., Hallberg, K.B. et al. Enumeration and characterisation of acidophilic microorganisms isolated from a pilot plant stirred-tank biooxidation operation. *Applied Environmental Microbiology* 69, 2003: 1936–1943.

Olson, G.J., Brierley, J.A. and Brierley, C.L. Bioleaching review part B: Progress in bioleaching: Applications of microbial processes by the minerals industries. *Applied Microbiology and Biotechnology* 63, 2004: 249–257.

Pani, C.K., Swain, S., Kar, R.N. et al. Bio-dissolution of zinc sulfide concentrate in 160 l 4-stage continuous bioreactor. *Minerals Engineering* 16, 2003: 1019–1021.

Pettit, L.D. and Powell, K.J. *SC-Database for Windows*. Academic Software Version 5.4. Sourby Old Farm, Timble, Otley, Yorks, UK, 2001.

Plumb, J.J., Hawkes, R.B. and Franzmann, P.D. *The Microbiology of Moderately Thermophilic and Transiently Thermophilic Ore Heaps*. In: D.E. Rawlings, D.B. Johnson (eds.), Biomining, Springer-Verlag, Berlin, Heidelberg, New York, 2007: 217–235.

Rawlings, D.E., Dew, D. and Plessis, C.D. Biomineralization of metal-containing ores and concentrates. *Trends in Biotechnology* 21, 2003: 38–44.

Rawlings, D.E. and Johnson, D.B. The microbiology of biomining: Development and optimization of mineral-oxidizing microbial consortia. *Microbiology* 53, 2007: 315–324.

Rawlings, D.E., Tributsch, H. and Hansford, G.S. Reasons why '*Leptospirillum*'-like species rather than *Thiobacillus ferrooxidans* are the dominant iron-oxidizing bacteria in many commercial processes for the biooxidation of pyrite and related ores. *Microbiology* 145, 1999: 5–13.

Readett, D.J. Biotechnology in the mining industry, Straits Resources Limited and the Industrial Practice of Copper Bioleaching in Heap. *Australasian Biotechnology* 11, 2001: 30–31.

Reith, F., Rogers, S.L., McPhail, D.C. et al. Potential for the utilisation of micro-organisms in gold processing. In: J. Avraamides, G. Deschênes, D. Tucker (eds.), World Gold 2007 by and co-products and the environment, ISBN-9781920806743, Australasian Institute of Mining and Metallurgy, World Gold 2007, Cairns, Australia, 2007: 1–8.

Ribbing, C. Environmentally friendly use of non-coal ashes in Sweden. *Waste Management* 27, 2007: 1428–1435.

Rodríguez, Y., Ballester, A., Blázquez, M.L. et al. New information on the sphalerite bioleaching mechanism at low and high temperature. *Hydrometallurgy* 71, 2003: 57–66.

Rohwerder, T., Gehrke, T., Kinzler, K. et al. Bioleaching review part A: Progress in bioleaching: Fundamentals and mechanisms of bacterial metal sulfide oxidation. *Applied Microbiology and Biotechnology* 63, 2003: 239–248.

Sandström, Å. and Petersson, S. Bioleaching of a complex sulphide ore with moderate thermophilic and extreme thermophilic microorganisms. *Hydrometallurgy* 46, 1997: 181–190.

Sandström, Å., Sundkvist, J.E. and Petersson, S. Bio-oxidation of a complex zinc sulphide ore: A study performed in continuous bench-and pilot scale. In: *Proceedings of Australian Mineral Foundation Conference, Biomine-97*, Glenside, Australia, 1997: M1.1.1–M1.1.11.

Schippers, A. and Sand, W. Bacterial leaching of metal sulfides proceeds by two indirect mechanisms via thiosulfate or via polysulfides and sulfur. *Applied and Environmental Microbiology* 65, 1999: 319–321.

Shi, S., Fang, Z. and Ni, J. Bioleaching of marmatite flotation concentrate with a moderately thermoacidophilic iron-oxidizing bacterial strain. *Minerals Engineering* 18, 2005: 1127–1129.

Shrader, V.J. and Su, S.X. Factors affecting elemental sulfur formation in biooxidized samples: Preliminay studies. In: *Biomine 97, International Biohydrometallurgy Symposium* IBS-97, Biotechnology Comes of Age, Australian Mineral Foundation, ISBN-0-908039-66-2, Sydney, Australia, 1997: M3.3.1–M3.3.10.

Steemson, M.L., Sheehan, G.J., Winborne, D.A. et al. An integrated bioleach/solvent extraction process for zinc metal production form zinc concentrates. PCT World Patent, WO 94/28184, 24 May, 1994.

Stephenson, D. and Kelson, R. Wiluna BIOX plant—Expansion and new developments. *Conference Proceedings of IBS-BIOMINE '97*, 4–6 August, Sydney, Australia, 1997: M4.1.1–M4.1.8.

Sundkvist, J.E., Gahan, C.S. and Sandström, Å. Modeling of ferrous iron oxidation by *Leptospirillum ferrooxidans*-dominated chemostat culture. *Biotechnology and Bioengineering* 99, 2008: 378–389.

Sundkvist, J.E., Sandström, Å., Gunneriusson, L. et al. Fluorine toxicity in bioleaching systems. *Proceedings of the XVI International Biohydrometallurgy Symposium*, Cape Town, South Africa, ISBN: 1-920051-17-1, 2005: 19–28.

Sweet, P.C. Virginias lime industry. *Quarterly Journal of Virginia*, Division of Mineral Resources, Charlottesville, Virginia, ISSN 0042-6652. 32, 1986: 33–43.

Tham, G. and Ifwer, K. Utilization of ashes as construction materials in landfills. Värmeforsk report 966, March 2006 (in Swedish).

The European Slag Association (EUROSLAG). Legal status of slags. Position paper. Available in http://www.euroslag.org/media/Position_paper_Jan_2006.pdf, March 2006.

U.S. EPA SW 846, method 1311; Title 40-261.24 of the Code of Federal Regulation, in Federal Register, 51, 1986: 21648–21693.

van Aswegen, P.C., Godfrey, M.W., Miller, D.M. et al. Developments and innovations in bacterial oxidation of refractory ores. *Minerals and Metallurgical Processing* 8, 1991: 188–191.

van Aswegen, P.C. and Marais, H.J. Advances in application of the BIOX® process for the refractory gold ores. *Minerals and Metallurgical Processing* 16, 1999: 61–68.

van Aswegen, P.C. and van Niekerk, J. New developments in the bacterial oxidation technology to enhance the efficiency of the BIOX process. In: *Proceedings of BacMin Conference*, Bendigo, Victoria, Australia, 2004: 181–189.

Van Aswegen, P.C., Van Niekerk, J. and Olivier, W. *The BIOX^{TM} Process for the Treatment of Refractory Gold Concentrates*. In: D.E. Rawlings, D.B. Johnson, (eds.), Biomining, ISBN-10 3-540-34909-X, Springer-Verlag, Berlin, Heidelberg, New York, 2007: 1–33.

Van Niekerk, J. Recent advances in the BIOX® technology. *Advanced Materials Research* 71–73, 2009: 465–468. www.talvivaara.com

Ziemkiewicz, P. Steel slag: Applications for AMD control. *Proceedings of the Conference on Hazardous Waste Research*, Snowbird, Utah, 1998: 44–62.

8

Application of Microbes for Metal Extraction from Mining Wastes

Sandeep Panda, Srabani Mishra, Nilotpala Pradhan, Umaballav
Mohaptra, Lala Behari Sukla and Barada Kanta Mishra

CONTENTS

8.1 Introduction

A variety of low-grade ore deposits such as copper, nickel, gold, uranium and others available throughout the world are dumped as wastes due to the unsuitability of conventional methods of metal recovery. As a result of the gradual depletion of the oxide as well as the high-grade ore deposits, particular concern is being raised to process the low-grade/run-off mine ores. Low-grade ore deposits are otherwise treated as waste due to the associated economic constraints in processing the conventional crush-grind and float technology that is adopted for the high-grade ores. On the other hand, dumping such material leads to several environmental issues. Since the metal content from such ores is quite low for processing through the conventional methods, the use of certain microorganisms have gained importance for the processing of low-grade ores and other industrial wastes (Baba et al. 2011; Panda et al. 2013a). The method of application of microorganisms to recover metal values is referred to as 'bioleaching' and the area of application technology is 'mineral biotechnology'.

Microorganisms involved in the leaching process have a number of characteristics in common that make them suitable for mineral solubilisation. These microorganisms act as catalysts to promote and enhance the dissolution rates of respective metals from ores and other mining wastes (Esther et al. 2013). Bioleaching offers several advantages to process the low and run-off mine ores and are considered as an eco-friendly and economic technology (Munoz et al. 2007; Panda et al. 2013a; Watling 2006). The most preferred methodology for recovery of metal values from low-grade ores through bioleaching is the heap bioleaching technology that offers several advantages. Furthermore, heap bioleaching coupled with hydrometallurgical methods of metal recovery such as those of the solvent extraction (SX) and electrowinning (EW) are gaining importance (Panda et al. 2012a). This method of bio-hydrometallurgical processing accounts for 20%–25% of copper production globally (Panda et al. 2014a and references therein).

Considering the importance of microbes in mining waste treatment, the chapter reviews and outlines some of the basic fundamentals of microbial processing of low-grade ores with special reference to copper ore. In addition, various methods of the microbial application toward recovery of metal values from the low-grade ores are discussed to give better insights into the mode of treatment.

8.2 Microorganisms in Waste Treatment and Their Area of Application

Microorganisms employed in leaching of metals from low grade and other industrial wastes have a number of characteristics in common that make them suitable for mineral solubilisation. The microorganisms capable of producing ferric iron and sulphuric acid are generally used for mineral bio-oxidation and bioleaching (Munoz et al. 2007; Sand et al. 2001). Microorganisms such as those of the iron oxidising chemolithotrophic bacteria possess the capability to convert soluble ferrous iron into ferric iron (Panda et al. 2014b). Ferric iron is a powerful oxidising agent that further gives an oxidative attack on the mineral moiety to release metal values.

On the other hand, sulphur oxidising chemolithotrophs are responsible to produce the necessary sulphuric acid that is used in bioleaching. In addition to the above, these microbes have the capability to grow autotrophically by fixing CO_2 from the atmosphere. Yet another advantageous characteristic of these acidophilic chemolithotrophs is the higher metal tolerance capability that enables them to survive under high metallic stress conditions (Hoque and Philip 2011). This unique chemolithoautotrophic metabolism seen in acidophiles makes the organisms industrially important. The following section outlines some of the microorganisms and their mode of operation.

8.2.1 Autotrophic

The microorganisms driving the mineral oxidation processes are autotrophic in character. They use CO_2 from the atmosphere to obtain their carbon for cell mass synthesis. To name a few, the chemolithoautotrophic metal solubilising bacteria such as *Acidithiobacillus ferrooxidans, Leptospirillum ferrooxidans* and *Leptospirillum ferriphilum* are iron-oxidising in nature while *Acidithiobacillus thiooxidans* and *Acidithiobacillus caldus* are sulphur-oxidising (Das et al. 1999). Bacteria such as the *Acidothiobacillus ferroxidans* and *A. thiooxidans* fix CO_2 by Calvin reductive pentose phosphate cycle using enzyme ribulose 1,5-bisphosphate carboxylase (Bergamo et al. 2004; Valdes et al. 2010). On the other hand, bacteria such as *L. ferriphilum* follow the reverse tricarboxylic acid (TCA) cycle for CO_2 assimilation (Valdes et al. 2010). *A. caldus* is believed to follow the Calvin or an incomplete TCA cycle (Valdes et al. 2010; Zhou et al. 2007) while least information is available regarding *L. ferrooxidans.*

The genus *Acidithiobacillus* and *Leptospirillum* represent a versatile group of chemolithoautotrophic microorganisms. *A. ferrooxidans* is a rod-shaped (usually occurs as single or in pairs), non-spore forming, Gram-negative, motile and single-pole flagellated bacterium. Despite its strictly aerobic nature, *A. ferrooxidans* has a unique character to utilise elemental sulphur or metal sulphides under anoxic conditions thereby facilitating microbial reduction reactions (Osorio et al. 2013). In an aerobic mode of respiration or, say, biooxidation reaction, this bacterium utilises ferrous iron as the electron donor and oxygen as the terminal electron acceptor. However, under oxygen-deficient conditions, elemental sulphur acts as the electron donor and ferric iron serves as the final electron acceptor (Osorio et al. 2013), thereby facilitating microbial reduction reactions. On the other hand, bacteria belonging to genus *Leptospirillum* are Gram-negative, vibrio or spiral shaped cells (Johnson 2001) that are obligately chemolithotrophic acidophilic microorganisms growing optimally in inorganic media within a pH range 1.3–2.0. Since, the only donor source of the electron being ferrous iron, these are among the most metabolically restricted microorganisms known. Such versatility makes these microorganisms most suitable for experimental studies and is widely applied at an industrial scale in consortia (Panda et al. 2012a).

8.2.2 Heterotrophic

In addition to the autotrophs, a group of heterotrophic microorganisms (bacteria and fungi) are also a part of the microbial bioleaching community. These microorganisms may not be directly efficient as the autotrophs towards leaching but can contribute a portion. Several heterotrophs contribute to metal solubilisation through the secretion of certain organic acids such as citrate, gluconate, oxalate or succinate. The source of carbon for the heterotrophs comes from the extracellular metabolites and cell lysates from autotrophs. This in turn results in the removal of carbon excess (which can be inhibitory) and stimulate the growth and iron oxidation of autotroph such as *A. ferrooxidans*. In certain microorganisms such as *Acidiphilium acidophilum* and *Acidimicrobium ferrooxidans*, an interesting feature is seen such as the capability to grow autotrophically with sulphur and iron (II) compounds, heterotrophically with glucose or yeast extract and mixotrophically with all of these substrates (Clark and Norris 1996; Hiraishi et al. 1998).

8.2.3 Thermophilic

Industrial applications of bioleaching usually prefer the use meso-acidophilic microorganisms (Panda et al. 2013b). Owing to the higher residence time, the process cannot be considered economically viable. Many researchers have thus proposed the use of thermo-acidophiles for faster recovery of metal values from low-grade ores (Liang et al. 2012; Zeng et al. 2010; Zhou et al. 2009). The thermophilic archea/bacteria grow at an optimal temperature between 60°C and 80°C and possess remarkable characteristics that make them especially suitable for bioleaching process. As a consequence of higher reaction temperature, the microorganisms increase the dissolution rates of some metallic sulphides such as molybdenite and chalcopyrite that are otherwise difficult to process using meso-acdiophiles.

A similar trend can be found in the bioleaching processes using certain type of microbes that operate at ambient 40°C to those operating within the temperature ranges 45–55°C and 75–80°C, irrespective of the mineral under attack. Generally, the Gram-positive leaching bacteria are moderately thermophilic in nature and are members of the genera *Acidimicrobium*, *Ferromicrobium* and *Sulfobacillus* (Clark and Norris 1996; Norris et al. 1996). Over recent years, some new species of thermophilic microorganisms such as *Metallosphaera sedula*, *Sulfolobus metallicus* or *Sulfurococcus* have been identified to have greater capacity to dissolve metallic sulphides (Huber et al. 1989). Archaebacteria involved in bioleaching belong to the group Sulfolobales (a group of extremely thermophilic, sulphur- and iron- (II) ion-oxidisers) including genera such as *Sulfolobus*, *Acidianus*, *Metallosphaera* and *Sulfurisphaera* (Fife et al. 2000; Fuchs et al. 1995; 1996; Kurosawa et al. 1998; Norris et al. 2000; Valdes et al. 2010). Certain archaea belonging to the genus

Acidianus such as the *Acidianus ambivalensi* are capable to grow on reduced sulphur and low pH conditions at high temperatures (Urich et al. 2004). However, the contribution of these organisms to industrial scale bioleaching is not well established. On the other hand, bioleaching tests carried out with bacteria of the genus *Sulfolobus* have provided high copper extraction yields from difficult-to-treat ore like chalcopyrite.

Owing to the importance of all the groups of microorganisms involved in bioleaching as discussed above, a summary of some of the microbial applications in bio-hydrometallurgical is shown in Table 8.1.

8.3 Mechanism of Microbial Action

To understand the underlying mechanisms of microbial action, Silverman and Ehrlich (1964) initiated attempts towards explaining a possible reaction mechanism for bioleaching. As an outcome of their attempt, two possible bioleaching routes were proposed that were named as 'Direct' and 'Indirect' bioleaching. According to a review performed by Tributsch (2001) on some of the existing proposals for chalcopyrite bioleaching, he proposed three different bioleaching mechanisms namely:

1. *Indirect bioleaching:* This mechanism points to the oxidative attack of the generated Fe^{3+} iron as a result of Fe^{2+} oxidation by the microorganisms in the leaching media to dissolute copper from low-grade ore.

2. *Contact bioleaching:* This mechanism demonstrates the direct attachment of the bacteria to the ore surfaces through a layer of extracellular polymeric substances (EPS) thereby preparing the medium for facilitating an electrochemical dissolution approach from the mineral surface.

3. *Cooperative bioleaching:* This mechanism combines both the indirect and direct or contact mechanism for microbial action on mineral surface. Both microorganisms directly attached to the ore and free bacteria in solution simultaneously affect the metal dissolution process. The attached microorganisms release certain chemicals to the solution that provide an energy source for the microbes free in solution.

Figure 8.1 shows the action of three different mechanisms proposed by Tributsch (2001).

With the basic knowledge on the mechanism of bioleaching, the advent of research to provide a clearer picture, the indirect mechanism via the

TABLE 8.1

Microorganisms in Biohydrometallurgy and Their Area of Application

Microorganisms	Process/Conditions	Area of Application	References
Bacteria of the genera *Thiobacillus* and *Leptospirillum*	• Oxidation of sulphide minerals, (S° and Fe(II)) • pH: 1.4–3.5 • Temp: 5°–35°C	• Dump or heap leaching, underground mining (in situ leaching) and tank leaching of metals from sulphidic and mixed-type ores • Concentrates from the wastes of pyrometallurgical industries • Desulphurisation of coal	Bosecker (2006), Mishra et al. (2014), Panda et al. (2012a)
Group of thermophilic bacteria	• Oxidation of sulphide minerals (S° and Fe(II)) • pH 1.1–3.5 • Temp: 30°–55°C	• Dump or heap leaching, underground mining (in situ leaching) and tank leaching of metals from sulphidic and mixed ores • Concentrates from the wastes of pyrometallurgical industries • Desulphurisation of coal.	Bosecker (2006), Liang et al. (2012), Mishra et al. (2014), Zeng et al. (2010), Zhou et al. (2009)
Organotrophic microorganisms such as fungi, bacteria, yeast, algae and their metabolites	• Destruction of sulphide minerals and aluminium silicates • Solubilisation of gold and biosorption of metal.	• Extraction of metals from carbonates and silicate-based ores • Leaching of gold	Bosecker (2006), Jena et al. (2012)

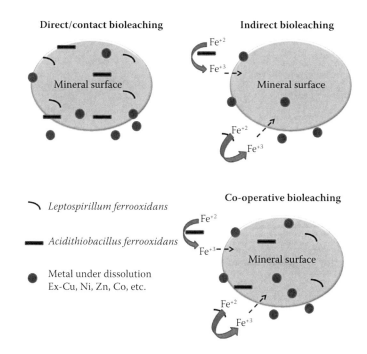

FIGURE 8.1
Mechanisms of bioleaching through direct/contact; indirect and co-operative means.

polysulphide and elemental sulphur production were proposed by Sand et al. (1995) and Schippers and Sand (1999). This is regarded as the most coherent attempt to explain the true bioleaching mechanism. The mechanism entails several steps that start with the dissolution of mineral sulphides with the protons present in the bioleaching medium, which is generated by the microorganisms through mineral hydrolysis.

As the meso-acidophilic, moderately thermo-acidophilic and/or extremo-philic (extreme thermo-acidophilic) bacteria attach to the ore surfaces, they start to catalyse the oxidation of ferrous to ferric ions using oxygen as an oxidant. This in turn results in sustaining the high oxidation potentials required to break down the sulphidic minerals.

The generalised biologically catalysed ferrous ion oxidation is shown in Equation 8.1.

$$4Fe^{2+} + 4H^+ + O_2 \rightarrow 4Fe^{3+} + 2H_2O \tag{8.1}$$

Simultaneously, microbial strains capable of oxidising sulphur catalyse the oxidation of sulphur to produce sulphuric acid that maintains the acidity of the medium required for metal dissolution and growth of the acidophiles.

The generalised biologically catalysed sulphur oxidation is shown in as Equation 8.2.

$$2S + 2H_2O + 3\,O_2 \rightarrow 2H_2SO_4 \tag{8.2}$$

The proposed mechanism of microbial action is based on the interdisciplinary studies that involved mineralogical and semiconducting properties of ore such as chalcopyrite ($CuFeS_2$), sphalerite (ZnS), galena (PbS), hauerite (MnS_2), orpiment (As_2S_3) and realgar (As_4S_4). The valance bonds of all these metal sulphides are formed by the atomic orbitals of both metallic and sulphur atoms.

8.4 Methods of Microbial Leaching

Microbial leaching of metals follows three basic methods. For an easy understanding of the methods and their relations to each other, a simplified flow diagram is outlined as Figure 8.2.

The details of the different microbial methods are discussed below.

8.4.1 Shake Flask Leaching

Shake flask leaching refers to the leaching or extraction of metals in shaking flask reactors (Figure 8.3) at lab scale, and hence referred to as 'lab scale

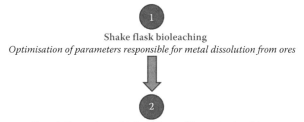

1

Shake flask bioleaching
Optimisation of parameters responsible for metal dissolution from ores

2

Percolation column bioleaching or bioreactor leaching
Optimisation of bench scale conditions responsible for metal dissolution from ores

3

Heap bioleaching or agitation (stirred) tank reactor leaching
Testing the feasibility and economics of bioleaching at pilot scale towards a commercial venture

FIGURE 8.2
Step-wise operation of three basic methods adopted for bioleaching experiments.

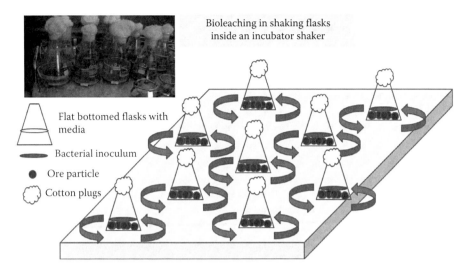

FIGURE 8.3
(**See colour insert.**) Showing a pictorial operation of shake flask leaching. A group of flat-bottomed flasks with bacterial inoculum and ore placed at top of a shaker. Arrow in the picture indicates direction of movement of shaking flasks.

bioleaching'. Some researchers have also reported a similar kind of method, 'roll bottles' (Robertson and Van Staden 2009), the objective being the same as shaking flasks. Shaking flask bioleaching is performed to study the amenability of a material towards bacterial leaching in view of optimising various physico-chemical as well as biological process parameters. Laboratory-scale bioleaching provides better insight into the scaling of the process in terms of studying its feasibility. The parameters that require optimisation are listed in Table 8.2.

8.4.2 Percolation Column Leaching

On the basis of the optimised parameters at lab scale, percolation column bioleaching is operated at a bench scale which further provides better

TABLE 8.2

Parameters Considered for Optimisation of Microbial Leaching of Metals

Nature	Parameters
Physico-chemical parameters	pH, temperature, redox potential, O_2 and CO_2 content, nutrient availability, Fe (III) concentration, etc.
Biological parameters of a leaching environment	Metal tolerance capacity, adaptation abilities of microorganisms to ore and other environmental conditions.
Ore/waste material	Ore composition, Type of mineral, size fraction (grain size), surface area, porosity, galvanic interactions, etc.

insights into the metal dissolution for scaling up to pilot scale. Column bio-leaching studies ensure optimization of bacterial lixiviant (bacterial solution that enables leaching of metals from ores) flow rates for dissolution of metal values (Panda et al., 2014a). In addition, this method provides a clear picture of the acid consumption patterns (Panda et al. 2013b), which helps in estimat-ing the costs for a scaled-up trial such as heap bioleaching (discussed in the next Section 8.3).

A column is a specially designed apparatus in which the low-grade ore or other industrial waste material of a particular size fraction of spe-cific weight per volume of microbial consortium is subjected to microbial leaching reactions. A percolation column is designed according to spe-cific measurements in height and diameter to accommodate a particular weight of the sample. The weight of the material is calculated based on the optimised pulp density (w/v) during lab-scale studies. Similarly, the total volume of bacterial lixiviant is likewise calculated with the optimised inoculum concentration. The bottom of the column is tightly fitted with a rubber cork that provides support for the material inside the column, hav-ing a hole in its middle built-in with an outlet pipe. A layer of glass wool is supplied at the top of the rubber cork that acts as a filter for any fine particles oozing out from the column outlet. The top surface of the column acts as the inlet for the bacterial lixiviant and as well enables diffusion of atmospheric O_2 and CO_2 into the column. The schematic diagram of a designed column bioleaching setup is shown in Figure 8.4a. An example of the experimental column bioleaching setup is shown in Figure 8.4b.

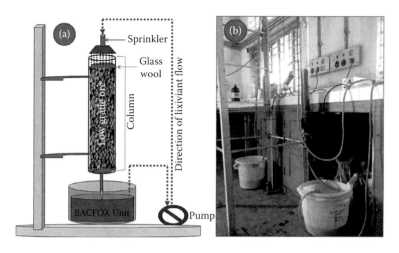

FIGURE 8.4
(**See colour insert.**) (a) Schematic of a percolation column bioleaching setup. (b) Experimental setup of a percolation column bioleaching for recovery of copper from low-grade ore. BACFOX Unit—Bacterial Film Oxidation Unit or Bacterial Growth Tank.

Preparation of the bacterial lixiviant is carried out in specially designed reactors from where the solution is pumped into the column through the inlet provided at the top. With gravity, the bacterial lixiviant slowly percolates through the column, filling up all the voids, facilitating the necessary bio-chemical reactions to occur. This in turn enables the dissolution of the respective metal values into the solution, which is collected at the bottom. This enriched solution is further subjected to metal analysis through atomic absorption spectroscopy in order to notice the recovery rates of metal from the ore.

Apart from percolation column leaching, stirred-tank reactors are also used in bioleaching. It can also be pointed out that bioleach processes can be categorised depending upon the type of resources to be processed. Based on the engineering aspects, biomining has been categorised as irrigation based (heap and/or dump leaching, *in situ* leaching) and stirred-tank or agitation-based leaching. Since percolation columns are the best simulation of heap leaching, materials such as waste rock, low-grade ores (oxides, primary and secondary sulphides or concentrator tailings) are preferred through this route. However, intermediate- to high-grade ores, chalcopyrite concentrates and gold ores are preferred through agitation or stirred-tank leaching. The material in this process is deposited in a tank and leaching is performed by mechanical agitation and aeration. In addition, thermophilic bacterial leaching can also be performed through stirred-tank leaching.

8.4.3 Heap Leaching

With all the optimised conditions at the bench scale, pilot-scale practices are initiated. An example of a pilot scale as well as commercial scale application is heap bioleaching. Heaps are a specially designed or engineered system that enables a particular weight of material (in tonnes) to be stacked in the form of a dump with specific dimensions of length, width, height and slant over a water/acid proof membrane. The heap is slightly slanted to enable the metal-rich effluent from the heap to be completely drained out to a reservoir, which acts as the storehouse of bacterial lixiviant and also as a bioreactor (Panda et al. 2012a).

The concept of bioleaching in heaps came from the earliest technologies that were very basic. These early technologies involved stacking the low or run-off mine ores in the form of a dump that was very irregular in shape and size or dimensions of length, breadth and height. Several boulder mounds of such waste material generated during mining excavations were dumped on a daily basis, which ultimately resulted in piling up the material over several hectares of land and a few meters in height. With time, during rainy seasons metals from the dumped material drained out as a result of microbial activity, which in turn was the cause of contaminating the ground water and soil. The result was also called acid mine drainage (AMD). The associated environmental problems with dumping called for the treatment of such

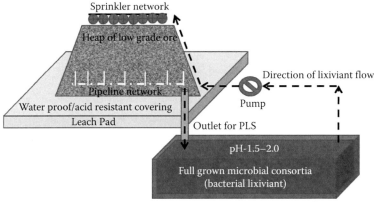

FIGURE 8.5
(**See colour insert.**) Schematic diagram of a heap bioleaching process.

wastes to recover metal values and as well check environmental pollution. The basic approach was to treat these huge stockpiles with dilute sulphuric acid that encouraged the growth of mineral oxidising acidophiles, primarily the meso-acidophiles contributing towards metal dissolution. However, the irregularity of the dumps in terms of dimensions and uneven sizes resulted in slower recovery of metals over years. To overcome such a problem, the efficiency of engineering aspects came into picture that modified the concept of dumps with heaps of proper dimensions, good reservoir facilities and pipeline networks. The outcome was exciting, resulting in enhanced recovery rates of various metals. The schematic of a heap bioleaching operation of a low-grade ore is pictorially shown in Figure 8.5.

8.5 Downstream Methods of Metal Extraction

As discussed in the previous sections, different methods are applied to optimise and scale up a bioleaching process. As a result of the (bio)-chemical reactions (discussed in Section 8.3), respective metal values dissolute into the bacterial solution. This metal-rich effluent is known as pregnant leach solution (PLS), which is subjected to two basic hydrometallurgical processing routes to recover pure metal. The methods are described below:

8.5.1 Solvent Extraction

The earliest used technologies preferred the process of cementation with scrap iron to recover copper from copper pregnant leach solutions from dumps/heaps and so on. The generalised chemical reaction is shown as Equation 8.3.

$$CuSO_4 + Fe^0 \rightarrow Cu^0 + FeSO_4 \tag{8.3}$$

The process had several drawbacks such as the copper metal was not absolutely pure and discarding of the iron sulphate solution resulted in several environmental issues. The use of cementation was discarded due to the following reasons:

a. The Fe^0 reagent used in the process is not recoverable.

b. In the presence of Fe^{3+} ions the consumption of Fe^0 consumption is increased.

c. The efficiency of the leach acid used during leaching is destroyed, which results in higher consumption of Fe^0.

d. The $FeSO_4$ generated is an additional environmental hazard.

e. Impure copper product which is typically about 7% pure.

f. The operation is labour intensive, so that alternative systems are desirable.

Later developments in hydrometallurgical processing resulted in the use of certain solvents to extract respective metals from the PLS with absolute efficiency and zero issues (Panda et al. 2012b, c). A typical process flow sheet of the SX of metal-rich bioleach liquor (PLS) is shown in Figure 8.6. SX units

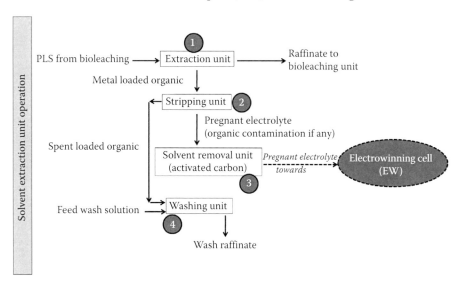

FIGURE 8.6
A typical SX process flow sheet for treatment of bioleach liquor (PLS).

basically consist of three different units of operation namely, the extraction unit, the stripping unit and washing unit. At first, the metal-rich bioleach liquor is subjected to the extraction unit with a desired concentration of organic solvent. The solvent is so chosen that it is very specific to the metal of interest. With time, all the metal in the PLS is transferred to the organic phase and the remaining aqueous phase is sent back to the bioleaching circuit. The metal devoid aqueous phase that is sent back to the bioleaching circuit is called raffinate. The metal-loaded organic phase is then subject to stripping of metal in the stripping unit. The pregnant electrolyte from the stripping unit may contain some contamination of the organic solvent, and hence it is further subject to organic removal through the use of activated carbon. Activated carbon adsorbs any organic in the pregnant electrolyte and produces an organic free electrolyte that is subjected to the EW cell for metal deposition.

8.5.2 Electrowinning

Electrowinning, also known an electroextraction, is a deposition phenomenon in which the organic free metal pregnant electrolyte from the SX unit is subjected to the application of electric current so that deposition of the desired metal from the aqueous solution is achieved. The current is passed from an inter anode through the pregnant electrolyte so that the metal is extracted and finally deposited on the cathode. The most common application of EW are depositions of various metal such as copper, nickel, lead, zinc, gold, silver, aluminium and so forth (Barbosa et al. 2001; Chatelut et al. 2000; Exposito et al. 2000; Haarberg et al. 2010; Lupi et al. 2006; Panda et al. 2014a; Saba and Elsherief 2000).

A generalised process flow sheet of biohydrometallurgical process for treatment of low-grade copper ores through heap bioleaching is shown in Figure 8.7.

8.6 Advantages and Disadvantages of Microbial Processing

Microbial processing of metal values from low-grade ores offers several advantages as well as disadvantages.

The advantages of the process are discussed below:

1. *Low-cost technology:* Microbial leaching is a low-cost technology. A heap bioleaching plant can be constructed at the site of a low-grade waste dump. Hence, transportation cost can be minimised. In addition, the setup does not need a sophisticated operation unit that adds to its benefit by avoiding large numbers of highly skilled manpower. The reagents used in the process are also cheap.

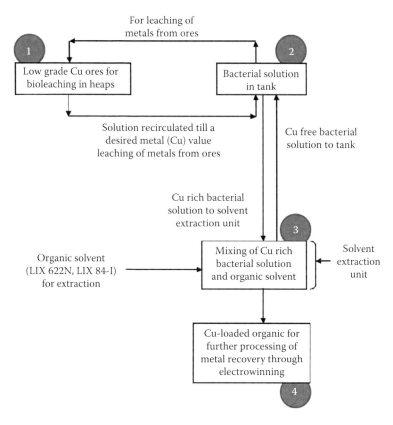

FIGURE 8.7
A generalised description of the biohydrometallurgical process of a typical Cu heap leach operation, processing of heap PLS through SX and EW for recovery of copper.

2. *Easy operation:* The bioleaching setup uses simple and easy method of operation.

3. *Quick start-up times:* Since, the method is simple and can be commissioned at the site of a dump; the processing starting from bioleaching to metal recovery does not need extensive start-up times.

4. *Uses indigenous microorganisms:* Another advantage of the process is the use of native microorganisms from the mine site. Indigenous microorganisms are well versed with the environmental conditions that enable them to the best agents of metal dissolution from low-grade ores.

5. *Eco-friendly:* The process uses less or no environmental hazardous chemicals that can in any way cause pollution. Another added advantage is that the entire system operates in a continuous circuit, which avoids dumping, or piling or reservoiring any end products. Also the microorganisms operating in the system grow at highly

acidic conditions; hence, the risk of contamination is quite low. Further, the microorganisms are non-pathogenic for humans and other life forms.

In addition to the several benefits of microbial processing at large scale, there are also several disadvantages of the method. The disadvantages are discussed below:

1. *Slower metal recoveries:* Unlike chemical leaching processes, bioleaching yields are very slow. It takes a few months to years to complete the entire process. Also, in order to have a higher yield or recovery of metal values, proper care needs to be taken regarding the various physico-chemical as well as biological parameters. Improper maintenance can lead to slower recoveries and deposition of passivation layers (Panda et al. 2013b).

2. *Environmental calamities:* At a large scale where heap bioleaching is carried out directly under atmospheric conditions, any changes to the climatic conditions such as heavy rainfall or sudden decrease or increase in temperatures can affect the bioleaching performance. An example of heap bioleaching of chalcopyrite with discussions to the effect of climatic changes on copper recovery has been discussed (Panda et al. 2012a).

8.7 Future Prospects

The future of microbial processing of various industrial wastes for the recovery of various metal values is quite encouraging. With the rapid depletion of high-grade ores, it has become evitable to process the low-grade ores to recovery metal values. So the future of a mining industry will be the processing of low-grade ores in an eco-friendly and economic way. Slowly, the method is gaining importance in the field of waste treatment (Bull 2001). Since microorganisms are the key agents responsible for metal dissolution, many attempts have been made to understand the underlying mechanisms of several groups of microorganisms. It has now been confirmed that a mixed bacterial consortium is more effective than a pure culture (Fu et al. 2008) for bioleaching reactions. Advent of genetically engineered microorganisms in consortia is also being proposed to enhance the capabilities of bioleaching and bioremediation (Brune and Bayer 2012). There has not been much study emphasised toward a direct prediction on development of a novel microbial consortium that can efficiently act on metal dissolution. Very recently, based on bioinformatics applications, a web server has been proposed to predict novel microorganisms to be used in bioleaching (Parida et al. 2014). The web server directly predicts the key proteins involved in iron and/or sulphur oxidation by several groups of bioleaching bacteria that can be very helpful toward development of a novel microbial consortia. Hence, the applications of such bioinformatics predictions can help shape the outlook of the bioleaching processing.

8.8 Conclusion

The current scenario of bioleaching is quite encouraging. In the next few years, bioleaching is expected to be a widely applied microbial technology to recover metal values in a more economic and eco-friendly way. The use of a percolation column and heap bioleaching is expected to be widely used to recover metal values from low-grade ores. Heap bioleaching is the most preferred technology today. Understanding a heap and its behaviour under different environmental conditions can open up new doors toward a more advanced industrial application of the technology. With the progress of the technology and a deeper understanding of the microbial aspects, it is expected that many new commercial plants will be installed and the mineral processing sector will see a more advanced and green mineral biotechnology.

Acknowledgements

The first two authors, Sandeep Panda and Srabani Mishra, are thankful to the Council of Scientific and Industrial Research (CSIR), New Delhi for the award of Senior Research Fellowships. The authors are also grateful to the director, CSIR-IMMT for granting permission to publish this chapter.

References

Baba, A.A., Adekola, F.A., Atata, R.F., Ahmed, R.N., Panda, S. Bioleaching of Zn (II) and Pb (II) from Nigerian sphalerite and galena ores by a mixed culture of acidophilic bacteria. *Transaction of Non-ferrous Metal Societies of China* 21, 2011: 2535–41.

Barbosa, L.A.D., Sobral, L.G.S., Dutra, A.J.B. Gold electrowinning from diluted cyanide liquors: Performance evaluation of different reaction systems. *Minerals Engineering* 14, 2001: 963–74.

Bergamo, R.F., Novo, M.T., Veríssimo, R.V., Paulino, L.C., Stoppe, N.C., Sato M.I., Manfio, G.P., Prado, P.I., Garcia, O., Ottoboni, L.M. Differentiation of *Acidithiobacillus ferrooxidans* and *A. thiooxidans* strains based on 16S–23S rDNA spacer polymorphism analysis. *Research in Microbiology* 155, 2004: 559–67.

Bosecker, K. Bioleaching: Metal solubilization by microorganisms. *FEMS Microbiology Reviews* 20, 2006: 59104.

Brune, K.D. and Bayer, T.S. Engineering microbial consortia to enhance biominig and bioremediation. *Frontiers in Microbiology* 3, 2012: 203.

Bull, A.T. Biotechnology for industrial sustainability. *The Korean Journal of Chemical Engineering* 18, 2001: 137–48.

Chatelut, M., Gobert, E., Vittori, O. Silver electrowinning from photographic fixing solutions using zirconium cathode. *Hydrometallurgy* 54, 2000: 79–90.

Clark, D.A. and Norris, P.R. *Acidimicrobium ferrooxidans* gen. nov., sp. nov.: Mixed-culture ferrous iron oxidation with *Sulfobacillus* species. *Microbiology* 142, 1996: 785–90.

Das, T., Subbanna, A., Chaudhury, G.R. Factors affecting bioleaching kinetics of sulfide ores using acidophilic micro-organisms. *Biometals* 12, 1999: 1–10.

Esther, J., Panda, S., Behera, S.K., Sukla, L.B., Pradhan, N., Mishra, B.K. Effect of dissimilatory Fe (III) reducers on bioreduction and nickel-cobalt recovery from Sukinda chromite-overburden. *Bioresource Technology* 146, 2013: 762–66.

Exposito, E., Gonzalez-Garcia, J., Bonete, P., Montiel, V., Aldaz, A. Lead electrowinning in a fluoborate medium. Use of hydrogen diffusion anodes. *Journal of Power Sources* 87, 2000: 137–43.

Fife, D.J., Bruhn, D.F., Miller, K.S., Stoner, D.L. Evaluation of a fluorescent lectin-based staining technique for some acidophilic mining bacteria. *Applied and Environmental Microbiology* 66, 2000: 2208–10.

Fu, B., Zhou, H., Zhang, R., Qiu, G. Bioleaching of chalcopyrite by pure and mixed cultures of *Acidithiobacillus* spp. and *Leptospirillum ferriphilum*. *International Biodeterioration and Biodegradation* 62, 2008: 109–15.

Fuchs, T., Huber, H., Burggraf, S., Stetter, K.O. 16S rDNA-based phylogeny of the archaeal order Sulfolobales and reclassification of *Desulfurolobus ambivalens* as *Acidianus ambivalens* comb. nov. *Systematic and Applied Microbiology* 19, 1996: 56–60.

Fuchs, T., Huber, H. Teiner, K., Burggraf, S., Stetter, K.O. *Metallosphaera prunae*, sp. nov., a novel metal-mobilizing, thermoacidophilic archaeum, isolated from a uranium mine in Germany. *Systematic and Applied Microbiology* 18, 1995: 560–66.

Haarberg, G.M., Kvalheim, E., Ratvik, A.P., Xiao, S.J., Mokkelbost, T. Depolarised gas anodes for aluminium electrowinning. *Transactions of Nonferrous Metals Society of China* 20, 2010: 2152–54.

Hiraishi, A., Nagashima, K.V., Matsuura, K. et al. Phylogeny and photosynthetic features of *Thiobacillus acidophilus* and related acidophilic bacteria: Its transfer to the genus *Acidiphilium* as *Acidiphilium acidophilum* comb. nov. *International Journal of Systematic and Evolutionary Microbiology* 48, 1998: 1389–98.

Hoque, M.E. and Philip, O.J. Biotechnological recovery of heavy metals from secondary sources—An overview. *Materials Science and Engineering C* 31, 2011: 57–66.

Huber, G., Spinnler, C., Gambacorta, A., Stetter, K.O. *Metallosphaera sedula* gen. and sp. nov. represents a new genus of aerobic, metal-mobilizing, thermoacidophilic archaebacteria. *Systematic and Applied Microbiology* 12, 1989: 38–47.

Jena, P.K, Mishra, C.S.K., Behera, D.K., Mishra, S., Sukla, L.B. Dissolution of heavy metals from (ESP) dust of a coal based sponge iron plant by fungal leaching. *African Journal of Environmental Science and Technology* 6, 2012: 208–13.

Johnson, D.B. Importance of microbial ecology in the development of new mineral technologies. *Hydrometallurgy* 59, 2001: 147–57.

Kurosawa, N., Itoh, Y.H., Iwai, T. et al. *Sulfurisphaera ohwakuensis* gen. nov., sp. nov., a novel extremely thermophilic acidophile of the order Sulfolobales. *International Journal of Systematic and Evolutionary Microbiology* 48, 1998: 451–56.

Liang, C.L., Xia. J.L., Nie, Z.Y., Yang, Y., Mac, C.Y. Effect of sodium chloride on sulfur speciation of chalcopyrite bioleached by the extreme thermophile *Acidianus manzaensis*. *Bioresource Technology* 110, 2012: 462–67.

Lupi, C., Pasquali, M., DellEra, A. Studies concerning nickel electrowinning from acidic and alkaline electrolytes. *Minerals Engineering* 19, 2006: 1246–50.

Mishra, S., Panda, P. P., Pradhan, N. et al. Effect of native bacteria *Sinomonas flava* 1C and *Acidithiobacillus ferrooxidans* on desulphurization of Meghalaya coal and its combustion properties. *Fuel* 117, 2014: 415–21.

Munoz, J.A., Dreisinger, D.B., Cooper W.C., Young, S.K. Silver catalyzed bioleaching of low grade ores. Part I. Shake flasks tests. *Hydrometallurgy* 88, 2007: 3–18.

Norris, P.R., Burton, N.P., Foulis, N.A.M. Acidophiles in bioreactor mineral processing. *Extremophiles* 4, 2000: 71–6.

Norris, P.R., Clark, D.A., Owen, J.P., Waterhouse, S. Characteristics of *Sulfobacillus acidophilus* sp. nov. and other moderately thermophilic mineral-sulphide-oxidizing bacteria. *Microbiology* 142, 1996: 775–83.

Osorio, H., Mangold, S., Denis, Y. et al. Anaerobic sulfur metabolism coupled to dissimilatory iron reduction in the extremophile *Acidithiobacillus ferrooxidans*. *Applied and Environmental Microbiology* 79, 2013: 2172–81.

Panda, S., Mishra, S., Pradhan, N., Mohapatra, U.B., Sukla, L.B. Application of some eco-diversified mineral oxidizers and reducers towards development of a sustainable biotechnological industry. *Current Biochemical Engineering* 1, 2014b: 117–124.

Panda, S., Parhi, P.K., Nayak, B.D., Pradhan, N., Mohapatra, U.B., Sukla, L.B. Two step meso-acidophilic bioleaching of chalcopyrite containing ball mill spillage and removal of the surface passivation layer. *Bioresource Technology* 130, 2013b: 332–38.

Panda, S., Parhi, P.K., Pradhan, N., Mohapatra, U.B., Sukla, L.B., Park, K.H. Extraction of copper from bacterial leach liquor of a low grade chalcopyrite test heap using LIX 984N-C. *Hydrometallurgy* 121–124, 2012b: 116–19.

Panda, S., Pradhan, N., Mohapatra, U.B. et al. Bioleaching studies for recovery of Copper values from pre and post thermally activated ball mill spillage samples. *Frontiers of Environmental Science and Engineering* 7, 2013a: 281–93.

Panda, S., Rout, P.C., Sarangi, C.K. et al. Recovery of copper from a surface altered chalcopyrite contained ball mill spillage through bio-hydrometallurgical route. *The Korean Journal of Chemical Engineering* 31, 2014a: 452–60.

Panda, S., Sanjay, K., Sukla, L.B. et al. Insights into heap bioleaching of low grade chalcopyrite ores—A pilot scale study. *Hydrometallurgy* 125–126, 2012a: 157–65.

Panda, S., Sarangi, C.K., Pradhan, N., Subbaiah, T., Sukla, L.B., Mishra, B.K., Bhatoa, G.L., Prasad, M.S.R., Ray, S.K. Bio-hydrometallurgical processing of low grade chalcopyrite ore for recovery of copper metal. *The Korean Journal of Chemical Engineering* 26, 2012c: 781–85.

Parida, B.K., Panda, S., Misra, N., Panda, P.K., Mishra, B.K. BBProF: An asynchronous application server for identification of potential bioleaching bacteria. *Geomicrobiology Journal* 31, 2014: 299–14.

Robertson, S.W. and Van Staden, P.J. 2009. The progression of metallurgical testwork during heap leach design. Hydrometallurgy Conference, The Southern African Institute of Mining and Metallurgy. http://www.saimm.co.za/Conferences/Hydro2009/031-042_Robertson.pdf (Accessed on 03.03.2014).

Saba, A.E. and Elsherief, A.E. Continuous electrowinning of zinc. *Hydrometallurgy* 54, 2000: 91–106.

Sand, W., Gehrke, T., Hallmann, R., Schippers, A. Sulfur chemistry, biofilm and the indirect attack mechanism—A critical evaluation of bacterial leaching. *Applied Microbiology and Biotechnology* 43, 1995: 961–66.

Sand, W., Gehrke, T., Jozsa, P.G., Schippers, A. Biochemistry of bacterial leaching—direct vs. indirect bioleaching. *Hydrometallurgy* 59, 2001: 159–75.

Schippers, A. and Sand, W. Bacterial leaching of metal sulfides proceeds by the indirect mechanisms via thiosulfate or via polysulfides and sulfur. *Applied and Environmental Microbiology* 65, 1999: 319–21.

Silverman, M.P. and Ehrlich, H.L. Microbial formation and degradation of minerals. in *Advances in Applied Microbiology.* New York: Wayne W. Umbreit, 1964, 153–98.

Tributsch, H. Direct versus indirect bioleaching. *Hydrometallurgy* 59, 2001: 177–85.

Urich, T., Bandeiras, T.M., Leal, S.S. et al. The sulphur oxygenase reductase from *Acidanus ambivalens* is a multimeric protein containing a low potential mononuclear non-heme iron centre. *Biochemical Journal* 381, 2004: 137–146.

Valdes, J., Cárdenas, J.P., Quatrini, R., Esparza, M., Osorio, H., Duarte, F., Lefimil, C., Sepulveda, R., Jedlicki, E., Holmes, D.S. Comparative genomics begins to unravel the ecophysiology of bioleaching. *Hydrometallurgy* 104, 2010: 471–76.

Watling, H.R. The bioleaching of sulphide minerals with emphasis on copper sulphide—A review. *Hydrometallurgy* 84, 2006: 81–108.

Zeng, W., Qiu, G., Zhou, H. et al. Community structure and dynamics of the free and attached microorganisms during moderately thermophilic bioleaching of chalcopyrite concentrate. *Bioresource Technology* 101, 2010: 7068–75.

Zhou, H.B., Zeng, W.M., Yang, Z.F., Xie, Y.J., Qiu, G.Z. Bioleaching of chalcopyrite concentrate by a moderately thermophilic culture in a stirred tank reactor. *Bioresource Technology* 100, 2009: 515–20.

Zhou, Q.G., Bo, F., Bo, Z.H., Xi, L., Jian, G., Fei, L.F., Hua, C.X. Isolation of a strain of *Acidithiobacillus caldus* and its role in bioleaching of chalcopyrite. *World Journal of Microbiology and Biotechnology* 23, 2007: 1217–25.

9

Bioleaching of Metals from Major Hazardous Solid Wastes

Gayathri Natarajan, Thulasya Ramanathan, Abhilasha Bharadwaj
and Yen-Peng Ting

CONTENTS

9.1 Introduction

Population growth, rapid industrialisation and rising standards of living have contributed to an increase in both the quantity and variety of solid wastes generated by industrial, mining, domestic and agricultural activities. This has also resulted in the depletion of natural resources. It has been estimated that about 19 billion tonnes of solid wastes will be generated annually by the year 2025 (Yoshizawa et al., 2004). Most of the solid waste generated ends up in open dumps and wetlands, polluting the environment, contaminating surface and ground water, and posing serious health hazards.

According to statistics, the economically exploitable metal minerals may be able to sustain between another 20 and 100 years (Yang, 2013). A recent publication indicates a very severe global problem of metal minerals scarcity (Diederen, 2009). For instance, copper prices have shown a steady increase because of increasing production costs due to the declining copper grades. Further with the growing population and urbanisation, the demand for copper in cars, appliances and electronics has shown an exponential increase (AQM Copper Inc., 2013). There is clearly an imperative need to find alternative sources of metals in addition to natural ores. A rational and sustainable utilisation of natural resources, recycling and resource recovery from solid waste and its protection from toxic releases is vital for sustainable socio-economic development. This chapter explores the use of biotechnological processes for the recovery of metals from major industrial and municipal wastes, including electronic wastes, battery wastes, spent catalysts and fly ash.

9.1.1 Electronic Waste

When millions of computers and mobiles purchased around the world every year become obsolete, they leave behind lead, cadmium, mercury and other hazardous wastes. According to a recent e-waste report by Environmental Protection Agency (EPA), more than 3.19 million tons of e-waste were discarded in the United States alone. Only 18% of the waste was recycled and the remaining was discarded in landfills or incinerators (ETBC, 2012). In recent years, potential environmental problems associated with electronic waste disposal have been gaining importance as the consumption of electronic devices becomes more prevalent, resulting in greater amounts of electronic waste to be handled and disposed (Gramatyka et al., 2007). Manufacturing mobile phones and personal computers consumes 3% of the gold and silver mined worldwide each year, 13% of the palladium and 15% of cobalt (UNEP, 2009). In other words, resources of raw materials, particularly metals are depleting at a higher rate to meet the demand.

In 2009, the United Nations Environment Program (UNEP) estimated that the global e-waste generated was around 40 million tons annually, not including the proportions that were not reported via official channels (UNEP, 2009). The same report projected that by 2020, the waste stream would double in Organisation for Economic Co-operation and Development (OECD) countries and surge up to 500% from 2007 levels in developing nations. E-waste shows a higher growth rate than any other category of municipal waste (Greenpeace, 2005), with the alarming rate due to increased consumption of electronic devices and shorter life span of mobile phones, computers, televisions and so on. Developing countries such as Africa, China and India face more problems as e-waste is illegally shipped to these countries where laws for the protection of workers and environment are more lax or less strictly enforced, and the cost of recycling is much less than in the developed countries. To reduce electronic waste disposal, countries have drafted regulations on the reuse and recycling of e-waste (Gramatyka et al., 2007). This can help to reduce the amount of electronic waste piling up in the landfills worldwide, reduce the volume of hazardous waste disposal and environmental damage caused by such methods of disposal.

The main economic driving force for recycling of electronic scrap is the recovery of precious metals such as silver, gold and palladium. The recovery of base metals such as copper, aluminium and iron is also of interest because of the huge quantities involved in the manufacturing of electronics (Sum, 1991). Compared with the metal content in natural ores, the metals contained in e-waste are higher (Li et al., 2007; Xiang et al., 2010), making it an economical and more attractive secondary source of metals. For instance, the gold content in natural gold ores is around 0.5–13.5 g of gold per ton (Korte et al., 2000), while the gold content in e-waste is significantly higher at around 10–10,000 g of gold per ton (Cui and Zhang, 2008; Pham and Ting, 2009), making it an attractive alternative source of gold compared to natural ores. Currently, the main source of gold in industries and other sectors comes from gold mining, which has been associated with many environmental problems. Toxic chemicals such as cyanide are often used in gold extraction. In addition, gold mining often leaves behind residual metal tailings that contain harmful metals such as lead and cadmium which, when leached into the ground, cause groundwater pollution. Besides, to produce 1 ton of gold, CO_2 emissions of up to 17,000 tons are generated (Mishra et al., 2008b). With the annual demand for gold in electrical and electronic equipment at 300 tonnes on average, this extraction alone produces 5.1 million tonnes of CO_2. Other metals such as copper, cobalt, tin, indium, silver, palladium, platinum and ruthenium used in electrical and electronic equipment account for an annual CO_2 emission level of 23.4 million tonnes, at almost 1/1000th of the world's CO_2 emissions (Mishra et al., 2008b). With the increasing demand for gold and the resultant depleting of natural ores, there is indeed an imperative to discover more alternative sources of gold, such as recycled e-waste.

9.1.2 Fly Ash

The ever-mounting consumption of energy and generation of wastes have also increased fly ash production. Fly ash refers to the residue that rises with flue gas during the combustion of coal or any solid waste. The two major types of fly ash include coal fly ash and municipal solid waste (MSW) incineration fly ash. These are usually generated in large quantities and hence demand huge landfill areas for disposal. More than 100 million tonnes of fly ash are produced annually worldwide. In the United States alone, fly ash production has increased from 49 million tons in 1990 to 67 million tons in 2010. Of this, only 37% is used and the rest is discarded in landfills (ACAA, 2010).

India produced about 175 million tonnes of fly ash in 2012, which required around 40,000 hectares of land for the construction of ash ponds. The projection that this figure would increase to 600 million tonnes by 2031–2032 highlights the need for greater utilisation of fly ash. Although the utilisation rate has increased from 1 million ton in 1993 to around 60 million tonnes in 2006–2007, it can be further improved with chemical treatment and detoxification of fly ash (Singh, 2010).

Fly ash may be utilised in several areas. For instance, it may potentially improve the physical health of soil and by serving as a soil modifier and enhance its water-retaining capacity. It improves water and nutrient uptake by plants, helps in the development of roots and soil binding and protects the soil from diseases and detoxifies contaminated soil (Singh, 2010). Fly ash can also be incorporated in Portland cement concrete to enhance its performance. One ton of Portland cement production typically generates 0.87 tonnes of carbon dioxide into the environment. The utilisation of fly ash in cement minimises carbon dioxide emission to the extent of its proportion in cement. Apart from these uses, fly ash has also been used in the production of fly ash–lime–gypsum bricks, cellular lightweight concrete blocks and fly ash-based polymers.

Fly ash is heavily loaded with toxic metals such as aluminium, zinc, cadmium, copper and lead, and is generally regarded as a hazardous waste. Potential leaching of these metals through groundwater and rainwater poses grave environmental problems (Lam et al., 2010; Singh, 2010). The removal of heavy metals from fly ash would alleviate such concerns. Although chemical treatment and detoxification of fly ash may improve the utilisation of ash, the traditional approach such as treatment using organic and inorganic acids or chloride evaporation, and thermal treatment has become obsolete due to the high cost capital and operating costs, increased energy consumption as well as negative environmental impact. On the other hand, with the high concentration of some heavy metals, fly ash may be considered an artificial ore and a resource from which metal values may be derived.

9.1.3 Spent Catalyst

The petroleum industry can be considered as the backbone of a modern economy because it provides the main source of energy to date. To meet increasingly stringent economic and environment regulations, petroleum refineries extensively utilise catalysts for the purification and upgrading of various petroleum streams and residues. Four major types of catalysts are commonly used in various processes: reforming, hydroprocessing, fluid catalytic cracking (FCC) and alkylation. Hydroprocessing consists of hydrotreating and hydrocracking catalysts. After repeated use, impurities such as coke, sulphur, nitrogen and heavy metals (Ni, V, Fe, etc.) from crude petroleum are deposited on the catalyst pores, resulting in catalyst deactivation. Although spent catalysts may be regenerated to extend their operational life, regeneration can only be applied for a few times, and on a limited number of catalytic systems. The U.S. EPA classifies these deactivated spent catalysts as hazardous waste due to their self-heating behaviour and toxic chemical content (Eijsbouts et al., 2008). With the global refinery catalyst market estimated to reach $4.3 billion in 2018 from about $3 billion in 2010, large quantities of hazardous spent catalysts will continue to be generated as solid wastes annually (Refinery Catalyst Market, 2012).

In most refineries, a major portion of the spent catalyst wastes comes from the hydroprocessing units. It has been reported that 150,000–170,000 tonnes of spent hydroprocessing catalyst are generated worldwide annually (Dufresne, 2007). These statistics have been increasing in the recent years due to the increasing demand for clean fuels with ultra-low sulphur levels together with the processing of low-quality feedstock containing higher contents of sulphur, nitrogen, asphaltene and so on. Currently, spent catalysts are managed industrially via chemical recovery and the recycling of valuable metals for different applications. While conferring an economic advantage, this approach unfortunately entails the use of acids in large-scale-processing operations, which generate large volumes of potentially hazardous wastes and gaseous emissions. The ultimate disposal of the spent catalyst in landfill sites also poses challenges, due to economic constraints, decreasing availability of landfill as well as the concern for pollution arising from the leaching of heavy metals into the environment (Marafi and Stanislaus, 2008b). Besides the formation of leachates, spent hydroprocessing catalysts liberate toxic gasses such as H_2S, HCN or NH_3 when they are in contact with water, oxygen or an inert environment (Furimsky, 1996). Indeed, environmental and economic constraints deter generators from the disposal of spent catalysts in landfills (ECMA, 2001; USEPA, 2003).

Spent catalysts discarded as solid wastes from industrial operations typically contain substantial amounts of alumina and heavy metals such as Mo, Co, V, Ni, W and Fe. These metals are highly valuable and are used extensively in the steel industry and in the manufacture of special alloys. With the increasing demand for heavy metals and the rapidly declining high-grade mineral ores, spent catalysts could serve as a cheap secondary source for these valuable metals. Microbial treatment and detoxification of the spent catalyst will reduce the environmental problems associated with landfilling as well as the amount of waste that requires landfilling (Marafi and Stanislaus, 2008a,b).

9.1.4 Battery Waste

The production and use of batteries have grown worldwide owing to the increased and extensive use in electrical and electronic devices, with lithium-ion batteries (LIBs), Zn–carbon batteries and Ni–Cd batteries as the most common portable batteries. World demand for primary and secondary batteries is forecast to rise 8.5% per year to $144 billion in 2016

(Refinery Catalyst Market, 2012). Primary Zn–carbon batteries occupy about 90% of the total sales annually because of low cost and rapid run out. Spent battery wastes are of environmental concern because they are toxic, abundant and prevalent for extended periods of time (Li and Xi, 2005). Ni–Cd batteries (estimated at around 3.9% of all discarded dry batteries) are hazardous, with Ni and Cd being the suspected carcinogens (Shapek, 1995). LIBs, on the other hand, are less toxic but are highly flammable due to the presence of metallic lithium which radically oxidises in the presence of moisture (Lee and Rhee, 2002). Although batteries generally make up only a tiny portion of MSW—(<1%)—they account for a disproportionate amount of the toxic heavy metals in MSW (USEPA, 2004). For example, the EPA has reported that nickel–cadmium batteries accounted for 75% of the cadmium found in MSW. Similarly lead acid batteries accounted for 65% of the lead found in MSW. Hence, it is important to recycle and recover the metals before they are disposed (USGS, 2008). The typical metal composition of Zn–Mn batteries is about 12%–28% Zn and 26%–45% Mn and the remaining components include graphite, K and Fe (Sayilgan et al., 2009). Ni–Cd batteries typically consist of 12%–15% Cd and 15%–20% Ni (Rydh and Svärd, 2003). LIBs are heavily loaded with metals with the composition of 5%–15% Co and 2%–7% Li.

Owing to the high metal concentrations in these waste materials, they may be considered as artificial ores (Brandl, 2001). With the increasing demand for metals and depleting natural ores, there is an imperative need to discover more alternative sources of metals, such as solid waste. Recycling and resource recovery can help to reduce the amount of waste piling up in the landfills worldwide, reduce the volume of hazardous waste disposal and also environmental damage caused by such methods of disposal. In view of the environmental and economic benefits, increasing attention has been paid to the development of processes for recovering metals and other valuable materials from these wastes. It is generally held that biotechnology offers promise in metallurgical processing of these waste materials. For instance, bioleaching has been used for the recovery of precious metals and copper from ores for many years. However, limited research has been carried out on the bioleaching of metals from solid waste. This chapter reviews the prospects of biotechnological recovery of four major solid wastes, namely electronic scrap, fly ash, spent catalyst and battery wastes.

9.2 Conventional Methods of Metal Extraction

Although the rich metal content in these wastes is appealing, metal-extracting operations on these wastes are scarce, and most of the waste generated is landfilled. Even if the wastes are landfilled, disposed in surface impoundments or recycled, their toxic effects are not completely eradicated.

Infiltration of rainwater into landfills results in the release of these toxic trace elements. Moreover, the problem of space still remains. Hence, wastes need to be treated appropriately in order to mitigate the toxic effects before disposal or reuse. Conventional metal extraction usually takes place through pyrometallurgical and hydrometallurgical processing.

9.2.1 Pyrometallurgy

Pyrometallurgical processing refers to the process of refining ores with heat to accelerate chemical reactions or to melt the metallic or non-metallic content. This includes incineration, smelting in blast furnaces, drossing, sintering, melting and gas-phase reactions at high temperatures. At elevated temperatures, physical and chemical transformation of the materials occurs which enables the recovery of the metals. This method is commonly used to recover non-ferrous metals such as copper, and precious metals such as silver from electronic waste over the past two decades (Zhou, 2007; Cui and Zhang, 2008). Pyrometallurgical processes have been used widely for the recycling of batteries where metals are selectively volatilised at high temperatures and then condensed. These processes are simple but employ extreme conditions and utilise enormous energy. The major limitations and disadvantages in the use of pyrometallurgical methods include the formation of dioxins from the presence of halogenated flame retardants in the smelter feed, limited separation of metals that needs further hydrometallurgical or electrochemical processing, requirement of high-grade feed and high-energy costs (Hageluken, 2006; Dalrymple et al., 2007; Cui and Zhang, 2008).

9.2.2 Hydrometallurgy

Hydrometallurgical methods involve the use of aqueous chemistry and exploit the chemical properties of materials to extract metals from ores, concentrates or solid materials. The solutions containing the metals of interest are then subjected to separation processes such as solvent extraction, precipitation, cementation, ion exchange, filtration and distillation to isolate and concentrate the metals of interest. Extensive research has been reported on the recovery of both precious metals such as gold, silver and platinum, and base metals such as copper, nickel, lead and zinc (Cui and Zhang, 2008; Pant et al., 2012; Tuncuk et al., 2012) from electronic scrap. Heavy metals from fly ash have been extracted using a wide variety of leaching agents such as organic and inorganic acids, alkalis and salts as well as chelating agents (Jadhav and Hocheng, 2012). Similar reports are available for battery waste and spent catalysts as well. This approach, while cheaper (e.g. lower power consumption and recycling of chemical agents) is usually not environmentally friendly. Moreover, heavy metals immobilised in chemically treated waste may leach when there is a change in the chemical environment, in particular pH (Shimaoka and Hanashima, 1996). Strict regulation on the equipment used, is also imposed, making the process unattractive.

9.3 What Is Bioleaching?

Bioleaching is a biomining process; it relies on the use of microorganisms to recover precious and base metals from mineral ores and concentrates. Traditional mining techniques are unable to economically extract metals from ores that have been worked and are lean. In biomining, microorganisms act as biocatalysts in extracting metals from metal-laden sources and can be used to exploit secondary sources. The process of biomining includes biooxidation and bioleaching (Brandl, 2001). In biooxidation, the microbes selectively dissolve undesired metals from the solid matrix, leaving the metal values of interest enriched in the solid phase to be recovered by other processes. The resulting supernatant is thereafter discarded. Bioleaching on the other hand refers to the solubilisation of the metals of interest from their unavailable solid state by microorganisms such as bacteria and fungi. Here, the metal of interest is recovered from the solution (Brandl, 2001). The discussion in this chapter focusses on bioleaching rather than biooxidation.

Bioleaching processes are based on the ability of microorganisms to transform solid compounds, resulting in soluble and extractable elements which can be recovered (Krebs et al., 1997). The earliest reports on bioleaching date back to the late nineteenth century. Rio Tinto mines in southwest Spain with low-grade Cu ores are regarded as the birthplace for commercial bioleaching practices (Brandl, 2001). Initially, bioleaching was employed only for ores. With relatively lower operating temperatures and lower consumption of reagents, bioleaching has been shown to be an economic alternative to conventional pyrometallurgical and hydrometallurgical methods for metal extraction. The biological process enables the recycling of metals by a process close to the natural biogeochemical cycles, thus reducing the demand for energy, and for natural resources such as ores and landfill space (Krebs et al., 1997). Indeed, bioleaching has emerged as an economical and simple technology for leaching of metals such as copper, gold, silver, uranium and so on from their respective ores and has opened up the possibility of extracting metals from natural ores and industrial residues not recoverable by conventional methods (Krebs et al., 1997). With the growing demand for metals, industrial wastes such as fly ash, electronic scrap and spent catalysts have also been considered for metal recovery. Indeed, with the increasing pressure for the industry to adopt environmentally friendly and sustainable processes, more attention has been focussed on bioleaching as an alternative clean and green technology (Brierley, 2008).

Interests in bioleaching intensified in 1998 when a study compared hydrochloric acid leaching, chloride evaporation, acetic acid leaching and bioleaching of fly ash by *Aspergillus niger* (Tateda et al., 1998). Chloride evaporation showed excellent extraction efficiency and a short reaction time but the process was uneconomical with high capital and operating cost and required harsh operating conditions and tremendous energy consumption. Inorganic and organic acid extractions consumed less energy and were more cost efficient but less process efficient than chloride evaporation. The study considered bioleaching inferior to the other technologies due to the extended reaction rate, increased cost and dearth of knowledge but the versatile, cost-efficient and eco-friendly nature of the bioleaching process encouraged research in the area.

9.3.1 Types of Bioleaching

Bioleaching can be conducted in one step or two steps. In one-step bioleaching, the microorganism is inoculated together with the solid waste in the microbial medium where microbial growth and metal leaching takes place together. While this technique is straightforward, the metal ions present in the waste may affect microbial growth, and thereby limit the efficiency of the bioleaching process. In two-step bioleaching, the microorganism is first cultured in the growth medium until it reaches the exponential phase (i.e. actively growing state), after which the solid waste is added. In another approach, termed spent medium leaching, the microorganism is cultured in its growth medium until the maximum production of metabolites occurs. The suspension and microorganisms are subsequently filtered to obtain the cell-free spent medium (containing only the biogenically produced metabolites) which is used for the subsequent leaching.

In general, two-step and spent medium leaching are more effective than one-step bioleaching (Aung and Ting, 2005; Mishra et al., 2007; Deng et al., 2012). This is due to the inhibition of microbial growth (leading to reduced metabolite secretion) in the presence of minerals and industrial waste containing high concentration of metals. Spent medium leaching is more advantageous, due to easier handling and a shorter processing time. More importantly, it also allows optimum operating conditions for both microbial growth and leaching, since the production of metabolites and the leaching process are decoupled. For example, the absence of viable microorganisms in spent medium leaching permits the use of more aggressive leaching conditions (low pH, high temperature, etc.) and easier scale-up and therefore results in higher metal recovery. In addition, as the microbes are not in contact with the toxic metals present in the waste, increase in metabolite production and continuous replenishment of metabolites can be achieved.

9.4 Mechanisms Involved in Bioleaching

A bioleaching process involves a three-phase system consisting of

- An aqueous phase, with the nutrient solution for the microorganisms.
- A solid phase, composed of the ore or industrial waste containing metal values.
- A gaseous phase, consisting of atmospheric oxygen and carbon dioxide.

In the aqueous phase, many physicochemical and biological interactions take place. Of primary importance are the growth of microorganisms, the adaptation to the toxic environment of metal concentrates and the interactions between the microbial consortium if more than one strain is present. The production of metabolites or leaching agents and the contact between the microbes and the solid residues are also major considerations. Although there are still debates on the actual mechanisms by which microorganisms carry out metal dissolution, two major mechanisms are widely reported. Leaching mechanisms are classified as direct and indirect bioleaching.

a. Direct leaching

Direct leaching refers to leaching mediated through the physical contact of the microorganisms with the solid substrate. Sometimes called the contact mechanism, direct leaching of metals from a solid structure occurs through oxidation or reduction, which involves the transfer of an electron either from the solid structure (oxidation) to an electron acceptor (normally oxygen) or the injection of electrons into the solid structure from an electron donor such as H_2 (reduction). The microbes form an extracellular polymeric substance (EPS) layer to which the metal sulphide surface is attached (Kinzler et al., 2003). Mineral dissolution (due to electrochemical processes) occurs at the interface between the bacterial cell wall and the metal sulphide surface. The attached microbes acquire energy during oxidation of the solid and produce leaching chemicals Fe^{3+} and H^+. The primary Fe^{3+} ions are supplied by the bacterial EPS, where they complex to glucuronic acid residues. Some reduced metal species serve as electron donors in the metabolism of chemolithotrophic bacteria (Figure 9.1) whereas some oxidised metal species serve as electron acceptors for heterotrophic bacteria.

b. Indirect leaching

Indirect leaching refers to the leaching caused by biogenically excreted metabolic products. These products act as chemical

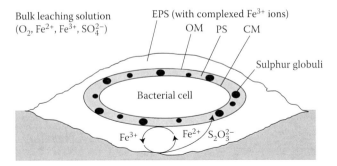

FIGURE 9.1
Model for the direct leaching mechanism of *A. ferrooxidans* showing the bacterial cell embedded in its EPS attached to the solid via electrostatic interactions. (CM, Cytoplasmic membrane; PS, periplasmic space; OM, outer membrane.) (With kind permission from Springer Science+Business Media, *Appl Microbiol Biotechnol*, Bioleaching review part A, 63(3), 2003, 239–248, Rohwerder, T. et al.)

oxidants or reductants to leach metals from the waste. The metabolites usually involved are organic compounds such as organic acids and ligands, or inorganic compounds such as carbonate, phosphate, acids and so on. Indirect leaching includes acidolysis, complexolysis, redoxolysis, ammonolysis or a combination of these mechanisms (Figure 9.2) (Ehrlich, 1992, 1997; Burgstaller and Schinner, 1993; Bosshard et al., 1996). In redoxolysis, redox reactions occur which are based either on electron transfer from minerals to microorganisms in the case of physical contact between organisms and solids, or on bacterial oxidation of Fe^{2+} to Fe^{3+} where ferric iron subsequently catalyses metal solubilisation as an oxidising agent (Krebs et al., 1997; Islam and Ting, 2009). Acidolysis occurs via the formation of organic or inorganic acids (protons) by the microorganisms. For instance, in the case of *Acidithiobacillus thiooxidans,* which produces extraneous sulphuric acid in the presence of sulphur, the acid leaches the metals from the solid waste (Bosecker, 1987). Complexolysis occurs via the excretion of metabolites as complexing agents that chelate metal ions and form stable co-ordinate complexes. For instance, cyanogenic bacteria such as *Chromobacterium violaceum* produce extracellular cyanide, which forms water-soluble complexes with many metals (Campbell et al., 2001). Recently, amino acids have also been identified as effective lixiviants. Ammonolysis refers to leaching by ammonia produced during urea hydrolysis or amino acid deamination, and has been observed in ammonifying bacteria (related to the genera *Bacillus, Acinetobacter* and *Vibrio*) and bacteria possessing urease enzymatic activity (*Corynebacterium*) (Groudeva et al., 2007). Figure 9.2 shows the major mechanisms of indirect leaching.

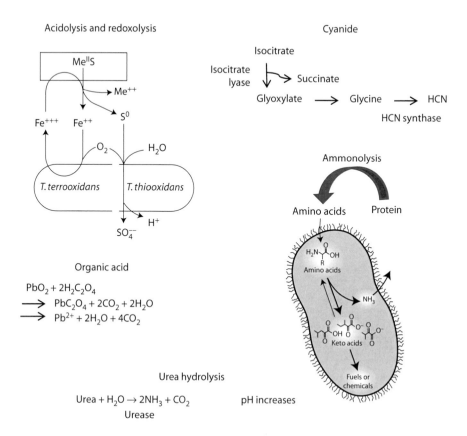

FIGURE 9.2
Indirect leaching mechanisms. (Reprinted by permission from Macmillan Publishers Ltd. *Geomaterials*, Das, A.P. et al. 2012. Reductive acid leaching of low grade manganese ores, 2: 70, copyright 2012; Hagins, J.M., Locy, R., and Silo-Suh, L. *J Bacteriol* 191: 6335–6339, copyright 2009; Moybel Research Laboratory (n.d.). *Helicobacter pylori* urease. University of Michigan Medical School. http://www.umich.edu/~hltmlab/research/pylori/. (Accessed 16 Feb, 2014.); Mielenz, J.R. *Nature Biotechnol* 29: 327–328, copyright 2011.)

9.5 Major Organisms and Bioleaching of Hazardous Waste

Tables 9.1 through 9.4 summarise significant work on the bioleaching of fly ash, electronic scrap, spent batteries and spent catalyst. For each of these solid wastes, a direct comparison of the reported work based on percentage recovery is not possible since metal recovery is highly dependent on the source and composition of the solid waste, as well as the acid-digestion protocol and its extent of digestion. The following sections review significant bioleaching

TABLE 9.1

Bioleaching of Fly Ash by Bacteria and Fungi

Organism	Metals Recovered	Reference
A. niger	Al	Singer et al. (1982)
A. niger	Cd, Zn, Cu, Pb, Mn, Al, Cr, Fe and Ni	Bosshard et al. (1996)
A. niger	Al, Cu, Pb, Fe and Zn	Xu and Ting (2004)
A. niger	Al, Cu, Pb, Fe, Mn and Zn	Wu and Ting (2006) Xu and Ting (2009)
A. niger	Cu, Pb, Fe, Mn, Zn, Cd and Cr	Yang et al. (2008)
A. niger	Cu, Pb, Cd and Zn	Wang et al. (2009b)
A. niger	Al	Torma and Singh (1993)
A. niger	Al, Pb, Fe and Zn	Yang et al. (2009a,b)
(Acidi)Thiobacillus strains	Zn	Bosecker (1987)
(Acidi)Thiobacillus thiooxidans ZYR1	Al, Ti	Fass et al. (1994)
(Acidi)Thiobacillus thiooxidans	Al, Fe	Brombacher et al. (1998)
Mixed culture: *(Acidi)Thiobacillus thiooxidans* and *(Acidi)Thiobacillus ferrooxidans*	Al, Zn, Fe, Cu, Mn, Cd, Ni, Cr and Pb	Seidel et al. (2001)
Mixed culture: *(Acidi)Thiobacillus thiooxidans* and *(Acidi)Thiobacillus ferrooxidans*	Zn, Cu, Cd, Cr and As	Ishigaki et al. (2005)
Sewage suspensions with four *(Acidi)Thiobacilli* strains, *T. neopolitanus, T. acidophilus* and two strains of *T. thiooxidans*	Al, Zn, Fe, Cu, Mn, Cd, Ni, Cr and Pb	Krebs et al. (2001)
A. brierleyi	Zn	Konishi et al. (2003)
Y. lipolytica	Zn, Ni, Cu and Cr	Bankar et al. (2012)

work reported and strategies employed in improving metal recovery from fly ash, electronic scrap, spent battery and spent catalyst.

9.5.1 Bioleaching of Metals from Fly Ash

A. niger is the most widely studied organism for the bioleaching of metals (such as Cd, Zn, Cu, Pb, Mn, Al, Cr, Fe and Ni) from fly ash. Two-step bioleaching with this fungus showed better metal recovery compared to one-step bioleaching. The growth of fungi in sugar-containing media results in the production of organic acids such as oxalic acid, citric acid and gluconic acid. *A. niger* produces citric acid at a higher concentration in the absence of fly ash, while gluconic acid is produced at a higher concentration in its presence (Bosshard et al., 1996). Citrate showed better leaching when compared to gluconate and hence, two-step bioleaching showed higher metal recovery (Bosshard et al., 1996). Metal tolerance was also higher in two-step

TABLE 9.2

Bioleaching of Electronic Waste by Bacteria and Fungi

Organism	Metals Recovered	Reference
Mixed culture: *A. thiooxidans, A. ferrooxidans, A. niger* and *P. simplicissimum*	Cu, Zn, Ni, Al and Sn	Brandl et al. (2001)
Mixed culture: *A. thiooxidans* and *A. ferrooxidans*	Cu, Pb and Zn	Wang et al. (2009a)
Mixed culture: *A. thiooxidans* and *A. ferrooxidans*	Cu, Zn, Ni, Al and Sn	Liang et al. (2010)
Mixed culture: *S. thermosulfidooxidans* and an unidentified acidophilic heterotroph (a local isolate)	Ni, Cu, Al and Zn	Ilyas et al. (2007)
Mixed culture: *Acidithiobacillus* sp., *Gallionella* sp. and *Leptospirillum* sp.	Cu	Xiang et al. (2010) Yang et al. (2009c)
A. ferrooxidans	Cu	Choi et al. (2005)
Cyanogenic microorganisms *P. fluorescens* and *C. violaceum*	Au	Faramarzi et al. (2004) Brandl et al. (2008) Chi et al. (2011) Pham and Ting (2009)
P. aeruginosa, P. fluorescens and *C. violaceum* and their mixed cultures	Au, Ag, Cu, Zn and Fe	Pradhan and Kumar (2012)
C. violaceum	Au, Cu	Natarajan and Ting (2014)
Genetically engineered *C. violaceum*	Au	Tay et al. (2013)

TABLE 9.3

Bioleaching of Battery Waste by Bacteria and Fungi

Organism	Type of Battery Waste	Metals Recovered	References
A. ferrooxidans	Li-ion battery waste	Co, Li	Mishra et al. (2008a)
A. ferrooxidans	Li-ion battery waste	Co	Zeng et al. (2012)
Mixed culture: *Thiobacillus thiooxidans* and *Thiobacillus ferrooxidans*	Li-ion battery waste	Co, Li	Xin et al. (2009)
Alicyclobacillus sp. and *Sulfobacillus* sp.	Zn–Mn battery waste	Zn, Mn	Xin et al. (2012a)
Alicyclobacillus sp. and *Sulfobacillus* sp.	Zn–Mn battery waste	Zn, Mn	Xin et al. (2012b)
A. ferrooxidans	Ni–Cd battery waste	Ni, Cd	Cerruti et al. (1998)
Indigenous *Thiobacilli* from sewage sludge	Ni–Cd battery waste	Ni, Cd and Co	Zhao et al. (2007, 2008a,b) Zhu et al. (2003)
Adapted and non-adapted *A. ferrooxidans*, Mixture of adapted and non-adapted *A. ferrooxidans* and *A. thiooxidans*	Ni–Cd battery waste	Ni	Velgosov et al. (2010)

TABLE 9.4

Bioleaching of Spent Catalyst by Bacteria and Fungi

Organism	Waste	Metals Recovered	Reference
A. ferrooxidans	Spent coal liquefaction catalyst	Mo	Blaustein et al. (1993)
A. ferrooxidans, denitrifying bacteria and Sulfolobus species	Spent Ni–Mo coal liquefaction catalyst	Mo, Ni	Joffe and Sperll (1993)
B. stearothermophilus and M. sedula	Spent Ni–Mo coal liquefaction catalyst	Mo	Sanback and Joffe (1993)
A. thiooxidans	Spent V–Ti and V–P catalyst	V	Briand et al. (1999)
A. niger	FCC catalyst	Ni, Al, Fe, V and Sb	Aung and Ting (2005)
A. niger	Spent hydroprocessing catalyst	Ni, Mo and Al	Santhiya and Ting (2005)
A. thiooxidans	Spent catalyst	Ni, Mo and V	Mishra et al. (2007)
A. thiooxidans, Desulfovibrio sp.	Spent catalyst	Ni	Bosio et al. (2007)
A. thiooxidans	Spent catalyst	Ni	Islam and Ting (2009)
Sulphur-oxidising bacteria and iron-oxidising bacteria	Spent hydroprocessing catalyst	Mo, Ni and V	Pradhan et al. (2009)
A. thiooxidans, A. ferrooxidans and L. ferrooxidans	Spent catalyst	Ni, Mo and V	Beolchini et al. (2010)
Adapted A. niger and P. simplicissimum	Spent hydrocracking catalyst	W, Fe, Mo, Ni and Al	Amiri et al. (2011a,b)
A. thiooxidans, A. ferrooxidans, A. niger, and P. simplicissimum	Spent hydroprocessing catalyst	Al, Co, Mo and Ni	Gholami et al. (2011a,b)
S. metallicus	Spent catalyst	Al, Ni, Mo and V	Kim et al. (2012)
A. brierleyi	Spent hydrotreating catalyst	Al, Ni, Mo and Fe	Bharadwaj and Ting (2013a)
Mixed culture of moderate thermophiles	Spent catalyst	Al, Ni, Mo and V	Srichandan et al. (2012)
A. brierleyi	Spent hydrocracking catalyst	Al, Mo and Ni	Gerayeli et al. (2013)

bioleaching; *A. niger* tolerates up to 5% (w/v) fly ash in two-step bioleaching but not in one-step bioleaching (Yang et al., 2008). In a study on fungal growth and metals inhibition kinetics for one-step bioleaching, a modified Gompertz model fits the experimental data for biomass concentration and citric acid production, while Han and Levenspiel model was found to be the best fit for inhibitory effects on fungi by the fly ash (Xu and Ting, 2009). Owing to a low tolerance for fly ash by the fungi, lower pulp densities showed higher-leaching efficiencies (Wu and Ting, 2006). Although this suggests that spent medium (cell-free filtrate) leaching would result in higher metal recovery than two-step bioleaching, it was not borne out; the higher metal recovery in two-step bioleaching suggested the involvement of direct leaching in the process (Wu and Ting, 2006). In contrast, another study found that indirect leaching by citrate was the dominant mechanism. The efficiency of the leaching agent was confirmed when biogenically produced; commercial citric acid gave similar extraction yields (Singer et al., 1982).

In a study identifying the major factors influencing two-step bioleaching, the pulp density and the concentration of the carbon source were found to be more important factors than the inoculum concentration and time of addition of fly ash in bioleaching (Xu and Ting, 2004). Apart from process conditions, the pretreatment of fly ash was found to enhance metal recovery. Washing with deionised water and calcination of fly ash resulted in higher extraction yields (Torma and Singh, 1993; Wang et al., 2009b). Another major factor in metal recovery is biosorption of the metals leached by the fungal biomass. Al, Fe and Zn leaching followed pseudo-first-order reaction kinetics whereas Pb leaching followed pseudo-second-order reaction kinetics due to chemisorption being the rate-controlling step. Sorption of Pb, Zn and Fe followed the Langmuir model which showed that monolayer biosorption took place at the functional groups/binding sites on the surface of the biomass while the sorption of Al followed Freundlich isotherm, suggesting multi-layer sorption on heterogeneous surfaces (Yang et al., 2009a,b). So far, no efforts have been reported on the biosorption of metals (in order to enhance metal recovery in the aqueous solution) in the bioleaching of fly ash.

Acidithiobacillus strains have also been used successfully in the bioleaching of fly ash. Indirect bioleaching with biogenically produced sulphuric acid was the major mechanism in bioleaching although *Acidithiobacillus ferroxidans* employed redoxolysis with Fe^{2+}/Fe^{3+} ions as well (Bosecker, 1987). Instances of direct bioleaching were also seen wherein enhanced EPSs production in the presence of fly ash was observed which promoted cell–solid surface interactions and thus, direct bioleaching (Brombacher et al., 1998). Since higher pulp densities were deleterious to bacteria (Bosecker, 1987), salt-tolerant and metal-tolerant bacteria were developed to combat this challenge and improve the extraction yield (Fass et al., 1994). Mixed culture leaching was also carried out to examine a symbiotic leaching effect shown by various *Acidithiobacillus* species. Mixed culture leaching with *A. ferroxidans* (an iron-oxidising bacteria [IOB]) and *A. thiooxidans* (a sulphur-oxidising bacteria [SOB]) was carried out. IOB uses contact leaching/redox mechanisms for metal leaching whereas SOB employed indirect leaching. IOB showed higher metal-leaching ability but poorer tolerance to heavy metals than SOB-mixed culture that showed enhanced metal leaching especially at higher pulp densities (3% w/v of fly ash) due to the high leaching potential of IOB and the high tolerance of SOB to ash addition. The coupling of redox mechanisms with sulphate leaching enhanced metal recovery (Ishigaki et al., 2005). In a scaled-up semi-continuous laboratory-scale-leaching unit using mixed culture, it was found that the extent of bioleaching followed the order: inoculated medium (i.e. one-step bioleaching) > filtered cell-free filter medium > autoclaved sterile spent medium > uninoculated medium (Seidel et al., 2001). Sewage suspensions with four *Thiobacilli* strains, namely *Thiobacillus neopolitanus, Thiobacillus acidophilus* and two strains of *Thiobacillus thiooxidans* were also examined for the bioleaching of fly ash. No external carbon source was needed as anaerobic sewage sludge provided the nutrients as well as the pH buffer. The pH of the suspension reduced more rapidly in the presence of sewage sludge compared to the presence of an external C source as sewage sludge enhanced the growth of SOB, and the metal solubilisation rate was increased (Krebs et al., 2001).

Acidianus brierleyi, a thermophilic acidophilic microbe was also used for bioleaching of fly ash. However, because of the poor fly ash and pH tolerance, bioleaching with this archaea strain was not studied in greater detail (Konishi et al., 2003).

A recent study on the bioleaching of fly ash by *Yarrowia lipolytica* showed that the main mechanism of leaching was via the production of citric acid, and that fly ash had no inhibitory effect on citric acid production. Furthermore, unlike *A. niger*, the yeast did not produce different metabolites in the presence of the ash. Apart from citric acid, two extracellular proteins induced in the presence of fly ash were speculated to be involved in bioleaching. Accumulation of metal ions at the cell wall, cell membrane and cytoplasm occurred. Cu ions were leached more effectively by the yeast cells than commercial citric acid, suggesting that other metabolites or mechanisms play a role in bioleaching. The attachment of *Y. lipolytica* cells to fly ash occurred due to the hydrophobic nature of fly ash and the cells. It was also observed that the increasing concentration of fly ash led to an increase in average cell length, cell volume and surface area of the yeast cells and a decrease in surface-to-volume ratio (Bankar et al., 2012).

pH and fly ash tolerance of the organism are among the most important factors influencing bioleaching, and are particularly critical when the process is scaled up, and operated at high pulp densities. As the pH of fly ash is typically around 11–13, the use of commonly examined acidophilic organisms requires considerable acidification of the ash. Temperature tolerance is less critical but nonetheless important, especially in heap leaching where the temperature in the interior of the heap rises above 50°C (and in some cases over 60°C) during bioleaching, since the process is exothermic. Organisms with extensive nutritional requirements increase the cost of the process, and hence are naturally not favoured. Indigenous microbial population isolated from heaps and dump sites have been examined extensively in order to identify waste-tolerant strains and better performers. Compared to acidophiles, heterotrophic fungi tolerate a wider range of pH and are employed for treating moderately alkaline wastes. Fungi produce extracellular metabolites that serve as complexing agents for metal ions in fly ash and thus reduce toxicity and bioleaching effects (Burgstaller and Schinner, 1993). Since most microbes reported for fly ash bioleaching tolerate not more than 10% (w/v) of fly ash, there is clearly a need to identify microorganisms with greater metal tolerance. One approach is the use of genetically modified strains, or strains selected by mutation, in order to accelerate the bioleaching process and enhance its efficiency.

9.5.2 Bioleaching of Metals from Electronic Waste

One of the earliest studies on the bioleaching of electronic waste material was reported by Brandl et al., who demonstrated that microbiological processes can be applied to mobilise metals from electronic waste materials, using a mixed culture of bacteria *A thiooxidans* and *A. ferroxidans* and individual fungal strains *A. niger and Penicillium simplicissimum* to mobilise copper, zinc, nickel, aluminium and tin from electronic scrap at 10 g/L pulp density (Brandl et al., 2001). The formation of inorganic and organic acids caused the mobilisation of metals. The study found that the growth of organisms in the presence of electronic scrap was poor, and a two-step process was proposed where microbial growth was decoupled from the bioleaching; toxicity of electronic scrap on the growth of microorganisms was significantly reduced, and higher concentrations of waste may be treated. Wang et al. reported that a mixed culture of *A. ferrooxidans* and *A. thiooxidans* performed better in the bioleaching of copper, lead and zinc than a pure culture of *A. ferroxidans,* which in turn showed better bioleaching than a pure culture of *A. thiooxidans* (Wang et al., 2009a). However, the synergistic effect in mixed bacterial leaching that led to increased bioleaching efficiency was not discussed. A new approach on multiple printed circuit board (PCB) additions (4 g/L at 48 h, 6 g/L at 96 h and 8 g/L at 144 h) was reportedly adopted, using the mixed culture of *A. thiooxidans* and *A. ferrooxidans* to operate at higher pulp densities (Liang et al., 2010). The bacteria were able to tolerate higher toxic scrap concentrations of up to 18 g/L and the bioleaching efficiencies obtained using this strategy were similar to the study by Brandl et al. which used a lesser pulp density of 10 g/L (Brandl et al., 2001).

Ilyas et al. demonstrated that the mixed culture of *Sulfobacillus thermosulfidooxidans* (a moderately thermophilic microorganism) and an unidentified acidophilic heterotroph (isolated from the local environment) leached more than 80% of base metals such as Ni, Cu, Al and Zn (Ilyas et al., 2007) from electronic waste. This work also shows enhanced leaching by pre-adaptation of microorganisms and the removal of potentially toxic non-metallic components by using high-density-saturated NaCl solution in the electronic scrap-washing step (to separate these components in the washing step). Xiang et al. and Yang et al. leached 95% copper from PCB using bacterial consortium (*Acidithiobacillus* sp., *Gallionella* sp. and *Leptospirillum* sp.) enriched from natural acid mine drainage. The influence of initial pH, initial Fe^{2+} concentration in the bioleaching solution and precipitate formation were discussed and optimum values for maximum recovery were determined (Yang et al., 2009c; Xiang et al., 2010). Choi et al. demonstrated that the addition of a complexing agent (citric acid) to the bioleaching solution of *A. ferroxidans* recovered the copper precipitated during bioleaching. The solubility of leached copper metal ions increased from 37 wt% to 80 wt% (Choi et al., 2005) with the addition of citric acid. However, the chemistry behind the process of complexing agent addition has not been discussed in depth.

Although it is known that a variety of bacteria, for example, *Chromobacterium violeaceum*, *Pseudomonas fluorescens*, *Pseudomonas aeruginosa* and *Bacillus megaterium* are capable of producing hydrogen cyanide (Knowles and Bunch, 1986), it was Faramarzi who demonstrated for the first time that cyanogenic microorganisms form water-soluble metal complexes with metal-containing solids such as PCB scrap (Faramarzi et al., 2004). Gold cyanidation reaction is summarised by Elsner's equation:

$$4Au + 8CN^- + O_2 + 2H_2O \rightarrow 4Au(CN)_2^- + 4OH^-$$

Cyanogenic microorganisms mobilise metals under alkaline conditions in contrast to acidophilic bacteria, which have so far been used in bioleaching of heavy metals from solid waste (Brandl et al., 2008). Further, it was also shown that copper present in electronic scrap competes with gold for the cyanide produced. A study on the effects of cyanide and dissolved oxygen concentration on gold-dissolution rate from Au powder revealed that dissolved oxygen exerts a greater effect on biological gold recovery. Dissolved oxygen may be enhanced by hydrogen peroxide; a study showed that 11% gold recovery from waste mobile phone PCB was achieved using *C. violaceum* with hydrogen peroxide supplementation (Chi et al., 2011).

Microbial oxidation using iron- and SOB for gold beneficiation from refractory gold ores prior to hydrometallurgical gold recovery has been extensively implemented in commercial processes worldwide (Lindström et al., 1992; Iglesias and Carranza, 1994; Olson, 1994; Bosecker, 1997; Brierley and Brierley, 2001; Campbell et al., 2001; Rawlings et al., 2003). In an e-waste bioleaching study, biooxidation as pretreatment removed about 80% of the copper, thereby increasing the gold/copper ratio and resulting in enhanced gold bioleaching by *C. violaceum* (Pham and Ting, 2009). Since higher pH increases the availabilty of free cyanide ions (CN^-) for metal complexation ($H^+ + CN^- \rightarrow HCN$), the work was extended using *C. violaceum* mutated to grow at pH 10 (Natarajan and Ting, 2014). In another attempt to increase cyanide concentration, Tay et al. demonstrated the utility of lixiviant metabolic engineering in the construction of cyanogenic bacteria with additional cyanide producing operon for enhancing bioleaching of precious metals from electronic waste by increased cyanide lixiviant production (Tay et al., 2013). A recent work by Pradhan and Kumar (2012) examined metal mobilisation by cyanogenic bacteria and showed that mixed culture of *C. violaceum* and *P. aeruginosa* had better ability to leach precious metals such as gold and silver and base metals such as iron, copper and zinc compared with the individual strains.

It is important to note that a direct comparison of the leaching efficiencies by different microorganisms is difficult since the electronic scrap used differs in composition and size (even for the same source of electronic waste). Other factors such as growth medium, bioleaching period and pulp density vary. To complicate matters, researchers used different acid-digestion protocols to determine the total metal composition of the electronic scrap (which is important in determining the bioleaching efficiency) due to the lack of a standard USEPA protocol, unlike in fly ash and spent catalyst leaching. In addition, most researchers do not report the presence of non-metallic components such as plastics, ceramics, refractory oxides and silica, which may affect the bacterial cells and lixiviant production.

Although the motivation and prime economic driver for the bioleaching of e-wastes was the recovery of precious metals such as gold, silver and platinum, unfortunately, bioleaching setups have not progressed past the laboratory-scale study. Indeed, bioleaching of metals from electronic scrap may find a potential industrial application only if the extraction efficiencies of the metals are improved by increased production of lixiviants, possibly through the optimisation of conditions such as aeration and pH or the use of genetically engineered strains.

9.5.3 Bioleaching of Metals from Battery Waste

In bioleaching experiments on LIBs using *A. ferroxidans*, it was reported that the presence of higher Fe (II) concentrations resulted in jarosite formation and the precipitation of Fe (III) ions reduced cobalt dissolution (Mishra et al., 2008a). Bioleaching of LIB with a mixed culture of *A. thiooxidans* and *A. ferroxidans* grown with different energy sources (elemental sulphur, FeS_2 and a combination of the two) showed that Li was bioleached by non-contact mechanism (i.e. indirect leaching) wherein sulphuric acid played a prominent role in the leaching process. Co^{3+} was first reduced to soluble Co^{2+} by Fe^{2+} and then released by acid dissolution. The dissolution of Co was higher in systems having FeS_2 as the energy source and was highest in $FeS_2 + S$ using the system while Li was best leached in systems using elemental sulphur as the energy source (Xin et al., 2009). In the case of Zn–Mn batteries, Zn leaching was chemical-reaction controlled while Mn leaching was diffusion controlled (Xin et al., 2012b). In a study on Ni–Cd battery, a two-stage system consisting of a bioreactor generating sulphuric acid (i.e. with *Thiobacillus ferroxidans* immobilised on elemental sulphur) and a percolation column (with the battery waste, with the liquid phase from sulphuric acid bioreactor) successfully recovered Cd and Ni (Cerruti et al., 1998). A three-stage system with a bioreactor/acidifying reactor, settling tank and leaching reactor was employed to leach nickel and cadmium completely from anodic and cathodic electrodes used in Ni–Cd batteries using a *Thiobacilli* species (Zhao et al., 2007). The authors also examined the effect of hydraulic retention time and process loading on the recovery of Ni, Cd and Co.

Ni leaching from spent Ni–Cd batteries is influenced by several factors. In a study on the effect of temperature, leaching medium and liquid-to-solid ratio, a mixture of adapted *A. ferroxidans* and *A. thiooxidans* at 30°C grown in 9 K medium and a L:S ratio of 100:1 gave the highest Ni recovery of 84% after 28 days in one-step bioleaching (Velgosov et al., 2010). Although the authors have identified the parameters influencing the bioleaching process, an important factor, pH was not discussed. One approach used by several researchers is multi-objective optimisation using factorial designs to bring out the interaction among different parameters (Xu and Ting, 2009; Amiri et al., 2011c).

9.5.4 Bioleaching of Metals from Spent Catalyst

Bioleaching of spent catalyst was first carried out in the early 1990s, for the extraction of Mo from a coal liquefaction catalyst residues using iron-oxidising cultures of *A. ferrooxidans* and *Leptospirillum ferrooxidans* (Blaustein et al., 1993). Low Mo recovery rates were reported, even after 42 days of bioleaching at 30°C. In another study, both *A. ferrooxidans* and a *Solfolobus* species efficiently leached Ni from Ni–Mo spent catalyst, but the presence of Mo had an adverse effect on microbial growth and subsequent extraction rates (Joffe and Sperll, 1993). It was concluded that only organisms tolerant to high Mo concentration could be used for bioleaching of the spent catalyst. For instance, denitrifying bacteria and thermophilic microbes growing at 60°C such as *Bacillus stearothermophilus* and *Metallosphaera sedula* were able to extract >80% of both Ni and Mo from the spent catalyst (Sanback and Joffe, 1993; Joffe and Sperll, 1993). Briand et al. utilised a sulphur-oxidising microbe, *A. thiooxidans* to recover 98% of vanadium from vanadium–titanium and spent vanadium–phosphorus catalyst (Briand et al., 1999). The main leaching agent was sulphuric acid produced by bacterial oxidation of elemental sulphur added as the energy source for *A. thiooxidans*. Apart from chemolithotrophic bacteria, heterotrophic fungus *A. niger* and *P. simplicissimum* have been extensively studied for the bioleaching of the spent hydroprocessing catalyst and FCC catalyst (Santhiya and Ting, 2005; Amiri et al., 2011a).

The effect of bioleaching methods, that is, one-step, two-step and spent medium leaching have been investigated for the bioleaching of the spent catalyst. Two-step bioleaching and spent medium leaching were invariably more effective than one-step bioleaching (Mishra et al., 2007; Deng et al., 2012) due to the inhibition of microbial growth (which leads to reduced metabolite secretion) in the presence of minerals and industrial waste containing high concentration of metals. The efficacy of two-step and spent medium leaching however was dependent on many factors including the microbial strain, type of spent catalyst and process parameters. For instance, the presence of the spent catalyst has been observed to induce organic acid production in fungal leaching, resulting in higher yields from two-step leaching (Aung and Ting, 2005; Amiri et al., 2011a). Results suggest that fungal leaching occurs via the indirect mechanism through biologically produced metabolites as well as direct interaction of the fungus through bioaccumulation (Burgstaller and Schinner, 1993). Unlike fungal leaching, spent medium leaching was found to be most effective for the leaching of the hydrotreating catalyst by thermophilic archaea *A. brierleyi* because the microbes were sensitive to the presence of the toxic spent catalyst (Bharadwaj and Ting, 2013a).

Higher optimum pulp density is desirable as it allows a smaller reactor volume and energy consumption per unit weight of the waste material. Microbial tolerance to high solid content can be improved by adapting it to high metal concentration prior to bioleaching. Adaptation experiments showed that *A. niger* was able to tolerate up to 1000 mg/L Ni, 1200 mg/L Mo and 2000 mg/L Al separately and a combination of all three metals up to 100 mg/L Ni, 200 mg/L Mo and 600 mg/L Al (with the metals added in a ratio of 1:2:6 to mimic the spent catalyst composition) (Santhiya and Ting, 2006). In contrast to un-adapted strains, *A. niger* adapted to all three metals in combination was able to grow in the presence of 3% pulp density and achieve the maximum leaching efficiency of 78.5% Ni, 82.3% Mo and 65.2% Al at 1% pulp density over 30 days. A similar approach was used for the bioleaching of spent hydrocracking catalyst with adapted *A. niger* and *P. simplicissimum* (Amiri et al., 2011a,b). The effect of pulp density was investigated using one-step, two-step and spent medium leaching of the spent catalyst. The adapted *P. simplicissimum* and *A. niger* was able to tolerate up to 3% and 5% pulp density, respectively. Amiri et al. reported that higher pulp densities (3%–5%) induced gluconic acid production (the main lixiviant in one-step and two-step leaching) and inhibited citric acid production (the main lixiviant in spent medium leaching) in *P. simplicissimum* and *A. niger*. Citric acid production was inhibited at higher pulp densities by the presence of Mn and Fe in the spent catalyst. The authors claimed that Mn strongly inhibits citric acid accumulation in the tricarboxylic acid (TCA) cycle.

Bacterial leaching of the spent catalyst has been extensively studied with mesophiles such as *A. ferrooxidans* and *A. thiooxidans*. Gholami et al. reported higher extraction rates by iron oxidising *A. ferrooxidans* (63% Al, 96% Co, 84% Mo and 99% Ni) than sulphur oxidising *A. thiooxidans* (2.4% Al, 83% Co, 95% Mo and 16% Ni) after 30 days of bioleaching (Gholami et al., 2011a).

Metal dissolution rate for *A. ferrooxidans* peaked during the first 5 days and remained stable for the remaining period, whereas the leaching rate for *A. thiooxidans* was slow during the initial days and increased sharply by the end of 30 days. Pradhan et al. studied the kinetic aspects of Ni, V and Mo leaching from spent catalyst by iron oxidising and sulphur oxidising mesophilic bacteria (Pradhan et al., 2009). Dual-leaching kinetics was observed, that is, an initial faster leaching rate within the first 24 h responsible for the majority of metal dissolution (80% leaching efficiency of Mo, Ni and V), followed by a slower rate. A dual-leaching rate was suggested to be due to surface and intraparticle diffusion resistance, resulting in faster and slower leaching rates, respectively. The mass transfer coefficients (surface and pore diffusivity) for surface- and intraparticle-mediated leaching were examined using Fick's law, with pore diffusion resulting in higher resistance to the leaching reaction compared to surface diffusion. Mo leaching efficiency was lowest in comparison to Ni and V, which was suggested to be due to the diffusion barrier from the sulphur layer and refractory nature of Mo sulphide dissolution. In a similar study by the same group, precipitation of Mo as molybdenum trioxide and the possibility of the pretreatment process exposing Mo phase to a different extent were cited as the reasons for low leaching efficiency (Mishra et al., 2008b).

More recently, thermophilic microbes have been utilised for the leaching of the spent catalyst. The major disadvantages with mesophile leaching are the slower leaching rates and long contact time required for efficient leaching. Previous studies on bioleaching of low-grade ores have reported higher metal extraction rates by thermophilic microbes than mesophiles (Dew et al., 1999; Mousavi et al., 2005). The extreme thermophile *A. brierleyi* demonstrated superior leaching kinetics (76% Al, 98% Ni and 93% Mo) compared to the mesophile *A. thiooxidans* (31% Al, 65% Ni and 44% Mo) after 9 days of spent medium leaching of the spent hydrotreating catalyst (Bharadwaj and Ting, 2013b). Both microbes solubilised metals through an indirect mechanism, where biogenically sulphuric acid produced from the oxidation of elemental sulphur was the primary leaching agent. In another study, pH, pulp density, inoculation volume and elemental sulphur concentration were identified as the significant parameters affecting the spent catalyst leaching by *A. brierleyi* using the Plackett–Burman factorial design (Gerayeli et al., 2013). The central composite design predicted the optimum conditions of pH 1.6, pulp density 0.6% (w/v), inoculation 4% (v/v) and elemental sulphur concentration 4 g/L for maximum metal extraction. A mixed culture of moderate thermophiles was able to efficiently leach Ni (up to 96%) and Al (up to 76%) although recovery rates for Mo and V remained low (<20% and <10%, respectively) irrespective of the varying pulp densities (5% and 10% w/v) and particle size (45–106, 106–212 and >212 μm) (Srichandan et al., 2012). Similar findings were also reported in the thermophile leaching of spent catalyst by *Sulfolobus metallicus* (Kim et al., 2012). Current research shows the distinct potential of improving leaching kinetics by using thermophilic microbes, but the higher energy input required for their growth cannot be ignored. It is therefore important to evaluate the economic trade-off between higher production cost (due to excess energy) and improvement in leaching efficiency.

In the preceding discussion on bioleaching studies, the common approach by most researchers has been to consider some of the factors affecting bioleaching in isolation, that is, using a 'one-factor-at-a-time' technique. This approach unfortunately does not pin-point the region of optimum response, since it fails to reveal any interactions that may occur among the factors during bioleaching. A useful scheme is to consider a matrix of all combinations of the operating variables as it would have the advantage of thoroughly exploring the experimental surface. Unfortunately, such an approach would require a large number of experimental measurements. A robust mathematical approach, called the response surface method (RSM) is superior as it is able to simultaneously consider several factors at different levels, and may reveal corresponding interactions among these factors, using a smaller number of experimental observations (Amiri et al., 2011c). Besides, the use of statistical inference techniques may reveal the importance of the individual factors, the appropriateness of their functional form and the sensitivity of the response to each factor (Xu and Ting, 2004).

If bioleaching yields greater metal extraction than chemical leaching, it may hold promise as a clean and more environmentally sustainable technology for the treatment of industrial waste. When chemical leaching (using citric acid, gluconic acid and oxalic acid) was compared with bioleaching at the same concentration of biogenically produced acids, bioleaching gave up to 20% higher metal extraction efficiency (Aung and Ting, 2005), which suggested the importance of secondary metabolites and direct leaching in the bioleaching process (Schinner and Burgstaller, 1989). Similar results have also been reported (Bosio et al., 2008). In a study comparing bioleaching efficiency of *A. thiooxidans* and chemical leaching performed using the same concentration of sulphuric acid produced in two-step bioleaching (200 mM), Ni-leaching efficiency was found comparable to bioleaching, whereas Mo and V extraction was higher in chemical leaching (Mishra et al., 2007).

9.6 Limitations of Bioleaching

Although considerable research has been done in bioleaching of metal-bearing waste materials, the technology still remains in its infancy. Despite its numerous advantages, it is recognised that the main limitation is the long leaching time required compared to chemical leaching, since lixiviants are produced in lower concentrations by the microorganism. Metal recovery by bioleaching requires anywhere from a few days in the case of cyanogenic bacteria for gold recovery from electronic scrap at the laboratory scale, to 3–18 months as in the case of heap bioleaching, depending on the metal of interest and the scale of operation. Since the microorganisms in most cases are unable to tolerate the high concentration of toxic metals present in solid waste, operation at high pulp densities is another major limitation. In the case of fly ash, electronic scrap and spent catalyst, the maximum pulp density reported is 10% w/v (Seidel et al., 2001), 18% w/v (Liang et al., 2010) and 5% w/v (Amiri et al., 2011b) respectively.

One has to note that the research on bioleaching of metal-bearing wastes has so far been carried out at bench scale and that more studies are needed to establish bioleaching as a suitable alternative to conventional techniques. Scaling up the bioleaching process is challenging due to issues on aeration, hydrodynamics and temperature control. Agglomeration and compaction of waste particles in a large-scale setup during bioleaching remain the major hindrances to scaling up.

9.7 Future of Solid Waste Bioleaching

Bioleaching offers significant advantages over the conventional chemical processes, being safer, operationally simpler, less energy intensive and requiring low capital and operating cost. Furthermore, the release of CO_2 and toxic chemicals is much lower in bioleaching, thus making it an environmentally friendly process (Moskvitch, 2012). There has been genuine growth in the commercialisation of low-grade ores and mine tailings, especially for copper recovery and pretreatment of refractory gold ores since the mid-1980s which reflects that these processes are robust and economically viable (Brierley and Brierley, 2001). Huge demand for base metals, precious metals and rare earths both in developed and developing countries and the rapid depeletion of primary resources by conventional pyro- and hydrometallurgical operations have motivated researchers and created opportunities to extend bioleaching operations to the recovery of metals from solid wastes.

There is an urgent need for lab-scale metal-recovery processes from solid wastes to be optimised and scaled up to the pilot plant to prove its potential for full-scale commercial operation. Indeed, full-scale commerical developments can reap huge benefits in meeting the global metal demand and protecting the environment from hazardous waste and operations. Column bioleaching of wastes, which mimics large-scale heap bioleaching, have to be studied extensively in order to identify problems facing commercial-scale operation. The economic feasibility of constructing bioreactors with improved aeration designs and corrosion-resistant materials should also be analysed, taking into account the value of recovered metal for advancement in this technology.

However, there are several major technical challenges that pose impediments to commercialisation, such as slow reaction rate and low process efficiency. Several studies have demonstrated the potential in the use of thermophiles for improved extraction and processing time (Konishi et al., 2003; Ilyas et al., 2007; Bharadwaj and Ting, 2013a). Further, research is needed on overcoming the challenges associated with high-temperature leaching such as gas transfer limitations, metal toxicity and economic constraints with slightly higher energy requirements. To enhance the bioleaching process, techniques such as sonobioleaching (Swamy et al., 2005), electrochemical bioleaching (Ho, 2011) and catalyst-driven bioleaching (Zeng et al., 2012) are being explored. It must be borne in mind, however, that the applicability of these techniques is specific to bioleaching reactions and conditions, the microorganisms used, metabolites produced by these microorganisms and the target metals to be recovered.

Higher leaching efficiency may also be obtained by using more tolerant strains of microorganisms. Indigenous microbial populations in bioleaching heaps and dump sites have been the source of waste-tolerant strains and better performers. For instance, a new indigenous strain of *A. ferroxidans* isolated from sulphide mine for the bioleaching application has shown high levels of tolerance to lead and zinc and has the ability to grow at very low pH, thereby significantly improving metal recovery yields (Nemati et al., 2014). The change in population dynamics should be closely monitored to study the effect of varying process parameters on microbial growth. Further studies need to be performed using genetically modified strains or mutated strains to enhance the efficiency of the bioleaching process. For instance, a metabolically engineered strain of *C. violaceum* produces a higher concentration of the cyanide lixiviant which leads to twice as much gold recovery from electronic waste (Tay et al., 2013). Rapid, accurate and simple techniques for monitoring the bioleach systems are needed for control of these processes by operators who are unlikely to be microbiologists.

In advancing biohydrometallurgy for metal recovery from solid waste, it is necessary to understand how the various components in the system interact with each other to bring about the bioleaching process. As waste generation continues to increase, the success of biological processing of wastes depends on the speed of integration of the developments in microbiological and hydrometallurgical engineering into a single system. Stringent government regulations and research policies that favour 'green' technologies could be key incentives for developing such processes. Indeed, solid wastes are considered as artificial ores and serve as secondary raw materials and hence reduce the demand for primary mineral resources. With increasing concern over the environmental impact of solid waste disposal as the driving force, as well as recognising the need for resource conservation in the modern era of environmental sustainability, the industry is likely to consider adopting bioleaching of solid waste for both detoxification and resource recovery.

References

ACAA. 2010. 1966–2010—Fly ash production and use charts. http://www.acaa-usa. org/Portals/9/Files/PDFs/1966-2010_FlyAsh_Prod_and_Use_Charts.pdf. (Accessed 20 May, 2014.)

Amiri, F., Yaghmaei, S., and Mousavi, S.M. 2011a. Bioleaching of tungsten-rich spent hydrocracking catalyst using *Penicillium simplicissimum*. *Bioresour Technol* 102:1567–1573.

Amiri, F., Yaghmaei, S., Mousavi, S.M., and Sheibani, S. 2011b. Recovery of metals from spent refinery hydrocracking catalyst using adapted *Aspergillus niger*. *Hydrometallurgy* 109: 65–71.

Amiri, F., Mousavi, S.M., and Yaghmaei, S. 2011c. Enhancement of bioleaching of a spent Ni/Mo hydroprocessing catalyst by *Penicillium simplicissimum*. *Sep Purif Technol* 80: 566–576.

AQM Copper Inc. 2013. Copper fundamentals. http://www.aqmcopper.com/s/ copperfundamentals.asp. (Accessed 15 Dec, 2013.)

Aung, K.M.M. and Ting, Y.P. 2005. Bioleaching of spent fluid catalytic cracking catalyst using *Aspergillus niger*. *J Biotechnol* 116: 159–170.

Bankar, A., Winey, M., Prakash, D., Kumar, A.R., Gosavi, S., Kapadnis, B., and Zinjarde, S. 2012. Bioleaching of fly ash by the tropical marine yeast, *Yarrowia lipolytica* NCIM 3589. *Appl Biochem Biotechnol* 168: 2205–2217.

Beolchini, F., Fonti, V., Ferella, F., and Veglio, F. 2010. Metal recovery from spent refinery catalysts by means of biotechnological strategies. *J Hazard Mater* 178: 529–534.

Bharadwaj, A. and Ting, Y.P. 2013a. Bioleaching of spent hydrotreating catalyst by acidophilic thermophile *Acidianus brierleyi*: Leaching mechanism and effect of decoking. *Bioresour Technol* 130: 673–680.

Bharadwaj, A. and Ting, Y.P. 2013b. Bioleaching of spent hydrotreating catalyst by thermophilic and mesophilic acidophiles: Effect of decoking. *Adv Mater Res* 825: 280–283.

Blaustein, B.D., Hauck, J.T., Olson, G.J., and Baltrus, J.P. 1993. Bioleaching of molybdenum from coal liquefaction catalyst residues. *Fuel* 72: 1613–1618.

Bosecker, K. 1987. Microbial recycling of mineral waste products. *Acta Biotechnol* 7: 487–497.

Bosecker, K. 1997. Bioleaching: Metal solubilization by microorganisms. *FEMS Microbiol Rev* 20: 591–604.

Bosio, V., Viera, M., and Donati, E. 2008. Integrated bacterial process for the treatment of a spent nickel catalyst. *J Hazard Mater* 154: 804–810.

Bosshard, P.P., Bachofen, R., and Brandl, H. 1996. Metal leaching of fly ash from municipal waste incineration by *Aspergillus niger*. *Environ Sci Technol* 30: 3066–3070.

Brandl, H. 2001. Microbial leaching of metals, In *Biotechnology Set*, Second Edition (eds H.-J. Rehm and G. Reed), Wiley-VCH Verlag GmbH, Weinheim, Germany. doi: 10.1002/9783527620999.ch8k.

Brandl, H., Bosshard, R., and Wegmann, M. 2001. Computer-munching microbes: Metal leaching from electronic scrap by bacteria and fungi. *Hydrometallurgy* 59: 319–326.

Brandl, H., Lehmann, S., Faramarzi, M.A., and Martinelli, D. 2008. Biomobilization of silver, gold, and platinum from solid waste materials by HCN-forming microorganisms. *Hydrometallurgy* 94: 14–17.

Briand, L., Thomas, H., de la Vega Alonso, A., and Donati, E. 1999. Vanadium recovery from solid catalysts by means of *Thiobacilli* action. *Process Metall* 9: 263–271.

Brierley, C. 2008. How will biomining be applied in future? *T Nonferr Metal Soc* 18: 1302–1310.

Brierley, J. and Brierley, C. 2001. Present and future commercial applications of biohydrometallurgy. *Hydrometallurgy* 59: 233–239.

Brombacher, C., Bachofen, R., and Brandl, H. 1998. Development of a laboratory-scale leaching plant for metal extraction from fly ash by *Thiobacillus* strains. *Applied Environ Microbiol* 64: 1237–1241.

Burgstaller, W. and Schinner, F. 1993. Leaching of metals with fungi. *J Biotechnol* 27: 91–116.

Campbell, S., Olson, G., Clark, T., and McFeters, G. 2001. Biogenic production of cyanide and its application to gold recovery. *J Ind Microbiol Biotechnol* 26: 134–139.

Cerruti, C., Curutchet, G., and Donati, E. 1998. Bio-dissolution of spent nickel–cadmium batteries using *Thiobacillus ferrooxidans*. *J Biotechnol* 62: 209–219.

Chi, T.D., Lee, J., Pandey, B., Yoo, K., and Jeong, J. 2011. Bioleaching of gold and copper from waste mobile phone PCBs by using a cyanogenic bacterium. *Miner Eng* 24: 1219–1222.

Choi, M.S., Cho, K.S., Kim, D.S., and Kim, D.J. 2005. Microbial recovery of copper from printed circuit boards of waste computer by *Acidithiobacillus ferrooxidans*. *J Environ Sci Health*, *Part A* 39: 2973–2982.

Cui, J. and Zhang, L. 2008. Metallurgical recovery of metals from electronic waste: A review. *J Hazard Mater* 158: 228–256.

Dalrymple, I., Wright, N., Kellner, R., Bains, N., Geraghty, K., Goosey, M., and Lightfoot, L. 2007. An integrated approach to electronic waste (WEEE) recycling. *Circuit World* 33: 52–58.

Das, A.P., Swain, S., Panda, S., Pradhan, N., and Sukla, L.B. 2012. Reductive acid leaching of low grade manganese ores. *Geomaterials* 2: 70.

Deng, X., Chai, L., Yang, Z., Tang, C., Tong, H., and Yuan, P. 2012. Bioleaching of heavy metals from a contaminated soil using indigenous *Penicillium chrysogenum* strain F1. *J Hazard Mater* 233–234: 25–32.

Dew, D., Van Buuren, C., McEwan, K., and Bowker, C. 1999. Bioleaching of base metal sulphide concentrates: A comparison of mesophile and thermophile bacterial cultures. *Process Metall* 9: 229–238.

Diederen, A. 2009. Metal minerals scarcity: A call for managed austerity and the elements of hope. TNO Defence, Security and Safety. http://europe.theoildrum.com/node/5239. (Accessed 10 May, 2014.)

Dufresne, P. 2007. Hydroprocessing catalysts regeneration and recycling. *Appl Catal A: General* 322: 67–75.

ECMA. 2001. Guidelines for the management of spent catalysts. European Chemical Industry Council. http://www.cefic.org/Documents/Industry%20sectors/ECMA/ECMA%20GUIDELINES%20FOR%20THE%20MANAGEMENT%-20OF%20SPENT%20CATALYSTS-201202.pdf. (Accessed 16 Jan, 2014.)

Ehrlich, H. 1992. Metal extraction and ore discovery. *Encycl Microbiol* 3: 75–80.

Ehrlich, H. 1997. Microbes and metals. *Appl Microbiol Biotechnol* 48: 687–692.

Eijsbouts, S., Battiston, A., and Vanleerdam, G. 2008. Life cycle of hydroprocessing catalysts and total catalyst management. *Catal Today* 130: 361–373.

ETBC. 2012. Facts and figures on E-waste and recycling. Electronics take back coalition. http://www.electronicstakeback.com/wp-content/uploads/Facts_and_Figures. (Accessed 12 Mar, 2014.)

Faramarzi, M.A., Stagars, M., Pensini, E., Krebs, W., and Brandl, H. 2004. Metal solubilization from metal-containing solid materials by cyanogenic *Chromobacterium violaceum*. *J Biotechnol* 113: 321–326.

Fass, R., Geva, J., Shalita, Z.P., White, M.D., and Fleming, J.C. 1994. Demetallization with biosynthetic sulfuric acid. Patent filed by The Israel Electric Corporation Ltd. (Patent number: US 5278069 A).

Furimsky, E. 1996. Spent refinery catalysts: Environment, safety and utilization. *Catal Today* 30: 223–286.

Gerayeli, F., Ghojavand, F., Mousavi, S.M., Yaghmaei, S., and Amiri, F. 2013. Screening and optimization of effective parameters in biological extraction of heavy metals from refinery spent catalysts using a thermophilic bacterium. *Sep Purif Technol* 118:151–161.

Gholami, R.M., Borghei, S.M., and Mousavi, S.M. 2011a. Bacterial leaching of a spent Mo–Co–Ni refinery catalyst using *Acidithiobacillus ferrooxidans* and *Acidithiobacillus thiooxidans*. *Hydrometallurgy* 106: 26–31.

Gholami, R.M., Borghei, S.M., and Mousavi, S.M. 2011b. Fungal leaching of hazardous heavy metals from a spent hydrotreating catalyst. *Proc World Acad Sci, Eng Technol* 76: 726–731.

Gramatyka, P., Nowosielski, R., and Sakiewicz, P. 2007. Recycling of waste electrical and electronic equipment. *J Achievements Mater Manuf Eng* 20: 535–538.

Groudeva, V., Krumova, K., and Groudev, S.N. 2007. Bioleaching of a rich-in-carbonates copper ore at alkaline pH. *Adv Mater Res* 20: 103–106.

Greenpeace. 2005. The e-waste problem. Greenpeace International. http://www.greenpeace.org/international/en/campaigns/toxics/electronics/the-e-waste-problem/. (Accessed 7 May, 2014.)

Hageluken, C. 2006. *Improving Metal Returns and Eco-Efficiency in Electronics Recycling—A Holistic Approach for Interface Optimisation between Pre-Processing and Integrated Metals Smelting and Refining*. Paper presented at the Electronics and the Environment, Scottsdale, Arizona.

Hagins, J.M., Locy, R., and Silo-Suh, L. 2009. Isocitrate lyase supplies precursors for hydrogen cyanide production in a cystic fibrosis isolate of *Pseudomonas aeruginosa*. *J Bacteriol* 191: 6335–6339.

Ho, K.Y. 2011. Electro-bioleaching of spent hydroprocessing Ni–Mo catalysts. Department of Chemical and Biomolecular Engineering (Meng thesis, National University of Singapore).

Iglesias, N. and Carranza, F. 1994. Refractory gold-bearing ores: A review of treatment methods and recent advances in biotechnological techniques. *Hydrometallurgy* 34: 383–395.

Ilyas, S., Anwar, M.A., Niazi, S.B., and Afzal Ghauri, M. 2007. Bioleaching of metals from electronic scrap by moderately thermophilic acidophilic bacteria. *Hydrometallurgy* 88: 180–188.

Ishigaki, T., Nakanishi, A., Tateda, M., Ike, M., and Fujita, M. 2005. Bioleaching of metal from municipal waste incineration fly ash using a mixed culture of sulfur-oxidizing and iron-oxidizing bacteria. *Chemosphere* 60: 1087–1094.

Islam, M. and Ting, Y.P. 2009. Fungal bioleaching of spent hydroprocessing catalyst: Effect of decoking and particle size. *Adv Mater Res* 71: 665–668.

Jadhav, U. and Hocheng, H. 2012. A review of recovery of metals from industrial waste. *J Achieve Mater Manufact Eng* 54(2): 10.

Joffe, P. and Sperll, G. 1993. Microbiological recovery of metals from spent coal lique-faction catalysts. SciTech Connect, DOE Report, PC-92119-T4. http://www.osti.gov/scitech/biblio/10150845. (Accessed 12 Dec, 2014.)

Kim, D.-J., Srichandan, H., Gahan, C.S., and Lee, S.-W. 2012. Thermophilic bioleaching of spent petroleum refinery catalyst using *Sulfolobus metallicus*. *Can Metall Q* 51: 403–412.

Kinzler, K., Gehrke, T., Telegdi, J., and Sand, W. 2003. Bioleaching—A result of interfacial processes caused by extracellular polymeric substances (EPS). *Hydrometallurgy* 71: 83–88.

Knowles, C.J. and Bunch, A.W. 1986. Microbial cyanide metabolism. *Adv Microb Physiol* 27: 73–111.

Konishi, Y., Matsui, M., Fujiwara, H., Nomura, T., and Nakahara, K. 2003. Zinc leaching from fly ash in municipal waste incineration by thermophilic archaean *Acidianus brierleyi* growing on elemental sulfur. *Sep Sci Technol* 38: 4117–4130.

Korte, F., Spiteller, M., and Coulston, F. 2000. The cyanide leaching gold recovery process is a nonsustainable technology with unacceptable impacts on ecosystems and humans: The disaster in Romania. *Ecotoxicol Environ Saf* 46: 241–245.

Krebs, W., Bachofen, R., and Brandl, H. 2001. Growth stimulation of sulfur oxidizing bacteria for optimization of metal leaching efficiency of fly ash from municipal solid waste incineration. *Hydrometallurgy* 59: 283–290.

Krebs, W., Brombacher, C., Bosshard, P.P., Bachofen, R., and Brandl, H. 1997. Microbial recovery of metals from solids. *FEMS Microbiol Rev* 20: 605–617.

Lam, C.H.K., Ip, A.W.M., Barford, J.P., and McKay, G. 2010. Use of incineration MSW ash: A review. *Sustainability* 2: 1943–1968.

Lee, C.K. and Rhee, K.I. 2002. Preparation of $LiCoO_2$ from spent lithium-ion batteries. *J Power Sources* 109: 17–21.

Li, J., Lu, H., Guo, J., Xu, Z., and Zhou, Y. 2007. Recycle technology for recovering resources and products from waste printed circuit boards. *Environ Sci Technol* 41: 1995–2000.

Li, Y. and Xi, G. 2005. The dissolution mechanism of cathodic active materials of spent Zn–Mn batteries in HCl. *J Hazard Mater* 127: 244–248.

Liang, G., Mo, Y., and Zhou, Q. 2010. Novel strategies of bioleaching metals from printed circuit boards (PCBs) in mixed cultivation of two acidophiles. *Enzyme Microb Technol* 47: 322–326.

Lindström, E.B., Gunneriusson, E., and Tuovinen, O.H. 1992. Bacterial oxidation of refractory sulfide ores for gold recovery. *Crit Rev Biotechnol* 12: 133–155.

Marafi, M. and Stanislaus, A. 2008a. Spent catalyst waste management: A review: Part I—Developments in hydroprocessing catalyst waste reduction and use. *Resour Conserv Recy* 52: 859–873.

Marafi, M. and Stanislaus, A. 2008b. Spent hydroprocessing catalyst management: A review: Part II. Advances in metal recovery and safe disposal methods. *Resour Conserv Recy* 53: 1–26.

Mielenz, J.R. 2011. Biofuels from protein. *Nature Biotechnol* 29: 327–328.

Mishra, D., Kim, D.J., Ralph, D.E., Ahn, J.G., and Rhee, Y.H. 2007. Bioleaching of vanadium rich spent refinery catalysts using sulfur oxidizing lithotrophs. *Hydrometallurgy* 88: 202–209.

Mishra, D., Kim, D.J., Ralph, D.E., Ahn, J.G., and Rhee, Y.H. 2008a. Bioleaching of metals from spent lithium ion secondary batteries using *Acidithiobacillus ferro-oxidans*. *Waste Manag* 28: 333–338.

Mishra, D., Kim, D.J., Ralph, D.E., Ahn, J.G., and Rhee, Y.H. 2008b. Bioleaching of spent hydro-processing catalyst using acidophilic bacteria and its kinetics aspect. *J Hazard Mater* 152: 1082–1091.

Moskvitch, K. 2012. Biomining: How microbes help to mine copper. BBC News Technology. http://www.bbc.com/news/technology-17406375. (Accessed 13 Jan, 2014.)

Mousavi, S.M., Yaghmaei, S., Vossoughi, M., Jafari, A., and Hoseini, S. 2005. Comparison of bioleaching ability of two native mesophilic and thermophilic bacteria on copper recovery from chalcopyrite concentrate in an airlift bioreactor. *Hydrometallurgy* 80: 139–144.

Moybel Research Laboratory (n.d.). *Helicobacter pylori* urease. University of Michigan Medical School. http://www.umich.edu/~hltmlab/research/pylori/. (Accessed 16 Feb, 2014.)

Natarajan, G. and Ting, Y.-P. 2014. Pretreatment of E-waste and mutation of alkali-tolerant cyanogenic bacteria promote gold biorecovery. *Bioresour Technol* 152: 80–85.

Näveke, R. 1986. Bacterial leaching of ores and other materials. Institut für Mikrobiologie, Technische Universität. http://www.spaceship-earth.org/REM/Naeveke.htm. (Accessed 10 Mar, 2014.)

Nemati, F., Arabian, D., Khalilzadeh, R., and Aghdam, F.A. 2014. Enhancing iron oxidation efficiency by a native strain of *Acidithiobacillus ferrooxidans* via response surface methodology, and characterization of proteins involved in metal resistance by proteomic approach. *J Appl Biotechnol Rep* 1: 1.

Olson, G.J. 1994. Microbial oxidation of gold ores and gold bioleaching. *FEMS Microbiol Lett* 119: 1–6.

Pant, D., Joshi, D., Upreti, M.K., and Kotnala, R.K. 2012. Chemical and biological extraction of metals present in E waste: A hybrid technology. *Waste Manag* 32: 979–990.

Pham, V. and Ting, Y.P. 2009. Gold bioleaching of electronic waste by cyanogenic bacteria and its enhancement with bio-oxidation. *Adv Mater Res* 71: 661–664.

Pradhan, D., Mishra, D., Kim, D.J., Chaudhury, G.R., and Lee, S.W. 2009. Dissolution kinetics of spent petroleum catalyst using two different acidophiles. *Hydrometallurgy* 99: 157–162.

Pradhan, J.K. and Kumar, S. 2012. Metals bioleaching from electronic waste by *Chromobacterium violaceum* and *Pseudomonas* sp. *Waste Manag Res* 30: 1151–1159.

Rawlings, D.E., Dew, D., and du Plessis, C. 2003. Biomineralization of metal-containing ores and concentrates. *Trends Biotechnol* 21: 38–44.

Refinery Catalyst Market Shares, Strategies, and Forecasts, Worldwide, 2012 to 2018, 2012. http://www.researchandmarkets.com/reports/2090113/refinery_catalyst_market_shares_strategies_and. (Accessed 10 Oct, 2014.)

Rohwerder, T., Gehrke, T., Kinzler, K., and Sand, W. 2003. Bioleaching review part A. *Appl Microbiol Biotechnol* 63(3): 239–248.

Rydh, C.J. and Svärd, B. 2003. Impact on global metal flows arising from the use of portable rechargeable batteries. *Science Total Environ* 302: 167–184.

Sanback, K. and Joffe, P. 1993. DOE Rep (PC-92119-T2 and T3).

Santhiya, D. and Ting, Y.P. 2005. Bioleaching of spent refinery processing catalyst using *Aspergillus niger* with high-yield oxalic acid. *J Biotechnol* 116: 171–184.

Santhiya, D. and Ting, Y.P. 2006. Use of adapted *Aspergillus niger* in the bioleaching of spent refinery processing catalyst. *J Biotechnol* 121: 62–74.

Sayilgan, E., Kukrer, T., Civelekoglu, G., Ferella, F., Akcil, A., Veglio, F., and Kitis, M. 2009. A review of technologies for the recovery of metals from spent alkaline and zinc–carbon batteries. *Hydrometallurgy* 97: 158–166.

Schinner, F. and Burgstaller, W. 1989. Extraction of zinc from industrial waste by a *Penicillium* sp. *Appl Environ Microbiol* 55: 1153.

Seidel, A., Zimmels, Y., and Armon, R. 2001. Mechanism of bioleaching of coal fly ash by *Thiobacillus thiooxidans*. *Chem Eng J* 83: 123–130.

Shapek, R.A. 1995. Local government household battery collection programs: Costs and benefits. *Resour Conserv Recy* 15: 1–19.

Shimaoka, T. and M. Hanashima (1996). Behavior of stabilized fly ashes in solid waste landfills. *Waste Manage* **16**(5): 545–554.

Singer, A., Navrot, J., and Shapira, R. 1982. Extraction of aluminum from fly-ash by commercial and microbiologically-produced citric acid. *Appl Microbiol Biotechnol* 16: 228–230.

Singh, Y. 2010. Fly ash utilization in India. Wealthy waste. http://www.wealthywaste.com/fly-ash-utilization-in-india. (Accessed 10 May, 2014.)

Srichandan, H., Gahan, C.S., Kim, D.-J., and Lee, S.-W. 2012. Bioleaching of spent catalyst using moderate thermophiles with different pulp densities and varying size fractions without Fe supplemented growth medium. *World Acad Sci Eng Technol* 61: 190.

Sum, E.Y.L. 1991. Recovery of metals from electronic scrap. *J Miner, Met Mater Soc* 43: 53–61.

Swamy, K., Narayana, K., and Misra, V.N. 2005. Bioleaching with ultrasound. *Ultrason Sonochem* 12: 301–306.

Tateda, M., Ike, M., and Fujita, M. 1998. Comparative evaluation of processes for heavy metal removal from municipal solid waste incineration fly ash. *J Environ Sci Chin-Eng Ed* 10: 458–465.

Tay, S.B., Natarajan, G., bin Abdul Rahim, M.N., Tan, H.T., Chung, M.C.M., Ting, Y.P., and Yew, W.S. 2013. Enhancing gold recovery from electronic waste via lixiviant metabolic engineering in *Chromobacterium violaceum*. *Sci Rep* 3: 2236.

Torma, A.E. and Singh, A.K. 1993. Acidolysis of coal fly ash by *Aspergillus niger*. *Fuel* 72: 1625–1630.

Tuncuk, A., Stazi, V., Akcil, A., Yazici, E.Y., and Deveci, H. 2012. Aqueous metal recovery techniques from e-scrap: Hydrometallurgy in recycling. *Miner Eng* 25: 28–37.

UNEP. 2009. Recycling—From E-waste to resources. Sustainable innovation and technology transfer industrial sector studies. http://www.unep.org/pdf/pressreleases/E-waste_publication_screen_finalversion-sml.pdf. (Accessed 12 Feb, 2014.)

USEPA. 2003. Hazardous waste management system. United States Environmental Protection Agency. In Federal Register vol 68. No. 202.

U.S. EPA. 2004. Product Stewardship. Batteries. http://permanent.access.gpo.gov/websites/epagov/www.epa.gov/epaoswer/non-hw/reduce/epr/products/batteries.html. (Accessed 12 Oct, 2014.)

USGS. 2008. Mineral Commodity Summaries, Cadmium: 42–43. http://minerals. usgs.gov/minerals/pubs/mcs/2008/mcs2008.pdf. (Accessed 10 Oct, 2014.)

Velgosov, O., Kadukov, J., Mražíkov, A., and Blaškov, A. 2010. Influence of selected parameters on nickel bioleaching from spent Ni–Cd batteries. *Miner Slov* 42: 365–368.

Wang, J., Bai, J., Xu, J., and Liang, B. 2009a. Bioleaching of metals from printed wire boards by *Acidithiobacillus ferrooxidans* and *Acidithiobacillus thiooxidans* and their mixture. *J Hazard Mater* 172: 1100–1105.

Wang, Q., Yang, J. and Wu, T. 2009b. Effects of water-washing pretreatment on bioleaching of heavy metals from municipal solid waste incinerator fly ash. *J Hazard Mater* 162: 812–818.

Wu, H.-Y. and Ting, Y.-P. 2006. Metal extraction from municipal solid waste (MSW) incinerator fly ash—Chemical leaching and fungal bioleaching. *Enzyme Microb Technol* 38: 839–847.

Xiang, Y., Wu, P., Zhu, N., Zhang, T., Liu, W., Wu, J., and Li, P. 2010. Bioleaching of copper from waste printed circuit boards by bacterial consortium enriched from acid mine drainage. *J Hazard Mater* 184: 812–818.

Xin, B., Jiang, W., Aslam, H., Zhang, K., Liu, C., Wang, R., and Wang, Y. 2012a. Bioleaching of zinc and manganese from spent Zn–Mn batteries and mechanism exploration. *Bioresour Technol* 106: 147–153.

Xin, B., Jiang, W., Li, X., Zhang, K., Liu, C., Wang, R., and Wang, Y. 2012b. Analysis of reasons for decline of bioleaching efficiency of spent Zn–Mn batteries at high pulp densities and exploration measure for improving performance. *Bioresour Technol* 112: 186–192.

Xin, B., Zhang, D., Zhang, X., Xia, Y., Wu, F., Chen, S., and Li, L. 2009. Bioleaching mechanism of Co and Li from spent lithium-ion battery by the mixed culture of acidophilic sulfur-oxidizing and iron-oxidizing bacteria. *Bioresour Technol* 100: 6163–6169.

Xu, T.-J. and Ting, Y.-P. 2004. Optimisation on bioleaching of incinerator fly ash by *Aspergillus niger*—Use of central composite design. *Enzyme Microb Technol* 35: 444–454.

Xu, T.-J. and Ting, Y.-P. 2009. Fungal bioleaching of incineration fly ash: Metal extraction and modeling growth kinetics. *Enzyme Microb Technol* 44: 323–328.

Yang, J., Wang, Q., Luo, Q., Wang, Q., and Wu, T. 2009a. Biosorption behavior of heavy metals in bioleaching process of MSWI fly ash by *Aspergillus niger*. *Biochem Eng J* 46: 294–299.

Yang, J., Wang, Q., and Wu, T. 2008. Comparisons of one-step and two-step bioleaching for heavy metals removed from municipal solid waste incineration fly ash. *Environ Eng Sci* 25: 783–789.

Yang, J., Wang, Q., and Wu, T. 2009b. Heavy metals extraction from municipal solid waste incineration fly ash using adapted metal tolerant *Aspergillus niger*. *Bioresour Technol* 100: 254–260.

Yang, T., Xu, Z., Wen, J., and Yang, L. 2009c. Factors influencing bioleaching copper from waste printed circuit boards by *Acidithiobacillus ferrooxidans*. *Hydrometallurgy* 97: 29–32.

Yang, Y. 2013. Metal recovery from industrial solid waste-contribution to resource sustainability, in *REWAS 2013: Enabling Materials Resource Sustainability*, doi: 10.1002/9781118679401.ch41.

Yoshizawa, S., Tanaka, M., and Shekdar, A.V. 2004. Global trends in waste generation. In: *Recycling, Waste Treatment and Clean Technology*, TMS Mineral, Metals and Materials Publishers, Spain, 1541–1552.

Zeng, G., Deng, X., Luo, S., Luo, X., and Zou, J. 2012. A copper-catalyzed bioleaching process for enhancement of cobalt dissolution from spent lithium-ion batteries. *J Hazard Mater* 199–200: 164–169.

Zhao, L., Wang, L., Yang, D., and Zhu, N. 2007. Bioleaching of spent Ni–Cd batteries and phylogenetic analysis of an acidophilic strain in acidified sludge. *Front Environ Sci Eng Chin* 1: 459–465.

Zhao, L., Yang, D., and Zhu, N.W. 2008a. Bioleaching of spent Ni–Cd batteries by continuous flow system: Effect of hydraulic retention time and process load. *J Hazard Mater* 160: 648–654.

Zhao, L., Zhu, N.W., and Wang, X.H. 2008b. Comparison of bio-dissolution of spent Ni–Cd batteries by sewage sludge using ferrous ions and elemental sulfur as substrate. *Chemosphere* 70: 974–981.

Zhou, G.M., Luo, Z.H., and Zhai, X.L. 2007. Experimental study on metal recycling from waste PCB. Paper presented at *International Conference on Sustainable Solid Waste Management*, Chennai, India.

Zhu, N., Zhang, L., Li, C., and Cai, C. 2003. Recycling of spent nickel–cadmium batteries based on bioleaching process. *Waste Manag* 23: 703–708.

(a)   (b)

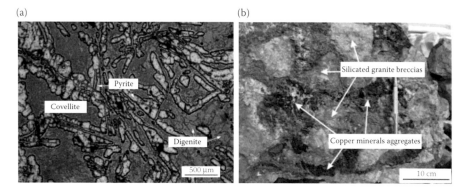

FIGURE 4.2
Optical microscope images showing coarse grains of sulphide aggregates (a) and copper sulphide aggregates associated with the matrix of minerals in the ores (b). (From Ruan, R. et al., 2011. *Hydrometallurgy*, 108(1): 130–135. With permission.)

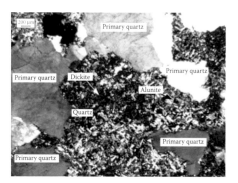

FIGURE 4.3
Optical microscope image showing the altered granite composed by quartz dickite and alunite in the ore body. (From Ruan, R. et al., 2011. *Hydrometallurgy*, 108(1): 130–135. With permission.)

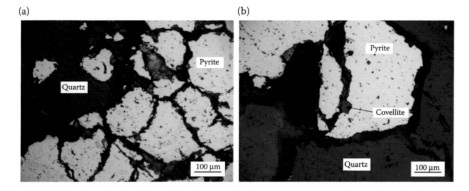

FIGURE 4.4
Optical microscope image showing pyrite in the bioleach residue of industrial heap. The voids between the pyrite grains are from leached copper sulphides (a). A large grain of copper sulphides was leached leaving unleached covellite (b). (From Ruan, R. et al., 2011. *Hydrometallurgy,* 108(1): 130–135. With permission.)

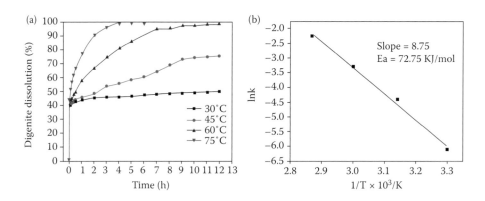

FIGURE 4.13
Effect of temperature on digenite dissolution: (a) Kinetic curve of the dissolution rate and (b) Arrhenius plot for the digenite second stage leaching. (From Ruan, R. et al., 2011. *Hydrometallurgy,* 108(1): 130–135. With permission.)

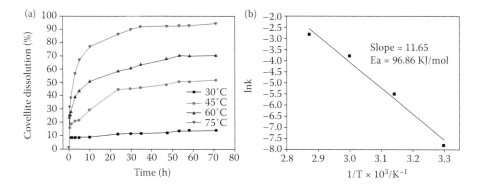

FIGURE 4.14
Effect of temperature on covellite dissolution: (a) Kinetic curve of the dissolution rate and (b) Arrhenius plot for the covellite leaching. (From Ruan, R. et al., 2011. *Hydrometallurgy*, 108(1): 130–135. With permission.)

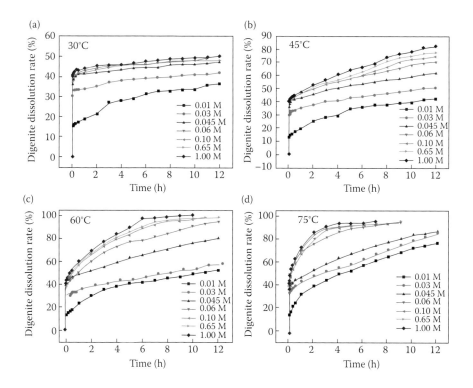

FIGURE 4.15
Effect of ferric concentration on digenite dissolution at (a) 30°C, (b) 45°C, (c) 60°C and (d) 75°C. (From Ruan, R. et al., 2011. *Hydrometallurgy*, 108(1): 130–135. With permission.)

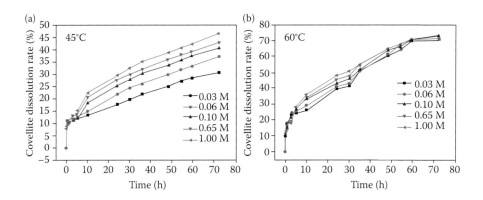

FIGURE 4.16
Effect of Fe^{3+} concentration on covellite dissolution at (a) 45°C and (b) 60°C. (From Ruan, R. et al., 2011. *Hydrometallurgy*, 108(1): 130–135. With permission.)

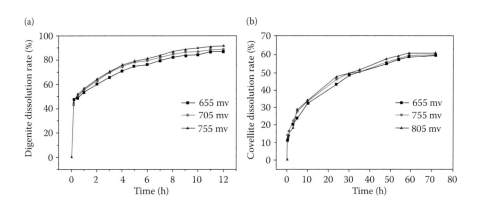

FIGURE 4.17
Effect of redox potential on (a) digenite dissolution and (b) covellite dissolution. (From Ruan, R. et al., 2011. *Hydrometallurgy*, 108(1): 130–135. With permission.)

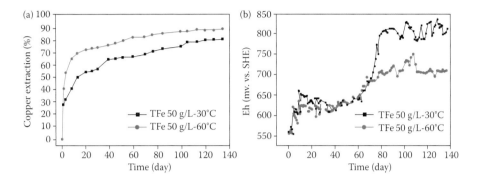

FIGURE 4.18
Column bioleaching at 30°C and 60°C. (a) Copper extraction from the solution and (b) changes of the redox potential. (From Ruan, R. et al., 2011. *Hydrometallurgy*, 108(1): 130–135. With permission.)

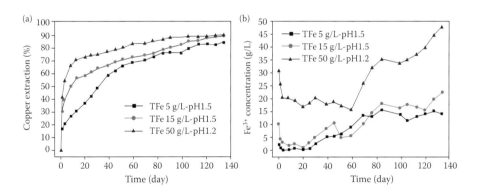

FIGURE 4.19
Effect of Fe^{3+} concentration on column bioleaching: (a) Copper extraction from the solution and (b) change of Fe^{3+} concentration during column leaching. (From Ruan, R. et al., 2011. *Hydrometallurgy*, 108(1): 130–135. With permission.)

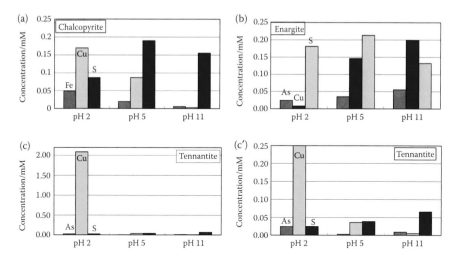

FIGURE 5.1
Dissolved Cu, S, Fe and As concentrations from (a) chalcopyrite, (b) enargite and (c) tennantite after 1 h oxidation by 0.013% H_2O_2 at pH 2, 5 and 11 under O_2 bubbling. Note the scale of the vertical axis in (c) and (c′). Mineral samples were ground to 74–105 μm. 0.5 g of each mineral sample was added into 50 mL of 0.013% H_2O_2 and 0.01 M KNO_3 at pH 2, 5 and 11 adjusted with 1 M HNO_3 and 1 M KOH. (Modified from Sasaki, K. et al., 2010a. *Hydrometal.*, 100, 144–151.)

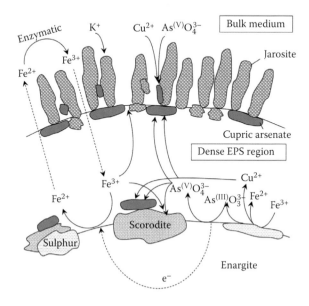

FIGURE 5.7
Schematic model of the bioleaching of enargite followed by the sequential formation of secondary minerals in the vicinity of the enargite surface. (Modified from Takatsugi, K., Sasaki, K., Hirajima, T., 2011. *Hydrometal.*, 109, 90–96.)

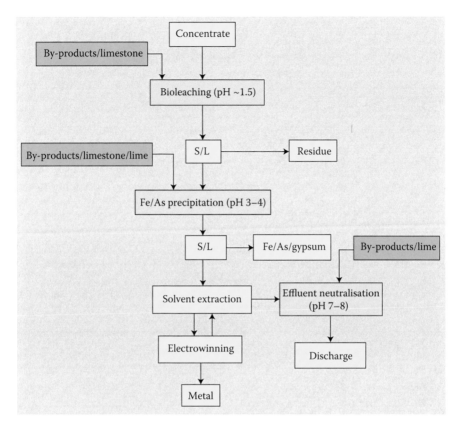

FIGURE 7.3
Flow sheet of a process for base metal production describing stages of neutralisation.

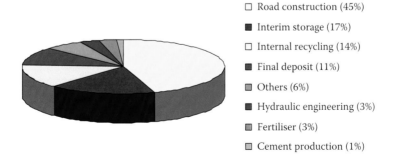

☐ Road construction (45%)

■ Interim storage (17%)

☐ Internal recycling (14%)

■ Final deposit (11%)

■ Others (6%)

■ Hydraulic engineering (3%)

■ Fertiliser (3%)

☐ Cement production (1%)

FIGURE 7.4
Use of steel slag produced in Europe. (Adapted from The European Slag Association (EUROSLAG). Legal status of slags. Position paper. Available in http://www.euroslag.org/media/Position_paper_Jan_2006.pdf, March 2006.)

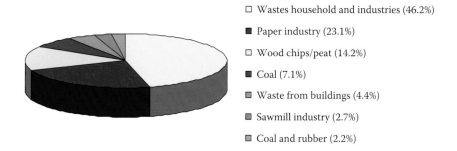

☐ Wastes household and industries (46.2%)

■ Paper industry (23.1%)

☐ Wood chips/peat (14.2%)

■ Coal (7.1%)

▨ Waste from buildings (4.4%)

▨ Sawmill industry (2.7%)

▨ Coal and rubber (2.2%)

FIGURE 7.5
Estimation of ash production per annum in Sweden 2003, estimation by C. Ribbing (2007).

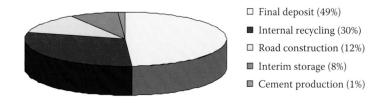

☐ Final deposit (49%)

■ Internal recycling (30%)

☐ Road construction (12%)

▨ Interim storage (8%)

▨ Cement production (1%)

FIGURE 7.6
Use of steel slag produced in Sweden. (Redrawn from Engström, F. Mineralogical influence of different cooling conditions on leaching behaviour of steel making slags. Licentiate thesis, Luleå University of Technology, ISSN: 1402-1757/ISRN LTU-LIC—07/58—SE/NR 2007:58.)

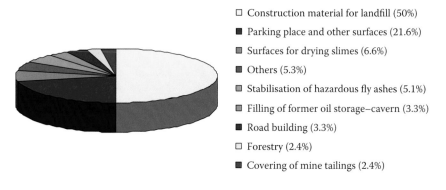

☐ Construction material for landfill (50%)

■ Parking place and other surfaces (21.6%)

▨ Surfaces for drying slimes (6.6%)

■ Others (5.3%)

▨ Stabilisation of hazardous fly ashes (5.1%)

▨ Filling of former oil storage–cavern (3.3%)

■ Road building (3.3%)

☐ Forestry (2.4%)

■ Covering of mine tailings (2.4%)

FIGURE 7.7
Uses of ashes in Sweden 2003, estimation by C. Ribbing (2007).

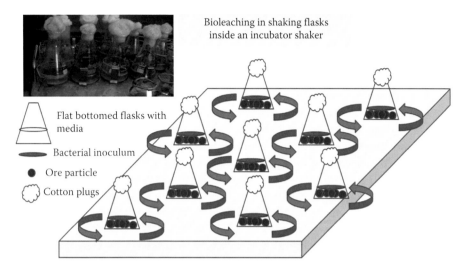

FIGURE 8.3
Showing a pictorial operation of shake flask leaching. A group of flat-bottomed flasks with bacterial inoculum and ore placed at top of a shaker. Arrow in the picture indicates direction of movement of shaking flasks.

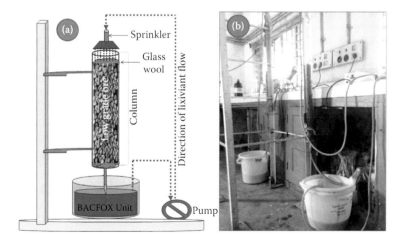

FIGURE 8.4
(a) Schematic of a percolation column bioleaching setup. (b) Experimental setup of a percolation column bioleaching for recovery of copper from low-grade ore. BACFOX Unit—Bacterial Film Oxidation Unit or Bacterial Growth Tank.

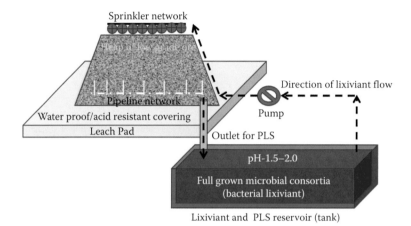

FIGURE 8.5
Schematic diagram of a heap bioleaching process.

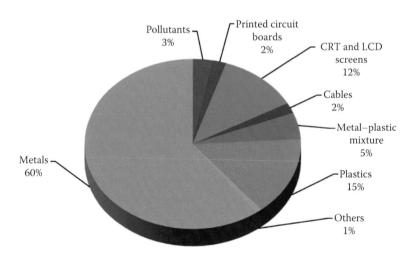

FIGURE 13.1
Typical material fractions in WEEE.

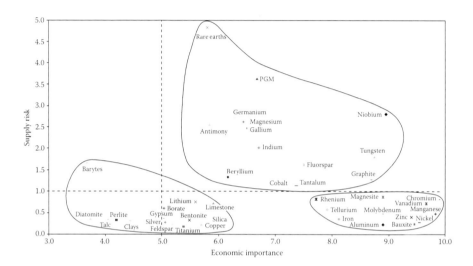

FIGURE 13.2
Critical raw materials in EU.

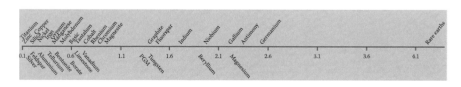

FIGURE 13.3
Critical raw materials according to their environmental country risk.

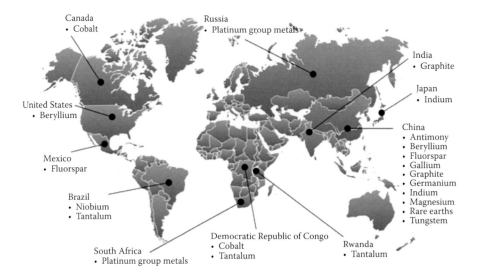

FIGURE 13.4
Production concentration of critical raw mineral materials.

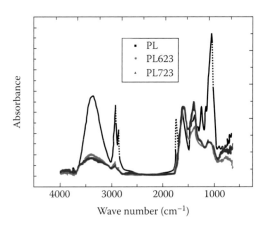

FIGURE 16.2
FT-IR spectra of pineapple leaves carbonised at 723 K (PL723) or 623 K (PL623) and dried pineapple leaves (PL). (Reprinted from *Chemical Engineering Journal* 172(2–3), Ponou, J. et al., Sorption of Cr(VI) anions in aqueous solution using carbonized or dried pineapple leaves, 906–13, Copyright 2011, with permission from Elsevier.)

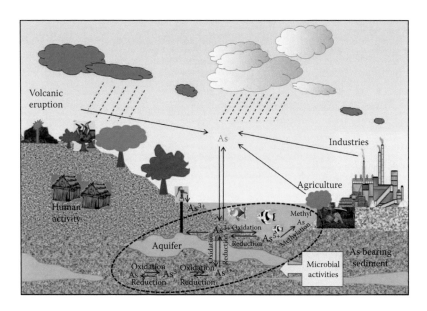

FIGURE 19.1
Global As biogeochemical cycle in environment. (Modified after Mukhopadhyay, R. et al. *FEMS Microbiology Reviews* 26(3), 2002: 311–325.)

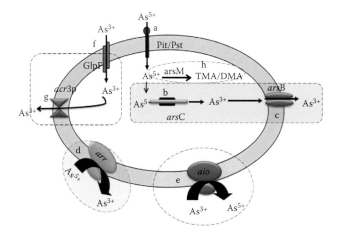

FIGURE 19.2
Schematic diagram of representing mechanisms of bacteria–As interaction. Arsenate (As^{5+}) enters the cells through the phosphate transporters (a); after entering the cells, As^{5+} is reduced to As^{3+} by cytosolic As^{5+} reductase, *arsC* (b); following reduction, As^{3+} is extruded out from the cell interior through efflux pump *arsB* (c). As^{5+} can be used as the terminal electron acceptor during respiration under anaerobic environment (d). As^{3+} can be oxidised to serve as an electron donor or a mechanism of resistance via periplasmic *aio* system (e). As^{3+} enters the cell through the aquaglyceroporin channel (f) and is directly extruded out from the cells via another As^{3+} transporter *acr3p* (g). Furthermore, inorganic As can also be transformed into organic species by methylation (h).

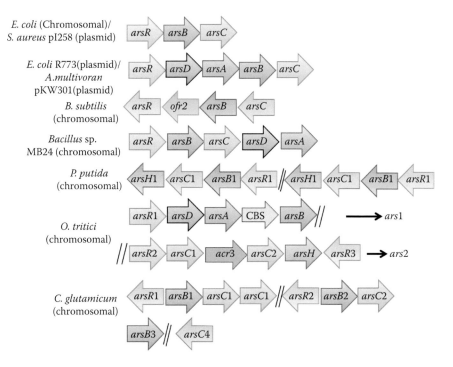

FIGURE 19.3d
Arrangement of gene clusters in *ars* operon in different bacteria.

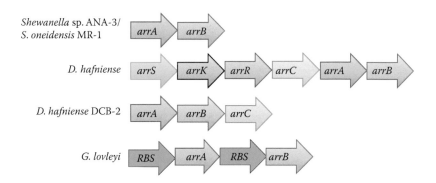

FIGURE 19.4b
Arrangement of gene clusters in *arr* operon in different bacteria.

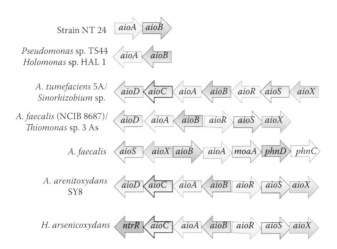

FIGURE 19.5b
Arrangement of gene clusters in *aio* operon in different bacteria.

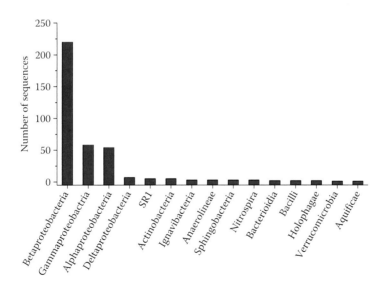

FIGURE 19.7
Distribution of major phylogenetic groups at phylum level retrieved from different sites of Bengal delta plain.

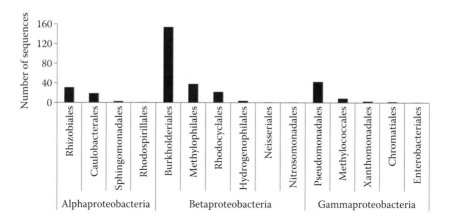

FIGURE 19.8
Distribution of major phylogenetic groups at genus level retrieved from different sites of Bengal delta plain.

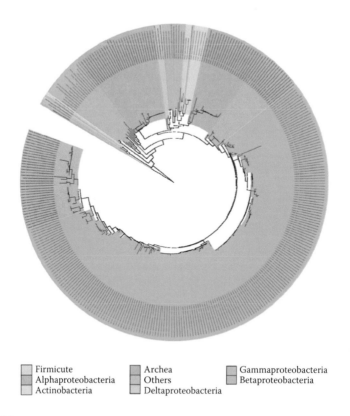

Firmicute
Alphaproteobacteria
Actinobacteria
Archea
Others
Deltaproteobacteria
Gammaproteobacteria
Betaproteobacteria

FIGURE 19.9
Phylogenetic tree of different bacterial genera identified from various As-contaminated sites of Bengal delta plain.

10

Biogeochemistry in Recycling of Trace Elements and Heavy Metals

Jian Chen, Fengxiang X. Han and Paul B. Tchounwou

CONTENTS

Human civilisation increasingly relies on the usage of various metal resources, including heavy metals. Heavy metals were originally sequestered inside the Earth as ore deposits. With industrialisation, great amounts of metals have been excavated and released on the Earth's surface and dissipated into the environment. The biosphere, atmosphere, hydrosphere, and pedosphere have been polluted by heavy metals or fingerprinted by human civilisation (Han et al. 2002, 2003). Heavy metals enter the pedosphere in various forms during metal mining, smelting, processing, recycling and the disposals of wastes (Han et al. 2000, 2001a,b,c). Heavy metals are emitted into the atmosphere through human activities, such as combustion of coal, mining and processing, and are transported globally and finally deposited by wet and dry processes into the pedosphere (Nriagu 1979; Nriagu and Pacyna 1988).

The inventory of heavy metals estimated since the industrial age clearly indicates a large amount of heavy metal sources produced and dissipated into the pedosphere (Han et al. 2002, 2003). In 2000, the cumulative industrial age anthropogenic global production levels of Cd, Cr, Cu, Hg, Ni, Pb and Zn were 1.1, 105, 451, 0.64, 36, 235 and 354 million metric tones, respectively. The ratios of the potential anthropogenic metal inputs to metal content in world soil in 2000 were 1.0, 0.4, 5.6, 8.7, 0.7, 4.2 and 3.5 for Cd, Cr, Cu, Hg, Ni, Pb and Zn, respectively, and global metal burdens per capita were 0.18, 17.3, 74.2, 0.10, 5.9, 38.6 and 58.2° kg for Cd, Cr, Cu, Hg, Ni, Pb and Zn, respectively. The most efficient approach to control heavy metal pollution in the pedosphere is to control anthropogenic pollution sources during heavy metal mining, processing, manufacturing, recycling and disposal of wastes. The technology of industrial processing is required to be continually improved to reduce the metal dissipation rates into the environment. Outdated heavy metal-containing products, by-products and wastes need to be continually recycled. In this chapter, we will discuss the life cycle and biogeochemical processes of trace elements and heavy metals in their recycling. Three case studies include biogeochemical cycling of trace elements in mine tailing, sewage-sludge and animal waste-applied soils

10.1 Life Cycle of Heavy Metals and Trace Elements

The recovery rate of metals from both mining and processing (primary and secondary) is estimated at 85% (Rudawsky 1986). The recycling efficiency with the modern technology varies with the metal type. Lead has the highest recycling efficiency (95%), followed by chromium (87%) and Zn the lowest (19%) (U.S. Dept. of Interior and U.S. Geological Survey 2002). However, the recovery rate of metals from mining depends highly on the technology used and the economy of the regions. In some undeveloped areas of China, the recovery rate of Hg and Pb–Zn from mining is 30%–45% (Chen et al. 1999). Some metals (like Zn) are partly dissipated into the environment during product use as coatings, even before recycling. The recovery of heavy metals in the form of first-generation products is approximately 80% for Pb, 74% for Cr and 16% for Zn. Most products containing heavy metals as a minor component often last from 20 to 50 years, in fewer cases from 10 to 100 years. After two to three recycling processes, metal recoveries for Zn and possibly Cd and Ni may be less than 10%. Non-recovered metals are mainly dissipated into the pedosphere as pollutants and wastes. The pedosphere is a major sink for heavy metals compared to the atmosphere. Heavy metals are almost completely dissipated into the pedosphere after about four generations of products for most metals, lasting for 80–200 years (complete-dissipation period), and in fewer cases lasting from 40 to 400 years. The complete-dissipation period has decreased in recent years since the life of each generation of metal-containing products has shortened. Heavy metals enter the pedosphere through direct dissipation during primary/secondary and recycling processes (as air pollutants, liquid and solid wastes), and disposal of various outdated products and wastes, in which most heavy metals are relatively concentrated. Further, agricultural practices including application of metal-containing pesticides, fertilisers, and wastes are also important pathways (Han et al. 2000, 2001a,b).

In earlier centuries, as today, most arsenic was a by-product from mining other metals (Loebenstein 1994). Arsenic is emitted into the air during extractive metallurgical processing of nonferrous metals. Since early last century, As compounds have been used as herbicides, pesticides, wood preservatives, animal feed additives, corrosion inhibitors, semi-conductors and metal alloys, and in glass manufacture. Of these, agricultural uses accounted for 90% of arsenic utilisation in the 1940s and then declined, while wood preservatives are the major current application of As (Loebenstein 1994). However, as of January 1, 2004, the USEPA will not allow chromated copper arsenate products to be used to treat wood intended for residential uses. This decision will decrease arsenic dissipation in both the manufacturing and retail sectors (USEPA 2002a,b). Arsenic-containing products have relatively short life periods. Wood preservatives are generally guaranteed to prevent lumber decay for a period of 30 years and other products last from 5 to 20 years (Loebenstein 1994). Most agricultural products containing arsenic have even shorter useful lives of less than 1 year and are dissipative (Loebenstein 1994). Compared to the complete dissipation periods of heavy metals-containing products (Han et al. 2002), As-containing products have much shorter useful lives (the short-life cycle), which makes our hypothesis (discussed above) and As sources-based approach more reasonable and rational. Non-recovered As is mainly dissipated into the pedosphere as pollutant and waste. Arsenic enters the pedosphere in various forms through direct dissipation during primary/secondary (metal mining, smelting and processing) and recycling processes, the disposal of various outdated products and wastes, and agricultural practices, such as application of As-containing pesticides, herbicides and fertilisers (Loebenstein 1994; Han et al. 2003). Arsenic is emitted into the atmosphere through human activities, such as combustion of coal and during mining and processing. It is transported globally and finally deposited by wet and dry processes into the pedosphere (Nriagu 1979; Nriagu and

Pacyna 1988). Thus, the pedosphere is a major sink for As compared to the atmosphere. Based on cumulative As production, we developed an estimate of the potential world-wide arable land loading and potential global As burden per capita to assess the potential environmental risks of As accumulation in the pedosphere.

10.2 Major Biogeochemical Processes Governing Recycling of Trace Elements

10.2.1 Oxidation–Reduction Processes

The redox process is one of the main factors governing the biogeochemical cycles of major heavy metals and trace elements. The chemical speciation, mobility, and bioavailability of trace elements are directly affected by reduction and oxidation in the environment. The globally biological redox activity driven by the anthropogenic activities has caused a highly oxidising surface environment, which promotes the release of mobile and bioavailable metals and trace elements (Borch et al. 2010). The oxidised trace metals can be transformed to reductive forms through both chemical- and biological-dependent processes. For instance, Cu(II) can be chemically transformed to Cu(I) by Fe^{2+} or H_2S (Pattrick et al. 1997; Matocha et al. 2005). For biological reduction, microorganisms can directly reduce many kinds of metals via dissimilatory pathways (Lovley 1993). The loss of sorption capacity may cause the reductive dissolution of metal oxides, which can be confirmed by the microbial reduction process of Co(III)- and Ni(II)-containing goethite (MZachara et al. 2011). Some Fe-reducing microorganisms may reduce Fe^{3+} to Fe^{2+} through intrinsic acetate metabolism (Hori et al. 2009). Trace metals may also be oxidised biogeochemically, driven by the entrance of O_2 into anoxic systems. However, this process is likely to occur rapidly in abiotic processes compared to the relatively slower processes driven by microorganisms (Lee et al. 2002). The redox processes of trace metals are always dependent on the interaction among different metal species. Borch et al. (2010) has proposed a conceptual model involved in Fe(II/III) mineral phases showing the effect of abiotic and biotic redox processes on the fate of As in the environment. For the abiotic processes, the formation of reactive Fe(II/III) mineral phase is indispensable for the As(V) reduction or As(III) oxidation. For the biotic processes, microbial-dependent release of Fe^{2+} or oxidation of Fe^{2+} can directly result in the oxidation of As(III) or reduction of As(V). The redox reactions of trace metals in the environment are complicated, but these reactions generally occur along redox gradients in time or space, which can be found in a wide range of environment conditions. The redox gradients are dependent on the reduction potentials, microbial composition and activity and chemical constitution (Borch et al. 2010). The anthropogenic activities and climate change need to be considered for the further modelling of the redox processes of biogeochemical recycling of trace metals in the environment.

10.2.2 Adsorption and Desorption Reactions

The biogeochemical recycle of trace metals can be significantly affected by their solution concentrations, which is largely dependent on the adsorption and desorption of trace metals in soil minerals and organic matters. The first factor for controlling the adsorption reaction is the pH of soil solution, which can affect the surface charges of hydroxyl groups on soil minerals. Cations are attracted by negatively charged sites while anions are adsorbed to positively charged sites. In a case where two kinds of adsorption sites are located very close on the surface, the anion may be adsorbed by electrostatic attraction to both sites (Amrhein and Doner 2014). The second factor affecting the adsorption is oxyanion. The replacement of surface hydroxyl by oxyanion-derived O^{2-} results in the formation of a new surface structure between mineral and oxyanion. This kind of complex is not easy to release trace metals through desorption. It is difficult to distinguish the controlling process between adsorption and precipitation of trace metals in a complicated soil environment; the adsorption of some trace metals to soil minerals may tend to cause further precipitation because trace metals can be incorporated into the newly formed chemical structures. The plant root exudates and soil microbial activity have significant effects on soil pH and oxygen environment (Blossfeld et al. 2011; Shi et al. 2011; Rudolph et al. 2013). Therefore, the application of some remediative technologies (including phytostabilisation) is based on the strategy that promotes the adsorption and the following participation of trace metals rather than desorption. This kind of strategy can reduce the mobility and bioavailability of trace metals (Anawar et al. 2013).

10.2.3 Volatilisation Reactions

The volatilisation reactions refers to some trace metals with volatilised forms, such as Hg, Se and As. Hg in soil can be in many forms, including elemental mercury (Hg^0), ionic mercury (Hg^{2+}), methyl mercury (MeHg), mercury hydroxide ($Hg(OH)_2$), and mercury sulphide (HgS). Among them, MeHg is the most toxic form while Hg^0 is the gaseous form. Soil microbes have developed specific mechanisms to resist Hg toxicity by transforming MeHg to Hg^0. One of the biological systems for detoxifying Hg^{2+} or MeHg compounds involves two enzymes referred to as mercuric ion reductase (MerA) and organomercurial lyase (MerB). In some Hg-resistant microbes, MerB catalyses the reaction of MeHg to Hg^{2+}, and then MerA transforms Hg^{2+} to Hg^0 that can volatilise out of cells (Nascimento and Chartone-Souza 2003). Unlike Hg, volatile Se compounds are always in organic form, such as dimethylselenide that are 1/600 to 1/500 as toxic as inorganic forms of Se

(Padmavathiamma and Li 2007). Dimethyl selenide (CH_3SeCH_3), dimethyl diselenide ($CH_3SeSeCH_3$), dimethyl selenenyl sulphide (CH_3SeSCH_3), methaneselenol (CH_3SeH), and hydrogen selenide (H_2Se) are the volatile Se species formed in soils and plants. Some plant species have the ability to metabolise Se compounds to volatile form. For instance, *Typha latifolia* and some members of the Brassicaceae are capable of metabolising various inorganic or organic species of Se (e.g. selenate, selenite and Se-methionine [Met]) into gaseous Se forms (e.g. dimethylselenide or dimethylselenide), which can be volatilised and released into the atmosphere (Bañuelos 2000; Pilon-Smits et al. 1999; Terry et al. 1999). S-adenosyl-L-Met:L-Met S-methyltransferase (MMT) is the key enzyme responsible for the process of Se phytovolatilisation (Tagmount et al. 2002). The volatile form of As is organic As compounds as well. Microbial reactions produce arsine (AsH_3), methylarsine (CH_3AsH_2), dimethylarsine ($(CH_3)_2AsH$) and trimethylarsine ($(CH_3)_3As$), which further transfer As in the soil to the atmosphere. These reactions involved in some microbial enzymes can occur in both aerobic and anaerobic conditions (Qin et al. 2006; Stolz et al. 2006; Turpeinen et al. 2002). In addition, plants have the capacity of volatilising As as well. When *Pteris vittata*, a famous As hyperaccumulator, grows in As-contaminated soil, vapour samples from chambers covering the shoots contain significantly volatilised As (37% for arsenite and 63% for arsenate) (Sakakibara et al. 2007). The above reports can help develop bioremediation strategies named biovolatilisation for toxic metals-contaminated soils. However, the efficiency of metal volatilisation depends on many factors (e.g. metal concentration, soil type, vegetation type, microbial community), which should be further investigated.

10.2.4 Dissolution and Precipitation

In addition, dissolution and precipitation processes as discussed earlier are also important for biogeochemistry of recycling of trace elements and heavy metals. In general, most of the cations and transitional heavy metals and trace elements tend to be precipitated and co-precipitated as hydroxides or oxides, carbonates and phosphates depending upon the high pH and presence of carbonates and phosphates. When pH becomes acid, the cations and traditional trace elements are dissolute into solution.

10.3 Biogeochemical Cycling of Trace Elements in Mine Tailing

Mine tailing poses a potential threat to the environment because it contains many kinds of toxic trace metals with variable biogeochemical characteristics. The trace metal concentration in the soils of mine tailing depends on several factors, such as distance and the proportions of releasable fractions. In the paddy soils of metal mine tailing in Korea, the trace metal contents were found to be proportional to both distance to the core mine tailing and higher proportions of relatively easily releasable fractions, such as exchangeable forms (Lee 2006). In addition, the metal status is various among different metal types in mine tailing. In the above Korea tailing, geochemical equilibrium modelling using PHREEQC indicates the presence of solubility controlling solid phases for Cd and Pb, whereas Zn and Cu seem to be controlled by adsorption/desorption processes (Lee 2006). Acidic mine drainage (AMD) is proposed to be the main source effect causing the dispersion of metal pollution from mine tailing, which may promote the reduction of trace metals. For example, the analysis of heavy metal distribution and chemical speciation in tailings and soils around a Pb–Zn mine in Spain indicates that Pb is associated with non-residual fractions, mainly in reducible form, while Zn appears mainly associated with the acid-extractable form in mine tailing samples (Rodriguez et al. 2009). The behaviour of the trace elements is closely associated with sulphide minerals in tailings of different age because AMD is rich in sulphide compounds playing important roles in the oxidation processes of trace metals. In the Guanajuato mining district of Mexico, the relative oxidation rate followed the order: $PbS > ZnS > FeAsS \approx CuFeS_2$, while the relative affinity of the elements with iron oxides followed the sequence $Cu \approx Zn > As > Pb$ (Ramos Arroyo and Siebe 2007).

Among the trace metals in AMD from mine tailing, the microbial-mediated oxidation of Fe^{2+} has been well investigated. Many microorganisms with rich biodiversity are involved in the transformation of Fe and S in mine tailings (Kock and Schippers 2008), *Acidithiobacillus ferrooxidans* is a major participant for the oxidation of Fe^{2+} in AMD. *A. ferrooxidans* oxidises Fe^{2+} in very high efficiency because it is a chemolithoautrophic, γ-proteobacterium using energy from the oxidation of Fe- and S-containing minerals for growth in extremely acidic condition (e.g. pH < 2) (Johnson and Hallberg 2005). The experimental and computational methodology for determining the kinetic equation and the extant kinetic constants of Fe^{2+} oxidation by *A. ferrooxidans* has been improved by Molchanov et al. (2007). The analysis from the genome sequence of *A. ferrooxidans* suggests that the genes encoding Fe^{2+} oxidation functions are organised in two transcriptional nits, the *petI* and *rus* operons (Valdes et al. 2008). Interestingly, the oxidation of Cu in AMD is dependent on *A. ferrooxidans*-mediated Fe^{2+} oxidation. First, Fe^{2+} is biologically oxidised to Fe^{3+}. Second, the chemical oxidation of Cu^+ to produce more soluble Cu^{2+} by Fe^{3+} that is reduced to Fe^{2+} in the process (Valdes et al. 2008).

The release of trace metals from mine tailing areas depends on their mobility. The organic matter maybe an important factor for determining the content, distribution and mobilisation of trace metals in mine sites (García-Sánchez et al. 2010). For example, the organic matter may play a positive role in enhancing the mobility of trace metals in mine tailing soils because of the possible formation of soluble organic ligands (Acosta et al. 2011). Humic acid (HA) seems to be the major carrier of metal ion mobilisation (Suteerapataranon et al. 2006). However, the organic carbon is not always essential for the mobility of trace metals. The sediment quality can impact the fate of trace metals in the environment. In the mine waters of Blesbokspruit located in South Africa, only Cu was primarily associated with the organic fraction whereas Ti and Zr were mostly found in the residual fraction (Roychoudhury and Starke 2006). The trace metals can be distributed through the carriage by some typical solid phases. In a mining-influenced basin in France, some secondary Fe–Mn oxyhydroxides carry very high concentrations of As_2O_5 and Cd, while the Al–Si fine-grained phase is the major carrier of Hg (Grosbois et al. 2007). Moreno-Jiménez et al. (2010) has also reported similar results, that up to 80% of the total As were retained by Al- and Fe-oxyhydroxides in soils adjacent to an old mine site in Spain. Compared to other metals, arsenic can be remobilised from sediments in the water bodies contaminated with mine tailings. The microbial-mediated process is involved in the transformation and remobilisation of As across the sediment–water interface. The microorganisms are capable of surviving in As-rich sediments and reduce As(V) to As(III), which further leads to a decrease in total As concentration in sediments and an increase in As(III) concentration in the aqueous media (Routh et al. 2007). Interestingly, the bacterial dissolution of Fe and Mn was found to occur simultaneously with the dissolution of As in mine tailing soils (Lee et al. 2010). Therefore, further studies are needed to test whether and how microorganisms affect the metal-oxyhydroxides to release both As and Fe in mine tailing. The different metallogenic mechanisms contribute to the mobility of trace metals over the long period of mining activity. Millions of tonnes of mine tailings have been dumped into a coastal bay in Spain over a long period. The fraction of metals associated with jarosites present a high stability under different physicochemical conditions, while metals associated with mineralogical phases that are undergoing supergenic alteration processes presented a high mobility (Martínez-Sánchez et al. 2008).

10.4 Biogeochemical Cycling of Trace Elements in Sewage/Sludge Applied Soils

Sewage sludge is an important source of recycling heavy metals and trace elements in soils. There is no doubt that heavy metals and trace elements have increased substantially along with population growth. Global sludge production has continued to increase. Sewage sludge may be a valuable source of essential nutritional elements (N, P, K and some micronutrients) for agricultural crops, especially in soils deficient in major, minor and micronutrients. Also, it may improve the physical properties of soils by addition of organic matter. The application of sewage sludge to agricultural land is considered the most preferred method for its disposal. Agricultural use of sewage sludge consumed approximately 44%; 20%–25% and more than 31% of sludge production in the United Kingdom, Germany and the United States, respectively (Bilitewski et al. 1994). However, sewage sludge has elevated levels of potentially toxic trace elements (Zn, Cd, Cu, Pb, Cr and Ni) (Han 2007), which may exceed natural soil concentrations by two orders of magnitude or more. The application of sewage sludge and irrigation with recycled sewage water for long periods of time has been shown to result in a substantial accumulation of heavy metals in the soils and crops grown on them (McGrath and Cegarra 1992).

The distribution of heavy metals among the solid-phase fractions of soils is affected by climate, soil formation processes, elemental make-up of the parent material, soil properties, such as pH and Eh, texture, mineralogical composition, organic matter and carbonate contents, as well as the chemical properties of the elements.

10.4.1 Distribution of Trace Elements in Sewage Sludge

In sludge and manure, metals are predominantly in carbonate bound, residual and organically bound fractions, except copper, which is mainly in organically bound fraction, and cadmium, which is in the exchangeable form. Exchangeable (EXC), carbonate (CARB) and easily reducible oxide-bound (ERO) fractions are the main forms for Cd in sludge (Banin et al. 1990). Sposito et al. (1983) found that most of Cd in the sludge was in the carbonate fraction, followed by the organically bound (OM) and residual fractions (RES). Copper in sludge prevailed in the organically bound (Banin et al. 1990), followed by the carbonate and residue fractions. Nickel has been found to be equally present in the carbonate (32%), residual (26%) and organically bound fractions (24%) in sludge (Emmerich et al. 1982). Banin et al. (1990) reported that most of the Pb was bound in the reducible oxide fraction, followed by the readily reducible oxides and carbonate fractions. Zinc in sludge has been reported to occur in the carbonate, easily reducible oxide and organically bound fractions (Banin et al. 1990).

10.4.2 Distribution and Redistribution of Trace Elements in Sludge-Amended Soils

In waste-amended soils, most of the Cd was in the exchangeable and carbonate fractions posing high potential availability. After the cessation of long term sludge applications, Cd still remained highly extractable by DTPA (Bell et al. 1991). The Cd uptake by corn decreased over time at all sludge rates and was inversely related (p > 0.01) to time in years following terminal application (Bidwel and Dowdy 1987). Copper in a sludge-treated Typic Udipsamment from England and Israeli arid-zone soils predominated in the organically bound fraction, followed by the carbonate fraction (Banin et al. 1990, McGrath and Cegarra 1992). With time, Cu was redistributed into the reducible oxide fractions. Nickel in a sludge-treated

Typic Udipsamment from England has been shown to occur mainly in the residual fraction, but the percentage of the residual form decreased and the carbonate and organically bound fractions increased with annual application of sludge (McGrath and Cegarra 1992). Another study also showed that Ni in California soils amended with sludge for 7 years was mostly present in the residual (64%) and less organically bound (12%) and carbonate fractions (18%) (Chang et al. 1984). During the first 10 years of an annual application of sludge, Ni in the carbonate, organically bound and exchangeable fractions increased and the percentage of the residual fraction decreased with time. During the following 20 years after the cessation of the sludge application, the organically bound and carbonate fractions declined slightly and the residual fraction increased with time (McGrath and Cegarra 1992). Most of the Pb was found to be in the carbonate fraction in a sludge-treated Typic Udipsamment from England during 20 years and after 25 years of applications of sewage sludge according to the findings of McGrath and Cegarra (1992). After the cessation of the sludge application, Pb in all fractions seems to remain the same (McGrath and Cegarra 1992). Zinc in a Typic Udipsamment amended with sewage sludge has been shown to occur mainly in the carbonate bound, followed by the residual fraction (McGrath and Cegarra 1992). The application of sludge increased the percentage of Zn in the carbonate fraction.

10.4.3 Transport and Bioavailability of Trace Elements in Sludge-Amended Soils

Transport of heavy metals to lower depths in the profile of sludge-amended soil, however, has been observed by several researchers. Sludge-borne Cd, Cr, Cu and Zn in a silt loam loessial soil were reported to move into the 16- to 30-cm soil layer, directly below the zone of incorporation of sludge after 10 years of application (Dowdy et al. 1994). After 17 years of sludge use, some further movement of Cr and Cu into the 45- to 60-cm layer was identified (Dowdy et al. 1994). Barbarick et al. (1998) also observed that Zn in sewage sludge-amended loam soils significantly and consistently increased in extractable levels (DTPA-Zn) below the plow layer.

LeClaire et al. (1984) reported that soluble and exchangeable Zn as extracted by HNO_3 and H_2O were highly labile; the organic matter fraction by NaOH was labile and the carbonate fraction extracted by EDTA represented a reservoir of potentially bio-available Zn to plants. Murthy (1982) found that the exchangeable and soluble organically complex Zn extracted by $Cu(OAc)_2$ were more available to rice in wetland soils, and amorphous iron and aluminium oxides by oxalate reagents also contributed to availability. The organically complex Zn played the most important role in Zn nutrition of rice in paddy soils (Han and Zhu 1992). The exchangeable and the chelated or soluble organically bound fractions were more available sources for plants (Han et al. 1995, Han, 2007). In calcareous soils, Zn and Cd in the

CARB fraction and Cu in the OM fraction were still available to plants (Banin et al. 1990; Han and Zhu 1992; Han et al. 1995).

10.5 Biogeochemical Cycling of Trace Elements in Animal Waste-Applied Soils

Poultry and pig swine wastes are the two most important animal wastes that received intensive study. During the past 40 years, U.S. poultry industry has increased rapidly. There is a clear trend of increase in the annual waste production. Furthermore, the major portion (90%) of poultry litter produced is applied to agricultural land (Moore et al. 1995). Several heavy metals including As, Co, Cu, Fe, Mn, Se and Zn are added to poultry diets for various purposes, resulting in fairly high concentrations of heavy metals in the waste (Sims and Wolf 1994). High concentrations of heavy metals in poultry waste-amended soils may result in environmental concerns due to the potential contamination of surface and groundwater via runoff and leaching (Han et al. 2001c). A number of different soils receiving repeated application of poultry litter for several years have exhibited high concentrations of extractable of Cu and Zn (Han et al. 2001).

10.5.1 Accumulation of Heavy Metals in Animal Waste-Amended Soils

Elevated levels of heavy metals were found in long-term poultry and pig manure waste-amended soils (Han et al. 2000, 2001c). However, most of the metal accumulation in the soils did not reach the toxicity levels to plants (Han et al. 2000, 2001c). Kornegay et al. (1976) reported that Cu levels in the plow layer of the soil increased after just 3 years of application of pig manure. Sutton et al. (1983) found that Cu-enriched swine manure increased the 1 N HCl-extractable Cu in the upper portion of the soil profile (0–31 cm) after five years of application, but not at lower depths. Mullins et al. (1982) indicated that the DTPA-extractable Cu in the surface soil increased after 3 years of application and Cu was found to move downward in the sandy soil. Payne et al. (1988) further pointed out that after 8 years of applications of swine manure, DTPA-extractable Cu continued to increase with the years of application and it was linearly related to the application amount. Some extent of the Cu downward movement was confirmed. The total concentrations of selected heavy metals (Cu, Zn, Pb, Ni, Cr and Mn) in a Mississippi surface soil receiving 25 years of the poultry waste (Han et al. 2000): Copper, Zn and Mn accumulated considerably in the poultry waste-amended soil over 25 years; other metals such as Ni, Cr and Pb increased slightly in the amended as compared to non-amended soil. Based on total concentrations, Cu and Zn in the amended soil increased at an average rate of approximately

2 mg kg^{-1}yr^{-1}. The difference in Mn represents an average rate of accumulation of 6.5 mg kg^{-1} yr^{-1}.

10.5.2 Fractionation Distribution of Heavy metals in Animal Waste-Amended Soils

Copper in the Coastal Plain (Aquic Hapludulfs), Ridge and Valley (Aquic Hapludalfs) and Piedmont regions (Typic Rhodudulfs) of Virginia following 3 and 8 years of applications of Cu-enriched manure was mostly bound in the organic matter fraction, followed by the oxide-occluded fraction (Mullins et al. 1982; Payne et al. 1988). Furthermore, these manure treatments resulted in higher Cu contents in the organically bound fraction than did sulphate salt application (Payne et al. 1988). In a 25-year poultry waste-amended Mississippi soil, copper in the amended soil was mostly present in the organic matter fraction (50%), whereas Zn was found in the easily reducible oxide fraction (50%) (Han et al. 2000). This suggests that Cu and Zn in this long-term amended soil are potentially bio-available. The mobility of Zn throughout soil profiles in the long-term waste-amended soil was observed. Copper was also transported to 60-cm depth. Zinc in the soil profile was mainly in the form of residual and crystalline iron oxide bound (CryFe) fractions, followed by the organic matter bound and exchangeable fractions, thus posing potential risks to contamination of underground water.

Compared to Cu, Zn concentration in the EXC fraction for both Mississippi soils were quite high, possibly due to equilibrium with a reservoir (ERO fraction) containing a high concentration of Zn in amended soil, and a relatively low pH in non-amended soil. This suggests a likelihood that Zn had high solubility and thus may be bioavailable. Zinc in the amended soil was predominantly in the ERO ($48 \pm 10\%$) and OM fractions ($23 \pm 2.8\%$), and to a lesser degree, in the amorphous iron oxide-bound (AmoFe) fraction ($12 \pm 6\%$). Zinc in the ERO fraction in amended soil was 20 times that in non-amended soil. It may then be suggested that adsorption–desorption and oxidation–reduction processes may govern Zn availability, phytotoxicity and mobility in this soil. Additionally, accumulated Zn in all fractions, especially the ERO, OM and AmoFe fractions indicates that Zn in this long-term waste-amended soil may pose potential environmental risks to surface and groundwater.

Our results indicate distinctly different distributions of Cu and Zn among the solid-phase components in poultry-amended soils (Han et al. 2000). The distribution patterns of these two heavy metals can be attributed to their chemical properties. Copper is known to bind strongly as an inner-sphere complex with organic matter (McBride 1981), and Zn may be bound by outer-sphere complexation with the metal retaining its inner hydration sphere. Due to its similarity with Fe and Mg ionic radii, Cu is capable of isomorphous substitution of Fe^{2+} and Mg^{2+} in layer silicates (Banin et al. 1990; Han and Banin 1997), which explains the concentrations of Cu observed in the RES fraction in non-amended soils. Copper incorporation into interlayer of alumino-silicate minerals in the amended soil, however, was perhaps a slow process. The significant proportion of metal (Cu and Zn) concentrations in the RES fraction in amended soil was, in part, due to metal incorporation into the interlayer of alumino-silicates and binding to humin, which was not completely removed in the OM fraction step. In addition, Cu and Zn may also co-precipitate with and form solid solutions of iron oxides (Lindsay, 1979).

10.5.3 Mobility in the Soil Profiles

In the soils receiving a long-term application of swine manure, DTPA-extractable Cu in the B horizons of the soils was found to increase, indicated some extent downward movement of the applied Cu in these soils, but larger increases in extractable Cu in one subsoil were attributed to downward movement caused by plowing (Mullins et al. 1982, Payne et al. 1988). Han et al. (2000) clearly demonstrate that long-term application of poultry waste resulted in considerable movement of Zn and to some extent Cu in the soil profiles.

Most heavy metals applied in poultry waste to a Mississippi soil accumulated in the upper 0–20 cm and to a lesser extent in the 20–40 cm depth (Han et al. 2000). In contrast, for the non-amended soil, the total concentrations of Cu and Zn showed a slight decrease close to the soil surface (0–20 cm). This decrease is perhaps due to plant uptake of these metals from the surface zone. For Cu, the total concentration was almost uniform with depth (below 40 cm) in both profiles, whereas total Zn, in the amended soil exhibited a noticeable increase with soil depth. Specifically, amended-soil Zn concentrations were nearly 1.5–2 times that of the non-amended soil. These results reflect the potential downward movement of Zn and to some extent Cu following long-term poultry-waste additions.

Bioavailability of heavy metals in poultry waste-amended soils has been shown to be strongly affected by soil pH. Van der Watt et al. (1994) found that plant uptake was positively correlated with the DTPA- and 0.1 M $NaNO_3$-extractable metal contents and negatively correlated with soil pH. They also found that 3%–5% of the Zn and 0.5% of the Cu in the added poultry waste were taken up by sudax (*Sorghum bicolor* (L.) Moench) in greenhouse experiments. Additions of poultry-litter significantly decreased both plant tissue Zn concentration and KNO_3-extractable Zn levels in amended soil (Pierzynski and Schwab 1993) and decreased Cd extracted in the exchangeable, complexed and HCl-extractable fractions.

10.6 Conclusion

In conclusion, trace elements and heavy metals have been dispersed and recycled through metal mining, smelting, processing, recycling and the disposals of wastes. Biogeochemical processes involved in recycling of trace elements and heavy metals include oxidation–reduction, adsorption–desorption, dissolution–precipitation/co-precipitation, and volatilisation processes. Through recycling of solid and liquid wastes, trace elements and heavy metals have been accumulated in waste-polluted ecosystems. This is exemplified by animal-waste-applied, sewage-sludge-applied, mining tailing lands with increasing mobility and bioavailability of trace elements and heavy metals. Therefore, understanding the biogeochemistry of these trace elements in recycling processes of various wastes is essential for preventing them from further pollution of the environments, their potential entrance into food chains, and protection of human health and ecosystem health.

Acknowledgement

This research study was supported by U.S. Nuclear Regulatory Commission (NRC–HQ-12-G-38-0038), U.S. Department of Commerce (NOAA) (NA11 SEC4810001-003499) and National Institutes of Health NIMHD-RCMI (G12MD007581).

References

Acosta, J. A., A. Faz, S. Martínez-Martínez, R. Zornoza, D. M. Carmona and S. Kabas. 2011. Multivariate statistical and GIS-based approach to evaluate heavy metals behavior in mine sites for future reclamation. *J Geochem Expl* 109:8–17.

Amrhein, C. and H. Doner. 2014. Biogeochemistry of trace elements: Reactions in soils. In *Salinity and Drainage in San Joaquin Valley, California*, eds. A. C. Chang and D. Brawer Silva, 123–146. The Netherlands: Springer.

Anawar, H., A. García-Sánchez and M. Zabed Hossain 2013. Biogeochemical cycling of arsenic in soil–plant continuum: Perspectives for phytoremediation. In *Heavy Metal Stress in Plants*, ed. D. K. Gupta, F. J. Corpas and J. M. Palma, 203–224. Berlin, Heidelberg: Springer.

Banin, A., Z. Gerstl, P. Fine, Z. Metzger and D. Newrzella. 1990. Minimizing soil contamination through control of sludge transformations in soil. Joint German–Israel research projects. Final report. No. of Project: Wt 8678/458. Hebrew University of Jerusalem.

Bañuelos, G. S. 2000. Phytoextraction of selenium from soils irrigated with selenium-laden effluent. *Plant Soil* 224:251–258.

Barbarick, K. A., J. A. Ippolito and D. G. Westfall. 1998. Extractable trace elements in the soil profile after years of biosolids application. *J Environ Qual* 27:801–805.

Bell, F. F., B. R. James and R. L. Chaney. 1991. Heavy metal extractability in long-term sewage sludge and metal salt-amended soils. *J Environ Qual* 20:481–486.

Bidwel, A. M. and R. H. Dowdy. 1987. Cadmium and zinc availability to corn following termination of sewage sludge application. *J Environ Qual* 16:438–442.

Bilitewski, B., G. Kardtle, K. Marek, A. Weissbach and H. Boeddicker. 1994. *Waste Management*. The Netherlands: Springer.

Blossfeld, S., D. Gansert, B. Thiele, A. J. Kuhn and R. Lösch. 2011. The dynamics of oxygen concentration, pH value and organic acids in the rhizosphere of *Juncus* spp. *Soil Biol Biochem* 43:1186–1197.

Borch, T., R. Kretzschmar, A. Kappler, P. V. Cappellen, M. Ginder-Vogel, A. Voegelin and K. Campbell. 2010. Biogeochemical redox processes and their impact on contaminant dynamics. *Environ Sci Technol* 44:15–23.

Chang, A. C., A. L. Page, J. E. Warneke and E. Grgurevice. 1984. Sequential extraction of soil heavy metals following a sludge applications. *J Environ Qual* 13:33–38.

Chen, H., C. Zheng, C. Tu and Y. Zhu. 1999. Heavy metal pollution in soils in China: Status and countermeasures. *AMBIO* 28:130–134.

Dowdy, R. H., C. E. Clapp, D. R. Linden, W. E. Larson, T. R. Halbach and R. C. Polta. 1994. Twenty years of trace metal partitioning on the Rosemount sewage sludge watershed, In *Sewage Sludge Land Utilization and the Environment*, ed. E. C. Clapp, W. E. Larson and R. H. Dowdy, 149–158, Madison, WI: Am Soc Agr.

Emmerich, W. E., L. J. Lund, A. L. Page and A. C. Chang. 1982. Solid phase forms of heavy metals in sewage sludge-treated soils. *J Environ Qual* 11:178–181.

García-Sánchez, A., P. Alonso-Rojo and F. Santos-Francés. 2010. Distribution and mobility of arsenic in soils of a mining area (Western Spain). *Sci Total Environ* 408:4194–4201.

Grosbois, C., A. Courtin-Nomade, F. Martin and H. Bril. 2007. Transportation and evolution of trace element bearing phases in stream sediments in a mining—Influenced basin (Upper Isle River, France). *Appl Geochem* 22:2362–2374.

Han, F. X. 2007. *Biogeochemistry of Trace Elements in Arid Environments*. The Netherlands: Springer.

Han, F. X. and A. Banin. 1997. Long-term transformations and redistribution of potentially toxic heavy metals in arid-zone soils. I: Under saturated conditions. *Water Air Soil Pollut* 95:399–423.

Han, F. X., A. Banin, Y. Su, D. L. Monts, M. J. Plodinec, W. L. Kingery and G. B. Triplett. 2002. Industrial age anthropogenic inputs of heavy metals into the pedosphere. *Naturwissenschaften* 89:497–504.

Han, F. X., A. T. Hu and H. Y. Qin. 1995. Transformation and distribution of forms of zinc in acid, neutral and calcareous soils of China. *Geoderma* 66:121–135.

Han, F. X. and Q. Q. Zhu. 1992. Fractionation of zinc in paddy soils of China. *Pedosphere* 2:283–288.

Han, F. X., J. Hargreaves, W. L. Kingery, D. B. Huggett and D. K. Schlenk. 2001a. Accumulation, distribution and toxicity of copper in soils of catfish pond receiving periodic copper sulfate applications. *J Environ Qual* 30:912–919.

Han, F. X., A. Banin and G. B. Triplett. 2001b. Redistribution of heavy metals in arid-zone soils under a wetting-drying soil moisture regime. *Soil Sci* 166:18–28.

Han, F. X., W. L. Kingery and H. M. Selim. 2001c. Accumulation, redistribution, transport and bioavailability of heavy metals in waste-amended soils. In *Trace Elements in Soil: Bioavailability, Fluxes and Transfer,* ed. I. K. Iskander and M. B. Kirkham, 141–168, Boca Raton, FL: Lewis Publishers.

Han, F. X., W. L. Kingery, H. M. Selim and P. Gerald. 2000. Accumulation of heavy metals in a long-term poultry waste-amended soil. *Soil Sci* 165:260–268.

Han, F. X., Y. Su, D. L. Monts, M. J. Plodinec, A. Banin and G. B. Triplett. 2003. Assessment of global industrial-age anthropogenic arsenic contamination. *Naturwissenschaften* 90:395–401.

Hori, T., A. Muller, Y. Igarashi, R. Conrad and M. W. Friedrich. 2009. Identification of iron-reducing microorganisms in anoxic rice paddy soil by ^{13}C-acetate probing. *ISME J* 4:267–278.

Johnson, D. B. and K. B. Hallberg. 2005. Acid mine drainage remediation options: A review. *Sci Total Environ* 338:3–14.

Kock, D. and A. Schippers. 2008. Quantitative microbial community analysis of three different sulfidic mine tailing dumps generating acid mine drainage. *Appl Environ Microbiol* 74:5211–5219.

Kornegay, E. T., J. D. Hedges, D. C. Martens and C. Y. Kramer. 1976. Effect on soil and plant mineral leaves following application of manures of different copper contents. *Plant Soil* 45:151–162.

LeClaire, J. P., A. C. Chang, C. S. Levesgue and G. Sposito. 1984. Trace metal chemistry in arid-zone field soils amended with sewage sludge. IV: Correlations between zinc uptake and extracted soil zinc fractions. *Soil Sci Soc Am J* 48:509–513.

Lee, G., J. M. Bigham and G. Faure. 2002. Removal of trace metals by coprecipitation with Fe, Al and Mn from natural waters contaminated with acid mine drainage in the Ducktown Mining District, Tennessee. *Appl Geochem* 17:569–581.

Lee, K.-Y., K.-W. Kim and S.-O. Kim. 2010. Geochemical and microbial effects on the mobilization of arsenic in mine tailing soils. *Environ Geochem Health* 32:31–44.

Lee, S. 2006. Geochemistry and partitioning of trace metals in paddy soils affected by metal mine tailings in Korea. *Geoderma* 135:26–37.

Lindsay, W. L. 1979. *Chemical Equilibria in Soils.* New York: John Wiley & Sons.

Loebenstein, J. R. 1994. *The Materials Flow of Arsenic in the United States.* U.S. Bureau of Mines Information. Washington, DC. Div of Mined Commodities, United States Department of the Interior.

Lovley, D. R. 1993. Dissimilatory metal reduction. *Annu Rev Microbiol* 47:263–290.

Martínez-Sánchez, M. J., M. C. Navarro, C. Pérez-Sirvent, J. Marimón, J. Vidal, M. L. García-Lorenzo and J. Bech. 2008. Assessment of the mobility of metals in a mining-impacted coastal area (Spain, Western Mediterranean). *J Geochem Expl* 96:171–182.

Matocha, C. J., A. D. Karathanasis, S. Rakshit and K. M. Wagner. 2005. Reduction of copper(II) by iron(II). *J Environ Qual* 34:1539–1546.

McBride, M. B. 1981. Forms and distribution of copper in solid and solution phases of soils, In *Copper in Soils and Plants*, ed. J. F. Loneragan, A. D. Robson and R. D. Graham, Sydney: Academic Press.

McGrath, S. P. and J. Cegarra. 1992. Chemical extractability of heavy metals during and after long-term applications of sewage sludge to soil. *J Soil Sci* 43:313–321.

Molchanov, S., Y. Gendel, I. Ioslvich and O. Lahav. 2007. Improved experimental and computational methodology for determining the kinetic equation and the extant kinetic constants of Fe(II) oxidation by *Acidithiobacillus ferrooxidans. Appl Environ Microbiol* 73:1742–1752.

Moreno-Jiménez, E., R. Manzano, E. Esteban and J. Peñalosa. 2010. The fate of arsenic in soils adjacent to an old mine site (Bustarviejo, Spain): Mobility and transfer to native flora. *J Soils Sediments* 10:301–312.

Moore, P. A, Jr., T. C. Daniel, A. N. Sharpley and C. W. Wood. 1995. Poultry manure management: Environmentally sound options. *J Soil Water Conserv* 50:321–327.

Mullins, C. L., D. C. Martens, W. P. Miller, E. T. Kornegay and D. L. Hallock. 1982. Copper availability, form and mobility in soils from three annual copper-enriched hog manure applications. *J Environ Qual* 11:316–320.

Murthy, A. S. P. 1982. Zinc fractions in wetland rice soils and their availability to rice. *Soil Sci* 133:150–154.

MZachara, J. M., J. K. Fredrickson, S. C. Smith and P. L. Gassman. 2011. Solubilization of Fe(III) oxide-bound trace metals by a dissimilatory Fe(III) reducing bacterium. *Geochim Cosmochim Acta* 65:75–93.

Nascimento, A. M. A. and E. Chartone-Souza. 2003. Operon *mer*: Bacterial resistance to mercury and potential for bioremediation of contaminated environments. *Genetics Mol Res* 2:92–101.

Nriagu, J. O. 1979. Global inventory of natural and anthropogenic emissions of trace metals to the atmosphere. *Nature* 279:409–411.

Nriagu, J. O. and J. M. Pacyna. 1988. Quantitative assessment of worldwide contamination of air, water and soils by trace metals. *Nature* 333:134–139.

Padmavathiamma, P. K. and L. Y. Li. 2007. Phytoremediation technology: Hyper-accumulation metals in plants. *Water Air Soil Pollut* 184:105–126.

Pattrick, R. A. D., J. F. W. Mosselmans, J. M. Charnock, K. E. R. England, G. R. Helz, C. D. Garner and D. J. Vaughan. 1997. The structure of amorphous copper sulfide precipitates: An X-ray absorption study. *Geochim Cosmochim Acta* 61:2023–2036.

Payne, G. G., D. C. Martens, C. Winarko and N. F. Perera. 1988. Availability and form of copper in three soils following eight annual applications of copper-enriched swine manure. *J Environ Qual* 17:740–746.

Pierzynski, G. M. and A. P. Schwab. 1993. Bioavailability of zinc, cadmium, and lead in a metal-contaminated alluvial soil. *J Environ Qual* 22:247–254.

Pilon-Smits, E. A. H., M. P. Desouza, G. Hong, A. Amini, R. C. Bravo, S. T. Payabyab, and N. Terry. 1999. Selenium volatilization and accumulation by twenty aquatic plant species. *J Environ Qual* 28:1011–1017.

Qin, J., B. P. Rosen, Y. Zhang, G. Wang, S. Franke and C. Rensing. 2006. Arsenic detoxification and evolution of trimethylarsine gas by a microbial arsenite S-adenosylmethionine methyltransferase. *Proc Natl Acad Sci USA* 103:2075–2080.

Ramos Arroyo, Y. R. and C. Siebe. 2007. Weathering of sulphide minerals and trace element speciation in tailings of various ages in the Guanajuato mining district, Mexico. *CATENA* 71:497–506.

Rodriguez, L., E. Ruiz, J. Alonso-Azcarate and J. Rincon. 2009. Heavy metal distribution and chemical speciation in tailings and soils around a Pb-Zn mine in Spain. *J Environ Manage* 90:1106–1116.

Routh, J., A. Bhattacharya, A. Saraswathy, G. Jacks and P. Bhattacharya. 2007. Arsenic remobilization from sediments contaminated with mine tailings near the Adak mine in Västerbotten district (northern Sweden). *J Geochem Expl* 92:43–54.

Roychoudhury, A. N. and M. F. Starke. 2006. Partitioning and mobility of trace metals in the Blesbokspruit: Impact assessment of dewatering of mine waters in the East Rand, South Africa. *Appl Geochem* 21:1044–1063.

Rudawsky, O. 1986. *Mineral Economics*. Amsterdam: Elsevier.

Rudolph, N., S. Voss, A. Moradi, S. Nagl and S. Oswald. 2013. Spatio-temporal mapping of local soil pH changes induced by roots of lupin and soft-rush. *Plant Soil* 369:669–680.

Sakakibara, M., A. Watanabe, M. Inoue, S. Sano and T. Kaise. 2007. Phytoextraction and phytovolatilization of arsenic from as-contaminated soils by *Pteris vittata*. In *Proceedings of the Annual International Conference on Soils, Sediments, Water and Energy*. 12: 26, Science Society of American Inc., Am. Soc. Agron. Inc., Madison, WI, USA.

Shi, S., A. E. Richardson, M. O'Callaghan, K. M. DeAngelis, E. E. Jones, A. Stewart, M. K. Firestone and L. M. Condron. 2011. Effects of selected root exudate components on soil bacterial communities. *FEMS Microbiol Ecol* 77:600–610.

Sims, J. T. and D. C. Wolf. 1994. Poultry waste management: Agricultural and environmental issues. *Adv Agron* 52:1–83.

Sposito, G., C. S. LeVesque, J. P. LeClaire and A. C. Chang. 1983. Trace elements chemistry in arid-zone field soils amended with sewage sludge: III. Effect of the time on the extraction of trace metals. *Soil Sci Soc Am J* 47:898–902.

Stolz, J. F., P. Basu, J. M. Santini and R. S. Oremland. 2006. Arsenic and selenium in microbial metabolism. *Annu Rev Microbiol* 60:107–130.

Suteerapataranon, S., M. Bouby, H. Geckeis, T. Fanghanel and K. Grudpan. 2006. Interaction of trace elements in acid mine drainage solution with humic acid. *Water Res* 40:2044–2054.

Sutton, A. L., D. W. Nelson, V. B. Mayrose and D. T. Kelly. 1983. Effect of copper levels in swine manure on corn and soil. *J Environ Qual* 12:198–203.

Tagmount, A., A. Berken and N. Terry. 2002. An essential role of *S*-adenosyl-l-methionine: l-Methionine *S*-methyltransferase in selenium volatilization by plants. Methylation of selenomethionine to selenium-methyl-l-selenium-methionine, the precursor of volatile selenium. *Plant Physiol* 130:847–856.

Terry, N., C. Carlson, T. K. Raab and A. Zayed. 1999. Rates of selenium volatilization among crop species. *J Environ Qual* 21:341–344.

Turpeinen, R., M. Pantsar-Kallio and T. Kairesalo. 2002. Role of microbes in controlling the speciation of arsenic and production of arsines in contaminated soils. *Sci Total Environ* 285:133–145.

U.S. Environmental Protection Agency (USEPA). 2002a. *Implementation Guidance for the Arsenic Rule*. U.S. Environmental Protection Agency, Washington, DC: U.S. Government Printing Office.

U.S. Environmental Protection Agency (USEPA). 2002b. Manufacturers to Use New Wood Preservatives, Replacing Most Residential Uses of CCA, http://www.epa.gov/pesticides/factsheets/chemicals/cca_transition.htm (accessed on 7 May 2003).

US Dept. of Interior, US Geol. Survey. 2002. Mercury in U.S. coal—Abundance, distribution, and modes of occurrence, http://pubs.usgs.gov/factsheet/fs095-01/fs095-01.html (accessed Aug. 28, 2002).

Valdes, J., I. Pedroso, R. Quatrini, R. J. Dodson, H. Tettelin, R. Blake, J. A. Eisen and D. S. Holmes. 2008. *Acidithiobacillus ferrooxidans* metabolism: From genome sequence to industrial applications. *BMC Genomics* 9:597.

Van der Watt, H. v. H., M. E. Sumner and M. L. Cabrera. 1994. Bioavailability of copper. Manganese, and zinc in poultry litter. *J Environ Qual* 23:43–49.

11

Role of Natural and Engineered Biofilms Composition in Toxic Inorganic Contaminants Immobilisation

Eric D. van Hullebusch and Yoan Pechaud

CONTENTS

11.1 What Is a Biofilm?

Most of the microorganisms on Earth live in aggregates such as films, flocs, granules and sludges (Costerton et al. 1995). In this chapter, biofilm is defined as a cluster of microorganisms able to multiply and to live in a slimy matrix composed of minerals and microbially synthesised extracellular polymeric substances (EPS).

There are several advantages of forming a biofilm versus living as individual cells. It is well known that microorganisms can adapt to different growth conditions (i.e. pH, redox, dissolved oxygen [DO] concentration, etc.). In biofilm matrices, the growth conditions are heterogeneous and therefore microbes usually express phenotypic peculiarities that are often distinct from those expressed during planktonic growth (Stewart and Franklin 2008). Consequently, microbial biofilm populations have emergent functions that their counterparts single planktonic cells lack (Stoodley et al. 2002). Biofilms are known to withstand the effect of antibiotics and/or toxic metals better than planktonic cultures of the same species (Harrison et al. 2007; Stewart and Franklin 2008). Several mechanisms have been proposed to explain these differences: (i) biofilm matrix represents a diffusional barrier for reactive, charged (heavy metals or metalloids), hydrophobic or large molecules that are either degraded or sorbed by the EPS matrix; (ii) microorganisms within biofilms can adapt to local environmental conditions by undergoing a process of cell specialisation, by changing physiological activities, by rising the population of particular microbial species that are well adapted to a particular condition and, by selecting mutant microbial strains better adapted under a given condition (Stewart and Franklin 2008).

11.1.1 Architecture and Dynamics of Biofilms

The recent use of new technologies such as confocal laser scanning microscopy (CLSM) has revealed a complex and diverse architecture and composition of biofilms. These heterogeneities are the result of biological, chemical and physical processes occurring simultaneously (Figure 11.1).

11.1.1.1 Processes Involved in Biofilm Development and Maturation

11.1.1.1.1 Biological Transformation Processes

In biofilms, several metabolic processes occur and interact with each other. Different electron donors/acceptors and nutrients are used for microbial growth and EPS production. The local microbial activity, local biomass production yield and EPS secretion strongly depend on microbial strains involved as well as on the local environmental conditions (electron donors and acceptors concentrations, nutrients concentrations, pH, redox, hydrodynamics, toxic metals concentrations, etc.) (Laspidou and Rittmann 2002). Biosynthesis is not the only relevant microbial process occurring in biofilms. The microorganisms undergo a number of loss processes including inactivation, decays or lysis, the latest inducing the release of the intracellular material in the EPS matrix (proteins, nucleic acids, lipids, etc.) (Laspidou and Rittmann 2002).

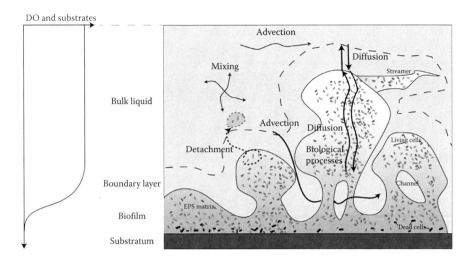

FIGURE 11.1
Comprehensive conceptual drawing showing biofilm structure and the main processes involved in biofilm development and maturation. The model divides the biofilm system into four compartments: the substratum, the biofilm, the boundary layer and the bulk liquid. The curve on the left illustrates relative changes in DO and substrates concentrations within the bulk water, the boundary layer and biofilm matrix of aerobic biofilms.

11.1.1.1.2 Transport Processes

Substrates, nutrients, heavy metals and so forth are generally transported by advection, molecular diffusion and turbulent dispersion. The dominant transport process for dissolved compounds within the biofilm depends on biofilm structural properties. For dense and homogeneous biofilms, molecular diffusion is the dominant transport process whereas for heterogeneous biofilms with large channels, advection may be the dominant transport process (Davit et al. 2013). Depending on the hydrodynamic regime and on the surface heterogeneities, spatial gradients in the boundary layer can occasionally be neglected when mixing is vigorous. In the bulk liquid, advection and turbulent dispersion are the main transport phenomena. The rates of these two processes depend on the hydrodynamic regime and the fluid velocity (Stewart 2012).

11.1.1.1.3 Detachment

Detachment is a key process in biofilms since it is the main process balancing biofilm growth and biofilm attachment and since it governs the retention time of particulate and soluble compounds (substrates, nutrients, heavy metals and metalloids, etc.). Moreover, this process influences the morphology of the biofilms and the distribution of microbial species. It has been highlighted that the size of detached particles varies significantly for a given biofilm and that the mean detachment extent depends on the environmental conditions (shear stress, nutrient limitation, etc.) as well as on the morphological and mechanical properties of biofilms (Derlon et al. 2013).

11.1.1.2 Overall Biofilm Composition

Biofilm, as a highly hydrated and porous structure, is mainly composed of water, in general up to 97% (Sutherland 2001). In most biofilm, microorganisms account for less than 10% of the dry mass, whereas the EPS matrix can account for over 90% (Flemming and Wingender 2010; Sutherland 2001). This matrix is composed of EPS produced by microorganisms, released from microorganisms or brought by the surrounding environment (Sutherland 2001). EPS are mainly represented by polysaccharides, proteins, glycoproteins while nucleic acids, phospholipids and humic acids are quantified to a less extent (Flemming and Wingender 2002; Marcato-Romain et al. 2012; van Hullebusch et al. 2003). The composition and quantity of the EPS depends on the type of microorganisms, the age of the biofilms and the environmental conditions. These include different levels of carbon, oxygen, nitrogen, phosphorus, temperature, pH and bioavailability of nutrients, toxic heavy metals, etc. (Flemming and Wingender 2010).

11.1.1.3 Chemical, Biological, Biochemical and Physical Heterogeneities

Biofilms are dynamic systems that continuously modify their local environment (physical, chemical and biological conditions) and their structure. In natural and engineered biofilms, the heterogeneities are characterised by a series of length scales that vary from nanometers (EPS, pore size, etc.) to millimeters or even centimeters for some natural biofilms (biofilm size) with an intermediate size of about several micrometers (channels, cells and cell clusters, etc.) (Davit et al. 2013).

Figure 11.2 shows various scanning electron microscope (SEM) images of biofilms at different spatial scales. Figure 11.2a shows an anaerobic granular biofilm developed in an upflow anaerobic sludge blanket (UASB) reactor. Physical and geometrical heterogeneities of the structure at the mesoscale can be observed with the presence of a rough surface with distinguishable channels. Figure 11.2b and c illustrates the heterogeneity of internal biofilm structure. One can observe voids, channels, pores and cells arranged in clusters or layers. Figure 11.2d illustrates the heterogeneous arrangement of bacteria and the presence of filaments involved in microbial aggregates.

Biofilms are heterogeneous in three main aspects (Bishop and Rittmann 1995):

- Chemical heterogeneities (de Beer and Schramm 1999): Diversity of chemical solutes (nutrients, metabolic products, inhibitors, metals, etc.), pH variations, diversity of chemical reactions (aerobic/anaerobic, etc.), redox conditions.

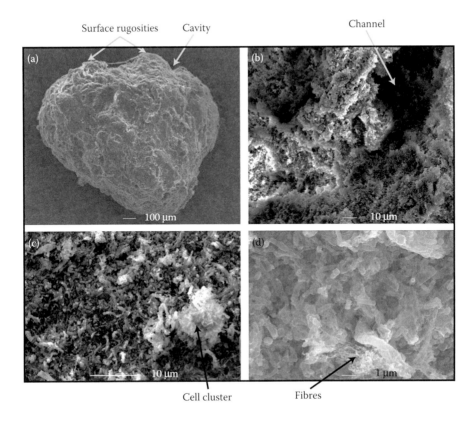

FIGURE 11.2
From mesoscopic to microscopic images of granular biofilms. (a) Overall image of an anaerobic granular biofilm, (b, c and d) SEM images of anaerobic granular biofilm surfaces.

- Microbial and biochemical heterogeneities (McSwain et al. 2005): Microbial diversity of species and their spatial distribution, differences in activity (growing cells, EPS producing cells, dead cells, etc.), EPS nature and quantity.
- Physical and geometrical heterogeneities (Paul et al. 2012): Biofilm density, rheology, soluble compounds diffusivity, biofilm thickness, roughness, porosity and so forth.

11.1.2 Distinction between Natural and Engineered Biofilms

The main difference between natural and engineered biofilms is the control of growth conditions. Engineering a biofilm consists of controlling the environmental conditions during its development to monitor the biofilm structure, composition and activity. In contrast, natural biofilms should cope with strong seasonal and Diel variations of environmental conditions.

11.1.2.1 Natural Biofilms (Epilithic Biofilm)

Natural biofilms can develop onto natural surfaces under all conditions facilitating microbial growth. Consequently, biofilms are ubiquitous in nature, developing on rocks, gravel, cobbles and on plants in freshwater and seawater. Stream biofilms are composed of eubacteria, archaea, fungi, filamentous and single cell microalgae, diatoms (Nimick et al. 2011). These biofilms are in general very heterogeneous and porous with the presence of large channels and of streamers under high water velocity (Graba et al. 2013) (Figure 11.3). They often contain mineral particles that can either be trapped by or precipitated within the EPS matrix or at the biofilm surfaces. As a result, metals and metalloids can interact with minerals at different locations (in mineral crusts, minerals entrapped in matrix biofilm or located within the surface and the interior of living cells) (Figure 11.3). Environmental conditions (water temperature, streamflow, dissolved gases concentrations, pH, substrate and nutrients, trace elements, etc.) in river beds vary significantly by day and season timescale. In addition, they are very complex with diverse substrates and electron donors (Nimick et al. 2011). Epilithic biofilms are thus dynamic structures that can create microscale variations of oxygen, pH, redox conditions and so on (Figure 11.3). For example, during the day, photosynthesis is the dominant metabolic process whereas during the night, in the absence of photosynthesis, respiration is the dominant process. This in turn influences significantly the distribution of pH, DO, redox conditions and so on within biofilms resulting, for example, in spatial and temporal variation of heavy metals concentration and speciation (Nimick et al. 2011).

11.1.2.2 Engineered Biofilms

Engineered biofilms are developed in reactors under well-controlled growth conditions. Depending on circulation and mixing of the solid and liquid phases, several biofilm reactor designs can be used (Nicolella et al. 2000).

For heavy metals or metalloid remediation studies, most research studies focused on mono-specie biofilms to simplify the system and better understand the mechanisms involved (see Section 11.2). However, for bioremediation purpose at industrial scales, engineered biofilms are in general multi-species and mainly composed of bacteria and also to some extent by fungi, protozoa and rotifers.

Biofilm structure and functionalities are significantly influenced by chemical, physical and hydrodynamic conditions in the surrounding environment. Thus, those factors must be controlled during biofilm growth in reactors.

Physico-chemical conditions (pH, loading rates, C/N/P ratios, electron donors and acceptors, ions, etc.) directly select bacterial populations and activity in biofilms. The pH, the substrate nature and nutrient conditions govern the nature of bacteria involved and their activity. Even if the majority of bacteria have an optimal pH around 7, some microorganisms need an acidic medium (acidophilic bacteria) or an alkaline medium (alkaliphilic bacteria) to

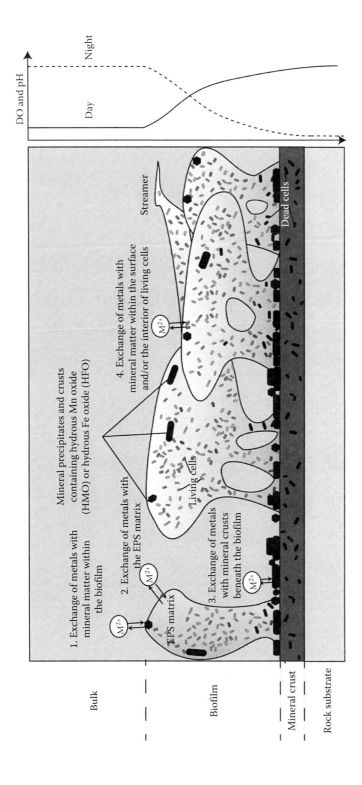

FIGURE 11.3

Possible exchange mechanisms of metals between water and components of biofilms developed on rock substrate. Distribution of living cells, mineral precipitates and crusts containing hydrous Mn oxide (HMO) or hydrous Fe oxide (HFO), EPS matrix, fossilised (dead) cells and rock substrate. Graph on the right shows relative changes in pH and DO concentration distribution within the biofilm for daytime (—) and nighttime (- -) conditions. (Modified from Nimick, D. A., C. H. Gammons, and S. R. Parker. 2011. *Chemical Geology* 283(1–2): 3–17.)

grow. Indirectly, as EPS are secreted by bacteria, chemical conditions govern EPS composition in biofilms. Substrate type and nutrients availability have an effect on the microbial metabolism, and thus influences the production of EPS (Durmaz and Sanin 2001). Moreover, the presence of toxic substances such as high concentrations in heavy metals can induce an increase in the production of EPS by microorganisms to protect themselves (Fang et al. 2002). Controlling temperature in bioreactors is also necessary since this operational parameter influences bacterial activities as well as transport phenomena and speciation of metals (Warren and Zimmerman 1994).

Hydrodynamic conditions play an important role on bacterial metabolism, on the detachment process and on transportation of soluble molecules (Stewart 2012). In general, an increase of hydrodynamic shear stresses or rates results in thinner, denser, more cohesive and less heterogeneous biofilms (Kwok et al. 1998; Paul et al. 2012). These observations at the meso-scale are certainly the results of processes that occur at microscale. Indeed, hydrodynamic conditions can influence the quantity and the type of EPS, secreted by bacteria and thus inorganic contaminants sorption and bioprecipitation phenomena (Ramasamy and Zhang 2005).

During the past decades, the main factors controlling biofilm structure and activities have been identified. However, the prediction of biofilm composition and structure still remains a difficult task. This is partly due to the lack of understanding on the influence of environmental conditions on EPS production in mixed-species biofilms and in turn on the relationship between biofilm composition and biofilm structure.

11.2 Toxic Inorganic Contaminants Removal Mechanisms in Biofilms

Biofilms may immobilise toxic heavy metals and metalloids (i.e. toxic inorganic contaminants) in the environment and thereby influence their migration behaviours. Over the past decades, the development of new technologies has improved our understanding on the biological, chemical and physicochemical mechanisms involved in metals and metalloid immobilisation and speciation in biofilms.

11.2.1 Case of Heavy Metals

In this section, only the main mechanisms involved in the removal of heavy metals will be discussed. Heavy metals accumulation in biofilms is mainly associated to metal biosorption, and to microbially induced metal precipitation (Figure 11.4). These mechanisms are described more specifically in the following subsections.

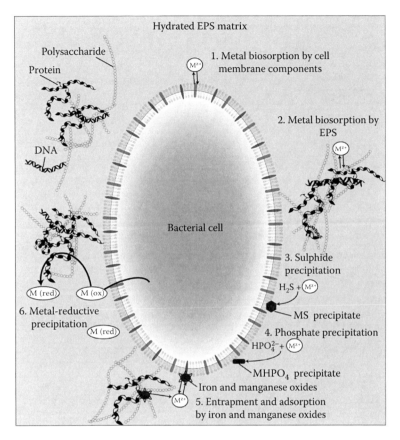

FIGURE 11.4
Schematic representation of the main interaction mechanisms between metals and a biofilm.

11.2.1.1 Biosorption

The biosorption of a wide range of metals onto bacterial cells and to the EPS has been studied (Fein et al. 1997; Guibaud et al. 2009). The cell wall of a bacterium and EPS contain functional groups that include carboxyl, phosphoryl, hydroxyl, amino and sulphhydryl groups (Flemming 1995). Due to the numerous sorption sites available in biofilms (Table 11.1), they can sorb water, inorganic and organic solutes, and particles. These sites display different sorption preferences, capacities and properties that depend on their chemical properties (ionic groups, hydrophobic groups, etc.). Many of these functional groups are negatively charged at neutral pH and are therefore able to sorb metals and/or to form organometallic complexes.

The ionisation states and sorption properties of the functional groups are pH dependent. In general, at a higher pH less protons are available to compete with metal for the EPS harboured binding sites leading to an enhanced

TABLE 11.1

Overall Composition of Microbial Biofilms and Possible Sorption Sites Available for Interactions with Metals and Metalloids

EPS
1.   EPS (including capsules) with:
• Cationic groups in amino sugars and proteins (e.g.: $-NH_3^+$)
• Anionic groups in uronic acids and proteins (e.g.: $-COO^-$; $-HPO_4^-$)
• Apolar groups from proteins (such as in aromatic amino acids)
• Groups with a high hydrogen bonding potential (such as polysaccharides)
Cells
2.   *Outer membrane* and:
• Lipopolysaccharides of Gram-negative cells with their lipid membrane
• The lipoteichoic acids in Gram-positive cells
3.   *Cell wall* consisting of N-acetylglucosamine and N-acetylmuramic acid, offering cationic and anionic sites
4.   *Cytoplasm membrane*, offering a lipophilic region
5.   *Cytoplasm*, as a water phase separated from the surrounding water
Minerals
6.   • Precipitates (sulphides, carbonates, phosphates, hydroxides, etc.)
• Free and bound metals

Source: From Flemming, H.-C. 1995. *Water Science and Technology* 32(8): 27–33; van Hullebusch, E D., M. H. Zandvoort, and P. N. L. Lens. 2003. *Reviews in Environmental Science and Biotechnology* 2(1): 9–33.

metal sorption in the exchange equilibrium (Comte et al. 2008; Li and Yu 2014). Several studies have also suggested that the deprotonated form of the reactive sites is primarily responsible for the binding of metal ions by biopolymers (van Hullebusch et al. 2005). For example, the carboxyl groups are negatively charged in neutral pH solution and attract positively charged ions through electrostatic interactions and form organometal complexes (Guibaud et al. 2005).

Moreover, a specificity associated to biofilm state is the possible pH gradients imposed by microbial metabolic activities and mass transport limitations. Indeed, bacterial metabolisms may influence biosorption by modifying local pH because proton exchanges in the environment generate some gradients within the bioaggregates (Batstone et al. 2006). In multi-species biofilms, depending on the environmental conditions, microbial composition is in general stratified. For example, in anaerobic biofilms, acidogens are located in the outer layers, while acetoclastic methanogens form clusters in the basal layers. Hydrolysis, acidogenesis and acetogenesis tend to release protons, while methanogenesis consumes protons. As a result, pH increase in the environment of methanogens clusters could generate pH gradient (Batstone et al. 2006) and thus stratification of sorption within biofilms. This distribution of pH conditions could in turn explain few experimental evidences of metals distribution in biofilms (Hu et al. 2007). However, new studies are required to better understand the sorption mechanisms within biofilm by measuring simultaneously pH distribution, biofilm composition and sorption of heavy metals locally in the biofilms.

11.2.1.2 Microbially Induced Metal Precipitation

Minerals formed within biofilms matrix are commonly denoted as biominerals, for example, sulphate, phosphate, carbonate, sulphide or silicate (van Hullebusch et al. 2003). In the case of metal sulphide precipitates, sulphate-reducing bacteria (SRB), anaerobic heterotrophic bacteria, utilise a range of organic substrates as electron donor and sulphate or other oxidised sulphur compounds as a terminal electron acceptor (Barton and Fauque 2009). The microbial reduction of sulphate by SRB in anaerobic environments is the major source of low-temperature sulphide in natural waters and is achieved by numerous bacterial strains (Labrenz and Banfield 2004). Then, sulphide can react with metals to form insoluble products (Lewis 2010). Immobilisation in a biofilm is an effective process to reduce the mobility of the metals, which is of value in bioremediation (Labrenz et al. 2000; van Hullebusch et al. 2003). Also, ancient microorganisms present as biofilm may have contributed to the low-temperature stages of formation of the large metal sulphide ore deposits (e.g. ZnS, Labrenz et al. 2000; Druschel et al. 2002). By combining scanning electron microscopy (SEM) and x-ray spectroscopy (EDX) analysis of a natural anaerobic biofilms, MacLean et al. (2007) highlighted that mineral formation within biofilms certainly depends on both physicochemical and biological phenomena. In this study, they observed that: ZnS minerals coated bacterial cells whereas FeS_2 minerals were not nucleated on the bacterial cell surfaces but were found within the biofilms matrix. In addition, they observed two pyrite mineral structures suggesting that distinct microenvironmental conditions have existed in the biofilm, leading to the formation of several forms of minerals.

Microbial metal sulphide precipitation reactions such as those considered here may be central to mine drainage remediation strategies (Johnson and Hallberg 2005). Bioprecipitation of metals by microbiologically produced sulphide is an effective way to remove and concentrate a range of metals, including Zn, Cu, Pb, Ni and Cd from wastewater and polluted streams (i.e. acid mine drainage [AMD], Papirio et al. 2013). The metals precipitate in bulk liquid as highly insoluble sulphides according to several physicochemical processes such as entrapment, nucleation and crystallisation of insoluble sulphides (e.g. ZnS, CdS, CuS and FeS) (Lewis, 2010). Various types of bioreactors have been used and generally the principal addition is an energy source for the SRB (Johnson and Hallberg 2005). One of the systems for immobilisation of toxic metals is the use of permeable reactive barriers. Permeable barrier technology is an *in situ* passive mitigation process designed to treat contaminated groundwaters. Metal sulphides are immobilised by SRB and the permeable barrier can retain pollutants without significantly altering the hydraulic gradient (Benner et al. 2000). Also, THIOPAQ technology developed and marketed by PAQUES Bio Systems in Balk, Netherlands, has been successfully used on a commercial scale at the Budelco zinc refinery in the Netherlands for the treatment of contaminated groundwater (Hockin and Gadd 2007). The treatment process uses H_2 as the electron source for the SRB to precipitate Zn^{2+} as ZnS.

Unfortunately, few studies report information regarding the influence of biofilm matrix composition on metal sulphide formation. Using both scanning transmission x-ray microscopy (STXM) and x-ray photoelectron emission microscopy (X-PEEM) to study natural anaerobic biofilms, MacLean et al. (2008) suggested that the pyrite crystals grew within an organic matrix composed mainly by polysaccharides and possibly extracellular DNA. This EPS matrix seems to play a key role in the growth and aggregation of the pyrite crystals by controlling crystal size and morphology.

In addition, no information is available regarding the influence of biofilm structure in the localisation of metal minerals in biofilm. However, one can suppose that in the same way than for sorption mechanisms, pH and supersaturation gradients will influence biomineralisation processes. For example, a pH increase in the environment of methanogen clusters could generate favourable conditions for precipitation of metals (Brown et al. 1998).

11.2.2 Metalloids Removal

Despite the importance of biofilms, there is limited fundamental detailed information on how they resist and transform potentially toxic elements such as selenium or uranium. The main microbial metalloid transformations in the environment are reduction and methylation (Wall and Krumholz 2006; Zannoni et al. 2007).

11.2.2.1 Case of Selenium

Selenium is found in five oxidation states. The soluble oxyanions selenate, Se(VI), and selenite, Se(IV), are poisonous in concentrations of ppm (Winkel et al. 2012). In contrast, elemental selenium Se(0) is highly insoluble and relatively non-toxic and occurs as a prevalent chemical species under anoxic conditions. Selenide, Se(-II), is both highly reactive and highly toxic, but is readily oxidised to Se(0) through several possible and energetically favourable inorganic and/or biochemical reactions (Fernández-Martínez and Charlet 2009).

The microbial reduction of selenate and selenite ions has been widely studied using planktonic bacteria, and these investigations are important to understand the fundamental interactions between microorganisms and metals. A variety of bacteria from soil and aquatic environments have the ability to reduce Se(VI) and Se(IV) oxyanions to insoluble Se(0). Some bacteria can use oxyanions of selenium (i.e. selenite or selenite) as terminal electron acceptors in dissimilatory reduction and also reduce and incorporate

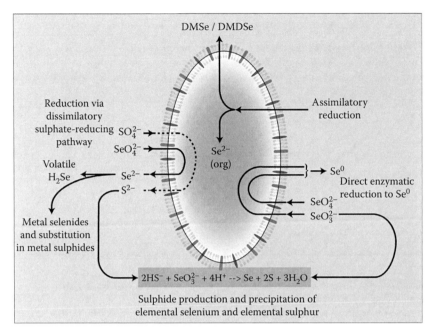

DMSe / DMDSe

Reduction via dissimilatory sulphate-reducing pathway

Assimilatory reduction

SO_4^{2-}
SeO_4^{2-}

Se^{2-} (org)

Volatile H_2Se

Se^{2-}
S^{2-}

$\} \rightarrow Se^0$
Direct enzymatic reduction to Se^0

SeO_4^{2-}
SeO_3^{2-}

Metal selenides and substitution in metal sulphides

$2HS^- + SeO_3^{2-} + 4H^+ \dashrightarrow Se + 2S + 3H_2O$

Sulphide production and precipitation of elemental selenium and elemental sulphur

FIGURE 11.5
Microbial transformation of selenium. DMSe, dimethyl selenide; DMDSe, dimethyl diselenide. (Modified from Hockin, S. L. and G. M. Gadd. 2003. *Applied and Environmental Microbiology* 69(12): 7063–7072.)

Se into organic compounds (e.g. selenoproteins) (assimilatory reduction). Selenate may also be reduced to selenide in nanomolar concentrations via the dissimilatory sulphate-reducing pathway, resulting in the production of volatile hydrogen selenide if no precipitation with metals such as Fe(II) or Cu(II) occurs (Figure 11.5). Assimilatory reduction of selenium by SRB is also required for the incorporation of selenide as an essential trace nutrient, and selenide may be released as the volatile alkylated species dimethyl selenide and dimethyl diselenide. A distinct pathway by which SRB enzymatically reduce selenium oxyanions to elemental selenium has also been established (Hockin and Gadd 2003 and references therein) (Figure 11.5).

Both aerobic and anaerobic reduction processes of selenium oxyanions are considered to be useful for removing Se toxic forms from Se-contaminated water. Table 11.2 reports the most recent studies regarding the role of biofilms in selenium removal. The reported studies are dealing with engineered or natural biofilms displaying the ability to reduce Se(VI) and/or Se(IV). Hockin and Gadd (2006) showed that Se removal is influenced by the presence of sulphate. A biofilm-selected strain of *Desulfomicrobium* sp. removed selenate from solution during growth on lactate (or hydrogen) and sulphate. When sulphate was not added in the medium, selenium was enzymatically reduced to selenide. However, when sulphate was added in

TABLE 11.2

Overview of the Literature Reporting on the Role of Engineered and Natural Biofilms in Selenium Removal

Biofilm Type	Environmental Conditions and Aims	Selenium Immobilisation Mechanisms	Authors
Engineered monospecific biofilm	Selenite reduction by biofilms of an iron-reducing bacterium, *Shewanella putrefaciens*, under anaerobic conditions	These findings indicate that biofilms of iron-reducing bacteria in the environment can immobilise selenium by reducing Se(IV) to Se(0), and Fe(III)-citrate complex promotes the reduction of Se(0) to Se(-II).	Suzuki et al. (2014)
Engineered monospecific biofilm	Study of the distribution and speciation of Se within aerobic *Burkholderia cepacia* biofilms formed on α-Al_2O_3 surfaces	Rapid reduction of Se(VI) by *B. cepacia* to Se(IV) and Se(0) subsequently results in a vertical segregation of Se species at the *B. cepacia*/α-Al_2O_3 interface. Elemental Se(0) accumulates within the biofilm with Se(VI), whereas Se(IV) intermediates preferentially sorb to the alumina surface.	Templeton et al. (2003)
Engineered monospecific biofilm	Selenite removal by a biofilm-forming strain of sulphate-reducing bacteria (*Desulfomicrobium norvegicum*) grown on lactate and sulphate	Elemental selenium and elemental sulphur were precipitated outside SRB cells. Precipitation occurred by an abiotic reaction of selenite with bacterially generated sulphide.	Hockin and Gadd (2003)
Engineered monospecific biofilm	A biofilm-selected strain of a *Desulfomicrobium* sp. grown on lactate (or hydrogen) and sulphate in order to remove selenate from solution	Under sulphate-limited growth conditions, selenium was enzymatically reduced to selenide. Under excess sulphate conditions, selenate removal was primarily by enzymatic reduction to elemental selenium. Sequestration by biofilms was greater under the latter condition.	Hockin and Gadd (2006)
Engineered multispecies biofilm	Removal of selenate from contaminated waters by using upflow anaerobic sludge bed (UASB) bioreactors either under methanogenic or sulphate-reducing conditions using lactate as electron donor	The selenate effluent concentrations in the sulphate-reducing and methanogenic reactor were 24 and 8 mg Se L^{-1}, corresponding to removal efficiencies of 97% and 99%, respectively. Selenium was mainly retained as elemental selenium in the biomass.	Lenz et al. (2008a)

Biofilm type	Description	Findings	Reference
Engineered multispecies biofilm	Removal of selenate from contaminated waters by using upflow anaerobic sludge bed (UASB) bioreactors under methanogenic using lactate as electron donor	This work showed a degree of complexity in the speciation of selenium treating anaerobic biofilms, with up to 4 selenium species contributing to the speciation, that is, different elemental, organic and metal- bound selenium species.	Lenz et al. (2008b)
Engineered multispecies biofilm	Selenate and selenite removal by UASB granular biofilms by using different electron donors (acetate or H_2/CO_2)	The high alkylation potential for selenite limits its bioremediation in selenium-laden waters involving H_2/CO_2 as the electron donor despite the fact that non-toxic elemental selenium and metal selenide species are formed.	Lenz et al. (2011)
Engineered multispecies biofilm	The authors focus on potential mechanisms involved in the removal of selenium from groundwater as cell-associated Se(0) nanoaggregates attached to tubing used to circulate acetate-amended groundwater.	Scanning and transmission electron microscopy revealed close association between Se(0) precipitates and cell surfaces, with Se(0) aggregates having a diameter of 50–60 nm.	Williams et al. (2013)
Natural multi-species biofilm	This study was aimed at characterising biofilms from a source with elevated levels of selenium in order to investigate the biotransformation of this element at high levels in the cultivated biofilms.	This study shows that selenium oxyanions significantly affect both the microbial diversity and structure in the multi-species biofilms studied. Investigations using CLSM showed a dominance of morphologically simplified populations in the multi-species biofilms in the presence of selenium species.	Yang et al. (2011)

excess, selenate removal occurred mostly by enzymatic reduction to elemental selenium. Sequestration by biofilms was greater under the latter condition. These results suggest that when a SRB biofilm bioreactor is considered for the treatment of acid mine drainage that contain selenium oxyanions, adequate selenate removal should be feasible under a wide range of environmental conditions as more recently demonstrated by Lenz et al. (2008a). However, the speciation and fate of the precipitated products is influenced by the dominant reduction pathway, which is controlled by environmental variables. Lenz et al. (2008a) showed that selenium was mainly immobilised as Se(0); however, advanced analysis of the biofilms by x-ray absorption spectroscopy showed that Se could also be significantly present as selenide and organic selenium compounds (Lenz et al. 2008b). Later on, Lenz et al. (2011) also showed that a significant quantity of volatile organic selenium compounds could be produced by anaerobic granular biofilm especially when the biofilm was fed with selenite as electron acceptor and hydrogen as electron donor. More recently, Suzuki et al. (2014) showed that biofilms of iron-reducing bacterial biofilms can immobilise selenium by reducing Se(IV) to Se(0), and Fe(III)-citrate complex promotes the reduction of Se(0) to Se(-II).

Yang et al. (2011) showed that selenium oxyanions significantly affect both the microbial diversity and structure in the multi-species biofilms studied. Investigations using CLSM showed a dominance of morphologically simplified populations in the multi-species biofilms in the presence of selenium species. The addition of selenium with or without sulphate in the influent was also shown to influence significantly the microbial diversity (Lenz et al. 2008a).

Templeton et al. (2003) studied a *B. cepacia* biofilm growing on α-Al$_2$O$_3$. The metabolically active bacteria were incubated with Se(VI) or Se(IV) leading to a significant reduction in the mobility of Se versus killed biofilms. Remobilisation experiments show that a large fraction of the insoluble Se(0) produced within the biofilm is retained during exchange with Se-free solutions. In addition, Se(IV) intermediates generated during Se(VI) reduction are preferentially bound to the alumina surface and do not fully desorb. In contrast, Se(VI) is rapidly and extensively remobilised (Templeton et al. 2003). This latter work shows the importance of the mineral and microbiological composition of the biofilm on possible selenium remobilisation. More work still needs to be done in order to understand the influence of water chemistry as well as biofilm mineral and microbial local composition on the removal and immobilisation of selenium species.

11.2.2.2 Case of Uranium

From all the uranium forms found in the environment, tetravalent uranium (U(IV)), which is often present as the mineral uraninite (UO$_2$), is the most stable and insoluble; whereas soluble U(VI), generally occurring as uranyl ion (UO$_2^{2+}$), is the most reactive species (Newsome et al. 2014). For remediation purposes, uranium bioimmobilisation may proceed through various processes: (i) direct or indirect reduction of U(VI) to U(IV) (Wall and Krumholz 2006); (ii) biosorption of uranium onto the cells surface; (iii) precipitation of uranium by organic complexing ligands released by the cells (Cao et al. 2011); and (iv) bioaccumulation of uranium in the cytoplasm through chelating to polyphosphate bodies or forming needle-like fibrils (Cao et al. 2010). A schematic illustration of the various mechanisms of U(VI) bioremediation using bacterial biofilm is shown in Figure 11.6 and the most recent studies are reported in Table 11.3.

Despite the potential benefits of using dissimilatory metal reducing microorganisms for U precipitation, the U(VI) reduction mechanisms using biofilms are still poorly understood. However, in the absence of bicarbonate, SRB biofilms have been shown to immobilise U(VI) for significant amounts of time as a result of both enzymatic and chemical reduction of U(VI) to insoluble uraninite (Beyenal et al. 2004; Marsili et al. 2007). In the presence of a carbonate buffer, the uranium removal extent and rate were satisfactory, but lower than in the absence of carbonate in SRB biofilm reactors over 5 months of operation (Marsili et al. 2005). In addition, the precipitated uranium has been shown to be stable as long as it is in sulphate-reducing conditions over a long period of time (4–5 months) (Marsili et al. 2007). More recently, well-controlled pure culture works have been applied to the field by studying uranium immobilisation using biofilms of indigenous microorganisms. Krawczyk-Bärsch et al. (2011) showed that *Ferrovum myxofaciens* dominated biofilms may have a substantial impact on the migration of uranium migration of uranium in acid mine drainage (AMD) waters. Krawczyk-Baärsch et al. (2012) indicated that uranium in the biofilm was immobilised intracellularly in microorganisms by the formation of metabolically mediated uranyl phosphate, similar to needle-shaped autunite ($Ca[UO_2]_2[PO_4]_2 \cdot 2\text{-}6H_2O$) or meta-autunite ($Ca[UO_2]_2[PO_4]_2 \cdot 10\text{-}12H_2O$) but aqueous calcium uranyl carbonate species formed in the groundwater contribute still to the migration of U(VI).

Knowing the complex biotic, abiotic and redox conditions in biofilms, it is challenging to predict the mobility of uranium because uranium immobilisation in biofilms with heterogeneous local conditions can be significantly different from that in bulk conditions (Cao et al. 2011, 2012; Nguyen et al. 2012). To efficiently stimulate natural multispecies biofilms to immobilise uranium at contaminated sites, many more studies are needed to elucidate the complex interactions among uranium, biofilms and various redox-sensitive minerals formed during *in situ* uranium bioremediation. Tapia-Rodríguez et al. (2012) reported the inhibitory impact of U(VI) towards different microbial populations in anaerobic biofilms, including methanogenic, denitrifying, and uranium-reducing microorganisms, which are commonly found at uranium bioremediation sites. The results suggest that microorganisms responsible for U(VI) reduction could tolerate much higher uranium concentrations compared to the other microbial populations investigated. Recently, Lee et al. (2014) showed that in the presence of biominerals such as FeS (mackinawite), a significant quantity of reduced uranium may remain in the form

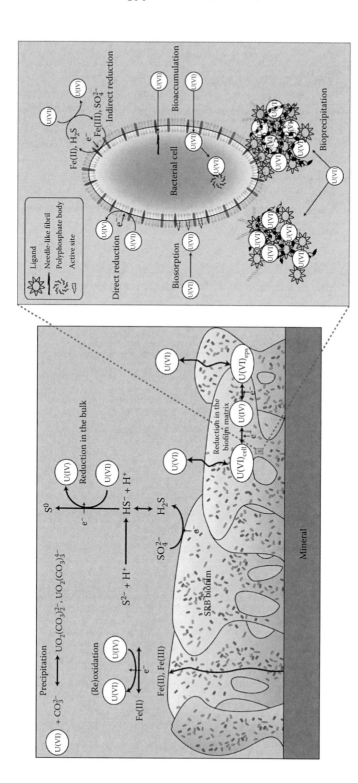

FIGURE 11.6
Uranium bioimmobilisation processes occurring when using microbial biofilms: direct and indirect reduction, biosorption, bioprecipitation and bioaccumulation. (Modified from Cao, B., B. Ahmed, and H. Beyenal. 2010. In *Emerging Environmental Technologies Volume II*, ed. V. Shah, 1–37. The Netherlands: Springer.)

Continued

TABLE 11.3

Overview of the Literature Reporting on the Role of Engineered and Natural Biofilms in Uranium Removal

Biofilm Type	Environmental Conditions and Aims	Uranium Immobilisation Mechanisms	Authors
Engineered monospecific biofilm	Biofilms of the sulphate-reducing bacterium (SRB) *Desulfovibrio desulfuricans* G20 was used to immobilise U(VI).	U was immobilised in the biofilms as a result of two processes: (1) enzymatic and (2) chemical reductions, by reacting with microbially generated H_2S.	Beyenal et al. (2004)
Engineered monospecific biofilm	Biofilms of sulphate-reducing bacteria *Desulfovibrio desulfuricans* G20 were used to immobilise U(VI).	Uranium accumulated mostly on microbial cell membranes and in the periplasmic space. The deposits had amorphous or poor nanocrystalline structures. Most likely nanocrystalline $U_x(PO_4)_y$.	Marsili et al. (2005)
Engineered monospecific biofilm	Biofilms of sulphate-reducing bacteria *Desulfovibrio desulfuricans* G20 were used to immobilise U(IV) in the presence of uranium-complexing carbonates.	This five-month study demonstrated that the sulphate-reducing biofilms were able to immobilise/reduce uranium efficiently, despite the presence of uranium-complexing carbonates.	Marsili et al. (2007)
Engineered monospecific biofilm	The goal of this study was to quantify the contribution of EPS to U(VI) immobilisation by *Shewanella* sp. HRCR-1.	The authors showed that (i) bound EPS from *Shewanella* sp. HRCR-1 biofilms contribute significantly to U(VI) immobilisation, through both sorption and reduction; (ii) bound EPS can be considered a functional extension of the cells for U(VI) immobilisation	Cao et al. (2011)
Engineered monospecific biofilm	The spatiotemporal metabolic responses of metabolically active *Shewanella oneidensis* MR-1 biofilms to U(VI) (UO_2^{2+}) was monitored by using non-invasive nuclear magnetic resonance imaging (MRI) and spectroscopy (MRS) approaches to obtain insights into adaptation in biofilms during biofilm–contaminant interactions.	The overall biomass distribution was not significantly altered upon exposure to U(VI). However, MRI and spatial mapping of the diffusion revealed localised changes in the water diffusion coefficients in the biofilms, suggesting significant contaminant-induced changes in structural or hydrodynamic properties during bioremediation	Cao et al. (2012)

TABLE 11.3 (Continued)

Overview of the Literature Reporting on the Role of Engineered and Natural Biofilms in Uranium Removal

Biofilm Type	Environmental Conditions and Aims	Uranium Immobilisation Mechanisms	Authors
Multispecies natural biofilm	*Ferrovum myxofaciens* dominated biofilms growing in acid mine drainage water as macroscopic streamers and as stalactite-like snottites hanging from the ceiling of the galleries.	Thermodynamic calculations and the plotting of the measured redox potential and pH into the pH–Eh diagram for the U–S–O–H–C system indicate that an aqueous U(VI) sulphate complex exists in the flooding water and a similar chemical composition, aqueous U(VI) sulphate complexation can be assumed in the biofilms.	Krawczyk-Bärsch et al. (2011)
Natural multispecies biofilms	Biofilm samples were taken from the walls of the ONKALO tunnel (Finland).	U(VI) was removed from solution and immobilised in biofilm microorganisms in the form of the U-phosphate minerals (autunite). In contrast, aqueous calcium uranyl carbonates species formed in the groundwater contribute to the migration of U(VI).	Krawczyk-Bärsch et al. (2012)
Multispecies natural sediment biofilm	This study aims at characterising, quantitatively, the microscale geochemical gradients (DO and H_2 concentration, pH and redox potential profiles) in Hanford 300A subsurface sediment biofilm and the influence of U(VI) on these gradients.	A biofilm reactor under air-saturated conditions was operated in the presence of U(VI) and characterised for U speciation in the sediment biofilm. X-ray absorption spectroscopy analysis showed that 80%–85% of the U was in the U(IV) valence state.	Nguyen et al. (2012)

of discrete UO$_2$ nanocrystals in the anaerobic groundwater, thereby compromising the effectiveness of bioreductive immobilisation as a remediation strategy. Finally, for effective long-term immobilisation of uranium, one should ensure that no biotic or abiotic reoxidation of the insoluble biogenic U(IV) could occur. It is therefore critical to understand the long-term stability of U(IV) under oxic- and nutrient-limited conditions at U-contaminated subsurface sites (Singh et al. 2014).

11.3 Current and Future Perspectives

Over the last decade, our understanding of metals and metalloids behaviour in natural and industrial biofilms has significantly improved. However, even though a large number of studies were carried out on the removal of metals and metalloids by biofilms at laboratory and pilot scales, as far as we know no industrial application of biofilm reactors still exist.

Several examples of studies close to industrial application have been presented by Diels et al. (2003, 2009), and are here shortly described. A moving-bed sand filter inoculated with heavy metal biosorbing and bioprecipitating bacteria has been used to remove metals from an industrial wastewater. In this configuration, a biofilm is developed on the sand grains after inoculation with heavy metal-resistant bacteria (*Ralstonia eutropha, Pseudomonas mendocina* and *Arthrobacter*). These bacteria have physico-chemical properties and secrete polymers that allow the biosorption and bioprecipitation of several heavy metals (Cd, Zn, Cu, Pb, Hg, Ni or Co). The wastewater containing metals goes through the biofilm sand filter. This leads to the binding of metals to the biofilms and thus to the removal of metals from the wastewater. Other studies have been performed at pilot scales to remove heavy metals from soils, groundwater and wastewater by using technologies such as bacteria immobilised composite membrane reactor to induce crystallisation of heavy metals (Diels et al. 2009). However, other large-scale studies should be performed to take advantage of the potential of these processes and to bring these technologies onto market.

However, even though most biogeochemical processes responsible for metals and metalloids removal by biofilms have been identified, still scientific and technical issues should be solved to understand the cycling and sequestration of toxic heavy metals and metalloids in natural biofilms and to develop industrial processes to treat contaminated wastewater, groundwater, soils and so forth. Therefore, additional research should be conducted in particular to understand and predict the long-term immobilisation of metals and metalloids in biofilms and the reversibility of the processes involved. This goes through

- A better understanding on the influence of environmental conditions (metals, metalloids, electron donor and acceptor quality and concentration, etc.) on bacterial metabolisms and thus biofilm composition (EPS composition, etc.).

- A better understanding on the influence of biofilm composition and stratification on the main mechanisms involved in immobilisation (sorption, precipitation, reduction, etc.).
- The development of mathematical models to describe the dynamics and the stratification of biofilms that should be included in addition to the processes conventionally considered in biofilm models, physicochemical and sorption processes.

References

Barton, L. L. and G. D. Fauque. 2009. Biochemistry, physiology and biotechnology of sulfate-reducing bacteria. *Advances in Applied Microbiology* 68: 41–98.

Batstone, D. J., C. Picioreanu, and M. C. M. van Loosdrecht. 2006. Multidimensional modelling to investigate interspecies hydrogen transfer in anaerobic biofilms. *Water Research* 40(16): 3099–3108.

Benner, S. G., W. D. Gould, and D. W. Blowes. 2000. Microbial populations associated with the generation and treatment of acid mine drainage. *Chemical Geology* 169(3–4): 435–448.

Beyenal, H., R. K. Sani, B. M. Peyton et al. 2004. Uranium immobilization by sulfate-reducing biofilms. *Environmental Science & Technology* 38(7): 2067–2074.

Bishop, P. L. and B. B. Rittmann. 1995. Modelling heterogeneity in biofilms: Report of the discussion session. *Water Science and Technology* 32(8): 263–265.

Brown, D. A., T. J. Beveridge, C. W. Keevil, and B. L. Sherriff. 1998. Evaluation of microscopic techniques to observe iron precipitation in a natural microbial biofilm. *FEMS Microbiology Ecology* 26(4): 297–310.

Cao, B., B. Ahmed, and H. Beyenal. 2010. Immobilization of uranium in groundwater using biofilms. In *Emerging Environmental Technologies Volume II*, ed. V. Shah, 1–37. The Netherlands: Springer.

Cao, B., B. Ahmed, D. W. Kennedy et al. 2011. Contribution of extracellular polymeric substances from *Shewanella* sp. HRCR-1 biofilms to U(VI) immobilization. *Environmental Science & Technology* 45(13): 5483–5490.

Cao, B., P. D. Majors, B. Ahmed et al. 2012. Biofilm shows spatially stratified metabolic responses to contaminant exposure. *Environmental Microbiology* 14(11): 2901–2910.

Comte, S., G. Guibaud, and M. Baudu. 2008. Biosorption properties of extracellular polymeric substances (EPS) towards Cd, Cu and Pb for different pH values. *Journal of Hazardous Materials* 151(1): 185–193.

Costerton, J. W., Z. Lewandowski, D. E. Caldwell et al. 1995. Microbial biofilms. *Annual Review of Microbiology* 49(1): 711–745.

Davit, Y., H. Byrne, J. Osborne et al. 2013. Hydrodynamic dispersion within porous biofilms. *Physical Review E* 87(1): 012718.

De Beer, D. and A. Schramm. 1999. Micro-environments and mass transfer phenomena in biofilms studied with microsensors. *Water Science and Technology* 39(7): 173–178.

Derlon, N., C. Coufort-Saudejaud, I. Queinnec, and E. Paul. 2013. Growth limiting conditions and denitrification govern extent and frequency of volume detachment of biofilms. *Chemical Engineering Journal* 218: 368–375.

Diels, L., S. V. Roy, S. Taghavi, and R. V. Houdt. 2009. From industrial sites to environmental applications with *Cupriavidus metallidurans*. *Antonie van Leeuwenhoek* 96(2): 247–258.

Diels, L., P. H. Spaans, S. Van Roy et al. 2003. Heavy metals removal by sand filters inoculated with metal sorbing and precipitating bacteria. *Hydrometallurgy* 71(1–2): 235–241.

Druschel, G. K., M. Labrenz, T. Thomsen-Ebert et al. 2002. Geochemical modeling of ZnS in biofilms: An example of ore depositional processes. *Economic Geology* 97(6): 1319–1329.

Durmaz, B. and F. D. Sanin. 2001. Effect of carbon to nitrogen ratio on the composition of microbial extracellular polymers in activated sludge. *Water Science and Technology* 44(10): 221–229.

Fang, H. H. P., L. C. Xu, and K. Y. Chan. 2002. Effects of toxic metals and chemicals on biofilm and biocorrosion. *Water Research* 36(19): 4709–4716.

Fein, J. B., C. J. Daughney, N. Yee, and T. A. Davis. 1997. A chemical equilibrium model for metal adsorption onto bacterial surfaces. *Geochimica et Cosmochimica Acta* 61(16): 3319–3328.

Fernández-Martínez, A. and L. Charlet. 2009. Selenium environmental cycling and bioavailability: A structural chemist point of view. *Reviews in Environmental Science and Bio/Technology* 8(1): 81–110.

Flemming, H.-C. 1995. Sorption sites in biofilms. *Water Science and Technology* 32(8): 27–33.

Flemming, H. C. and J. Wingender. 2002. What biofilms contain—Proteins, polysaccharides, etc. *Chemie in Unserer Zeit* 36(1): 30–42.

Flemming, H.-C. and J. Wingender. 2010. The biofilm matrix. *Nature Reviews Microbiology* 8(9): 623–633.

Graba, M., S. Sauvage, F. Y. Moulin et al. 2013. Interaction between local hydrodynamics and algal community in epilithic biofilm. *Water Research* 47(7): 2153–2163.

Guibaud, G., S. Comte, F. Bordas et al. 2005. Comparison of the complexation potential of extracellular polymeric substances (EPS), extracted from activated sludges and produced by pure bacteria strains, for cadmium, lead and nickel. *Chemosphere* 59(5): 629–638.

Guibaud, G., E. D. van Hullebusch, F. Bordas et al. 2009. Sorption of Cd(II) and Pb(II) by exopolymeric substances (EPS) extracted from activated sludges and pure bacterial strains: Modeling of the metal/ligand ratio effect and role of the mineral fraction. *Bioresource Technology* 100(12): 2959–2968.

Harrison, J. J., H. Ceri, and R. J. Turner. 2007. Multimetal resistance and tolerance in microbial biofilms. *Nature Reviews Microbiology* 5(12): 928–938.

Hockin, S. L. and G. M. Gadd. 2003. Linked redox precipitation of sulfur and selenium under anaerobic conditions by sulfate-reducing bacterial biofilms. *Applied and Environmental Microbiology* 69(12): 7063–7072.

Hockin, S. L. and G. M. Gadd. 2006. Removal of selenate from sulfate-containing media by sulfate-reducing bacterial biofilms. *Environmental Microbiology* 8(5): 816–826.

Hockin, S. L. and G. M. Gadd. 2007. Bioremediation of metals and metalloids by precipitation and cellular binding. In *Sulphate-Reducing Bacteria: Environmental and Engineered Systems*, ed. L. L. Barton and W. A. Hamilton, 405–434. Cambridge: Cambridge University Press.

Hu, Z., J. Jin, H. D. Abruña et al. 2007. Spatial distributions of copper in microbial biofilms by scanning electrochemical microscopy. *Environmental Science & Technology* 41(3): 936–941.

Johnson, D. B. and K. B. Hallberg. 2005. Acid mine drainage remediation options: A review. *Science of the Total Environment* 338(1–2): 3–14.

Krawczyk-Bärsch, E., H. Lünsdorf, T. Arnold et al. 2011. The influence of biofilms on the migration of uranium in acid mine drainage (AMD) waters. *Science of the Total Environment* 409(16): 3059–3065.

Krawczyk-Bärsch, E., H. Lünsdorf, K. Pedersen et al. 2012. Immobilization of uranium in biofilm microorganisms exposed to groundwater seeps over granitic rock tunnel walls in Olkiluoto, Finland. *Geochimica et Cosmochimica Acta* 96: 94–104.

Kwok, W. K., C. Picioreanu, S. L. Ong et al. 1998. Influence of biomass production and detachment forces on biofilm structures in a biofilm airlift suspension reactor. *Biotechnology and Bioengineering* 58(4): 400–407.

Labrenz, M. and J. F. Banfield. 2004. Sulfate-reducing bacteria-dominated biofilms that precipitate ZnS in a subsurface circumneutral-pH mine drainage system. *Microbial Ecology* 47(3): 205–217.

Labrenz, M., G. K. Druschel, T. Thomsen-Ebert et al. 2000. Formation of sphalerite (ZnS) deposits in natural biofilms of sulfate-reducing bacteria. *Science* 290(5497): 1744–1747.

Laspidou, C. S. and B. E. Rittmann. 2002. Non-steady state modeling of extracellular polymeric substances, soluble microbial products, and active and inert biomass. *Water Research* 36(8): 1983–1992.

Lee, S. Y., W. S. Cha, J. G. Kim et al. 2014. Uranium(IV) remobilization under sulfate reducing conditions. *Chemical Geology* 370: 40–48.

Lenz, M., E. D. van Hullebusch, F. Farges et al. 2008b. Selenium speciation assessed by X-ray absorption spectroscopy of sequentially extracted anaerobic biofilms. *Environmental Science & Technology* 42(20): 7587–7593.

Lenz, M., E. D. van Hullebusch, F. Farges et al. 2011. Combined speciation analysis by X-ray absorption near-edge structure spectroscopy, ion chromatography, and solid-phase microextraction gas chromatography-mass spectrometry to evaluate biotreatment of concentrated selenium wastewaters. *Environmental Science & Technology* 45(3): 1067–1073.

Lenz, M., E. D. van Hullebusch, G. Hommes et al. 2008a. Selenate removal in methanogenic and sulfate-reducing upflow anaerobic sludge bed reactors. *Water Research* 42(8–9): 2184–2194.

Lewis, A. E. 2010. Review of metal sulphide precipitation. *Hydrometallurgy* 104(2): 222–234.

Li, W. W. and H. Q. Yu. 2014. Insight into the roles of microbial extracellular polymer substances in metal biosorption. *Bioresource Technology* 160: 15–23.

MacLean, L. C. W., T. J. Pray, T. C. Onstott et al. 2007. Mineralogical, chemical and biological characterization of an anaerobic biofilm collected from a borehole in a deep gold mine in South Africa. *Geomicrobiology Journal* 24(6): 491–504.

MacLean, L. C. W., T. Tyliszczak, P. U. Gilbert et al. 2008. A high-resolution chemical and structural study of framboidal pyrite formed within a low-temperature bacterial biofilm. *Geobiology* 6(5): 471–480.

Marcato-Romain, C. E., Y. Pechaud, E. Paul et al. 2012. Removal of microbial multispecies biofilms from the paper industry by enzymatic treatments. *Biofouling* 28(3): 305–314.

Marsili, E., H. Beyenal, L. Di Palma et al. 2007. Uranium immobilization by sulfate-reducing biofilms grown on hematite, dolomite, and calcite. *Environmental Science & Technology* 41(24): 8349–8354.

Marsili, E., H. Beyenal, L. Di Palma et al. 2005. Uranium removal by sulfate reducing biofilms in the presence of carbonates. *Water Science & Technology* 52(7): 49–55.

McSwain, B. S., R. L. Irvine, M. Hausner, and P. A. Wilderer. 2005. Composition and distribution of extracellular polymeric substances in aerobic flocs and granular sludge. *Applied and Environmental Microbiology* 71(2): 1051–1057.

Newsome, L., K. Morris, and J. R. Lloyd. 2014. The biogeochemistry and bioremediation of uranium and other priority radionuclides. *Chemical Geology* 363: 164–184.

Nguyen, H. D., B. Cao, B. Mishra et al. 2012. Microscale geochemical gradients in Hanford 300 area sediment biofilms and influence of uranium. *Water Research* 46(1): 227–234.

Nicolella, C., M. C. M. van Loosdrecht, and J. J. Heijnen. 2000. Wastewater treatment with particulate biofilm reactors. *Journal of Biotechnology* 80(1): 1–33.

Nimick, D. A., C. H. Gammons, and S. R. Parker. 2011. Diel biogeochemical processes and their effect on the aqueous chemistry of streams: A review. *Chemical Geology* 283(1–2): 3–17.

Papirio, S., D. K. Villa-Gomez, G. Esposito et al. 2013. Acid mine drainage treatment in fluidized-bed bioreactors by sulfate-reducing bacteria: A critical review. *Critical Reviews in Environmental Science and Technology* 43(23): 2545–2580.

Paul, E., J. C. Ochoa, Y. Pechaud et al. 2012. Effect of shear stress and growth conditions on detachment and physical properties of biofilms. *Water Research* 46(17): 5499–5508.

Ramasamy, P. and X. Zhang. 2005. Effects of shear stress on the secretion of extracellular polymeric substances in biofilms. *Water Science and Technology* 52(7): 217–223.

Singh, G., S. S. Sengör, A. Bhalla et al. 2014. Reoxidation of biogenic reduced uranium: A challenge toward bioremediation. *Critical Reviews in Environmental Science and Technology* 44(4): 391–415.

Stewart, P. S. 2012. Mini-review: Convection around biofilms. *Biofouling* 28(2): 187–198.

Stewart, P. S. and M. J. Franklin. 2008. Physiological heterogeneity in biofilms. *Nature Reviews Microbiology* 6(3): 199–210.

Stoodley, P., K. Sauer, D. G. Davies, and J. W. Costerton. 2002. Biofilms as complex differentiated communities. *Annual Review of Microbiology* 56(1): 187–209.

Sutherland, I. W. 2001. Biofilm exopolysaccharides: A strong and sticky framework. *Microbiology-Uk* 147: 3–9.

Suzuki, Y., Y. Sakama, H. Saiki et al. 2014. Immobilization of selenium by biofilm of *Shewanella putrefaciens* with and without Fe(III)-citrate complex. *Journal of Nuclear Science and Technology* 51(1): 108–115.

Tapia-Rodríguez, A., A. Luna-Velasco, J. A. Field, and R. Sierra-Alvarez. 2012. Toxicity of uranium to microbial communities in anaerobic biofilms. *Water, Air, & Soil Pollution* 223(7): 3859–3868.

Templeton, A. S., T. P. Trainor, A. M. Spormann, and G. E. Brown Jr. 2003. Selenium speciation and partitioning within *Burkholderia cepacia* biofilms formed on α-Al$_2$O$_3$ surfaces. *Geochimica et Cosmochimica Acta* 67(19): 3547–3557.

van Hullebusch, E. D., A. Peerbolte, M. H. Zandvoort, and P. N. L. Lens. 2005. Sorption of cobalt and nickel on anaerobic granular sludges: Isotherms and sequential extraction. *Chemosphere* 58(4): 493–505.

van Hullebusch, E D., M. H. Zandvoort, and P. N. L. Lens. 2003. Metal immobilisation by biofilms: Mechanisms and analytical tools. *Reviews in Environmental Science and Biotechnology* 2(1): 9–33.

Wall, J. D. and L. R. Krumholz. 2006. Uranium reduction. *Annual Review of Microbiology* 60: 149–166.

Warren, L. A. and Zimmerman, A. P. 1994. The influence of temperature and NaCl on cadmium, copper and zinc partitioning among suspended particulate and dissolved phases in an urban river. *Water Research* 28(9): 1921–1931.

Williams, K. H., M. J. Wilkins, A. L. N'Guessan et al. 2013. Field evidence of selenium bioreduction in a uranium-contaminated aquifer. *Environmental Microbiology Reports* 5(3): 444–452.

Winkel, L. H. E., C. A. Johnson, M. Lenz et al. 2012. Environmental selenium research: From microscopic processes to global understanding. *Environmental Science & Technology* 46(2): 571–579.

Yang, S. I., J. R. Lawrence, G. D. W. Swerhone, and I. J. Pickering. 2011. Biotransformation of selenium and arsenic in multi-species biofilm. *Environmental Chemistry* 8(6): 543–551.

Zannoni, D., F. Borsetti, J. J. Harrison, and R. J. Turner. 2007. The bacterial response to the chalcogen metalloids Se and Te. In *Advances in Microbial Physiology Volume 53*, ed. R. K. Poole, 1–312. Oxford, UK: Academic Press.

12

Bioprocessing of Electronic Scraps

Sadia Ilyas and Jae-Chun Lee

CONTENTS

12.1 Introduction

The generational growth of technologically sound electrical and electronic products continues to revolutionise the various modes of communication, entertainment, transportation, education and health care globally, and affects human life significantly. There is no sign of abating such a revolution soon in the near future. This leads to a significant increase in the generation of electronic wastes after the product's end of life (EOL) or, getting outdated technologically. According to the directive of the European Community, goods which can function by supplying the electrical current or electromagnetic field as well; can generate, transfer and measure such a current and/or field that has been defined as the electrical and electronic equipment, is electronic scrap at the EOL (Tuncuk et al., 2012).

In this advanced technological era, the global generation of electronic scrap has been estimated up to 50 million tons. By 2005, the European Union has estimated the generation of 8.3–9.1 million tons of electronic scrap. It has been predicted that with an annual growth rate of 2.5%–2.7%, this figure will reach at least 12.3 million tons by 2020 (Ongondo et al., 2011). The quantum of such waste generation presents a challenge in terms of sustainability for both the materials and environment. A variety of hazardous organic and inorganic substances present in e-waste may cause severe environmental problems if not handled in an effective manner. In contrast, even after the EOL, e-waste contains several valuable, precious and rare metals. The metal contents in e-waste are even higher than those of the present in their primary ores for some of the metals. Owing to the fast depletion of primary reserves and by considering the contents of metallic values in electronic scrap, the recycling of electronic scrap is of prime importance to minimise the volume of landfill and mitigate/prevent the pollution of soil, groundwater and air due to the release of hazardous components. Furthermore, the treatment of electronic scrap for the recovery of metals appears to have important economic and environmental prospects (Tuncuk et al., 2012; Ilyas et al., 2013).

Among the several recycling routes for the treatment of the electronic wastes, biohydrometallurgy has been considered to be a better green technology for the metal recovery. Biohydrometallurgy offers many attractive features such as operational flexibility, low cost and low-energy consumption, besides being environmentally benign. Microbes encounter metals and metalloids of various kinds in the environment, and hence, they interact with them indifferently either for their benefit or detriment. The interaction of microbes with metals depends in part on whether the organism is prokaryotic or eukaryotic and with which metal or multi-metal system they are interacting. Both types of microbes have the ability to bind the metal ions present in the external environment at the cell surface or, to transport them into the cell for various intercellular functions. This metal–microbes interaction provides the possibility of selective or non-selective recovery of metals from electronic scrap, particularly that of copper and gold, which is the main concern in this chapter.

12.2 Leaching of Gold and Copper from Electronic Wastes by Biometallurgy

Electronic wastes are rich sources of base and precious metals. Besides the high-value share of most of these metals, copper has a high carbon footprint; nearly 4 kg of CO_2 produced for 1 kg of copper. Recycled gold represents a lower carbon footprint compared to that of primary-mined gold. Although only small amounts of gold are found in these scraps, it has the largest material footprint, for example, nearly 50 and 5000 times to that of copper and aluminum (U.S. Geological Survey, 2007; Groot and Pistorius, 2008). Based on the material footprint data for electronic scrap (PC, printer and CRC screen), Groot and Pistorius (2008) suggested that recycling efforts should primarily focus on the metals such as gold and tin followed by nickel, copper and lead for sustainability.

Currently, metallic scraps containing copper and gold are generally treated by pyrometallurgical and/or hydrometallurgical methods. Most methods involving pyrometallurgical processing of electronic waste give rise to the following limitations:

- Integrated smelters cannot recover aluminum and iron as metals; so, these metals are transferred into the slag.
- The presence of halogenated flame retardants in the smelter feed can lead to the formation of dioxins, unless special installations and measures are present.
- Ceramic components and glass in electronic scrap increase the amount of slag in the smelting furnace, which leads to the loss of precious metals.
- Management of energy recovery from the associated organic constitutes is at its stage of infancy.

However, in the case of hydrometallurgical routes, special stainless-steel and rubber-lined equipment is required to resist the highly corrosive acidic and oxidising conditions, and proper control of the emitted gases to avoid health risks. Similarly, the excessive use of cyanide for the dissolution of gold is also associated with the environmental risk.

In the current scenario, biological methods are considered to be environmentally friendly, simple and self-sustaining and have low manpower requirement, thereby cutting the labour cost. Biological gold-dissolution methods produce little cyanide and when coupled with subsequent detoxification, they are particularly environment friendly. Biological cyanidation has an economic advantage for processing ores and electronic scrap because cyanide-producing bacteria (*Chromobacterium violaceum*, *Bacillus megaterium*, *Pseudomonas fluorescens*, and *Pseudomonas plecoglossicida*) produce cyanide and autonomously detoxify cyanide at the end of the process by converting it into simple compounds (Brandl et al., 2001; Ilyas et al., 2007).

12.2.1 Bioleaching/Biooxidation of Metals from Electronic Scrap

The mobilisation of metal cations from an often almost insoluble material by biological oxidation and complexation processes is referred to as bioleaching. However, for the recovery of precious metals (Au, Pt, Pd, etc.), the leaching bacteria are applied only to dissolve interfering metals prior to the cyanidation treatment. Here, the term biooxidation is used preferentially because the bioleached metals, in most cases iron and arsenic, are not intended to be recovered. In order to extract metals from the electronic scrap by bioleaching, much attention was focussed on the printed circuit boards (PCBs). The metal contents of the PCBs are quite variable depending on the origin of the material, and to some extent on the sampling procedure followed and analytical methods used. The composition is generally in the range: 10%–27% Cu, 8%–38% Fe, 2%–19% Al, 0.3%–2% Ni, 1%–3% Pb, 200–3000 ppm Ag, 20–500 ppm Au, 10–200 ppm Pd and so on (Ilyas et al., 2007). In general, the values of precious metals (Au, Pt, Pd, etc.) and some base metals (Cu, Pb) are the prime economic drivers to recycle the electronic scrap. Rapid increases in production costs and finally a sustained period of uncertainty in metal prices, have forced companies for the contraction of the recycling industries. Hence, the recovery of metals such as Au and Cu from electronic scrap is mostly stressed upon, which is associated with the high market price as well as reducing the carbon and material footprint.

12.2.2 Principles of Microbial Interaction with Metals

Microbes encounter metals and metalloids of various kinds in different ways. In particular, metals of interest in the e-waste processing could be the base metals such as Co, Cu, Cd, Cr, Fe, Mn, Ni, V, Zn, Mo and Pb and the precious metals such as Au, Ag, Pt and Pd. Since the metal content in many of the solid wastes/secondary resources is in the form of inorganic sulphides/oxides or pure metals and sometimes as organo-metallic complexes, the metal of interest can be solubilised by microorganisms. As mentioned above, the way microbes interact with metals depends in part on whether the organisms are prokaryotic or eukaryotic. Both types of microbes have the ability to bind metal ions present in the external environment at the cell surface or to transport them into the cell for various intracellular functions (Rossi, 1990). On the other hand, only the prokaryotes (eubacteria and archaea) include organisms that are able to oxidise Mn^{2+}, Fe^{2+}, Co^{2+} and Cu^{1+} or reduce Mn^{4+}, Fe^{3+} and Co^{3+} on a large scale and conserve energy in these reactions. Some prokaryotes and eukaryotes may form metabolic products such as acids or ligands that dissolve base metals contained in minerals, such as Fe, Cu, Zn, Ni, Co and so on. Others may form anions such as sulphides or carbonate, that precipitate dissolved metal ions (Ehrlich, 1997; Ilyas et al., 2013). There can be several levels and modes of the interaction of microbes with metals such as (Ehrlich, 1997; Brierley and Brierley, 1999)

- Metabolic/enzymatic interaction—where the uptake of trace metals and their subsequent incorporation into metalloenzymes (nitrogenase, cytochromes, cytochrome oxidase, superoxide dismutases, bacterio-chlorophyll and formate dehydrogenase) takes place.

- Metabolic/enzymatic interaction—where the uptake of trace metals may be fast, non-specific and constitutive, as for instance, with the CorA Mg^{2+} exchange system.

- Metabolic/enzymatic interaction—where some eubacteria (*Acidithiobacillus ferrooxidans*, *Leptospirillum ferrooxidans*) and archaea (*Acidianus brierleyi*, *Sulfolobus acidocaldarius*) use metals or metalloids as electron donors or acceptors in energy metabolism.

- Enzymatic microbial detoxification of harmful metals or metalloids.

- Metabolic/non-enzymatic interaction: Prokaryotic and eukaryotic microbes are capable of accumulating metals by binding them as cations to the cell surface in a passive process.

- In nature, a noticeable microbial interaction with metals frequently manifests itself through metal immobilisation or mobilisation.

- Mineralytic interactions of prokaryotic or eukaryotic microorganisms with minerals/material are mainly based on three principles, namely, acidolysis, complexolysis and redoxolysis.

Microorganisms are able to mobilise metals in various ways. These are mobilisation by formation of organic or inorganic acids (protons), mobilisation involving oxidation and reduction reactions and mobilisation due to complexation with secreted chelating agents. Sulphuric acid is the main inorganic acid found in leaching environments. It is formed by sulphur-oxidising microorganisms such as *Thiobacilli*. A series of organic acids are formed by bacterial as well as fungal metabolism resulting in organic acidolysis, complex and chelate formation (Ehrlich, 1997).

12.2.3 Diversity among Leaching Microorganisms

Microbial diversity within mining biotopes is very broad and complex. A heterogeneous and complex microflora consists of both acidophilic heterotrophic and autotrophic microorganisms that exist with the commercial bioprocessing systems. Leaching bacteria are distributed among the Proteobacteria (*Acidithiobacillus*, *Acidiphilium*, *Acidiferrobacter* and *Ferrovum*); Nitrospirae (*Leptospirillum*); Firmicutes (*Alicyclobacillus*, *Sulfobacillus*) and Actinobacteria (*Ferrimicrobium*, *Acidimicrobium* and *Ferrithrix*). Within all groups, mesophilic as well as moderately thermophilic microorganisms can be found. Leaching archaea mostly belong to the Sulfolobales, a group of extremely thermophilic, sulphur and Fe^{2+} oxidisers including genera such as *Sulfolobus*, *Acidianus*, *Metallosphaera* and *Sulfurisphaera*. Also, within the Thermoplasmales, two Fe^{2+}-oxidising species, *Ferroplasma acidiphilum* and *Ferroplasma acidarmanus* are known (Brierley and Brierley, 1999; Norris et al., 2000; Vera et al., 2013). Furthermore, a large diversity among leaching bacteria has been found with respect to carbon assimilation pathways. *Acidithiobacillus* sp. and *Leptospirillum* sp. can grow only chemolithoautotrophically. In contrast, *Acidiphilium acidophilum* and *Acidimicrobium ferrooxidans* are able to grow autotrophically with reduced sulphur compounds and Fe^{2+}, heterotrophically with glucose or yeast extract and mixotrophically with all these substrates (Norris et al., 2000; Vera et al., 2013). In addition, several *Acidiphilium* sp. and *Acidisphaera rubrifaciens* possess pigments that may confer the ability for some photosynthetic activity (Vera et al., 2013).

Some heterotrophic bacteria (*C. violaceum, P. fluorescens*) and fungi (*Aspergillus niger, Penicillium simplicissimum*) are also considered among the potentially important and diversified bioleaching organisms (Faramarzi et al., 2004).

12.2.4 Bacterial Cultures and Consortia as Bioleaching Tools

Among the acidophilic prokaryotes applied mostly for bioleaching of copper from electronic scrap, the mesophilic (20–35°C) or moderately thermophilic (40–55°C) cultures exhibit optimum growth in the pH range 1.5–2.5. The most dominant mesophiles used in bioleaching of copper from electronic scrap belong to the genus *Acidithiobacillus* and *Leptospirillum*, which are essentially the iron and sulphur oxidisers (*A. ferrooxidans*), sulphur oxidisers (*Acidithiobacillus thiooxidans, Acidithiobacillus caldus*) and iron oxidisers (*L. ferrooxidans, Leptospirillum ferriphilum*). They are autotrophic in nature as they use inorganic carbon (CO_2) as the carbon source and are strictly chemolithotrophic, that is, they derive energy for growth and other metabolic functions from the oxidation of ferrous iron, reduced sulphur compounds, metal sulphides and some species also derive energy from the oxidation of hydrogen sulphide (Brierley, 1997).

Numerous investigations on the bioleaching of metals were carried out with wild cultures of *A. thiooxidans, A. caldus* and *A. ferrooxidans*. The description of *L. ferrooxidans* shows that it is not as easily enriched as *A. ferrooxidans* from the samples containing both organisms. But, the propensity of *L. ferrooxidans* to attach to sulphide minerals, its high affinity towards ferrous iron and low sensitivity to inhibit ferric ions compared to *A. ferrooxidans*, are the additional evidence of its increasing importance in the bioleaching. The complete biooxidation of sulphide minerals or the added energy source involves the oxidation of both iron and sulphur. Iron-oxidising bacteria produce ferric ions. Oxidation of metal sulphides is reported to occur through thiosulphate and the polysulphide pathways. The thiosulphate is oxidised to sulphuric acid biologically and/or chemically while elemental sulphur is produced through polysulphide intermediates and the sulphur-oxidising microorganisms such as *A. thiooxidans* and *A. caldus* reduce the accumulation of sulphur efficiently, compared to *A. ferrooxidans* while improving the bioleaching efficiency (Brierley, 1997).

With a view to improve the leaching efficiency with enhanced oxidation of sulphur and iron, studies have often been conducted with the consortia of microorganisms. Brandl et al. (2001) reported the preliminary investigation on the feasibility of the recovery of base metals from the dust of the e-waste shredding with the consortium of *A. ferrooxidans* and *A. thiooxidans*. The leaching results show that the addition of larger amounts of scrap leads to an increase of the initial pH due to the alkalinity of electronic scrap. In view of the alkalinity of the electronic scrap and acid consumption at higher pulp density and to reduce the toxic effects on the microbes, the biomass is reported to be produced in the first stage in the absence of electronic scrap over a period of 7 days. Electronic scrap is subsequently added in different concentrations and the cultures are incubated for an additional period of 10 days in the second stage. The consortium could leach more than 90% of the available Al, Cu, Ni and Zn. At higher concentration, metal mobilisation is reduced, especially for Al and Cu. Ni and Zn show much better results with the mobilisation of 60% and 95%, respectively. In all cultures, Pb and Sn are not detected in the leachate. It is proposed that Pb is precipitated as $PbSO_4$ and Sn is precipitated probably as SnO. Because of the precipitation of the metals during the microbial leaching process, subsequent treatment of the leach residue is essential to recover the metals already leached.

Overall, we can assume that the mechanism of metals bioleaching from electronic scrap involving bacteria is the same, as in the case of metal sulphides leaching. Fe^{3+} created by *A. ferrooxidans* oxidises elemental copper and/or other metals contained in the e-waste to an ionic form, according to the reactions:

$$M + Fe_2(SO_4)_3 \rightarrow 2Fe^{3+} + H_2O \tag{12.1}$$

$$2FeSO_4 + H_2SO_4 + 0.5O_2 + \text{bacteria} \rightarrow Fe_2(SO_4)_3 + H_2O \tag{12.2}$$

Moderately thermophilic microorganisms consist of the genera *Acidimicrobium*, *Ferromicrobium* and *Sulfobacillus*. *Sulfobacillus* belongs to mixotrophic organisms that grow on media containing both inorganic (reduced iron and sulphur compounds, sulphide minerals) and the organic compounds (yeast extract, casein hydrolysate, peptone, some sugars and amino acids). In heterotrophic conditions, *Sulfobacilli* grow in the presence of Fe^{2+} and yeast extract. In the autotrophic conditions, the presence of reduced sulphur compounds (thiosulphate or tetrathionate) or yeast extract which are added to the medium along with Fe^{2+} (Brierley, 1997), these being the moderate thermophiles permit the use of a higher temperature for a faster bioleaching rate.

Ilyas et al. (2007) used wild cultures, adapted cultures and the consortium of moderately thermophilic bacteria such as *Sulfobacillus thermosulfidooxidans* and an unidentified acidophilic heterotroph (code A1 TSB) isolated from the local environment to leach metals from washed and un-washed electronic scrap. At a scrap (PCBs) concentration of 10 g/L, the metal-adapted consortium leached more than 89% Cu, 81% Ni, 79% Al and 83% Zn in 18 days in the presence of 10 g/L sulphur. The secreted carboxylic acids such as citric, oxalic and succinic acids from the heterotroph were able to increase the leaching rate in this case which could be due to the synergistic effect of the heterotroph on the growth of *S. thermosulfidooxidans*.

A detailed study by Ilyas et al. (2010), with an attempt to upscale the process from shake flasks to lab-scale column reactors, indicates that the effect of the non-metallic portion of scrap contributing towards the alkalinity can be minimised by the washing treatment without the alteration of metal concentration. Furthermore, the mixed culture of acidophilic chemolithotrophic bacteria and acidophilic heterotrophic bacteria exhibited a greater bioleaching potential than individual cultures due to the synergistic effect. The pulp density of wastes can be increased by increasing the tolerance of the bacterial culture to mixed metal ions. The results from these studies demonstrate that the acid pre-leach for 27 days followed by bioleaching for 280 days can dissolve 86% Cu, 74% Ni, 80% Zn and 64% Al at a fixed temperature column reactor level. Recent studies by Ilyas et al. (2013) in the shake flask and column reactor demonstrate that supplementation of FeS_2 and S^0 as an additional energy source to the bacterial consortium during the growth and mixing while preparing the column charge, has a multi-fold effect which includes enhanced bacterial growth, increased leaching efficiencies, reduction of the bioleaching period and stabilisation of the optimum pH conditions without the pre-leaching step.

Apart from the basic and associated metals (e.g. Cu, Fe, Ni, Al, Sn, Zn and Pb) contained in electronic waste materials, precious metals (Ag, Au and Pt) were taken into consideration. The presence of these metals has a decisive influence on the value of electronic scrap and the cost of its processing. Some of the earliest works on the possibility of extracting gold from printed electronic circuits by bioleaching were done by Faramarzi et al. (2004). In the experiments, cut pieces of PCBs (5 × 10 mm) were used. In the presence of *C. violaceum*, gold was microbiologically dissolved in the form of dicyanoaurate. The maximum concentration of dicyanoaurate corresponds to the dissolution of initially added gold at the level of 14.9%. Glycine is commonly used as a precursor of cyanide, which is produced by oxidative decarboxylation of glycine as in the reaction:

$$NH_2CH_2COOH \rightarrow HCN + CO_2 + 4H^+ \qquad (12.3)$$

The conversion of glycine into hydrogen cyanide takes place by the associated enzymes under the metabolic activity of HCN synthase which is followed by the reaction of cyanide ions with the substrate to dissolve gold/copper.

12.2.5 Fungal Biogenic Acids/Metabolites as Bioleaching Tools

Some species of heterotrophic fungi, such as *Aspergillus* and *Penicillium*, have shown the potential for metal bioleaching (Table 12.1). Metal solubilisation may be due to the enzymatic reduction of highly oxidised metal compounds or by the production of organic acids (e.g. lactic acid, oxalic acid, citric acid and gluconic acid) and by compounds with hydrophilic reactive groups (Brandl et al., 2001). Fungal metabolites usually have the dual effect of dissolving insoluble metal compounds from the material by increasing metal dissolution with the lowering of the pH and increasing the concentration of soluble metals by complexing/chelating with soluble organo-metallic complexes. Acidolysis, complexolysis and redoxolysis are three mechanisms in bioleaching processes. In the acidolysis process, solubilisation occurs via protonation of the metal compound, making it less available to the metal cation and promotes dissolution of the solid where MO is the metal oxide.

$$MO + 2H^+ \rightarrow M^{2+} + H_2O \qquad (12.4)$$

The term complexolysis refers to the solubilisation of a metal ion due to the complexing capacity of a molecule, such as organic acids or amino acids. Complex formation between acid anions with metal cations decreases the concentration of the free metal ions, and thus allows more solid to go into the solution. This mechanism is slower than solubilisation with protons (acidolysis), but plays an important role in enhancing the solubility of metal ions, which has been solubilised via acidolysis. The mediation by redox reaction is based on electron transfer between excreted metabolites with metals on the material. The reduction of insoluble metals on the material therefore leads to the metal solubilisation. An example of redoxolysis in fungal bioleaching is the reduction of ferric iron and manganese mediated by oxalic acid in an acidic environment.

The feasibility for using fungi (*A. niger* and *P. simplicissimum*) to leach metals from electronic scrap has also been investigated by Brandl et al. (2001) in a one-step- and two-step-leaching mode. The microbial growth is found to be inhibited in the one-step process with the scrap concentration of above 10 g/L in the medium. However, almost complete dissolution of the metals such as Cu, Pb, Sn and Zn was noticed in a two-step-leaching process using a commercial gluconic acid solution (Nagluso™, 2.5 M) produced by *A. niger*.

Recently, Ilyas and Lee (2013) carried out studies with a view to develop an economically feasible and environmentally friendly technique to extract metals such as Al, Fe, Cu, Zn and Ni from electronic scrap with *Penicillium chrysogenum* in various leaching modes. This includes one-step leaching with combined growth and leaching phases (mode-1), two-step leaching with alternative growth and leaching phases (mode-2), spent medium leaching with separated growth and leaching phases (mode-3), leaching with fresh growth medium without fungal culture (mode-4), leaching with deionised

TABLE 12.1

Active Bio-Tops Used as a Tool in Bioleaching and the Level or Research

Source	Microorganisms	Leaching Efficiency	Level of Research	References
Mesophilic Bacteria				
Autotrophs				
PCBs/electronic scrap/e-wastes	*A. ferrooxidans + A. thiooxidans*	Cu; 90%	Shake flasks	Brandl et al. (2001)
	A. ferrooxidans	Cu; 37%	Shake flasks	Choi et al. (2004)
	Acidithiobacillus sp. + *Leptospirillum* sp.	Cu; 100%	Shake flasks	Vestola et al. (2010)
	A. ferrooxidans	Cu; 98%	Shake flasks	Yang et al. (2009)
TV circuit boards	*A. ferrooxidans + L. ferrooxidans + A. thiooxidans*	Cu; 89%	Shake flasks	Bas et al. (2013)
Heterotrophs				
PCBs/electronic scrap/e-wastes	*C. violaceum, P. fluorescens* and *P. plecoglossicida*	Au; 68.5%	Shake flasks	Brandl et al. (2008)
Waste mobile phone PCBs	*C. violaceum*	Au; 13%, Cu; 37%	Shake flasks	Chi et al. (2011)
	C. violaceum	Au; insignificant	Shake flasks	Kita et al. (2006)
Moderately Thermophilic Bacteria				
Electronic scrap	*S. thermosulfidooxidans + Thermoplasma acidophilum*	Cu; 86%	Column	Ilyas et al. (2010)
	S. thermosulfidooxidans + Sulfobacillus acidiphilus	Cu; 78%	Column	Ilyas et al. (2013)
	S. thermosulfidooxidans + Acidophilic hetrotroph	Cu; 89%	Shake flasks	Ilyas et al. (2010)
Fungi				
PCBs/electronic scrap/e-wastes	*A. niger + P. simplicissimum*	Cu;100%	Shake flasks	Brandl et al. (2001)
Electronic scrap	*P. chrysogenum*	Cu; 97%–98%	Shake flasks	Ilyas and Lee (2013)

water (mode-5) and leaching with commercial organic acids (mode-6). The main lixiviants produced in leaching mode-1 are found to be citric, tartaric, oxalic and gluconic acids in concentrations of 12, 2.5, 1.8 and 152 mM, whereas in leaching mode-2, the concentrations are 15, 0.5, 1.0 and 1.162 mM, respectively. In leaching mode-3, 63 mM citric acid, 23 mM tartaric acid and 29 mM oxalic acid have been observed. At 5% pulp density, both leaching mode-1 and mode-2 show similar metal extraction yields from the electronic scrap at 35°C; the extraction yield of Cu is much lower in leaching mode-1 and mode-2 (48%–50%) than in leaching mode-3 (97%) and mode-6 (98%). Thus, leaching mode-2 and mode-3 are proven to be techno-economically feasible for metal extraction with separate stages for the metabolic and leaching reactors.

12.2.6 Practical Applications of Bioleaching

In the scenario of commercial application of biohydrometallurgy, heap or dump leaching and stirred-tank leaching are the ways of applying the bioleaching technology for the recovery of copper and gold. Heap or dump leaching has been applied as a controlled process to low-grade copper ores containing secondary minerals such as covellite and chalcocite for about 40 years. The static bioleaching techniques are based on the principle of circulating water and air through heaps of ore coarsely fragmented to activate the growth of microorganisms that amplify the oxidation of the sulphidic material. Bio-heap leaching for the recovery of copper and biooxidation, as a pre-treatment, for the recovery of gold has now become a common practice in the industry (Olson et al., 2003). Since the 1980s, at least 13 copper bioleach operations have been commissioned.

The first commercial stirred-tank bioleach plant was commissioned in 1986 to pre-treat a sulphidic gold concentrate to enhance gold recovery. Typically, refractory gold plants operate with a 15%–20% slurry density. The slurry is continuously fed to a primary reactor. Most of the microbial growth occurs in the primary reactor, which has a typical slurry residence time of 2–2.5 days. The primary reactor overflows to a series of smaller secondary reactors connected in series. This design increases the efficiency of sulphide oxidation by reducing the short circuiting of sulphide particles. The total residence time in the circuit is about 4–6 days (Olson et al., 2003). In the case of electronic scrap, all these processes are only at the trial stage.

The prediction of future long-term commercial applications of bio-hydrometallurgical processing of secondary resources in the metal extraction industry is quite difficult. Nevertheless, it is almost certain that some of today's research will lead to innovative processes for commercial application. The reason is because biohydrometallurgy offers advantages such as operational simplicity, low capital and operating cost and shorter construction times that no other alternative process can provide. However, there are several specific needs for advancing the commercialisation of biohydrometallurgy.

- An iterative relationship is required with the metallurgical engineers to maximise the understanding of operational needs and to help the engineers understand the role and requirements of the microorganisms in their operations.
- The dynamics of the microbial population also change with time and conditions in the bioprocessing system. There is a need to both define and understand the potential interactions among the components of the microflora. This has the potential for improving bioleaching and mineral biooxidation through how the components of the system interact to bring about bioleach processes.
- Rapid, accurate and simple techniques for monitoring the microbial activity in bioleach mineral biooxidation systems are needed for control of these processes by operators.
- New materials for high temperature, highly corrosive conditions that are of relatively low cost, would also advance the technology.

12.3 Recovery of Gold and Copper from Leachate by Biosorption

The biosorption/bioreduction of metals has attracted interest recently as these transformations can play crucial roles in the recycling of both inorganic and organic species and can be of particular interest from recovering metals (Cu, Au) from electronic scrap-leached liquors. The elucidation of the mechanism of metal uptake is essential to develop technologies related to the metal recovery. In general, microbes-mediated metal recovery from waste electrical and electronic equipment (WEEE) leach solutions may involve the following pathways (Giller et al., 2009):

- Metal cations can bind on cell surfaces by biosorption or within the cell wall by bioaccumulation and in turn, augment metal uptake by microprecipitation
- Metal ions can be actively translocated inside the cell through metal-binding proteins
- Metal precipitation can occur during the interaction of heavy metals with extracellular polymers or with anions such as sulphide or phosphate produced by microbes
- Metal volatilisation through enzymes mediated biotransformation

Among the general mechanisms of microbes-mediated metal recovery, the mechanism of biosorption is very much intricate and therefore, a complete account concerning the biosorption processes is not available, but the key factors which control and characterise these mechanisms are as follows:

- The type of biological ligands accessible for metal binding
- Type of the biosorbent (living, non-living)
- Chemical, stereo-chemical and co-ordination characteristics of the targeted metals
- Characteristics of the metal solution such as pH and the competing ions

On the basis of dependence of cell metabolism, the biosorption mechanism can be divided into metabolism dependent and metabolism independent. According to the location where the metal recovered from the solution, biosorption may be classified as extracellular accumulation/precipitation, cell surface sorption/precipitation and intracellular accumulation. Physicochemical mechanisms such as ion exchange, complexation, coordination and chelation between metal ions and ligands, depend on the specific properties of the biomass (alive, or dead, or as a derived product). Other metal-recovery mechanisms dependent on metabolism include metal precipitation as sulphides or phosphates, sequestration by metal-binding proteins, peptides or siderophores, transport and internal compartmentalisation as indicated in Figure 12.1 (Ilyas and Lee, 2014).

Microorganisms used for biosorption processes include bacteria, fungi, yeasts and algae. The differences in metal uptake are due to the properties of each microorganism such as cell wall structure, nature of the functional groups and surface area. Several functional groups are present on the bacterial, fungal and algal cell wall including carboxyl, phosphonate, amine and hydroxyl groups (Vijayaraghavan et al., 2004).

12.3.1 Algae as Biosorbents

Chojnacka et al. (2005) reported the biosorption of Cu^{2+} ions by blue-green algae *Spirulina* sp. These algae are found to be capable of adsorbing one or more heavy metals including K, Mg, Ca, Fe, Sr, Co, Cu, Mn, Ni, V, Zn, As, Cd, Mo, Pb, Se and Al. Some results are summarised in Table 12.2.

A detailed investigation of gold binding in an algal biomass is carried out by Watkins et al. (1987). It is proposed that the mechanism of tetrachloroaurate interaction with *Chlorella vulgaris* involves the rapid reduction of Au^{3+} to Au^+, followed by a slow reduction to Au. It is believed that the ligand exchanges reactions, leading to the formation of bonds between Au^+ and sulphur and/or nitrogen contained in the algae.

Spectroscopic studies of the biosorption of Au^{3+} by de-alginated seaweed waste have been carried out by Romero-Gonz´alez et al. (2003) who reported that the colloidal Au is formed on the surface of de-alginated seaweed by

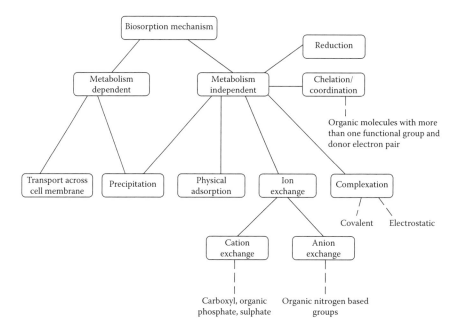

FIGURE 12.1
Mechanisms of biosorption.

reduction of Au^{3+} to Au. Extended x-ray absorption fine structure (EXAFS) measurements show that colloidal Au is present on the surface of the biosorbent. The adsorption mechanism of gold and palladium within condensed-tannin gel has also been reported (Kim and Nakano, 2005). Tannin gel particles that adsorbed palladium have been sampled for x-ray diffraction (XRD) analysis to find out the chemical form of the adsorbed palladium. The presence of metallic palladium on the tannin gel particles confirms that the Pd^{2+} can be reduced to Pd, while the hydroxyl groups of tannin gel are oxidised by redox reaction during the adsorption. It is also pointed out that the adsorption mechanism of palladium by condensed-tannin gel is more complicated, because palladium exists as various chemical forms, by which the adsorption rate is affected.

12.3.2 Fungi as Biosorbents

Studies have been conducted with fungal biomass by various researchers for heavy metals uptake. It is pointed out that several types of ionisable sites that influence the metal uptake efficiency of the fungal cell wall included phosphate, carboxyl groups on uranic acids and proteins, nitrogen-containing ligands on protein and chitin or chitosan. The feasibility of many algal species (*Halimeda opuntia*, *C. vulgaris* and *Ascophyllum nodosum*) has been evaluated to recover metals (Table 12.2).

TABLE 12.2

Bioleaching of Metals from Electronic Wastes and Level of Research

Algae	Metal	pH	Uptake (mmol)	References
C. vulgaris	Au^{3+}	2	0.5	Darnall et al. (1986)
Sargassum natans	Au^{3+}	2.5	2.1	Kuyucak and Volesky (1988)
A. nodosum	Au^{3+}	2.5	0.15	Kuyucak and Volesky (1988)
Sargassum fluitans	$Au(CN)_2^-$	2.0	0.0032	Niu and Volesky (1999)
Alginate cross-linked with $CaCl_2$	Au^{3+}	2.0	1.47	Torres et al. (2005)
De-alginated seaweed waste	Au^{3+}	3	0.4	Romero-Gonz´alez et al. (2003)
Laminaria japonica	Cu^{2+}	4.5	4.5 (mg/g)	Kuyucak and Volesky (1988)
Sargassum filipendula	Cu^{2+}	4.5	0.59 (mg/g)	Kuyucak and Volesky (1988)
A. nodosum	Cu^{2+}	6	70 (mg/g)	Kuyucak and Volesky (1988)
Fungi				
Saccharomyces cerevisiae	Cu^{2+}	6.5	2.4 (mg/g)	Qian et al. (2012)
Zoogloea ramigera	Cu^{2+}	4.0	29 (mg/g)	Kuyucak and Volesky (1988)
Pseudomonas syringae	Cu^{2+}	4.5	25	Kuyucak and Volesky (1988)
S. cerevisiae	Au^{3+}	5.0	0.026	Savvaidis (1998)
C. cladosporioides	Au^{3+}	4.0	0.5	Pethkar and Paknikar (1998)
C. cladosporioides	Au, waste	4.0	0.18	Pethkar and Paknikar (1998)
C. cladosporioides-1	Au^{3+}	4.0	0.4	Søbjerg et al. (2011)
C. cladosporioides-2	Au^{3+}	4.0	0.5	Søbjerg et al. (2011)
C. cladosporioides-1	Ag^+	4.0	0.4	Søbjerg et al. (2011)
C. cladosporioides-2	Ag^+	4.0	0.12	Søbjerg et al. (2011)
A. niger	Au^{3+}	2.5	1.0	Kuyucak and Volesky (1988)
Rhizopus arrhizus	Au^{3+}	2.5	0.8	Kuyucak and Volesky (1988)

Continued

TABLE 12.2 (Continued)

Bioleaching of Metals from Electronic Wastes and Level of Research

Algae	Metal	pH	Uptake (mmol)	References
Agro/Food/Plant				
Hen eggshell membrane	$Au(CN)_2^-$	3	0.67	Schiewer and Volesky (1996)
Hen eggshell membrane	$AuCl_4^-$	3	3.1	Schiewer and Volesky (1996)
Lysozyme	Au^{3+}, Pd^{2+}, Pt^{4+}	–	–	Maruyama et al. (2007)
Bovine serum albumin	Au^{3+}, Pd^{2+}, Pt^{4+}	–	–	Maruyama et al. (2007)
Alfalfa	Au^{3+}	5.0	0.18	Gamez et al. (2003)
Condensed-tannin gel	Au^{3+}	2.0	1.0	Kim and Nakano (2005)
Bayberry tannin-immobilised collagen fibre membrane	Au^{3+}	3.0	0.23	Ma et al. (2006)
Glutaraldehyde cross-linked chitosan	$Au(CN)_2^-$	1.6	2.9	Arrascue et al. (2003)
Sulphur derivative of chitosan	Au^{3+}	3.2	3.2	Arrascue et al. (2003)
Glutaraldehyde cross-linked chitosan	Au^{3+}	2.0	2.44	Guibal et al. (2002)
Bacteria				
Streptomyces erythraeus	Au^{3+}	4.0	0.03	Ishikawa et al. (2002)
Spirulina platensis	Au^{3+}	4.0	0.026	Ishikawa et al. (2002)
Bacillus subtilis	$Au(CN)_2^-$	2.0	0.008	Niu and Volesky (1999)
Arthrobacter sp.	Cu^{2+}	3.5–6	40 (mg/g)	Bai et al. (2012)
Sphaerotilus natans	Cu^{2+}	6	3.5 (mg/g)	Beolchini et al. (2006)
S. natans	Cu^{2+}	5.5	70 (mg/g)	Beolchini et al. (2006)
Streptomyces coelicolor	Cu^{2+}	5.0	195 (mg/g)	Öztürk et al. (2004)
Thiobacillus ferrooxidans	Cu^{2+}	6.0	40 (mg/g)	Ruiz-Manriquez et al. (1997)
B. subtilis	Cu^{2+}	5	3.5 (mg/g)	Nakajima et al. (2001)

Pethkar and Paknikar (1998) elucidated the binding mechanisms involved in the adsorption of Au and Ag ions to two strains of a fungus, *Cladosporium cladosporioides*. These methods confirm that no chemical change to the biosorbent takes place after metal loading, suggesting that the acidic conditions merely favour the electrostatic interaction between gold anions ($AuCl^{4-}$) and protonated biomass.

12.3.3 Bacteria as Biosorbents

Many bacterial species (e.g. *Bacillus, Pseudomonas, Streptomyces, Escherichia, Micrococcus*, etc.) have been tested for metal uptake by several researchers. It is concluded that the bacterial cell walls are efficient metal-chelating agents. Moreover, bacteria possess polysaccharide slime layers which readily offer amino, carboxyl, phosphate and sulphate groups for metals binding (Vijayaraghavan et al., 2004). Furthermore, studies by Vasudevan et al. (2001) indicate that heavy metal binding onto the surface of the bacterial cell wall is generally a two-stage process. The first stage involves the interaction between metal ions and reactive groups on the cell surface and the second stage includes deposition of successive metal species in greater concentrations. Qian et al. (2012) reported the simultaneous biodegradation of Ni–citrate complexes and recovery of Ni from solutions by *Pseudomonas alcaliphila*. They inferred that the addition of an excess amount of citrate to Ni–citrate complexes encourages the complex degradation as well as Ni recovery. They suggested that the possible mechanism for this change might be the generation of an alkaline pH by the metabolism of an excess citrate which results in the dissociation of citrate from the Ni–citrate complexes, thereby facilitating the recovery of Ni.

Creamer et al. (2006) studied the biometallurgical recovery of precious metals (Au, Pd) from electronic scrap leachates by *Desulfovibrio desulfuricans* in a columnar electro-bioreactor (2 L) that was supplied with either nitrogen or hydrogen through an external electrochemical cell from the bottom of the vessel and a glass-fritted gas diffuser. It is pointed out that the addition of *D. desulfuricans* biomass and bubbling of H_2 through the electronic scrap leachate samples, supplemented with 2 mM of Pd^{2+} insignificantly contribute to the deposition of Pd while the presence of copper inhibits hydrogenase activity and hinders Pd^{2+} recovery. This study further evaluates the effect of Cu^{2+} on the removal of Pd^{2+} and Au^{3+} from test solutions and leachate. The results show that Cu^{2+} inhibits the recovery of Pd^{2+}, but not of Au^{3+}. It is concluded that a bio-separation method is feasible for Au^{3+}, Pd^{2+} and Cu^{2+} and a three-step process is proposed for the selective recovery of Cu^{2+} and platinum group metals from electronic scrap leachate. Supplementation with palladised biomass catalyses the recovery of Pd^{2+} auto-catalytically and the Cu^{2+} can be finally recovered by a bio-precipitation method using bacterial off-gas (Table 12.2).

12.3.4 Agro-/Food Wastes as Biosorbents

Ishikawa et al. (2002) studied the recovery of gold from electroplating wastes using an eggshell membrane and proposed the physical sorption mechanism for the sorption of dicyanoaurate by an eggshell membrane since the Au^+ sorption reaction by an eggshell membrane is exothermic. In contrast, Pethkar and Paknikar (1998) indicated that biosorption of Au^{3+} by an eggshell membrane is a chemical reaction involving the dissociation of Au^{3+} from tetrachloroaurate. They confirmed their findings by XPS (x-ray photoelectron spectroscopy) and FT-IR (Fourier transform infra-red spectroscopy) for the elucidation of binding mechanisms involved in the sorption of Au and Ag ions onto two strains of a fungus, *C. cladosporioides* (Table 12.2).

12.4 Prospective on Bio-Hydrometallurgical Processing of Electronic Scrap

Biohydrometallurgy will continue to play an important role in mining and metal industries due to certain merits over the conventional technologies:

- Bioleaching lends itself economically and technically to the processing of small deposits and also the electronic scraps that cannot support high capital costs.
- Bioleaching, in some cases, may be more environmentally acceptable than other technologies, resulting in faster permitting of operations.

Some technical challenges and their possible solution can be as follows, for bioprocessing metals from electronic scrap:

- The widespread industrial use of bioleaching microbes susceptible for leaching precious metals has been constrained by the limited cyanogenic capabilities of lixiviant-producing microorganisms such as *C. violaceum*. So, the development of a metabolically engineered strain of *C. violaceum* or related cultures can produce more cyanide lixiviants and recover more than twice the amount of gold from electronic waste compared to wild culture. With further enhancement in cyanogenesis through subsequent metabolic engineering, the production of biogenic acids can be enhanced and bioleaching efficiency of metals can be improved.
- For efficient metal recovery from the solution, bacterial cells can be immobilised on supports such as polyurethane foam and/or an integrated flow through column reactors can be designed. Metal precipitation and crystallisation can be local through the production of high concentrations of phosphate ligand, which exceeds the

solubility product of the metal phosphate in the vicinity of nucleation sites on the cell surface that is most probably a lipid—a component of cell surface lipopolysaccharide. All metals having insoluble phosphates can be amenable to bio-recovery in this way.

- Biomass may be derived from activated sludge or fermentation wastes from food industries. Microorganisms such as bacteria, fungi, yeast and algae from their natural habitats are excellent sources of a biosorbent. The fast-growing organisms, for example, crab shells and seaweeds can also be used as biosorbents. In addition to the microbial sources, the agricultural products such as wool, rice, straw, coconut husks, peat moss, exhausted coffee, waste tea, walnut skin, coconut fibre, cork biomass, rice hulls, soybean hulls and cotton seed hulls, wheat bran, hardwood sawdust, pea pod, cotton and mustard seeds cake can serve as low-cost biosorbent sources.

- As an alternative approach to the economic provision of phosphate ligand into the crystalline metal phosphate, biological phosphate-removing bacteria (e.g. *Acinetobacter* sp.), mostly used in the waste water treatment process, grow aerobically and deposit intracellular reserves of polyphosphate (polyP). Upon transfer to anaerobic conditions (which can be spatial or temporal), the polyP can be mobilised and inorganic phosphate can be released from the cells. This can harness to the deposition of uranium and lanthanum phosphates similar to those of bio-manufacture using the phosphatase route. As a second approach, the use of phytic acid, inositol phosphate, a component of plant wastes and a by-product from biodiesel production, as the phosphate donor molecule can have a good economic potential for metal bio-recovery.

Acknowledgement

The work was supported by the Korea Institute of Geoscience and Mineral Resources (KIGAM), Daejeon, Korea.

References

Arrascue, M. L., Garcia, H. M., Horna, O., Guibal, E. 2003. Gold sorption on chitosan derivatives. *Hydrometallurgy* 71(1–2): 191–200.
Bai, J., Wu, X., Fan, F., Tian, W., Yin, X., Zhao, L., Fan, F., Li, Z., Tian, L., Qin, Z., Guo, J. 2012. Biosorption of uranium by magnetically modified *Rhodotorula glutinis*. *Enzyme Microb. Tech.* 51: 382–387.

Bas, A. D., Deveci, H., Yazici, E. Y. 2013. Bioleaching of copper from low grade scrap TV circuit boards using mesophilic bacteria. *Hydrometallurgy* 138: 65–70.

Beolchini, F., Pagnanelli, F., Toro, L., Vegliò, F. 2006. Ionic strength effect on copper biosorption by *Sphaerotilus natans*: Equilibrium study and dynamic modeling in membrane reactor. *Water Res.* 40: 144–152.

Brandl, H., Bosshard, R., Wegmann, M. 2001. Computer-munching microbes: Metal leaching from electronic scrap by bacteria and fungi. *Hydrometallurgy* 59(2–3): 319–326.

Brandl, H., Lehman, S., Faramarzi, M. A., Martinelli, D. 2008. Biomobilization of silver, gold, and platinum from solid waste materials by HCN-forming. *Hydrometallurgy* 94: 14–17.

Brierley, J. A. 1997. Heap leaching of gold bearing deposits, theory and operational description. In: *Biomining: Theory, Microbes and Industrial Processes*, ed., D. Rawlings, 103–115. New York: Springer.

Brierley, J. A., Brierley, C. L. 1999. Present and future commercial applications of biohydrometallurgy. In: *Biohydrometallurgy and the Environment toward the Mining of the 21st Century*, eds., R. Amils and A. Ballester, 81–89. Amsterdam: Elsevier.

Choi, M. S., Cho, K. S., Kim, D. S. 2004. Microbial recovery of copper from printed circuit boards of waste computer by *Acidithiobacillus ferrooxidans*. *J. Environ. Sci. Health A Tox. Hazard Subst. Environ. Eng.* 39(11–12): 2973–2982.

Chi, T. D., Lee, J.-c., Pandey, B. D., Jeong, J., Yoo, K.-k., Huyung, T. H. 2011. Bacterial cyanide generation in presence of metal ions (Na^{1+}, Mg^{2+}, Fe^{2+}, Pb^{2+}) and gold bioleaching from waste PCBs. *J. Chem. Eng. Japan* 44: 692–700.

Chojnacka, K., Chojnacki, A., Gorecka, H. 2005. Biosorption of Cr^{3+}, Cd^{2+} and Cu^{2+} ions by blue-green algae *Spirulina* sp. kinetics, equilibrium and the mechanism of the process. *Chemosphere* 59: 75–84.

Creamer, N. J., Baxter-Plant, V. S., Henderson, M., Macaskie, L. E. 2006. Palladium and gold recovery from precious metal solutions and electronic scrap leachates by *D. desulfuricans*. *Biotechnol. Lett.* 28: 1475–1484.

Darnall, D. W., Greene, B., Henzl, M. T., Hosea, J. M., McPherson, R. A., Sneddon, J., Alexander, M. D. 1986. Selective recovery of gold and other metal ions from an algal biomass. *Environ. Sci. Technol.* 20(2): 206–208.

Ehrlich, H. L. 1997. Microbes and metals. *Appl. Microbiol. Biotechnol.* 48: 687–692.

Faramarzi, M. A., Stagars, M., Pensini, E., Krebs, W., Brandl, H. 2004. Metal solubilization from metal-containing solid materials by cyanogenic *Chromobacterium violaceum*. *J. Biotechnol.* 113: 321–326.

Gamez, G., Gardea-Torresdey, J. L., Tiemann, K. J., Parsons, J., Dokken, K., Yacaman, M. J. 2003. Recovery of gold from multi-elemenal solution by alfalfa biomass. *Adv. Environ. Res.* 7: 563–571.

Giller, K. E., Witter, E., McGrath, S. P. 2009. Heavy metals and soil microbes. *Soil Biol. Biochem.* 41: 2031–2037.

Groot, D. R., Pistorius, P. C. 2008. Can we decrease the ecological footprint of base metal production by recycling? *J. S. Afr. I Min. Metall.* 108: 161–169.

Guibal, E., Sweeney, N. V. O., Vincent, T., Tobin, J. M. 2002. Sulfur derivatives of chitosan for palladium sorption. *React. Funct. Polym.* 50(2): 149–163.

Ilyas, S., Anwar, M. A., Niazi, S. B., Ghauri, A. 2007. Bioleaching of metals from electronic scrap by moderately thermophilic acidophilic bacteria. *Hydrometallurgy* 88: 180–188.

Ilyas, S., Lee, J.-c., Chi, R. 2013. Bioleaching of metals from electronic scrap and its potential for commercial exploitation. *Hydrometallurgy* 131–132: 138–143.

Ilyas, S., Lee, J.-c. 2013. Fungal leaching of metals from electronic scrap. *Miner. Metall. Proc.* 30: 151–156.

Ilyas, S., Lee, J.-c. 2014. Biometallurgical recovery of metals from waste electric and eletronic equipments: A review. *ChemBioEng Rev.* 1: 148–169.

Ilyas, S., Ruan, C., Bhatti, H. N., Ghauri, M. A., Anwar, M. A., 2010. Column bioleaching of metals from electronic scrap. *Hydrometallurgy* 101: 135–140.

Ishikawa, S., Suyama, K., Arihara, K., Itoh, M. 2002. Uptake and recovery of gold ions from electroplating wastes using eggshell membrane. *Bioresour. Technol.* 81(3): 201–206.

Kim, Y. H., Nakano, Y. 2005. Adsorption mechanism of palladium by redox within condensed-tannin gel. *Water Res.* 39(7): 1324–1330.

Kita, Y., Nishikawa, H., Takemoto, T. 2006. Effects of cyanide and dissolved oxygen concentration on biological Au recovery. *J. Biotechnol.* 124: 545–551.

Kuyucak, N., Volesky, B. 1988. Biosorbents for recovery of metals from industrial solutions. *J. Biotechnol. Lett.* 10(2): 137–142.

Ma, H.-w., Liao, X.-p., Liu, X., Shi, B. 2006. Recovery of platinum(IV) and palladium(II) by bayberry tannin immobilized collagen fiber membrane from water solution. *J. Membr. Sci.* 278(1–2): 373–380.

Maruyama, T., Matsushita, H., Shimada, Y., Kamata, I., Hanaki, M., Sonokawa, S., Kamiya, N., Goto, M. 2007. Proteins and protein-rich biomass as environmentally friendly adsorbents selective for precious metal ions. *Environ. Sci. Technol.* 41(4): 1359–1364.

Nakajima, A., Yasuda, M., Yokoyama, H., Ohya-Nishiguchi, H., Kamada, H. 2001. Copper biosorption by chemically treated *Micrococcus luteus* cells. *World J. Microbiol. Biotechnol.* 17: 343–347.

Niu, H., Volesky, B. 1999. Characteristics of gold biosorption from cyanide solution. *J. Chem. Technol. Biotechnol.* 74: 778–784.

Norris, P. R., Burton, N. P., Foulis, N. A. 2000. Acidophiles in bioreactor mineral processing. *Extremophiles* 4: 71–76.

Olson, G. J., Brierley, J. A., Brierley, C. L. 2003. Progress in bioleaching: Applications of microbial processes by the minerals industries. *Appl. Microbiol. Biotechnol.* 63: 249–257.

Ongondo, F. O., Williams, I. D., Cherrett, T. J. 2011. How are WEEE doing? A global review of the management of electrical and electronic wastes. *Waste Manag.* 31: 714 – 730.

Öztürk, A., Artan, T., Ayar, A. 2004. Biosorption of nickel(II) and copper(II) ions from aqueous solution by *Streptomyces coelicolor* A3(2). *Colloids Surf. B Bio-Interfaces* 34: 105–111.

Pethkar, A. V., Paknikar, K. M. 1998. Recovery of gold from solutions using *Cladosporium cladosporioides* biomass beads. *J. Biotechnol.* 63(2): 121–136.

Qian, J., Li, D., Zhan, G., Zhang, L., Su, W., Gao, P. 2012. Simultaneous biodegradation of Ni–citrate complexes and removal of nickel from solutions by *Pseudomonas alcaliphila*. *Bioresour. Technol.* 116: 66–73.

Romero-Gonz´alez, M. E., Williams, C. J., Gardiner, P. H. E., Gurman, S. J., Habesh, S. 2003. Spectroscopic studies of the biosorption of gold(III) by dealginated seaweed waste. *Environ. Sci. Technol.* 37(18): 4163–4169.

Rossi, G. 1990. *Biohydrometallurgy*. Hamburg, New York: McGraw-Hill.

Ruiz-Manriquez, A., Magana, P. I., López, V., Guzmán, R. 1997. Biosorption of Cu by *Thiobacillus ferrooxidans*. *Bioprocess Biosyst. Eng.* 18: 113–118.

Savvaidis, I. 1998. Recovery of gold from thiourea solutions using microorganism. *BioMetals* 11: 145–151.

Schiewer, S., Volesky, B. 1996. Modelling multi-metal ion exchange in biosorption. *Environ. Sci. Technol.* 30(10): 2921–2927.

Søbjerg, S. L., Lindhardt, A. T., Skrydstrup, T., Finster, K., Meyer, R. L. 2011. Size control and catalytic activity of bio-supported palladium nanoparticles. *Colloids Surf. B: Biointerfaces* 85: 373–378.

Torres, E., Mata, Y. N., Bl´azquez, M. L., Munoz, J. A., Gonzalez, F., Ballester, A. 2005. Gold and silver uptake and nanoprecipitation on calcium alginate beads. *Langmuir* 21(17): 7951–7958.

Tuncuk, A., Stazi, V., Akcil, A., Yazici, E. Y., Deveci, H. 2012. Aqueous metal recovery techniques from e-scrap: Hydrometallurgy in recycling. *Miner. Eng.* 25: 28–37.

U.S. Geological Survey: Mineral Commodity Summaries (Ed.), 2007. http://www.minerals.usgs.gov/minerals/pubs/mcs.

Vasudevan, P., Padmavathy, V., Tewari, N., Dhingra, S. C. 2001. Biosorption of heavy metal ions. *J. Sci. Ind. Res.* 60: 112–120.

Vestola, E. A., Kusenaho, M. A., Narhi, H. M. 2010. Acid bioleaching of solid waste materials from copper, steel and recycling industries. *Hydrometallurgy* 103: 74–79.

Vera, M., Schippers, A., Sand, W. 2013. Progress in bioleaching: Fundamentals and mechanisms of bacterial metal sulfide oxidation—Part A. *Appl. Microbiol. Biotechnol.* 97: 7529–7541.

Vijayaraghavan, K., Jegan, J. R., Palanivelu, K., Velan, M. 2004. Copper removal from aqueous solution by marine green alga Ulva reticulata. *Electron. J. Biotechnol.* 7: 61– 71.

Watkins II, J. W., Elder, R. C., Greene, B., Darnall, D. W. 1987. *Inorg. Chem.* 26: 1147–1151.

Watling, H. R. 2006. The bioleaching of sulphide minerals with emphasis on copper sulphides—A review. *Hydrometallurgy* 84: 81–108.

13

Urban Biomining: New Challenges for a Successful Exploitation of WEEE by Means of a Biotechnological Approach

Viviana Fonti, Alessia Amato and Francesca Beolchini

CONTENTS

13.1 Introduction

The production of end-of-life equipment, known as waste of electrical and electronic equipment (WEEE) or e-waste, is a direct consequence of the modern revolution of the electronic industry and of the constant evolution of technology. According to the United Nations Environment Programme (UNEP), a dramatic increase in the illegal import of e-waste, 200%–400% in

South Africa and 500% in India, can be estimated for the period between 2017 and 2020 (Schluep et al. 2009). Organic and inorganic components in WEEE can cause environmental problems if not properly managed. Today, in the United States and Europe, the main WEEE management strategies are incinerators and landfills and these can represent a serious threat to the environment and human health, due to contaminant leaching in soils and ground waters, and to a release of potentially hazardous by-products in the atmosphere (Robinson 2009; Tsydenova and Bengtsson 2011). On the other hand, base (e.g. Cu) and precious metals (e.g. Au, Ag and Pd) in WEEE could be employed as raw materials, making WEEE an attractive secondary resource of valuable elements (Oguchi 2013). In the last decade, the scientific community and industrial research have made a significant effort in developing techniques for the recovery of metal components from WEEE, especially by pyrometallurgical and hydrometallurgical approaches. Nevertheless, such techniques may be extremely polluting and not environmentally sustainable (Korte et al. 2000; Mecucci and Scott 2002; Cui and Zhang 2008; Tsydenova and Bengtsson 2011). Biohydrometallurgical strategies, based mainly on the ability of microorganisms to produce leaching agents, are gaining increasing prominence in this field. For instance, the technique known in mining activities as bioleaching (i.e. biological leaching) is considered a novel approach for metal mobilisation from various types of solids (Beolchini et al. 2012). The main advantages of biohydrometallurgical methods would be low operating costs (due to a low energy input), reduced environmental impact and a general minimisation of the end product (Ehrlich 2001; C. L. Brierley 2010).

This chapter aims to define where we are in the application of biotechnological strategies for the exploitation of WEEE and the recovery of high-added-value metals. In the first part, WEEE is described within its regulatory scenario and within its potential as a source of important critical metals. Then conventional treatments for metals recovery are briefly described, with their corresponding advantages and disadvantages. Finally, the main scientific knowledge in this field about the exploitation of biotechnological strategies is summarised and discussed, with a particular focus on the potential of different microbial strains in the application of *urban biomining* for an efficient use of e-waste as a secondary resource.

13.2 Electric and Electronic Waste as an Important Source of Secondary Raw Materials

13.2.1 Waste Production

WEEE is defined by the EU directive 2002/96/EC (European Parliament, 2002) as electrical or electronic equipment that is a waste within the meaning of Article 1(a) of Directive 75/442/EEC (Council Directive, 1975), including all its components, sub-assemblies and consumables, which are part of this product at the time of discarding (Li et al. 2013). In general, electronic waste comprises a broad range of electronic and electrical products, including large household appliances (e.g. washing machines and television sets) and hand-held equipment (e.g. cellular phones and personal computers). The economic growth increases the ownership of electronics, decreases simultaneously the life span of the electronic goods and eventually leads to rapid growth in the amount of unwanted and obsolete electronics, so that WEEE can be considered as one of the fastest-growing waste streams (Bigum et al. 2013). According to Huismann et al. (2008), the estimated rate of global WEEE generated is ~40 million tons per year; in 2007, India produced 439 kton of computers, printers, washing machines, mobile phones and TVs. Assuming an Indian population of ~1.12 billion in 2007, this equates to 0.4 kg per capita (Ongondo et al. 2011).

The composition of e-waste is extremely variable. Many studies have examined WEEE composition, dividing its components into five categories, namely, ferrous metals, non-ferrous metals, glass, plastics and other materials, but others classification could be made. A typical composition of WEEE is given in Figure 13.1 (full dataset in Ongondo et al. 2011). The metallic component itself can be very different due to the collected equipment; for instance, typical metals in printed circuit boards (PCBs) are copper (20%), iron (8%), tin (4%), nickel (2%), lead (2%), zinc (1%), silver (0.2%), gold (0.1%) and palladium (0.005%) (He et al. 2006). Depending on the type of equipment from which it originated, WEEE may contain substances that are particularly hazardous, such as Hg in fluorescent tubes, CFCs in freezers and refrigerators or Cd in mobile phones (Crowe et al. 2003). Owing to its content

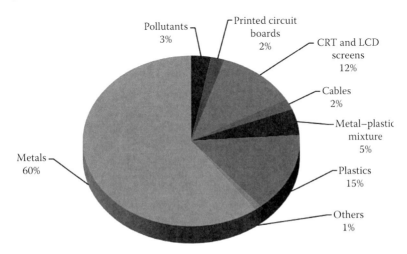

FIGURE 13.1
(See colour insert.) Typical material fractions in WEEE.

TABLE 13.1

Annual Import and Export of WEEE in the World

Country/Region	Annual Household Production (Million Tons)	Annual Export (Million Tons)	Annual Import (Million Tons)
USA	6.6	1.3	–
EU-25	7	1.9	–
Japan	3.1	0.62	–
China	3.1	–	2.0
India	0.36	–	0.85
West Africa	0.05	–	0.57
Total	20.21	3.82	3.42

Source: Adapted from Zoeteman, B. C. J., H. R. Krikke and J. Venselaar. 2009. *The International Journal of Advanced Manufacturing Technology* 47(5–8): 415–436.

in hazardous substances and to the volumes involved, the management of WEEE is an environmental issue of high priority; big amounts of WEEE are often exported from one country to another where disposal and treatment costs are lower or legislation is less restrictive (Table 13.1). The transboundary movement of WEEE has been particularly controversial. Initially, the main WEEE traffic routes were towards Asia (especially China) but, after the introduction of a tighter legislation in China, new destinations are chosen, such as Ghana, Nigeria, South Africa, Vietnam, India and the Philippines (Li et al. 2013). In addition, data about real waste flux among nations are very difficult to obtain because a very high percentage of WEEE disappear, leaving no official trace: a Greenpeace report (2009) states that 75% of European WEEE and 80% of American WEEE are illegally treated in developing countries. In order to control the mechanisms of waste import and export, some countries have adhered to the Organisation for Economic Co-operation and Development (OECD).

13.2.2 Metal and Rare Earth Elements

The diffusion of new technologies and economic growth causes a continuous request of raw material, especially metals, since they are key elements in the production processes of goods such as mobile phones, liquid crystal displays and other electrical and electronic equipment (EEE). Fourteen classes of raw elements have been identified as a very high risk of supply interruption for EU (Figure 13.2). The importance of these materials is connected to their growing demand, which, in turn, is driven both by the rate of production of new emerging technologies and by the growing rate of developing economies; on the contrary, geological scarcity is not considered in determining which raw elements are critical or not, at least within the next 10 years. Criteria for the identification of critical raw materials for EU have

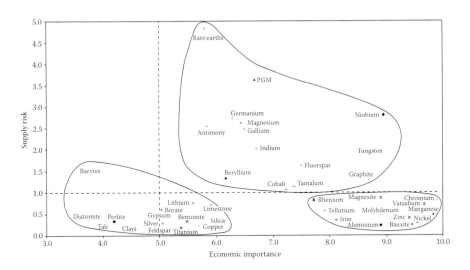

FIGURE 13.2
(See colour insert.) Critical raw materials in EU.

been recently established by the European Commission (2010); in particular, three aggregated indicators are used:

- Economic importance
- Supply risk (associated with political and socio-economic factors)
- Environmental risk (Figure 13.3)

According to such criteria, rare earth elements (REEs) are the most critical materials for EU. It is a group of 17 chemically similar metallic elements: the 15 lanthanides, plus scandium and yttrium. The demand for REEs derives from their key role in the production of electric and electronic equipment; for example, REEs are important constituents of permanent magnets, lamp phosphors, rechargeable NiMH batteries, catalysts and other applications. World demand for REEs has been estimated as around 133,600 tons per year, but the request is increasing and a demand of 210,000 tons is expected for 2015 (Humphries 2012). At the moment, China provides more than 90% of REEs and owns little less than 40% of the proven reserves (Binnemans et al.

FIGURE 13.3
(See colour insert.) Critical raw materials according to their environmental country risk.

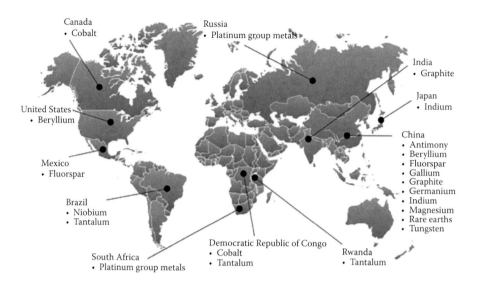

FIGURE 13.4
(See colour insert.) Production concentration of critical raw mineral materials.

2013). Moreover, China is the main provider in the market of other important elements, such as antimony, beryllium, fluorspar, gallium, graphite, germanium, indium, magnesium and tungsten (Figure 13.4).

The described situation is of concern for the whole world market and also for the environment. Using less critical metals as an alternative to critical REEs and investing in sustainable primary mining from old or new REEs deposits are two of the most promising strategies for facing the problems of resource depletion and the dependence of the European market on non-European countries (Binnemans et al. 2013). 'Urban mining' could be an additional alternative: the recovery of critical metals from a particular of type of waste, such as WEEE; this would allow to solve simultaneously both the economic issues, associated with the production of EEE, and the environmental issues, associated with their disposal.

13.2.3 Conventional Technologies for the Recovery of Metals

As mentioned above, the final disposal of electrical and electronic devices is a current issue of worldwide concern. WEEE management follows the principles set by waste hierarchy, reported within the Directive 2008/98/EC:

a. Prevention

b. Preparing for reuse

c. Recycling

d. Other recovery, for example, energy recovery

e. Disposal

Landfill disposal and incineration are the main practices in waste management, despite some differences among countries, but they can pose serious threats to human health and the environment. Metals can be released from WEEE disposed in landfill sites by leaching processes, with consequences on the whole ecosystem, from the atmospheric to the aquatic and terrestrial compartments. Incineration reduces waste volumes but it is a source of very hazardous pollutants, such as dioxins and furans, which can be released into the environment if adequate flue gas cleaning systems are not implemented. In addition, incineration contributes significantly to the annual emissions of Cd and Hg (Crowe et al. 2003). In this context, the recovery of base valuable and precious metals (e.g. Au, Ag, Cu, Zn, Co, Y) from WEEE represents a valid alternative to landfills and incinerators. Indeed, WEEE can represent a concentrated source of metals, which were mined from ore minerals where their concentrations were lower. In addition, the extraction of such elements from ores consumes land and energy, produces wastewaters and releases huge amounts of sulphur dioxide and carbon dioxide, while metal reclamation from waste would reduce the phenomenon, with a substantial decrease in the environmental impacts. For example, the energy used to obtain 1 kg of Al by reclamation from waste is equal or less than 1/10 of that required for it extraction; recycling e-waste reduces the bauxite residues of about 1.3 kg and the emission of carbon dioxide and sulphur dioxide of 2 kg and 0.011 kg, respectively (Zhang et al. 2012). In this scenario, recycling policies are gaining more attention, and many countries have drawn up regulations for the management of WEEE. In particular, the European WEEE Directive (Directive 2012/19) aims at WEEE recycling to reduce the disposal of waste and 'to contribute to the efficient use of resources and the retrieval of valuable secondary raw materials'. Considering that WEEE can work as a source of 14 economically important raw materials identified by the European Commission, 'urban mining' looks more attractive than conventional management strategies.

The first step in recycling treatment schemes is the pre-treatment, by manual de-pollution, shredding, air classifiers/hoods, magnetic sorting and/or eddy-current separation, aimed at removing dangerous substances (CFCs, Hg, PCB) and optical systems. According to Bigum et al. (2012), the major pre-treatment outputs are manually sorted components (29%), a magnetic-iron (33%) and a residual plastic fraction (26%); any other fraction constitutes about 2%–3% of the flow. The overall output from a pre-treatment plant is (per 1000 kg of received high-grade WEEE):

- 114 kg of substances requiring special treatment according to regulation
- 165 kg of copper and precious metal fraction

- 381 kg of iron and magnetic steel
- 22 kg of aluminium
- 53 kg of fluff and residual waste
- 265 kg of plastic

Today, the subsequent treatment steps are based on pyrometallurgical, hydrometallurgical and electro-metallurgical processes. Pyrometallurgical processes require the heating of WEEE at high temperatures (often greater than 1000°C) to separate metals and impurities into different phases, so that valuable metals can be recovered. As a consequence, high energy requirements and the production of hazardous gases are associated with such strategy. Hydrometallurgical treatments are based on the use of leaching agents in aqueous solutions, such as strong acids and bases, often applied together with other complexing agents, such as oxalic acid, acetic acid, cyanide, halide, thiourea and thiosulphate. Compared to pyrometallurgical treatments, hydrometallurgical processes are less energy and cost demanding and require plants with relatively small capacities. Electro-metallurgy processes use electrical current to recover metals (e.g. electro-winning and electro-refining of copper, zinc and other elements), with high energy consumption. Recently, hybrid technologies have also been applied, which integrate the chemical approach (more efficient) with biological strategies (more environmentally friendly), thus taking advantage of the benefits of both chemical and biological leaching (Rocchetti et al. 2013).

13.3 Biohydrometallurgy: Main Mechanisms and Involved Microorganisms

Among the technologies for metal reclamation from WEEE, biohydrometallurgy is gaining increasing attention within scientific and industrial research. In biohydrometallurgy, the metabolic products of key microorganisms are the driving force in dissolving metal species into aqueous solutions, which are easier to manage or treat (Erüst et al. 2013). The specific knowledge about the metabolic mechanisms underlying biohydrometallurgical processes was first developed in the mining field, where microorganisms are exploited for large-scale operations of metal recovery from ores ('biomining'). Nevertheless, 'biohydrometallurgy' covers all the processes in which the mobilisation of elements from solid materials is mediated by microorganisms.

The first scientific evidence about the role of microorganisms in metal extraction (Colmer and Hinkle 1947) has led to the description and isolation from acid mine drainage (AMD) of the acidophilic, autotrophic, iron- and sulphur-oxidising bacterium *Acidithiobacillus ferrooxidans* (formerly, *Thiobacillus ferrooxidans*; Temple and Colmer, 1951). Although it was not the first acidophilic strain with sulphur-oxidising activity to be described (namely, *Acidithiobacillus thiooxidans*; Waksman and Joffe, 1922), biohydrometallurgy originated from Colmer's discoveries, with the first industrial-scale extraction of copper from mine dumps through microbial action (early 1950s, Kennecott Copper Corporation; Zimmerley et al. 1958). Today, minerals/ores containing copper, gold and cobalt are processed on an industrial scale and promising results have been obtained for the industrial processing of sulphides of Ni, Zn, Mo Co, Ga, Pb and the platinum group metals (Ehrlich 2001; J. A. Brierley 2008; Ndlovu 2008; C. L. Brierley 2010; Lee and Pandey 2012).

In parallel with the increasing application of biohydrometallurgy, microorganisms and biochemical processes involved in metal mobilisation have been the object of growing investigations. The exploration of extreme environments (hot springs, volcanic regions, etc.) and the development of molecular biology techniques for detecting and identifying organisms have led to a broader knowledge of the microorganisms that interact with metals which that could be exploited in biohydrometallurgy. Today, the microorganisms that find real or potential application in biohydrometallurgy belong to all three domains of the tree of life: archaea, bacteria and eucarya (Ehrlich 2001; Olson et al. 2003; Norris 2007; Schippers 2007; Vera et al. 2013). An overview of the main strains involved (or potentially involved) in the biological treatment of WEEE for metal recovery and their mechanisms are provided in the next sections.

13.3.1 Fe/S-Oxidising Bacteria and Archaea

The transformation of base metal sulphides into water-soluble metal sulphates mediated by extremely acidophilic microorganisms (i.e. bioleaching) is the most known and commercially applied biohydrometallurgical process. Another successful and extensive commercial application of biohydrometallurgy is the bio-oxidation pre-treatment of refractory sulphide gold ores, where microorganisms are used to oxidise pyrites and other sulphides to expose gold occluded within the mineral matrix (Olson, Brierley, and Brierley 2003). Microorganisms involved in such processes belong to a non-phylogenetic group of acidophilic, aerobic and chemolithotrophic strains in the bacteria domain, often known as Fe/S-oxidising bacteria, due to their metabolism based on the oxidation of reduced sulphur compounds (and/or elemental sulphur) and ferrous ions. The main products are sulphuric acid (and a list of sulphur oxidation intermediates) and ferric ions. The most known and studied strains in this group are *At. ferrooxidans*, *At. thiooxidans* and *Leptospirillum ferrooxidans* but many other species are known today. In addition, iron-oxidising and sulphur-oxidising strains among archaea have been identified, so the microbial diversity in bioleaching and bio-oxidation processes is much wider than previously hypothesised (Johnson 2012; Dopson and Johnson 2012; C. L. Brierley and Brierley

2013; Vera et al. 2013). Indeed, leaching bacteria are distributed among α-, β- and γ-proteobacteria (*Acidithiobacillus, Acidiphilium, Acidiferrobacter, Ferrovum*), Nitrospirae (*Leptospirillum*), Firmicutes (*Alicyclobacillus, Sulfobacillus*) and Actinobacteria (*Ferrimicrobium, Acidimicrobium, Ferrithrix*); leaching archaea belong mostly to Crenarchaeota phylum (*Sulfolobus, Acidianus, Metallosphaera, Sulfurisphaera*), although within Euryarchaeota there are two acidophilic iron(II)-oxidising strains (*Ferroplasma acidiphilum* and *Ferroplasma acidarmanus*; Vera et al. 2013).

These microorganisms can be roughly organised into three broad groups based on the temperature ranges for iron and sulphur oxidation: mesophiles (below 40°C), moderate thermophiles (about 45°C or more) and thermophiles (about 70°C or more). Obviously, there is not a precise temperature threshold that divides these groups; for example, *Acidithiobacillus caldus* and *Acidimicrobium ferrooxidans* are quite active from about 25°C to almost 55°C. The group of mesophiles is constituted by Gram-negative strains within bacteria domains: *Acidithiobacillus, Thiobacillus* and *Leptospirillum* are the main genera. *At. ferrooxidans* has been an object of particular interest, owing to its remarkably broad metabolic capacity: it can live aerobically either on the oxidation of iron(II) or reduced inorganic sulphur compounds (RISCs) to elemental sulphur, but the anaerobic growth is possible by the oxidation of hydrogen or sulphur coupled with iron(III) reduction (Pronk et al. 1992; Johnson 2012). The other species within the *Acidithiobacillus* genus can oxidise sulphur but not iron. Mesophilic iron oxidisers are mainly affiliated with the *Leptospirillum* genus (*L. ferrooxidans, L. ferriphilum*). Among moderate thermophiles, sulphur oxidisers are affiliated with Gram-positive genera within Firmicutes (*Sulfobacillus* and *Alicyclobacillus*), while iron oxidisers belong to Actinobacteria (*Ferrimicrobium, Acidimicrobium* and *Ferrithrix*). The thermophile group is formed by strains belonging to the archaea domain. Sulphur-oxidising archaea are affiliated mainly with the genera *Sulfolobus* and *Metallosphaera*, while iron oxidisers are affiliated with the genus *Ferroplasma*. Although relatively few works have been published about the thermophile group, the number of archaea known to be directly involved in mineral sulphide oxidation appears to be comparable with that of mesophiles and moderate thermophiles (Norris 2007). Indeed, novel isolates have been shown to grow on pyrite and chalcopyrite at about 90°C (Plumb et al. 2002). No psychrophilic sulphur-oxidising acidophiles have been described, although the sulphur- and iron-oxidising proteobacterium *Acidithiobacillus ferrivorans* is psychro-tolerant, and grows at 4–35°C (Dopson and Johnson 2012).

Sulphur oxidation follows different pathways in the three groups. Curiously, *At. ferrooxidans* can be considered as the most studied acidophile, but its sulphur-oxidising pathways are not so well known. Owing to difficulties in developing genetic techniques in acidophiles, a large proportion of the hypotheses regarding RISCs metabolic pathways in these procaryotes are based on systems biology: putative genes have been assigned in several microorganisms. As best described by Dopson and Johnson (2012) and other review works (Rohwerder and Sand 2007; Barrie Johnson and Hallberg 2008; Jones et al. 2012), the oxidation of elemental sulphur and RISCs in Gram-negative bacteria (such as *At. ferrooxidans* and related species) appears to be due to a dioxygenase enzyme (rather than a dehydrogenase as observed in neutrophilic bacteria), termed sulphur oxygenase reductase (SOR). The production of ATP is due to an F_0F_1 ATPase; the process includes the final electron acceptors cytochrome bo_3 and cytochrome c oxidases. In contrast, many enzymes as well as their encoding genes, have been characterised in *Acidianus ambivalens* (archaea), so our image of sulphur metabolism for archaeal acidophiles is much clearer than that in the bacterial acidophiles. In spite of this, little is known about RISC metabolism in Gram-positive acidophiles (e.g. *Sulfobacillus* spp.). The primary difference between archaea and Gram-negative bacteria is that S° is disproportionated in the former, as opposed to being oxidised (Dopson and Johnson 2012).

While many sulphur-oxidising acidophiles appear to be obligate aerobes (e.g. *At. thiooxidans* and *At. caldus*), others (e.g. *At. ferrooxidans* and *Acidiferrobacter thiooxydans*) can use ferric iron as an alternative electron acceptor and grow in anoxic environments. Very interesting mutualistic interactions have been described. For example, neither the sulphur-oxidising autotroph *At. thiooxidans* nor the iron-oxidising heterotroph *Ferrimicrobium acidophilum* can oxidise pyrite when grown in pure culture, but they can do so when grown in co-culture: ferric iron is generated by bio-oxidation of ferrous iron by *Fm. acidophilum*, which uses organic carbon (as C source) provided by *At. thiooxidans*; the dissolution of pyrite occurs for abiotic oxidation by Fe(III) ions produced (Okibe and Johnson 2004).

Although involved molecules are often not well characterised, the overall mechanisms of metal solubilisation can been considered as well established. Nevertheless, such knowledge refers almost exclusively to sulphide ores (Sand et al. 2001; Rohwerder and Sand 2007). Although metals in PCBs are in metallic form (zero-valence, with the exception of ferric iron), there is no reason to suppose that Fe/S-oxidising bacteria can mobilise metals from WEEE by a sort of direct mechanism. Owing to its unique ability to use both sulphur and iron, *At. ferrooxidans* has been broadly used as a model microorganism to study and describe the mechanisms mediated by leaching bacteria in solubilising metals. The solubilisation of metal sulphides by *At. ferrooxidans* and related strains has long been described as a process based on two independent mechanisms: a 'direct mechanism' (i.e. the direct enzymatic oxidation of the sulphur moiety of the metal sulphide) and an 'indirect mechanism' (i.e. the non-enzymatic metal sulphide oxidation by Fe(III) ions combined with enzymatic (re)-oxidation of the resulting Fe(II) ions; Sand et al. 2001). However, it is now completely accepted that the 'direct mechanism' of biological metal sulphide oxidation does not exist: the 'indirect mechanism' has been singled out as the sole mechanism and can be more appropriately defined as 'the non-enzymatic metal sulphide oxidation by Fe(III) ions combined with the enzymatic (re)oxidation of the resulting

Fe(II) ions' (Rohwerder et al. 2003). Anyway, a microbe attachment to the ore really occurs and enhances the rate of mineral dissolution, so both 'contact' and 'non-contact' mechanisms occur (Sand et al. 2001; D. E. Rawlings 2002; Rohwerder et al. 2003; Rohwerder and Sand, 2007). Fe/S-oxidising bacteria approach the mineral surface by creating a biofilm, whereas the majority of cells attach to the sulphide surface, and planktonic bacterial cells remain floating in the bulk solution. The attachment process is predominantly mediated by the extracellular polymeric substances (EPS) that create a complex slime and fill the space between the cell wall and the surface. In bioleaching with Fe/S-oxidising bacteria, metal dissolution occurs for acid leaching and/or oxidation attack; the main role of leaching bacteria consists of the generation and regeneration of leaching agents, mainly Fe(III) ions and protons.

Some authors have proposed reactions describing the main mechanisms of copper bioleaching from PCBs (Choi et al. 2004; Cui and Zhang 2008; Xiang et al. 2010; Zhu et al. 2011; Ilyaset al. 2013). In the presence of iron as the energy source for leaching strains, the bio-oxidation of Fe(II) produces Fe(III), which is responsible for the oxidation of Cu^0 (insoluble) in Cu^{2+} (soluble), according to Equations 13.1 and 13.2:

$$4Fe^{2+} + O_2 \xrightarrow{\text{Fe-oxidisers}} 4Fe^{3+} + H_2O \qquad (13.1)$$

$$2Fe^{3+} + Cu^0 \rightarrow 2Fe^{2+} + Cu^{2+} \quad \Delta G^0 = -82.9\,kL/mol \qquad (13.2)$$

Cu leaching was also observed in the absence of iron, with elemental sulphur as the energy source. This suggests that part of zero-valence copper is solubilised by protons, although in such cases, molecular oxygen is involved:

$$2Cu^0 + 4H^+ + O_2 \rightarrow 2Cu^{2+} + 2H_2O \qquad (13.3)$$

During the dissolution of copper, Fe(III) ions are reasonably released from the PCB metallic component; Fe(III) ions will participate in the reaction (Equation 13.2) producing new protons by hydrolysis, and enhance copper solubilisation in Equation 13.3. The solubilisation of Zn, Ni and Al would follow the same mechanisms, according to their thermodynamic feasibility reactions:

$$2Fe^{3+} + Zn^0 \rightarrow 2Fe^{2+} + Zn^{2+} \quad \Delta G^0 = -295.4\,kL/mol \qquad (13.4)$$

$$2Fe^{3+} + Ni^0 \rightarrow 2Fe^{2+} + Ni^{2+} \quad \Delta G^0 = -196.6\,kL/mol \qquad (13.5)$$

$$3Fe^{3+} + Al^0 \rightarrow 3Fe^{2+} + Al^{3+} \quad \Delta G^0 = -1085.2\,kL/mol \qquad (13.6)$$

13.3.2 Cyanogenic Microorganisms

In the mining industry, bio-oxidation by Fe/S-oxidising bacteria is used to oxidise sulphides surrounding gold in refractory ores, in order to improve its accessibility for subsequent steps in the gold extraction circuit. Bio-oxidation is usually followed by chemical extraction of gold by complexation with cyanide ions (cyanidation) in alkaline conditions. The application of the cyanidation process for the extraction of gold and silver from ores has been known since 1898 (Smith and Mudder 1991). Cyanide is one of the few compounds (e.g. chlorides, other halides and thiourea) able to form water-soluble complexes with gold, with high extraction efficiencies even in gold amounts as small as 0.25% (Smith and Mudder 1991; Syed 2012). Recently, the biological production of cyanide by microorganisms and its potential application are gaining new attention by the scientific community. The use of biogenic cyanide (i.e. biocyanidation) may offer a valid alternative to conventional gold extraction techniques; the expressions 'alkaline bioleaching' or 'heterotrophic bioleaching' are often used to indicate the solubilisation of metals by cyanogenic microorganisms (G M Gadd 2000; Cui and Zhang 2008; Hennebel et al. 2013; Mishra and Rhee 2014), although the latter is also used for bioleaching by fungi (Aung and Ting 2005; Santhiya and Ting 2006; Sabra et al. 2011). Moreover, Au is not the sole target in biocyanidation, since other metals (such as Ni, Fe, Ag and Zn) can form stable complexes with CN⁻ (A. Smith and Mudder 1991). The use of cyanogenic strains would allow the mobilisation of metals from solids under alkaline conditions and this might be advantageous in view of commercial-scale operations. For instance, copper mobilisation from carbonate-rich rocks is known to be very acid consuming (Krebs et al. 1997; Dopson and Johnson 2012), so, the application of autotrophic Fe/S-oxidising microorganisms or heterotrophic microorganisms forming organic acids is very likely unfeasible. Moreover, environmental problems due to a potential release of cyanide in natural waters may be significantly reduced by exploiting the capability of cyanogenic bacteria in detoxifying cyanide by the enzyme β-cyanoalanine synthase (Macadam and Knowles 1984).

Hydrocyanic acid (HCN) is formed by a variety of heterotrophic bacteria (e.g. *Chromobacterium violaceum*, *Pseudomonas fluorescens*, *P. aeruginosa*) and fungi (e.g. *Marasmius oreades*, *Clitocybe* sp., *Polysporus* sp.); the production of HCN by *Pseudomonas plecoglossicida* was just known in 2006 (Brandl and Faramarzi 2006). In general, cyanide formation (i.e. cyanogenesis) by strains in bacteria domain has attracted larger attention, compared to fungi (Campbell et al. 2001; Faramarzi et al. 2004). Although archaea–gold interactions are known (Reith et al. 2007), cyanogenesis in archaea has been not reported yet.

C. violaceum is a mesophilic, Gram-negative β-proteobacterium, described at the end of the nineteenth century, and it is probably the most studied microorganism for potential application in low environmental impact processes of Au recovery from ores (Lawson et al. 1999; de Vasconcelos et al. 2003). HCN is produced as a secondary product of the oxidative decarboxylation in the catabolism of glycine, the main precursor for cyanide. The production of HCN occurs in a short period of the growth curve: between the end of the exponential phase and the early stationary phase in *C. violaceum*, just at the beginning of the stationary phase in the genus *Pseudomonas*. The involved enzyme is called HCN synthase (encoded by the *hcnABC* operon in *C. violaceum*); the amount of cyanide produced typically ranges around 2–50 mg/L (Brandl et al. 2008; Pham and Ting 2009). In the late stationary phase, *C. violaceum* detoxifies cyanide by converting it to β-cyanoalanine (Knowles and Bunch 1986).

Au dissolution by (bio-)cyanidation consists of an anodic reaction (Equation 13.7) and a cathodic one (Equation 13.8):

$$4Au + 8CN^- \rightarrow 4Au(CN)_2^- + 4e^- \tag{13.7}$$

$$O_2 + 2H_2O + 4e^- \rightarrow 4OH^- \tag{13.8}$$

The overall reaction is known as Elsner's equation, as shown in Equation 13.9 (Hedley and Tabachnick, 1958; Smith and Mudder, 1991; Kita et al. 2006):

$$4Au + 8CN^- + O_2 + 2H_2O \rightarrow 4Au(CN)_2^- + 4OH^- \tag{13.9}$$

Cyanidation of other metals follows very similar reactions (Campbell et al. 2001; Faramarzi et al. 2004). HCN pK_a is 9.3 but optimal pH for cyanogenesis in *C. violaceum*, and other cyanogenic bacteria, ranges between 7 and 8 (Lear et al. 2010). At physiological pHs, the main compound is HCN, of which a large amount is lost via volatilisation and little cyanide is available for metal complexation; despite this, cyanide anions are the most available form at a pH equal or greater than 10.5, values that compromise bacteria activity and life (Knowles and Bunch 1986). Another issue is related to oxygen availability. In the growth phases in which cyanogenesis occurs, *C. violaceum* rapidly consumes dissolved oxygen for respiration; therefore, little molecular oxygen is available for Au dissolution (Equations 13.6 to 13.8). In congruence, Kita and co-authors (2006) have described a decrease in Au dissolution from solids due to low dissolved oxygen concentrations.

13.3.3 Potential of Fungi in Metal Recovery from WEEE

Many fungal species can survive and grow in adverse conditions (e.g. low pH, low temperature) and show high tolerance levels to various types of contaminants (Gadd 2010). Bioremediation strategies based on the exploitation of fungi have also been proposed. In particular, bioleaching of metals by fungi (i.e. 'fungal leaching' or 'heterotrophic leaching') has been mostly investigated for metal extraction from low-grade ores and mine tailings (Mulligan et al. 2004), industrial waste (e.g. spent refinery catalysts; Aung and Ting 2005; Santhiya and Ting 2006) and wastewater sludge (More et al. 2010). Considering the ability of fungi to modify metal speciation and mobility, such microorganisms could also offer a potential alternative in the recovery of critical metals from WEEE, where base valuable and precious metals are in zero-valence form (with the exception of ferric iron). Fungi can solubilise metals through several mechanisms, but the production of weak organic acids is considered to be the most important mechanism of metal solubilisation. Indeed, citric, oxalic and gluconic acids can solubilise metals by forming water-soluble complexes (complexolysis; Burgstaller and Schinner 1993; Bosecker 1997). In addition, carboxylic acids produced by fungi can attack the mineral surface by their protons and lead to a release of associated metals (acidolysis; Gadd 2007). Metal complexation can occur also with functional groups on the cell wall surface (e.g. carboxyl, carbonyl, amine, amide, hydroxyl and phosphate groups; Baldrian 2003).

The main scientific contribution of fungal leaching on WEEE is by Brandl and co-authors (2001), where *Aspergillus niger* and *Penicillium simplicissimum* were investigated. *A. niger* showed the best efficiency in Cu mobilisation. Compared to lithotrophic bacterial leaching, fungal leaching offers the advantage of operating at mildly acidic conditions, which may minimise the eventual phenomenon of H_2S production that results from the addition of strong inorganic acids (Sabra et al. 2012).

13.4 State of the Art of Biohydrometallurgy in Metal Extraction from WEEE

The early works about the recovery of metals from e-waste by means of biological strategies have involved the exploitation of either of leaching bacteria/archaea (*Sulfolobus* sp., *At. ferrooxidans* and *At. thiooxidans*; Bowers-Irons et al. 1993; Brandl et al. 2001) or of bacteria and fungi that are able to produce 'biological surface active compounds' (BSAC; *Bacillus* sp., *Saccharomyces cerevisiae* and *Yarrowia lipolytica*; Hahn et al. 1993). These studies have demonstrated that biomobilisation of metals from WEEE is possible and may be advantageous in view of commercial applications. New research has been carried out subsequently and the number of papers in this field is increasing; nevertheless, the number of studies is still limited and new research is needed to reach a full-scale application. The biohydrometallurgical studies, published in international scientific journals between 2001 and early 2014, have focussed on the recovery of base valuable and precious metals from PCBs (also known as PWBs, printed wire boards): Fe/S-oxidising bacteria (i.e. leaching bacteria) can be exploited to recover base valuable metals such as Cu and Zn (Table 13.2), while cyanogenic strains are suitable for the recovery of precious metals, such as Au (Table 13.3).

TABLE 13.2

Summary of the Main Scientific Paper within Base-Valuable Metal Recovery by Bioleaching

Type of E-Waste	Microorganisms	Scale[a]	Energy Source	Temperature	Metals	References
PCBs (Tv)	*At. ferrooxidans + At. thiooxidans + L. ferrooxidans*	Flask (10 g/L)	Pyrite, Fe(II), S^0	35°C	Cu	Bas et al. (2013)
PCBs (unspecified)	AMD[b] (*Acidithiobacillus* spp., *Galianella* spp., *Leptospirillum* spp.)	Flask (20 g/L)	Fe(II)	30°C	Cu	Xiang et al. (2010)
PCBs (unspecified)	*S. thermosulfidooxidans + T. acidophilum*	Reactor (150 g/L)	S^0 (biogenic)	45°C	Cu, Zn, Ni, Cd, Al, Cr, Pb	Ilyas et al. (2014)
PCBs (unspecified)	*S. thermosulfidooxidans + T. acidophilum*	Column (10 kg)	Fe(II), S^0	45°C	Cu, Zn, Ni, Al	Ilyas et al. (2010)
PCBs (unspecified)	Municipal activated sludge	Flask (6.7 g/L)	S^0	22°C	Cu, Zn, Ni, Cd, Al, Cr, Pb	Karwowska et al. (2014)
PCBs (unspecified)	*At. ferrooxidans + At. thiooxidans* isolated from sewage treatment plant (China)	Flask (18 g/L)	Fe(II), S^0	32°C	Cu, Zn, Ni, Pb	Liang et al. (2010)
Poly-metallic concentrate (PCBs)	AMD[b] (China)	Flask (12 g/L)	Fe(II)	30°C	Cu, Zn, Ni, Al	Zhu et al. (2011)
Poly-metallic concentrate (unspecified WEEE and end-of-life vehicles)	*At. ferrooxidans + At. thiooxidans + L. ferrooxidans*	Reactor (10 g/L)	Fe(II)	35°C	Cu, Zn	Lewis et al. (2011)
Unspecified WEEE	*At. ferrooxidans + At. thiooxidans*	Flask (50 g/L)	Fe(II), S^0	30°C	Cu, Zn, Ni, Al	Brandl et al. (2001)

Note: Additional studies are discussed in the main text.
[a] Optimal pulp density for Cu recovery in brackets.
[b] Bacteria enriched from acid mine drainage (AMD).

TABLE 13.3

Summary of the Main Scientific Paper within Gold Recovery
by Cyanogenic Bacteria

Type of E-Waste	Microorganisms	Scale[a]	Energy Source	Temperature	Metals	References
PCBs (PC)	*C. violaceum* + *P. aeruginosa*	Flask (10 g/L)	Luria Broth	30°C	Au, Cu, Fe, Zn, Ag	Pradhan and Kumar (2012)
Unspecified WEEE (mainly PCBs)	*C. violaceum* (mutation for alkaline conditions)	Flask (5 g/L)	Luria Broth	30°C	Au, Cu	Natarajan and Ting (2014)
Unspecified WEEE (mainly PCBs)	*C. violaceum* (engineered strains)	Flask (5 g/L)	Luria Broth	30°C	Au	Tay et al. (2013)
Unspecified WEEE	*C. violaceum, P. fluorescens migula*	Flask (5 g/L)	n/a	n/a	Au, Cu	Pham and Ting (2009)

Note: Additional studies are discussed in the main text.
[a] Optimal pulp density for Cu recovery in brackets.

13.4.1 Recovery of Cu and Other Base Valuable Metals from WEEE by Bioleaching with Fe/S-Oxidising Strains

A first study by Brandl and co-authors (2001) has compared the capability of two microbial groups in the recovery of base valuable metals (i.e. Al, Cu, Ni, Pb and Sn) from powdered e-waste: mesophilic Fe/S-oxidising bacteria (*At. thiooxidans* and *At. ferrooxidans*) and fungi (*Penicillium simplicissimum* and *A. niger*). Their results have suggested that Al and Cu biomobilisation efficiencies are highly affected by the concentrations of solids to be treated (i.e. pulp density), while Ni and Zn are more easily recovered by bioleaching processes: with 100 g/L pulp density, the solubilisation decreased from >90% to 0%–5% and from 65 to 20% for Al and Cu, respectively, while Zn and Ni removal efficiencies were still high (>95% for Zn, about 60% for Ni). The same study showed that bioleaching with adapted fungi allows operating with very low amounts of e-waste (1–10 g/L). We have found references to another scientific work dealing with the application of fungi in the recovery of metals from WEEE (Hahn et al. 1993). Other works have confirmed the problems associated with a high solid concentration and have also dealt with other important constraints. According to the scientific literature, the main aspects that need a deep investigation and optimisation are particle size, solid concentration, kind and dosage of energy source, bacterial inoculum size and style, initial pH and temperature. Mainly, elemental sulphur and/or ferrous ions have been tested as suitable energy sources for leaching bacteria, although the use of pyrite (FeS$_2$), as a source of both iron and sulphur, has been recently reported (Ilyas et al. 2010; Bas et al. 2013). Studies have reported that the lower the particle size, the higher the metal removal rate, unless the particle size is too small (Wang et al. 2009; Zhu et al. 2011; Ilyas et al. 2014). In this case, particle–particle collisions would lead to bacterial cell damage. Nevertheless, it is not possible to fix a specific limit because such a phenomenon also depends upon shaking conditions and reactor type.

The majority of the papers about the recovery of base valuable metals from WEEE by bioleaching address the application of mesophilic Fe/S-oxidising bacteria, such as *At. ferrooxidans*, *At. thiooxidans* and *L. ferrooxidans* (11 papers out of 16), while a smaller number deals with the exploitation of moderate thermophilic strains, such as the bacterium *Sulfobacillus thermosulfidooxidans* or, among archaea, *Sulfolobus*-like organisms (4 papers out of 16; see also Table 13.2). Cu is one of the most studied metals; some papers have addressed other metals, too, but optimisation studies have focussed mainly either on Cu or on the total amount of metals (Liang et al. 2013; Ilyas et al. 2014). According to the scientific literature, Zn, Cu, Cd and Ni can be removed with high efficiency, although precipitation and re-complexation phenomena can occur (in the presence of either S^0 or Fe^{2+}, as energy source). The metal precipitation phenomenon may be favoured when Fe(II) is used as the energy source, since jarosite is produced in acidic environments (Fonti et al. 2013). Zhu and co-authors (2011) have observed that a longer incubation time determines a high jarosite precipitation, with slowdown in Cu removal. Indeed, solubilised metals can co-precipitate with Fe(III) in jarosites, and/or jarosite may form a layer on the surface of PCB crumbs and cause passivation (Choi et al. 2004; Zhu et al. 2011). Ni and Cd appear to be removed in a shorter time than Zn and Cu, which could require several days of treatment. In addition, the solubilisation of Cu, Pb, Ni and Cr improves when the temperature is increased to 37°C; air bubbling appears to be insignificant (Karwowska et al. 2014). Very interestingly, the solubilisation of Cr is favoured in the co-presence with microorganisms able to produce BSAC, such as *Bacillus subtilis* and *Bacillus cereus* (Karwowska et al. 2014).

The analysis of the scientific literature indicates that the direct growth of microorganisms in the presence of e-waste is not advisable because of toxic effects on living cells that would allow processing just small concentrations of waste materials (i.e. 1–5 g/L). Although Fe/S-oxidising bacteria are known to tolerate high metal concentrations (Tuovinen et al. 1971; Das et al. 1997; Leduc et al. 1997; Nies 1999; Dopson et al. 2003; Watkin et al. 2009), a solid concentration equal/greater than 10 g/L can cause an activity depletion in *At. ferrooxidans* and *At. thiooxidans* (Brandl et al. 2001; Vestola et al. 2010). Such a toxic effect occurs not only due to the high metal concentration but also due to the non-metallic component of PCBs and other e-waste (i.e. plastic and organic compounds, such as isocyanates, acrylic and phenolic resins, epoxides and phenols; Ludwig et al. 2002). The majority of the authors have tackled this problem using various strategies (one of the following or more than one): (1) two-step bioleaching, in which the biomass is produced in the absence of electronic scraps and the solid to be treated is added after growing (Brandl et al. 2001; Liang et al. 2010; Xiang et al. 2010; Karwowska et al. 2014); (2) pre-adaptation of microorganisms (Ilyas et al. 2007; Yang et al. 2009; Ilyas et al. 2010; Liang et al. 2010; Ilyas et al. 2014); (3) pre-treatment for removing the non-metallic component or to stabilise the pH at optimal values for bacteria (Ilyas et al. 2010; Zhu et al. 2011); and (4) high inoculum size (Yang et al. 2009; Zhu et al. 2011). A study by Liang et al. (2010) have demonstrated that in a bioleaching treatment with a two-step strategy, multiple PCB additions (4–8 g/L) at different times can improve the feasibility of Cu, Ni, Zn and Pb removal.

The use of indigenous Fe/S-oxidising bacteria appears to be more advantageous than type or purchased strains, and mixed cultures are shown to be more efficient than single cultures (Ilyas et al. 2007; Wang et al. 2009; Liang et al. 2010; Zhu et al. 2011; Karwowska et al. 2014). In an interesting study by Karwowska and co-authors (2014), Zn, Cu, Cd, Ni and Cr have been solubilised from crumbled PCBs (two main fractions were considered: ø = 0.1 ÷ 0.25 and ø > 2 mm) by a mixture of municipal activated sludge and municipal wastewater (containing *Thiobacillus thioparus*, *Thiobacillus denitrificans*, *At. thiooxidans* and *At. ferrooxidans*, according to PCR-based techniques). A maximum efficiency (of about 70%, 90%, 100%, 60% and 15%) has been obtained in the presence of S^0 1% and solid 6.7% (w/v). Bacteria enriched from AMD (identified genera *Acidithiobacillus*, *Galionella* and *Leptospirillum*) have been reported to remove about 95% Cu from PCB powder in 5 days without the adaptation step, under the initial conditions of pH 1.5 and 9 g/L Fe(II) (Xiang et al. 2010). Similarly, Zhu and co-authors (2011) reported that unidentified leaching bacteria enriched from AMD can solubilise about 90% Cu from metal concentrates (PCBs pre-treated to remove the whole non-metallic component), in the presence of ferrous ions (optimum: 9–12 g/L) and with a solid concentration up to 12 g/L. At higher solid concentrations, Cu extraction efficiencies decreased, but considering that this study was performed only on the metallic part of PCBs, the actual metal content was very high. Lewis and co-authors (2011) have carried out studies on the recovery of Zn and Cu from polymetallic concentrates (obtained by physical pre-treatment of a mixed feed of metallic scraps, WEEE and end-of-life vehicles) by bioleaching with a copper-adapted consortium of mesophilic Fe/S-oxidising bacteria (*At. ferrooxidans*, *At. thiooxidans* and *L. ferrooxidans*). Pure cultures of *At. ferrooxidans* can remove about 100% Cu from PCBs powder in 36 h, in the presence of Fe(II) 9 g/L and pulp density 25 g/L (Yang et al. 2009); a mixed culture of non-indigenous *At. ferrooxidans*, *At. thiooxidans* and *L. ferrooxidans* can remove about 90% Cu from powdered Tv PCBs in the presence of Fe(II) 9 g/L (80% Cu with 50 g/L pyrite) and solid 10 g/L (Bas et al. 2013). Nevertheless, no information about the mobilisation of the other metals is given in such papers. In addition, many studies with type/wild strains and/

or monocultures have reported that the pH was maintained low (around 2 or less) by adding concentrated sulphuric acid. It is not always clear if these operations for pH controlling were carried out also for the abiotic control tests as this could significantly change the scientific impact of the results and should be taken into account when bioleaching studies with different characteristics are compared.

According to Ilyas and co-authors, moderate thermophilic Fe/S-oxidising bacteria (working at 45°C or more, vs. 32–35°C optimal for mesophiles) would determine higher rates of metal solubilisation than mesophilic and extremely thermophilic strains (Ilyas et al. 2007; Ilyas et al. 2010; Ilyas et al. 2013; Ilyas et al. 2014). Mixed cultures of chemolithotrophic (e.g. *S. thermosulfidooxidans*) and heterotrophic acidophiles (e.g. *Thermoplasma acidophilum*) would offer greater bioleaching potential, although to gain high metal bioleaching efficiencies (about 70%–80% for Ni, Al and Zn, 90% for Cu, with 150 g/L pulp density; Ilyas et al. 2014) a preadaptation step for the microorganisms involved is needed. Mesophilic Fe/S-oxidising strains have been tested at 50°C and an increase in the Cu solubilisation rate was observed, but it was confirmed to be due to abiotic processes (Lewis et al. 2011). Ilyas and co-authors have also demonstrated that sources of biogenic S^0 from desulphurisation refinery plants can be a suitable growth substrate (greater sulphur oxidation in a shorter time period). This could contribute to making the bioremediation process more economical, and avoid the addition of ferrous ions in the medium (Ilyas et al. 2014).

Bioleaching strategies are not suitable when the main objective is the recovery of Pb and Sn, since the bioleaching approach appears to suffer of Pb and Sn precipitation phenomena (Pb would precipitate as $PbSO_4$, Sn as SnO). All the scientific articles have reported Pb and Sn precipitation, when these two metals were investigated, there were just two exceptions; some authors had also reported a high level of Cu precipitation (Choi et al. 2004; Cui and Zhang 2008). Low iron concentrations could reduce the phenomenon, but concerns about the application of bioleaching strategies are still reasonable: further steps of treatment could be required, with an increase in cost and environmental impact. Although the study by Brandl and co-authors, which we discussed above, would suggest that bioleaching with Fe/S-oxidising bacteria (i.e. 'lithotrophic leaching') could represent a more advantageous strategy, the exploitation of fungi allowed the mobilisation of Sn and Pb, which usually precipitate in bioleaching (Sn: 60% with *P. simplicissimum*, 40% with *A. niger*; Pb: 100% with *P. simplicissimum*, 40% with *A. niger*; 1 g/L pulp density). In addition, bioleaching techniques may be characterised by a high environmental impact, if a high amount of acid and a long period of treatment are need (Beolchini et al. 2013), so a combined approach could improve the feasibility of an eventual bioleaching process; live fungi could be substituted by a direct use of organic acids produced in optimal conditions (gluconic, oxalic, citric and ascorbic acids). For instance, commercially available gluconic acid (Naglusol™) is known to solubilise >97% of Pb, Ni, Sn and Zn with a scrap concentration of 100 g/L (Raimann 1996). However, in this regard, we have to mention that studies of toxicity assessment by leaching procedures (e.g. toxic characteristic leaching procedure (TCLP), synthetic precipitation leaching procedure (SPLP) and waste extraction test (WET)) have shown that environmental risks due to Pb appear to be relatively low (Ilyas et al. 2014). A similar approach has been studied by Choi et al. (2004) to improve copper solubilisation and increase the feasibility of a bioleaching process by adding citric acid during the process.

Solid concentration is a very important constraint in the development of commercial-scale processes. At present, among the papers analysed, only three have attempted to scale up from flask experimentation. 150 g/L of powdered PCBs (pre-treated by acid leaching to stabilise pH and reduce toxic effects) can be processed in a 2.5 L stirred-tank reactor (STR) with biogenic S^0 as a sole energy source for bacteria, while 10 kg can be treated by column bioleaching (about 1.2 L volume) in the presence of both Fe(II) and S^0 powder, after removing non-metallic components by a high-density saturated solution of NaCl (Ilyas et al. 2010; Ilyas et al. 2014). The recovery of more than 95% of Cu and Zn is possible in 50 h with 100 g/L pulp density and in the presence of Fe(II) 3 g/L as the sole energy source, by a two-step bioleaching strategy in a batch-mode reactor with ceramic rings as biofilm carriers (Lewis et al. 2011): in these conditions, mesophilic bacteria have been demonstrated to favour the solubilisation of Cu (Zn solubilisation was probably abiotic) and regenerate ferric ions.

13.4.2 Recovery of Au and Other Metals from WEEE by Cyanogenic Bacteria

A relevant interest in the potential utilisation of WEEE as a secondary source of raw materials is related to its content in gold and other precious metals. Au is used in electronics for its excellent resistance to corrosion and high electrical conductivity. Existing processes for recovering Au from electronic waste are based on a pyrometallurgical or hydrometallurgical approach and are extremely polluting and not environmentally sustainable (Korte et al. 2000; Mecucci and Scott 2002; Cui and Zhang 2008; Tsydenova and Bengtsson 2011). Although bacterial cyanide production has been known for many years and gold mobilisation by *C. violaceum* has already been reported (A. D. Smith and Hunt 1985; Campbell et al. 2001; Faramarzi et al. 2004; Faramarzi and Brandl 2006; Kita et al. 2006; Lear et al. 2010), very few works in the scientific literature address the extraction of gold (and/or other metals) from WEEE by cyanogenic strains. Among these, just four papers have addressed the topic with a quantitative approach (Table 13.3).

An analysis of the scientific literature available has revealed that the efficiency of gold recovery from WEEE usually ranges from 5% to 30%, with the exception of the study by Pradhan and Kumar (2012) that has reported 10 times higher efficiencies; on the contrary, in the early reports about the exploitation of cyanogenic microorganisms from minerals and other solids (e.g. gold powder) gold solubilisation efficiencies reached even 100% (Campbell et al. 2001; Faramarzi et al. 2004; Kita et al. 2006; Brandl et al. 2008). The rate of Au dissolution is known to depend upon several factors, including cyanide concentration, particle size, dissolved oxygen concentration, T, pH and competition with other metals for cyanide complexation. In biological cyanidation processes, pH is probably the most important constrain because it significantly affects both the bacterial growth rate and the availability of cyanide ions. Indeed, the recovery of gold by cyanidation requires very alkaline conditions (Windom et al. 1989), which are disadvantageous for the growth of *C. violaceum* (Knowles and Bunch 1986). In turn, a depletion of bacterial activity determines low or no cyanogenesis; therefore, low or no CN$^-$ ions are available for gold complexation. Researchers have faced this issue by two-step bioleaching strategies (all the studies in Table 13.3) or by using strains adapted to high pH values. Natarajan and Ting (2014) induced a strategic mutation in *C. violaceum* by exposing cells to N-nitroso-N-ethyl urea (100 mM), which allowed the selection of *C. violaceum* cells capable of growth in alkaline media (pH 9–10). Without mutation, pH 10 was inhibitory for the wild strain to grow. In 2013, Tay and co-authors have reported on Scientific Reports (*Nature*) the construction of two metabolically engineered *C. violaceum* strains, able to produce more cyanide than wild-type bacteria. The two engineered strains were allowed to recover from WEEE more than twice the gold dissolved by wild-type strains.

The best gold recovery efficiencies were obtained with a very low e-waste pulp density. In the study by Natarajan and Ting (2014), cells of *C. violaceum* (mutated for alkaline growth) solubilised about 5% of gold in the presence of 5 g/L of non-pre-treated powdered WEEE; an acid pre-leaching allowed to increase the efficiency up to 22.5% of total Au. Pham and co-authors (2009) have obtained slightly higher yields (about 10%) from non-pre-treated e-waste, either by wild strains of *C. violaceum* or by *P. fluorescens*. However, all the studies considered here have reported a decrease in Au solubilisation efficiency with a higher solid concentration. Biocyanidation of powdered WEEE previously treated by bioleaching with *At. ferrooxidans* has allowed working with solid up to 40 g/L, with gold extraction values comparable to those with un-pre-treated solid 5 g/L (Pham and Ting 2009). Pradhan and Kumar have obtained very high values of gold extraction from powdered PCBs in the presence of mixed culture of *C. violaceum* and *P. aeruginosa*; with a 10 g/L pulp density, about 85%, 75% and 50% of Cu, Au and Zn, respectively, were solubilised. Nevertheless, in the presence of the sole *C. violaceum*, extraction yields were comparable or just slightly lower. However, observed differences in gold solubilisation efficiencies appear to be closely related to the property and composition of the e-waste to be treated, but no study has addressed this point.

C. violaceum appears to be the most suitable strain for gold recovery by biocyanidation. Brandl et al. (2008) compared *C. violaceum* and *P. fluorescens* for the production of cyanide and subsequent formation of dicyanoaurate complex and reported a higher solubilisation of gold for *C. violaceum*. Other information about the performances of different cyanogenic microorganisms has been obtained from research on minerals or pure metal powder (Campbell et al. 2001; Faramarzi et al. 2004; Faramarzi and Brandl 2006; Kita et al. 2006).

In general, the identification of the best operating conditions is the main challenge to face toward the real application of biocyanidation to WEEE. Another important issue to face is related to the presence of other metals than Au. Cu and other base valuable metals in WEEE compete with Au in the complexation with CN^- anions. For this reason, some authors suggest to pre-treat powdered WEEE in order to remove most metals. For instance, Natarajan and Ting (2014) used 6 M nitric acid, which resulted in a removal of nearly 80% of Cu, between 55% and 70% for Ag, Zn, Ni, Pb, Fe and Al, and about 90% for Sn. Moreover, there is a need for studies that deal with the effect of WEEE heterogeneity in structure and composition. No study has addressed a scale-up of the process yet.

13.5 New Challenges for the Development of Sustainable Processes Based on a Biotechnological Approach

Incineration or landfill disposal of WEEE represent both an environmental threat and an important economic loss. Today, it is clear that WEEE is a very promising secondary source of critical metals. Various chemical and physical techniques have been proposed for the reclamations of these elements; nevertheless, high costs and environmental impacts are often associated with such techniques, due to metal speciation and other characteristics of e-waste. Bacteria, archaea and eucarya domains offer a variety of microorganisms that can find (or have already found) real application in biohydrometallurgy. This could offer alternative, cost-efficient and environmentally sustainable strategies for the recovery of base valuable and precious metals from WEEE. Despite the wide heterogeneity in WEEE composition, microorganisms with particular metabolic capabilities (e.g. Fe/S-oxidising strains, cyanogenic bacteria) have shown to be successfully applied for the extraction of various metals from this waste. In the case of bioleaching by Fe/S-oxidising strains, scale-up attempts have already been performed. In general, however, full-scale applications require further investigations; an optimisation of the operating conditions is needed and many aspects still must be addressed, such as the evaluation of costs and actual environmental impacts. At the same time, various insights and possible solutions have recently been suggested. The application of more than one biological strategy could lead to high solubilisation efficiencies for many more elements compared to a singular approach, and it could also offer the possibility to face issues highlighted by researchers and still just partially solved, such as the acid-neutralising power of WEEE and the toxicity of its components. Biohydrometallurgical extraction of metals from WEEE could be coupled with biological strategies for metal recovery from the derived leachates, such as biosorption, bioprecipitation and extraction by biogas. In addition, the use of other types of waste as an energy source for the microorganism involved has been demonstrated as a valid option and could offer the possibility to face two environmental problems contemporaneously, with several advantages.

References

Aung, K. M. M. and Y.-P. Ting. 2005. Bioleaching of spent fluid catalytic cracking catalyst using aspergillus niger. *Journal of Biotechnology* 116(2);(March 16): 159–170.

Baldrian, P. 2003. Interactions of heavy metals with white-rot fungi. *Enzyme and Microbial Technology* 32(1);(January): 78–91.

Barrie Johnson, D and K. B. Hallberg. 2008. Carbon, iron and sulfur metabolism in acidophilic micro-organisms. *Advances in Microbial Physiology* 54: 201–255.

Bas, a. D., H. Deveci and E.Y. Yazici. 2013. Bioleaching of copper from low grade scrap TV circuit boards using mesophilic bacteria. *Hydrometallurgy* 138;(June): 65–70.

Beolchini, F., V. Fonti, A. Dell'Anno, L. Rocchetti and F. Vegliò. 2012. Assessment of biotechnological strategies for the valorization of metal bearing wastes. *Waste Management* 32(5);(May): 949–956.

Beolchini, F., V. Fonti, L. Rocchetti, G. Saraceni, B. Pietrangeli and A. Dell'Anno. 2013. Chemical and biological strategies for the mobilization of metals/semi-metals in contaminated dredged sediments: Experimental analysis and environmental impact assessment. *Chemistry and Ecology* 29(5): 415–426.

Bigum, M., L. Brogaard and T. H. Christensen. 2012. Metal recovery from high-grade WEEE: A life cycle assessment. *Journal Hazardous Material* 207–208: 8–14.

Bigum, M., C. Petersen, T. H. Christensen and C. Scheutz. 2013. WEEE and portable batteries in residual household waste: Quantification and characterisation of misplaced waste. *Waste Management* 33(11): 2372–2380.

Binnemans, K., P. Tom Jones, B. Blanpain, T. Van Gerven, Y. Yang, A. Walton and M. Buchert. 2013. Recycling of rare earths: A critical review. *Journal of Cleaner Production* 51(July): 1–22.

Bosecker, K. 1997. Bioleaching: Metal solubilization by microorganisms. FEMS *Microbiology Reviews* 20: 591–604.

Bowers-Irons, G., R. Pryor, T. Bowers-Irons, M. Glass, C. Welsh and R. Blake. 1993. The bio-liberation of gallium and associated metals from gallium arsenide ore and semiconductor wafers. *Biohydrometallurgical Technologies* 1: 335–342.

Brandl, H., R. Bosshard and M. Wegmann. 2001. Computer-munching microbes: Metal leaching from electronic scrap by bacteria and fungi. *Hydrometallurgy* 59(2): 319–326.

Brandl, H. and M. A. Faramarzi. 2006. Microbe-metal interactions for the biotechnological treatment of metal-containing solid waste. *China Particuology* 4(2): 93–97.

Brandl, H., S. Lehmann, M. A. Faramarzi and D. Martinelli. 2008. Biomobilization of silver, gold, and platinum from solid waste materials by HCN-Forming microorganisms. *Hydrometallurgy* 94(1–4) (November): 14–17.

Brierley, C. L. and J. A. Brierley. 2013. Progress in bioleaching: Part B: Applications of microbial processes by the minerals industries. *Applied Microbiology* and *Biotechnology* 97(17) (September): 7543–7552.

Brierley, C. L. 2010. Biohydrometallurgical prospects. *Hydrometallurgy* 104 (3–4) (October): 324–328.

Brierley, J. A. 2008. A perspective on developments in biohydrometallurgy. *Hydrometallurgy* 94(1–4) (November): 2–7.

Burgstaller, W. and F. Schinner. 1993. Leaching of metals with fungi. *Journal of Biotechnology* 27(2): 91–116.

Campbell, S. C., G. J. Olson, T. R. Clark and G. McFeters. 2001. Biogenic production of cyanide and its application to gold recovery. *Journal of Industrial Microbiology and Biotechnology* 26(3) (March): 134–139.

Choi, M.-S., K.-S. Cho, D.-S. Kim and D.-J. Kim. 2004. Microbial recovery of copper from printed circuit boards of waste computer by *Acidithiobacillus ferrooxidans*. *Journal of Environmental Science and Health, Part A* 39(11–12): 2973–2982.

Colmer, A. R. and M. E. Hinkle. 1947. The role of microorganisms in acid mine drainage: A preliminary report. *Science* 106(2751): 253–256.

Council Directive, 1975. 75/442/EEC of 15 July 1975 on waste. *Official Journal of the European Union* 194: 39–41.

Crowe, M., A. Elser, B. Göpfert, L. Mertins, Th Meyer, J. Schmid, A. Spillner and R. Ströbel. 2003. Waste from electrical and electronic equipment (WEEE)-quantities, dangerous substances and treatment methods. D. Tsotsos, European Topic Centre on Waste, European Environmental Agency (EEA).

Cui, J. and L. Zhang. 2008. Metallurgical recovery of metals from electronic waste: A review. *Journal of Hazardous Materials* 158(2–3) (October 30): 228–256.

Das, A., J. M. Modak and K. A. Natarajan. 1997. Technical note studies on multi-metal ion tolerance of *Thiobacillus ferrooxidans*. *Minerals Engineering* 10(7): 743–749.

De Vasconcelos, A. T. R., D. F. de Almeida, M. Hungria, C. T. Guimaraes, R. V. Antônio, F. C. Almeida, L. G P de Almeida, R. de Almeida, J. A. Alves-Gomes and E. M. Andrade. 2003. The complete genome sequence of *Chromobacterium violaceum* reveals remarkable and exploitable bacterial adaptability. *Proceedings of the National Academy of Sciences of the United States of America*: 11660–11665.

Dopson, M., C. Baker-Austin, P. R. Koppineedi and P. L. Bond. 2003. Growth in sulfidic mineral environments: Metal resistance mechanisms in acidophilic microorganisms. *Microbiology* 149(8);(August 1): 1959–1970.

Dopson, M. and D. B. Johnson. 2012. Biodiversity, metabolism and applications of acidophilic sulfur-metabolizing microorganisms. *Environmental Microbiology* 14(10);(October): 2620–2631.

Ehrlich, H. L. 2001. Past, present and future of biohydrometallurgy. *Hydrometallurgy* 59(2–3);(February): 127–134.

Erüst, C., A. Akcil, C. S. Gahan, A. Tuncuk and H. Deveci. 2013. Biohydrometallurgy of secondary metal resources: A potential alternative approach for metal recovery. *Journal of Chemical Technology & Biotechnology* 88(12);(December 31): 2115–2132.

European Commission, 2010. Critical raw materials for the EU. DG Enterprise and Business.

European Parliament, 2002. Directive 2002/95/EC, "Restriction of the use of certain hazardous substances in electrical and electronic equipment". *Official Journal of the European Union* L37: 19–23.

European Parliament, 2012. Directive 2012/19/EU of the European Parliament and of the Council of 4 July 2012 on waste electrical and electronic equipment (WEEE). *Official Journal of the European Union* 197: 38–70.

Faramarzi, M. a. and H. Brandl. 2006. Formation of water-soluble metal cyanide complexes from solid minerals by *Pseudomonas plecoglossicida*. *FEMS Microbiology Letters* 259(1);(June): 47–52.

Faramarzi, M. A., M. Stagars, E. Pensini, W. Krebs and H. Brandl. 2004. Metal solubilization from metal-containing solid materials by Cyanogenic *Chromobacterium violaceum*. *Journal of Biotechnology* 113(1–3);(September 30): 321–326.

Fonti, V., A. Dell'Anno and F. Beolchini. 2013. Influence of biogeochemical interactions on metal bioleaching performance in contaminated marine sediment. *Water Research* 47(14);(September 15): 5139–5152.

Gadd, G. M. 2000. Bioremedial potential of microbial mechanisms of metal mobilization and immobilization. *Current Opinion in Biotechnology* 11(3);(June): 271–279.

Gadd, G. M. 2007. Geomycology: Biogeochemical transformations of rocks, minerals, metals and radionuclides by fungi, bioweathering and bioremediation. *Mycological Research* 111:(Pt 1);(January): 3–49.

Gadd, G. M. 2010. Metals, minerals and microbes: Geomicrobiology and bioremediation. *Microbiology* 156;(March): 609–643.

Greenpeace. Rifiuti elettronici—La campagna di Greenpeace. http://www.greenpeace.org/italy/Global/italy/report/2009/8/rifiuti-elettronici2.pdf (acceses December 15th 2013).

Hahn, M., S. Willscher and G. Straube. 1993. Copper leaching from industrial wastes by heterotrophic microorganisms. *Biohydrometallurgical Technologies*. 1: 99–108.

He, W., G. Li, X. Ma, H. Wang, J. Huang, M. Xu and C. Huang. 2006. WEEE recovery strategies and the WEEE treatment status in China. *Journal of Hazardous Materials* 136(3);(August 25): 502–512.

Hedley, H. and H. Tabachnick. 1958. Chemistry of cyanidation. American Cyanamid Company, Explosives and Mining Chemicals Department. New York.

Hennebel, T., N. Boon, S. Maes and M. Lenz. 2013. Biotechnologies for critical raw material recovery from primary and secondary sources: R&D priorities and future perspectives. *New Biotechnology* 00;(August 28): 1–7.

Huismann, J., F. Magalini, R. Kuehr, C. Maurer, S. Ogilvoe, J. Poll, C. Delgado, E. Artim, J. Szlezak and A. Stevels. 2008. Review of directive 2002/96 on waste electrical and electronic equipment (WEEE). Final report. http://ec.europa.eu/environment/waste/weee/pdf/final_rep_unu.pdf (accessed December 15, 2013).

Humphries, M. 2012. Rare earth elements: The global supply chain. *Congressional Research Service* 2011: 7–5700.

Ilyas, S., M. A. Anwar, S. B. Niazi and M. A. Ghauri. 2007. Bioleaching of metals from electronic scrap by moderately thermophilic acidophilic bacteria. *Hydrometallurgy* 88(1–4);(August): 180–188.

Ilyas, S., J.-C. Lee and R.-A. Chi. 2013. Bioleaching of metals from electronic scrap and its potential for commercial exploitation. *Hydrometallurgy* 131–132;(January): 138–143.

Ilyas, S., J.-C. Lee and B.-S. Kim. 2014. Bioremoval of heavy metals from recycling industry electronic waste by a consortium of moderate thermophiles: Process development and optimization. *Journal of Cleaner Production* 70(1): 192–202.

Ilyas, S., C. Ruan, H. N. Bhatti, M. A. Ghauri and M. A. Anwar. 2010. Column bioleaching of metals from electronic scrap. *Hydrometallurgy* 101(3–4);(March): 135–140.

Johnson, D. B. 2012. Geomicrobiology of extremely acidic subsurface environments. *FEMS Microbiology Ecology* 81(1);(July): 2–12.

Jones, D. S., H. L. Albrecht, K. S. Dawson, I. Schaperdoth, K. H. Freeman, Y. Pi, A. Pearson and J. L. Macalady. 2012. Community genomic analysis of an extremely acidophilic sulfur-oxidizing biofilm. *The ISME Journal* 6(1);(January): 158–170.

Karwowska, E., D. Andrzejewska-Morzuch, M. Łebkowska, A. Tabernacka, M. Wojtkowska, A. Telepko and A. Konarzewska. 2014. Bioleaching of metals from printed circuit boards supported with surfactant-producing bacteria. *Journal of Hazardous Materials* 264;(January 15): 203–210.

Kita, Y., H. Nishikawa and T. Takemoto. 2006. Effects of cyanide and dissolved oxygen concentration on biological Au recovery. *Journal of Biotechnology* 124(3);(July 25): 545–551.

Knowles, C. J. and A. W. Bunch. 1986. Microbial cyanide metabolism. *Advanced Microbial Physiology* 27: 73–111.

Korte, F., M. Spiteller and F. Coulston. 2000. The cyanide leaching gold recovery process is a nonsustainable technology with unacceptable impacts on Ecosystems and Humans: The Disaster in Romania. *Ecotoxicology and Environmental Safety* 46(3);(July): 241–245.

Krebs, W., C. Brombacher, P. P. Bosshard, R. Bachofen and H. Brandl. 1997. Microbial Recovery of Metals from Solids. *FEMS Microbiology Reviews* 20: 605–617.

Lawson, E. N., M. Barkhuizen and D. W. Dew. 1999. Gold solubilisation by cyanide producing bacteria *Chromobacterium violaceum*. In *Biohydrometallurgy and the Environment Toward the Mining of the 21st Century*. eds. Amils R., Ballester A. 9A: 239–246. Elsevier, Amsterdam.

Lear, G., J. M. McBeth, C. Boothman, D. J. Gunning, B. L. Ellis, R. S. Lawson, K. Morris et al. 2010. Probing the biogeochemical behavior of technetium using a novel nuclear imaging approach. *Environmental Science & Technology* 44(1);(January 1): 156–162.

Leduc, L. G., G. D. Ferroni and J. T. Trevors. 1997. Resistance to heavy metals in different strains of *Thiobacillus ferrooxidans* 13: 453–455.

Lee, J.-C. and B. D. Pandey. 2012. Bio-processing of solid wastes and secondary resources for metal extraction—A review. *Waste Management* (New York, NY) 32(1);(January): 3–18.

Lewis, G., S. Gaydardzhiev, D. Bastin and P.-F. Bareel. 2011. Bio hydrometallurgical recovery of metals from fine shredder residues. *Minerals Engineering* 24(11);(September): 1166–1171.

Li, J., B. N. Lopez, L. Liu, N. Zhao, K. Yu and L. Zheng. 2013. Regional or global WEEE recycling. Where to go? *Waste Management* 33(4): 923–934.

Liang, G., Y. Mo and Q. Zhou. 2010. Novel strategies of bioleaching metals from printed circuit boards (PCBs) in mixed cultivation of two acidophiles. *Enzyme and Microbial Technology* 47(7);(December): 322–326.

Liang, G., J. Tang, W. Liu and Q. Zhou. 2013. Optimizing mixed culture of two acidophiles to improve copper recovery from printed circuit boards (PCBs). *Journal of Hazardous Materials* 250–251;(April 15): 238–245.

Ludwig, C., S. Hellweg and S. Stucki. 2002. *Municipal Solid Waste Management*. Springer-Verlag, Berlin.

Macadam, A. M. and C. J. Knowles. 1984. Purification and Properties of β-Cyano-L-Alanine Synthase from the Cyanide-Producing Bacterium, *Chromobacterium violaceum*. *Biochimica et Biophysica Acta (BBA)—Protein Structure and Molecular Enzymology* 786(3): 123–132.

Mecucci, A. and K. Scott. 2002. Leaching and electrochemical recovery of copper, lead and tin from scrap printed circuit boards. *Journal of Chemical Technology & Biotechnology* 77(4);(April): 449–457.

Mishra, D. and Y. H. Rhee. 2014. Microbial Leaching of Metals from Solid Industrial Wastes. *Journal of Microbiology* 52(January): 1–7.

More, T. T., S. Yan, R. D. Tyagi and R. Y. Surampalli. 2010. Potential use of filamentous fungi for wastewater sludge treatment. *Bioresource Technology* 101(20);(October): 7691–700.

Mulligan, C. N., M. Kamali and B. F. Gibbs. 2004. Bioleaching of heavy metals from a low-grade mining ore using *Aspergillus niger*. *Journal of Hazardous Materials* 110(1–3);(July 5): 77–84.

Natarajan, G. and Y.-P. Ting. 2014. Pretreatment of e-waste and mutation of alkali-tolerant cyanogenic bacteria promote gold biorecovery. *Bioresource Technology* 152;(January): 80–85.

Ndlovu, S. 2008. Biohydrometallurgy for sustainable development in the African Minerals Industry. *Hydrometallurgy* 91(1–4);(March): 20–27.

Nies, D. H. 1999. Microbial heavy-metal resistance. *Applied Microbiology and Biotechnology* 51;(June): 730–750.

Norris, P. R. 2007. Acidophile diversity in mineral sulfide oxidation. In *Biomining*, 199–216. Springer.

Oguchi, M., H. Sakanakura and A. Terazono. 2013. Toxic metals in WEEE: Characterization and substance flow analysis in waste treatment processes. *The Science of the Total Environment* 463–464;(October 1): 1124–1132.

Okibe, N. and D. B. Johnson. 2004. Biooxidation of pyrite by defined mixed cultures of moderately thermophilic acidophiles in pH-controlled bioreactors: Significance of microbial interactions. *Biotechnology and Bioengineering* 87(5);(September 5): 574–583.

Olson, G. J., J. A. Brierley and C. L. Brierley. 2003. Bioleaching review part B: Progress in bioleaching: Applications of microbial processes by the minerals industries. *Applied Microbiology and Biotechnology* 63(3);(December): 249–257.

Ongondo, F. O., I. D. Williams and T. J. Cherrett. 2011. How are WEEE doing? A global review of the management of electrical and electronic wastes. *Waste Management* 31(4): 714–730.

Pham, V. A. and Y. P. Ting. 2009. Gold bioleaching of electronic waste by cyanogenic bacteria and its enhancement with bio-oxidation. *Advanced Materials Research* 73: 661–664.

Plumb, J. J., B. Gibbs, M. B. Stott, W. J. Robertson, J. A. E. Gibson, P. D. Nichols, H. R. Watling and P. D. Franzmann. 2002. Enrichment and characterisation of thermophilic acidophiles for the bioleaching of mineral sulphides. *Minerals Engineering* 15(11): 787–794.

Pradhan, J. K. and S. Kumar. 2012. Metals bioleaching from electronic waste by *Chromobacterium violaceum* and *Pseudomonads* sp. *Waste Management & Research* 30(11);(November): 1151–1159.

Pronk, J. T., J. C. de Bruyn, P. Bos and J. G. Kuenen. 1992. Anaerobic growth of *Thiobacillus ferrooxidans*. *Applied and Environmental Microbiology* 58(7);(July): 2227–2230.

Raimann, W. 1996. Bleaching regulator compositions and bleaching processes using them. Google Patents.

Rawlings, D. E. 2002. Heavy metal mining using microbes. *Annual Review of Microbiology* 56;(January): 65–91.

Reith, F., M. F. Lengke, D. Falconer, D. Craw and G. Southam. 2007. The geomicrobiology of gold. *The ISME Journal* 1(7);(November): 567–584.

Robinson, B. H. 2009. E-waste: An assessment of global production and environmental impacts. *The Science of the Total Environment* 408(2);(December 20): 183–191.

Rocchetti, L. et al., 2013. Environmental impact assessment of hydrometallurgical processes for metal recovery from WEEE residues using a portable prototype plant. *Environmental Science & Technology*, 47(3), 1581–1588.

Rohwerder, T., T. Gehrke, K. Kinzler and W. Sand. 2003. Bioleaching review part A: Progress in bioleaching: Fundamentals and mechanisms of bacterial metal sulfide oxidation. *Applied Microbiology and Biotechnology* 63(3);(December): 239–248.

Rohwerder, T. and W. Sand. 2007. Oxidation of inorganic sulfur compounds in acidophilic prokaryotes. *Engineering in Life Sciences* 7(4);(July): 301–309.

Sabra, N., H. C Dubourguier, M. N. Duval and T. Hamieh. 2011. Study of canal sediments contaminated with heavy metals: Fungal versus bacterial bioleaching techniques. *Environmental Technology* 32: 1307–1324.

Sabra, N., H.-C. Dubourguier and T. Hamieh. 2012. Fungal leaching of heavy metals from sediments dredged from the deûle canal, France. *Advances in Chemical Engineering and Science* 02(01): 1–8.

Sand, W., T. Gehrke, P.-G. Jozsa and A. Schippers. 2001. (Bio) chemistry of bacterial leaching—direct vs. indirect bioleaching. *Hydrometallurgy*: 159–175.

Santhiya, D. and Y.-P. Ting. 2006. Use of adapted *Aspergillus niger* in the bioleaching of spent refinery processing catalyst. *Journal of Biotechnology* 121(1);(January 2): 62–74.

Schippers, A. 2007. Microorganisms involved in bioleaching and nucleic acid-based molecular methods for their identification and quantification. In *Microbial Processing of Metal Sulfides*, 3–33. Springer.

Schluep, M., C. Hagelueken, R. Kuehr, F. Magalini, C. Maurer, C. Meskers, E. Mueller and F. Wang. 2009. Sustainable innovation and technology transfer industrial sector studies: Recycling–from E-waste to resources. United Nations Environment Programme & United Nations University, Bonn, Germany.

Smith, A. and T. Mudder. 1991. *Chemistry and Treatment of Cyanidation Wastes*. Mining Journal Books Ltd. (UK), 345.

Smith, A. D. and R. J. Hunt. 1985. Solubilisation of gold by *Chromobacterium violaceum*. *Journal of Chemical Technology and Biotechnology* 35(2): 110–116.

Syed, S. 2012. Recovery of gold from secondary sources—A review. *Hydrometallurgy* 115–116;(March): 30–51.

Tay, S. B., G. Natarajan, M. N. B. A. Rahim, H. T. Tan, M. C. M. Chung, Y. P. Ting and W. S. Yew. 2013. Enhancing gold recovery from electronic waste via lixiviant metabolic engineering in *Chromobacterium violaceum*. *Scientific Reports* 3;(January): 2236.

Temple, K. L. and A. R. Colmer. 1951. The autotrophic oxidation of iron by a new bacterium: *Thiobacillus ferrooxidans*. *Journal of Bacteriology* 62(5): 605.

Tsydenova, O. and M. Bengtsson. 2011. Chemical hazards associated with treatment of waste electrical and electronic equipment. *Waste Management* 31(1);(January): 45–58.

Tuovinen, O. H., S. I. Niemelä and H. G. Gyllenberg. 1971. Tolerance of *Thiobacillus ferrooxidans* to some metals. *Antonie van Leeuwenhoek* 37(1): 489–496.

Vera, M., A. Schippers and W. Sand. 2013. Progress in bioleaching: Fundamentals and mechanisms of bacterial metal sulfide oxidation—Part A. *Applied Microbiology and Biotechnology* 97(17);(September): 7529–7541.

Vestola, E. A., M. K. Kuusenaho, H. M. Närhi, O. H. Tuovinen, J. A. Puhakka, J. J. Plumb and A. H. Kaksonen. 2010. Acid bioleaching of solid waste materials from copper, steel and recycling industries. *Hydrometallurgy* 103(1–4);(June): 74–79.

Wang, J., J. Bai, J. Xu and B. Liang. 2009. Bioleaching of metals from printed wire boards by *Acidithiobacillus ferrooxidans* and *Acidithiobacillus thiooxidans* and their mixture. *Journal of Hazardous Materials* 172(2–3);(December 30): 1100–1105.

Waksman, S. A. and J. S. Joffe. 1922. Microörganisms concerned in the oxidation of sulfur in the soil: II. *Thiobacillus thiooxidans*, a new sulfur-oxidizing organism isolated from the soil. *Journal of Bacteriology* 7(2): 239–256.

Watkin, E. L. J., S. E. Keeling, F. A. Perrot, D. W. Shiers, M.-L. Palmer and H. R. Watling. 2009. Metals tolerance in moderately thermophilic isolates from a spent copper sulfide heap, closely related to *Acidithiobacillus caldus*, *Acidimicrobium ferrooxidans* and *Sulfobacillus thermosulfidooxidans*. *Journal of Industrial Microbiology & Biotechnology* 36(3);(March): 461–465.

Windom, H. L., S. J. Schropp, F. D. Calder, J. D. Ryan, R. G. Smith, L. C. Burney, F. G. Lewis and C. H. Rawlinson. 1989. Natural trace metal concentrations in estuarine and coastal marine sediments of the southeastern united states. *Environmental Science & Technology* 23(3);(March): 314–320.

Xiang, Y., P. Wu, N. Zhu, T. Zhang, W. Liu, J. Wu and P. Li. 2010. Bioleaching of copper from waste printed circuit boards by bacterial consortium enriched from acid mine drainage. *Journal of Hazardous Materials* 184(1–3);(December 15): 812–818.

Yang, T., Z. Xu, J. Wen and L. Yang. 2009. Factors influencing bioleaching copper from waste printed circuit boards by *Acidithiobacillus ferrooxidans*. *Hydrometallurgy* 97(1–2);(June): 29–32.

Zhang, K., J. L. Schnoor and E. Y. Zeng. 2012. E-waste recycling: Where does it go from here? *Environmental Science and Technology*, 46;(September): 10861–10867.

Zhu, N., Y. Xiang, T. Zhang, P. Wu, Z. Dang, P. Li and J. Wu. 2011. Bioleaching of metal concentrates of waste printed circuit boards by mixed culture of acidophilic bacteria. *Journal of Hazardous Materials* 192(2);(August 30): 614–619.

Zimmerley, S. R., D. G. Wilson and J. D. Prater. 1958. Cyclic leaching process employing iron oxidizing bacteria. U.S. Patent 2,829,964.

Zoeteman, B. C. J., H. R. Krikke and J. Venselaar. 2009. Handling WEEE waste flows: On the effectiveness of producer responsibility in a globalizing world. *The International Journal of Advanced Manufacturing Technology* 47 (5–8): 415–436.

14

Hybrid Leaching: An Emerging Trend in Bioprocessing of Secondary Resources

Sadia Ilyas and Jae-Chun Lee

CONTENTS

14.1 Introduction

Bio-hydrometallurgy is the hybridisation of biotechnology and metallurgy. From unique properties of some potentially significant biotopes and active principles of interactions between microbial metabolisms and materials, efficient integrated bio-metallurgical processes have been developed to extract metals (Morin et al., 2006). The hybrid process may involve a sequence in which a substance (ore/mineral/waste and by-product) is treated by applying a bio-leaching/bio-beneficiation approach followed by chemical processing or vice versa. Early practical applications of bio-hydrometallurgy entailed dump leaching of low-grade, low-value and run-of-mine material. Modern practical applications of bio-hydrometallurgy became a reality in the 1950s with the advent of copper bio-leaching at the Kennecott Copper Bingham

Mine, Utah. Paradoxically, integrated bio-hydrometallurgical practices for processing high-grade refractory gold ores began using tank bio-oxidation followed by cyanide leaching of the bio-treated ore to extract gold, subsequent to the processing of low-grade, lower-value ores of copper in heaps. As the confidence in practical bio-hydrometallurgical processing grows, the experience extends to the applications with knowledge base, innovations and newer practices with diversified materials (Brierley and Brierley, 1999). Considerable efforts have been made to develop an integrated approach for bio-metallurgical processing of wastes and by-products generated from metallurgical and industrial processes and manmade resources due to the rapid growth and access to the new-generation materials. Several options have been tested until now for the management of these secondary resources in view of efficient metal recovery and environmental safety (Brierley and Brierley, 1999).

The metal contents in these secondary resources are in the form of inorganic sulphides/oxides, pure metallic forms and sometimes as organo-metallic complexes. From these resources the metal of interest can be leached out by autotrophic or heterotrophic microorganisms supplemented with suitable energy sources. Autotrophic microorganisms such as bacteria (e.g. *Acidithiobacillus* sp.) and archaea (e.g. *Ferroplasma* sp.); heterotrophic microorganisms such as bacteria (e.g. *Pseudomonas* sp., *Bacillus* sp.) and fungi (e.g. *Aspergillus* sp., *Penicillium* sp.) are the major groups of microbes involved in the bio-leaching of metals (Schinner and Burgstaller, 1989). The ability of these microbes to solubilise metals from secondary raw materials has accelerated the growth of hybrid/integrated bio-hydrometallurgy. Metals that are present in different types of waste streams, including smelter dusts, converter/smelter slags, metallurgical sludges, spent catalysts/lithium ion batteries and electronic scraps, can be leached and/or immobilised by hybrid bio-hydrometallurgical processing routes with twin objectives of resource reclamation and low environmental degradation.

14.2 Fundamental and Alternative Routes of Hybrid Bio-Leaching

Bioprocessing of secondary resources by bacteria and fungi, with integrated/hybrid approach is mainly based on three fundamental routes, namely acidolysis, complexolysis and redoxolysis due to the diversity of material composition. Microorganisms are able to mobilise metals by the formation of organic or inorganic acids (protons), oxidation and reduction reactions and the secretion of complexing agents.

Mineral composition of these materials mainly consists of phases such as $Cu–Fe–Sn–Sb$, $FeO(OH)$, $Mn_3O_3(OH)_6$, M_2SiO_4 (M; Fe, Zn, Al), Cu_9S_5, Cu_5FeS_4,

$ZnFe_2O_4$, $FeSiO_4$, $FeCr_2O_4$, Fe_3O_4, Al_2O_3, MnO, Ni/Mo/P/Al_2O_3, FeS_2, MoS_2 and WS_2), As_2S_3, $CuFeS_2$, Cu_2S, FeS, Fe_7S_8, MnS_2, PbS, ZnS, Cu^0, Zn^0, Ni^0, Fe^0, Al^0, Au^0, Ag^0, Pt^0, Pd^0 and so on. Metal oxides in these materials can be leached out by proton-induced and ligand-induced reactions. Proton-induced and ligand-induced mineral/material solubilisation occurs simultaneously in the presence of ligands under acidic conditions.

In regard to metal sulphides or metals in a zero-valent state with material supplemented with reduced sulphur compounds, sulphur/or iron sources, generally direct or indirect leaching are assumed to operate. Direct leaching means a direct electron transfer from the metal sulphide to the cell attached to the material surface. The reactions in simplified form may be presented as (Bosecker, 1997; Ilyas et al., 2013)

$$MS + H_2SO_4 + \frac{1}{2}O_2 \rightarrow MSO_4 + S^0 + H_2O \tag{14.1}$$

$$S^0 + \frac{3}{2}O_2 + H_2O \rightarrow H_2SO_4 \tag{14.2}$$

Indirect leaching proceeds through metal sulphide oxidising agent, Fe^{3+}, which is generated by Fe^{2+} oxidising bacteria either planktonic or attached to the material surface.

$$MS + 2Fe^{3+} \rightarrow M^{2+} + 2Fe^{2+} + S^0 \tag{14.3}$$

$$2Fe^{2+} + \frac{1}{2}O_2 + 2H^+ \rightarrow 2Fe^{3+} + H_2O \tag{14.4}$$

$$M + 2Fe^{3+} \rightarrow M^{2+} + 2Fe^{2+} + S^0 \tag{14.5}$$

Here, M can be Cu^0, Zn^0, Ni^0, Al^0, Fe^0, Co^0 and so on. Since a direct electron transfer between the metal sulphide and the attached cell via enzymes, nanowires and so on has not been demonstrated, instead, attached cells provide an efficient extracellular polymeric substance (EPS)-filled reaction compartment for indirect leaching with Fe^{3+} ions (Sand et al., 1995). Thus, the terms 'contact leaching' and 'non-contact leaching' have been proposed for bio-leaching by attached and planktonic cells. A third term, 'cooperative leaching', has been proposed for dissolution of sulphur colloids, sulphur intermediates and mineral fragments by planktonic cells (Rawlings, 2002). These terms describe the physical status of cells without any information about the underlying chemical mechanisms.

For a comprehensive understanding of the chemical aspect of biological metal sulphide dissolution, two different pathways, namely, the thiosulphate mechanism and the polysulphide mechanism were proposed by Schippers and Sand (1999). In general, dissolution was achieved by a combination of proton attack and oxidation processes. However, instead of crystal structure, the reactivity of metal sulphides with protons (acid solubility) was considered the relevant criterion that was determined by the electronic configuration, valence bond and molecular orbital theory. The metal sulphides FeS_2, MoS_2 and WS_2 consist of pairs of sulphur atoms (Vaughan and Craig, 1978) that form non-bonding orbitals. Consequently, the valence bonds of these metal sulphides are only derived from the orbitals of metal atoms. Thus, the valence bonds of FeS_2, MoS_2 and WS_2 do not contribute to the bonding between the metal and the sulphur moiety of the metal sulphide. Consequently, these metal sulphides show resistance against a proton attack. The bonds can only be broken via multistep electron transfers with an oxidant such as the Fe^{3+}. After the initial attack of the oxidant Fe^{3+}, the sulphur moiety of pyrite can be oxidised to soluble sulphur intermediates. Here, the main sulphur intermediate is thiosulphate. The thiosulphate mechanism can be simplified by the following equations (Schippers and Sand, 1999):

$$FeS_2 + 6Fe^{3+} + 3H_2O \rightarrow S_2O_3^{2-} + 7Fe^{2+} + 6H^+ \tag{14.6}$$

$$2Fe^{2+} + \frac{1}{2}O_2 + 2H^+ \rightarrow 2Fe^{3+} + H_2O \tag{14.7}$$

The valence bond of As_2S_3, $CuFeS_2$, FeS, Fe_7S_8, MnS_2, PbS, ZnS and so on are derived from both metal and sulphur orbitals. So, in addition to an oxidant such as Fe^{3+}, protons can remove electrons from the valence bond, causing a cleavage of the bonds between the metal and the sulphur moiety of the metal sulphide. Consequently, these metal sulphides are relatively soluble in acid. In this case, the main sulphur intermediate is polysulphide and a series of reactions inherently explains the formation of elemental sulphur via polysulphides. The polysulphide mechanism can be simplified by the following equations:

$$MS + Fe^{3+} + H^+ \rightarrow M^{2+} + \frac{1}{2}H_2S_n + Fe^{2+} \quad (n \geq 2) \tag{14.8}$$

$$\frac{1}{2}H_2S_n + Fe^{3+} \rightarrow \frac{1}{8}S_8 + Fe^{2+} + H^+ \tag{14.9}$$

$$\frac{1}{8}S_8 + \frac{3}{2}O_2 + H_2O \rightarrow SO_4^{2-} + 2H^+ \tag{14.10}$$

Although elemental sulphur is chemically inert in the natural environment, it can be biologically oxidised to sulphuric acid.

14.2.1 Potentially Important Microbes

Potentially important leaching bacteria, with bio-metallurgical perspective, belong to the genus *Acidithiobacillus*, *Acidiphilium*, *Leptospirillum*, *Sulfobacillus* and *Sulfolobus* (Norris et al., 2000).

The most important bio-leaching mesophiles belong to the genus *Acidithiobacillus*, which includes the iron- and sulphur-oxidising *Acidithiobacillus ferrooxidans* and the sulphur-oxidising *Acidithiobacillus thiooxidans*. *Acidithiobacillus ferrooxidans* is a Gram-negative, acidophilic, aerobic, rod-shaped bacterium and is obligately chemolithotrophic for nutrition. The optimum pH for growth is 2.5 and the optimum temperature range is 30–35°C. The G + C content of the DNA is 58–59 mol% (Temple and Colmer, 1951). *Acidithiobacillus thiooxidans* is motile, with a polar flagellum as the locomotory organelle. It derives energy solely through the oxidation of reduced sulphur compounds, and cannot oxidise iron or pyrite, but has been able to grow on sulphur from pyrite in co-culture with *Leptospirillum ferrooxidans*, which is an iron-oxidising bacterium. They are also Gram-negative, acidophilic, aerobic, rod-shaped bacteria and obligately chemolithotrophic for nutrition. The optimal pH for growth is 2–3 and the temperature range for optimal growth is between 28°C and 30°C. The G + C content of the DNA is 52 mol% (Waksman and Joffe, 1922).

Acidithiobacillus ferrooxidans was endowed with a remarkably broad metabolic capacity and was the first acidophilic bacterium that was investigated from bio-metallurgical perspective. The study conducted at the putative boundary of mesophilic and moderately thermophilic conditions (45°C) shows that *Acidithiobacillus caldus* acts as the dominant sulphur oxidiser and *Leptospirillum* as the dominant iron oxidiser. *Leptospirillum ferrooxidans* is a Gram-negative, motile, acidophilic, aerobic, small curved rod-shaped bacteria. The pH range for their ambient growth ranges from 2.5 to 3.0. It can only oxidise ferrous iron to derive energy for their growth. The G + C content of the DNA is 51.7 mol%. *Acidithiobacillus caldus* is a Gram-negative, motile bacterium, having an optimum pH of 2–2.5 and an optimum growth temperature of 45°C. It is chemolithotrophic and grows on reduced sulphur substrates and can use molecular hydrogen as the electron donor. The G + C content of this species is 63.1–63.9 mol% (Markosyan, 1972).

The role of iron-oxidising *Leptospirillum ferrooxidans* in bio-leaching was recognised to be one with a slow attack due to its poor enrichment from samples compared to *Acidithiobacillus ferrooxidans*. An early indication of the importance of *Leptospirillum ferrooxidans* in bio-leaching was that a mixed culture of *Leptospirillum ferrooxidans* and *Acidithiobacillus thiooxidans* could oxidise pyrite faster than *Acidithiobacillus ferrooxidans* due to its high propensity to attach to leaching material, its high affinity for ferrous iron and low sensitivity to inhibition by ferric iron (Sand et al., 1992).

The relation between microbial activity and heat generation in metal bearing sulphidic materials points out the significance of thermophiles in bio-leaching units. Moderately thermophilic iron and sulphur-oxidising bacteria were initially cultured from mining environments (Brierley, 1978), hot springs (Brierley, 1978) and later from iron ore deposits (Ilyas et al., 2010). The development of moderate thermophiles include *Sulfobacillus thermosulfidooxidans*, which continues to be observed in recent mining operations, including heaps of refractory gold ore undergoing bio-oxidation (Brierley, 1997) and in bio-leaching of electronic scraps (Ilyas et al., 2007, 2010).

More recent discovery of extremely thermophilic microorganisms in acidic, sulphidic hot springs led to evaluations of these microorganisms for their ability to oxidise difficult materials such as MoS_2 and $CuFeS_2$. These microorganisms have expanded bioprocessing options in terms of temperature and metal tolerance.

Besides the role of acidophilic autotrophs in bio-leaching, some heterotrophic bacteria (*Chromobacterium violaceum*, *Pseudomonas fluorescens*) and fungi (*Aspergillus niger*, *Penicillium simplicissimum*) are also considered potentially important. Metals from these heterotrophic microorganisms are generally leached out by displacement with hydrogen ions, or by formation of soluble metal complexes and chelates (Bosshard et al., 1996).

14.2.2 Interaction of Microbe with Metals

Microbes encounter metals in the environment either for their benefit or to develop defenses against them. Their small size, high surface-to-volume ratio and negative net charge of the cell envelope makes these organisms prone to accumulate metal cations from the environment. Microbes can potentially accumulate metals by either a metabolism-independent passive process or a metabolism-dependent active process. Thus, overall accumulation is determined by the sorptivity of the cell envelope and capacity for taking up metals into the cytosol.

The uptake of trace metals and their subsequent incorporation into metalloenzymes or utilisation in enzyme activation occurs in all microbes (Wackett et al., 1989). Some examples of metalloenzymes are nitrogenase (Mo/Fe or sometimes V/Fe, or Fe only), cytochromes (Fe) and cytochrome oxidase *aa*3 (Fe, Cu), superoxide dismutases (Fe, Mn, Cu or Zn), bacteriochlorophyll (Mg), iron-sulphur proteins, CO dehydrogenase with Ni in anaerobic bacteria and Mo in aerobic bacteria, NADP-dependent formate dehydrogenase (W/Se/Fe) and formate dehydrogenase H (Mo/Se/Fe). For uptake, these metals must be in ionic form and uptake may require genetically determined and controlled transport mechanisms (Ehrlich and Newman, 1996).

Some eubacteria (*Acidithiobacillus ferrooxidans*, *Leptospirillum ferrooxidans*) and archaea (*Acidianus brierleyi*, *Sulfolobus acidocaldarius*) use metals or metalloids as electron donors or acceptors in energy metabolism. As energy sources, oxidisable metals or metalloids may satisfy the entire energy demand of an organism (chemolithotrophs). Some hydrogen-oxidising autotrophic bacteria also use oxidised metal species in their respiration, where an oxidised metal or metalloid species serves as the terminal electron acceptor that includes Fe^{3+} reduction to Fe^{2+}.

Enzymatic microbial detoxification of harmful metals or metalloids is another type of interaction. The bacterial oxidation of AsO^{2-} to AsO_4^{3-} by a strain of *Alcaligenes faecalis*, and the reduction of CrO_4^{2-} to $Cr(OH)_3$ by *Pseudomonas fluorescens* or *Enterobacter clocae* are examples of such redox reactions (Ehrlich, 1997).

Procaryotic and eucaryotic microbes are capable of accumulating metals by binding them as cations to the cell surface in a passive process. Even dead cells can bind metal ions. Depending on the conditions, such binding may be selective or non-selective. In some cases, if the cell surface becomes saturated by a metal species, the cell may subsequently act as a nucleus in the formation of a mineral containing the metal (Macaskie et al., 1987). Some bacteria and fungi can promote selective and non-selective leaching of one or more metal constituents with metabolic products such as organic or inorganic acids and/ or ligands produced by them (Ehrlich, 1997). Microbes may secrete inorganic metabolic products such as sulphide, carbonate or phosphate ions in their respiratory metabolism and with them toxic metal ions may precipitate in the process of non-enzymatic detoxification (Macaskie et al., 1987; Ehrlich, 1997).

14.2.3 Chemical–Biological Couple

The predominant route for processing secondary resources in the mid-1980s was hydrometallurgical/chemical leaching. Chemical leaching can be performed by utilising various inorganic acids or organic/inorganic reagents (H_2SO_4, HCl, HNO_3, $FeCl_3$, $CuCl_2$, $S_2O_3^{2-}$, CN^-, $(NH_2)_2CS$) and various chelating agents/ligands (ethylene diamine tetraacetic acid (EDTA), diethylene triamine penta acetic acid (DTPA), nitrilo tri acetic acid (NTA), oxalate, citrate, acetate, etc.) to form complexes. Well-known examples of chemical leaching are cyanide, halide, thiourea and thiosulphate leaching.

A series of environmental accidents at various gold mine sites has precipitated widespread concern over the use of cyanide as a leaching reagent, while in the case of thiourea, it is more expensive than cyanide and its consumption in gold processing is high because it is readily oxidised in a solution. The principal problem with thiosulphate leaching is also the high consumption of reagent during extraction. In comparison, bio-leaching is a cost effective, environmentally friendly and self-sustaining approach. Bio-leaching is based on the natural ability of microbes (bacteria, fungi) to transform solid metallic compounds to its soluble and extractable form by acidolysis, complexolysis, redoxolysis and bioaccumulation (Bosshard et al., 1996). In the current scenario, several biological and chemical couples/hybrids can be profitable for future applications.

Table 14.1 presents various chemical and biological routes and chemical–biological couples for the processing of secondary resources. A perusal of this table reveals that the use of such coupled processes can reasonably enhance

TABLE 14.1

Chemical, Biological and Hybrid Coupling for Bioprocessing Secondary Resources

Chemical Leaching: Ligands Used	Biological Leaching: Microbes Used	Hybrid Leaching: Combinations	Metals to Extract	References
EDTA	Acidithiobacillus ferrooxidans, bacterial strain	EDTA + Acidithiobacillus ferrooxidans / bacterial strain	Zn, Pb, Cd, Cu	Cheikh et al. (2010); Wasay et al. (1998); Satroutdinov et al. (2000)
Citrate	Aspergillus niger, Penicillium bilaiae, Penicillium sp.	Citrate + Aspergillus niger / Penicillium bilaiae / Penicillium sp.	Cd, Cu, Pb	Arwidsson and Allard (2009); Wasay et al. (1998)
Tartrate	Aspergillus niger, Penicillium bilaiae, Penicillium sp.	Tartrate + Aspergillus niger / Penicillium bilaiae / Penicillium sp.	Cd, Pb, Zn	Arwidsson and Allard (2009); Wasay et al. (1998)
Oxalate	Aspergillus niger, Penicillium sp.	Oxalate + Aspergillus niger / Penicillium sp.	Zn	Arwidsson and Allard (2009); Elliott and Shastri (1999)
Oxalate + ammonium citrate	Aspergillus niger, Penicillium sp.	(Oxalate + ammonium citrate) + Aspergillus niger / Penicillium sp.	Cd, Pb, Cu, Zn	Arwidsson and Allard (2009); Wasay et al. (1998)
DTPA	Candida albicans	DTPA + Candida albicans	Cr, Pb, Cu, Zn	Hong et al. (2000); Sohnle et al. (2001)
Sulphides	Acidithiobacillus ferrooxidans, Sulfolobus spp., Phanerochaete chrysosporium	Sulfides + Acidithiobacillus ferrooxidans / Sulfolobus spp. / Phanerochaete chrysosporium	Au	Bosecker (1997); Kohr (1995); Ofori-Sarpong et al. (2011)
EDDS	Amycolatopsis orientalis	EDDS + Amycolatopsis orientalis	Cu	Zwicker et al. (1997)
Nitrilotriacetic acid (NTA)	Acidithiobacillus caldus strain BC13	NTA + Acidithiobacillus caldus strain BC13	Zn, Fe	Hong et al. (2000); He et al. (2011); Aston et al. (2010)

the efficiency of metal extraction. Enhanced leaching of Zn, Cd, Pb and Cu involving EDTA (Wasay et al., 1998) coupled with either *Acidithiobacillus ferrooxidans* (Cheikh et al., 2010) or *Shewanella putrefaciens* (Dobbin et al., 1995) can be obtained. Many fungi, including *Aspergillus niger* and *Penicillium simplicissimum*, secrete various organic acids such as citric, tartaric and oxalic acids, which can act as chelating agents and hence are employed for the extraction of Cu, Cd, Zn and Pb (Mulligan and Kamali, 2003). Oxalate along with ammonium citrate can be used for the extraction of Cd, Pb, Cu and Zn (Wasay et al., 1998) and the efficiency of this process can reasonably be enhanced by adding *Aspergillus niger* or *Penicillium* sp.

Precious metals such as Au and Cu can be extracted by sulphide ligands (Ofori-Sarpong et al., 2011), along with microbes such as *Acidithiobacillus ferrooxidans*, *Sulfolobus* spp. (Kohr, 1995), *Acidithiobacillus thiooxidans* (Bosecker, 1997) and *Phanerochaete chrysosporium* (Ofori-Sarpong et al., 2011), which can accelerate the rate of metal extraction. Environmental issues associated with ligands are also a matter of concern that can be resolved by using various biodegradable ligands such as ethylenediaminedisuccinic acid (EDDS), DTPA and NTA, which are well known for their use in the extraction of Cd, Pb, Cu, Zn and Fe from their ores (Hong et al., 2000).

14.2.4 Oxidative–Reductive/Aerobic–Anaerobic Couple

Existing bio-hydrometallurgical processes use acidophilic procaryotes (bacteria and archaea) to degrade reduced (sulphide) minerals/materials with oxidative dissolution by direct attach, indirect attach, thiosulphate and polysulphide pathway (Equations 14.1 through 14.10) depending upon the properties of solid to be leached.

The metals contained in most of the oxide minerals are generally not present as discrete minerals, but entrapped within the structures of host minerals, most often ferric iron oxides or manganese oxyhydroxides (FeO(OH), $Mn_3O_3(OH)_6$, etc.). In order to solubilise metals from this type of host mineral, the strong bond between oxygen and ferric iron has to be broken. The economics of metal processing from this type of material can be improved significantly by breakage of the iron/manganese and oxygen bond by reductive dissolution. These biocatalytic process may enable potential bio-hydrometallurgical opportunities because of the fact that FeO(OH) is one of the major host minerals of various metals (Ni, Co, Cr, etc.) and its reduction can be feasible by utilising a low-cost electron donor (Table 14.2). All the fundamental process steps, apart from the metal removal step, can be biologically catalysed. The overall process can make use of low-intensity integrated processing enabled by biocatalysed reactions to, first, leach the oxide material, subsequently remove iron from the solution and finally regenerate a significant portion of the acid requirement for reuse in the leaching step. Theoretical mineral composition, dissolution reactions, indicative reaction extents and resulting acid and sulphur consumptions have been summarised

TABLE 14.2

Reagent Consumption as per the Extent of Reaction and Possible Cost of Electron Donor for Reductive Bioprocessing

Mineral	Chemical Reactions	% Extent of Reaction	Acid Consumption (kg/t)	Sulphur Consumption (kg/t)
(a) Reagent Consumption as per the Extent of Reaction:				
Goethite	$6FeOOH + 5H_2SO_4 + S° \rightarrow 6FeSO_4 + 8H_2O$	85	487	31.9
Boehmite	$6AlOOH + 9H_2SO_4 \rightarrow 3Al_2(SO_4)_3 + 12H_2O$	85	40	0
Serpentine	$Mg_3Si_2O_5(OH)_4 + 3H_2SO_4 \rightarrow 3MgSO_4 + 2SiO_2 + 8H_2O$	80	60	0
Nontronite	$Al_4Si_4O_{10}(OH)_8 + 6H_2SO_4 \rightarrow 4H_4SiO_4 + 2Al_2(SO_4)_3 + 2H_2O$	30	2	0
Asbolan	$Mn_3O_3(OH)_6 + 2H_2SO_4 + S° \rightarrow 3FeSO_4 + 5H_2O$	100	15	2.5
Chromite	$FeCr_2O_4 + 4H_2SO_4 \rightarrow FeSO_4 + Cr_2(SO_4)_3 + 4H_2O$	100	49	0
Nickel oxide	$NiO + H_2SO_4 \rightarrow NiSO_4 + H_2O$	90	21	0
Cobalt oxide	$Co_2O_3 + 3H_2SO_4 \rightarrow Co_2(SO_4)_3 + 3H_2O$	90	3	0

Donor Compound	Chemical Reactions	Electron Mole^{-1}	Cost (US$/t of Material)	Cost (US$/kmol Electrons)
(b) Electron Donor Costs as per the Participatory Electron During Reaction:				
Sodium dithionite	$Na_2S_2O_4 + 4H_2O \rightarrow 2Na^+ + 2SO_4^{2-} + 8H^+ + 6e^-$	6	1600	46.4
Glycerol	$C_3H_5(OH)_3 + 3H_2O \rightarrow 3CO_2 + 14H^+ + 14e^-$	14	800	5.3
Ethanol	$C_2H_6O + 3H_2O \rightarrow 2CO_2 + 12H^+ + 12e^-$	12	1200	4.6
Sulphur dioxide	$SO_2 + 2H_2O \rightarrow SO_4^{2-} + 4H^+ + 2e^-$	2	100	3.2
Gaseous hydrogen	$H_2 \rightarrow 2H^+ + 2e^-$	2	2000	2.0
Elemental sulphur	$S° + 4H_2O \rightarrow 2SO_4^{2-} + 8H^+ + 6e^-$	6	200	1.1

Source: Redrawn from Hallberg, K. et al. 2011. *Miner Eng* 24: 620–624.

in Table 14.2. The overall process involved (Equation 14.11) the catalytic transfer of electrons from sulphur (Equation 14.12) and the electron donor half-reaction to ferric iron contained in mineral form (Equation 14.13); the electron acceptor half-reaction is, therefore, an important step in the development of such processing options (Hallberg et al., 2011).

$$S^0 + 6FeO(OH) + 10H^+ \rightarrow SO_4^{2-} + 6Fe^{2+} + 8H_2O \tag{14.11}$$

$$S^0 + 4H_2O \rightarrow 6e^- + SO_4^{2-} + 8H^+ \tag{14.12}$$

$$6FeO(OH) + 6e^- + 18H^+ \rightarrow 6Fe^{2+} + 12H_2O \tag{14.13}$$

Similarly, cobalt is often associated with manganese oxyhydroxides of the asbolane-lithiophorite group. Since the manganese in this mineral group is in an oxidised form, acid dissolution of asbolane is enhanced under reducing conditions. Bacteria use a similar energy metabolism for dissimilatory iron and manganese reduction (Lovley and Phillips, 1988). Using a sulphur electron donor, the overall reductive dissolution of asbolane-lithiophorite group minerals under reductive conditions can be described by Equations 14.14 and 14.15).

$$Mn_3O_3(OH)_6 + S^0 + 4H^+ \rightarrow 3Mn^{2+} + 5H_2O + SO_4^{2-} \tag{14.14}$$

$$Mn_3O_3(OH)_6 + 6Fe^{2+} + 12H^+ \rightarrow 3Mn^{2+} + 6Fe^{3+} + 9H_2O \tag{14.15}$$

This reaction may either occur directly (Equation 14.14) or via the reduction of Mn^{4+} by Fe^{2+} (Equation 14.15), with ferrous iron derived from reductive dissolution of FeO(OH). The resulting ferric iron from Equation 14.3 may then be reduced (in a reaction similar to Equation 14.13). However, the overall net reaction (Equation 14.14) is the same, independent of whether iron is involved as an intermediary. In the current scenario, secondary resources containing sulphidic and oxide minerals and/or minerals in oxidised and reduced forms (spent catalysts, slags, ocean manganese nodules and contaminated sediments), a multistage bioprocessing unit can be established with alternative oxidative or reductive dissolutions.

Another possible redox/aerobic–anaerobic couple, of particular interest can be the case of bio-leaching and bio-recovery of metals from secondary resources (slags, dusts, fly ashes). After oxidative dissolution of metals with acidophilic, iron and sulphur oxidising microorganisms, metals could be subsequently precipitated from the bio-leach solution using biogenic H_2S produced by sulphur-reducing microorganisms (SBR) as follows:

$$SO_4^{2-} + 2CH_2O \rightarrow H_2S + 2HCO^{3-} \tag{14.16}$$

$$H_2S + M^{2+} \rightarrow MS + 2H^+ \qquad (14.17)$$

$$2HCO^{3-} + H^+ \rightarrow CO_2 + H_2O \qquad (14.18)$$

Precipitation of metals as sulphides using biogenic H_2S offers several advantages over chemical hydroxide precipitation: lower effluent concentrations of metals, better thickening characteristics of the metal sludge and the possibility to recover valuable metals (Kaksonen et al. 2011).

14.3 Hybrid Bio-Leaching of Secondary Resources

Integration of innovative biotechnological processes for valuable metal recovery from secondary resources is the absolute necessity of time. As most of the secondary resources have a complex and heterogeneous nature, hybrid bio-leaching of these resources, including alternative chemical–biological stages and successive oxidative–reductive leaching stages, can reduce leaching time, reduce the volume of wastes and increase the rate of extraction of metals, and can also be eco-friendly. The hybrid bio-hydrometallurgical approach can provide an opportunity to conduct advanced research in order to process a number of secondary resources, such as smelter dusts, slags, sludges, electronic wastes and spent catalyst.

14.3.1 Processing of Smelter Dust

Dusts mainly from the reverberatory/flash and converter furnaces are often recycled in the plant after blending with the concentrate to recover copper at the cost of plant productivity and causing environmental pollution, besides damaging the refractory bricks. The main copper sulphide in a typical smelting dust have been identified as chalcocite (16%), chalcopyrite (2%–3%), bornite (2%–3%) and covellite (1.0%), and 13% copper oxide (Oliazadeh et al., 2006). Some R&D work for bioprocessing of theses dusts has been summarised in Table 14.3. The high leaching of copper from the dust may be attributed to the presence of significant amount of secondary sulphides such as chalcocite and bornite, which are easily leached by such bacteria (Watling, 2006). The bio-leaching of copper from the Iranian (Sarcheshmeh) smelter dust was examined using integrated GeoCoat technology (Bakhtiari et al., 2008). In order to develop the leaching process in a continuous mode, two-stage stirred-tank bioreactors and two-stage airlift reactors (Bakhtiari et al., 2008) were applied.

Apart from the *Acidithiobacillus ferrooxidans* and *Acidithiobacillus thiooxidans*, the mixed culture also had *Leptospirillum ferrooxidans* and was inoculated after adaptation over the dust feed. The operation of laboratory (2.0 L)-scale

TABLE 14.3

R&D Work for Bioprocessing of Secondary Resources (Smelter Dusts, Slags, Sludges, Li Ion Batteries, Spent Catalyst and Electronic Scrap)

Source Material	Microbes Used	Metal Recovery	References
Waste PCBs	*Acidithiobacillus ferrooxidans* and 6.66 g/L Fe(III)	72%–98% Cu	Yang et al. (2009)
	Acidithiobacillus sp. and *Leptospirillum* sp.	100% Cu and Ni	Vestola et al. (2010)
	Sulfobacillus thermosulfidooxidans and acidophilic heterotroph	89% Cu; >81% Ni and Zn; 79% Al	Ilyas et al. (2007)
	Sulfobacillus thermosulfidooxidans and *Thermoplasma acidiphilum*	865 Cu; 74% Ni; 80% Zn; 64% Al	Ilyas et al. (2010)
	Chromobacterium violaceum	14.9% Au[a]; 13% Au[b]	[a]Faramarzi et al. (2004); [b]Chi et al. (2010)
Waste e-scrap	*Chromobacterium violaceum*	68.5% Au	Brandl and Faramarzi (2006)
Printed wire boards	*Acidithiobacillus ferrooxidans*, *Acidithiobacillus thiooxidans* and mixed culture	Up to 99.9% Cu	Wang et al. (2009)
Spent petro-refinery catalysts	*Aspergillus niger*	>80% Mo; 65%–88% Ni	Islam and Ting (2009); Santhiya and Ting (2005)
	Acidithiobacillus thiooxidans	94.8% V; 46.3% Mo; 88.3% Ni	Mishra et al. (2007)
	Mixed acidophilic culture bacteria	92% V; 85% Ni; 26% Mo	Kim et al. (2008)
	Acidithiobacillus ferrooxidans	84% Mo; 99% Ni; 96% Co; 63% Al; 97% V	Gholami et al. (2010)
	Penicillium simplicissimum	100% W and Fe; 92.7% Mo; 66.4% Ni; 25% Al	Amiri et al. (2011)
Spent Li-ion batteries	*Acidithiobacillus ferrooxidans* and *Acidithiobacillus thiooxidans* in the presence of sulphur	90% Co; 80% Li	Mishra et al. (2008); Xu et al. (2008)

Continued

TABLE 14.3 (Continued)

R&D Work for Bioprocessing of Secondary Resources (Smelter Dusts, Slags, Sludges, Li Ion Batteries, Spent Catalyst and Electronic Scrap)

Source Material	Microbes Used	Metal Recovery	References
Spent Ni–Cd batteries	*Acidithiobacillus ferrooxidans*	100% Cd; 96.5% Ni; 95% Fe	Cerruti et al. (1998)
	Acidithiobacillus ferrooxidans and *Acidithiobacillus thiooxidans*	96.1% Cu; 94.6% Zn; 81.7% Cd; 71.6% Ni; 61.5% Cr; 48.5% Pb	Zhu et al. (2003)
NiMH spent batteries	*Acidithiobacillus ferrooxidans* and *A. thiooxidans*	95.7% Ni; 72.4% Co; 10% Li	Wu et al. (2008)
Zn–Mn spent batteries	*Alicyclobacillus* sp. and *Sulfobacillus* sp.	100% Zn; 94% Mn	Xin et al. (2012)
Pb/Zn smelting slag	Mixed culture isolate of moderate thermophilic bacteria	> 82% Al; >85% As; >86% Cu; >85% Mn; > 85% Fe; >95% Zn	Cheng et al. (2009); Guo et al. (2010)
Copper smelter flue dust	*Acidithiobacillus ferrooxidans* and *Acidithiobacillus thiooxidans*	81%–87% Cu	Massianaie et al. (2006); Oliazadeh et al. (2006)
Copper converter slag	*Acidithiobacillus ferrooxidans*	93% Cu	Carranza et al. (2009)
Copper dump slag	Coryneform bacteria	38% Mn; 46% Mg; 68% Ca; 26% Ni; 40% Co; 80% Pb	Willscher and Bosecker (2003)

bioreactors (two-stage) arranged in a series for 180 days in continuous mode demonstrated the recovery of 86.8% Cu in 6 days at 70 g/L pulp density and 32°C. Investigation in an airlift reactor that was operated for 150 days to leach the acid-treated flue dust using the adapted mixed culture further confirmed the increased rate of dissolution (86% Cu) in 5 days at 34°C and 70 g/L pulp density. These results have a clear potential for bio-leaching copper from the copper smelter flue dusts if tested and implemented on a large scale (Lee and Pandey, 2012).

14.3.2 Processing of Converter/Smelter Slags

Slags are among the fastest-growing category of solid waste streams generated from the metallurgical industries. These wastes are harmful if released into the environment, but can be potentially valuable sources of metals if processed efficiently. In Korea, slag from copper smelting industry is considered to be one of the important metals containing waste materials. Approximately 1,850,000 tons of slag is generated annually and only 30% is recycled for cement production and the remaining is disposed-off in landfills. If the copper of slag could be recovered, it would supply 1.5% of annual copper production. Some R&D work for bioprocessing of theses slags has been summarised in Table 14.3. Recently, an integrated (chemical–biological couple) approach was adopted by Ilyas et al. (2013) for processing slag. A moderately thermophilic bacterial culture *Sulfobacillus thermosulfidooxidans* was used as inoculum for biogenic Fe^{3+} containing H_2SO_4 solution and then chemical leaching was carried out in a 2.0 dm^3 stirred-tank reactor with biogenic Fe^{3+} solution. The effect of various factors (pH, Fe^{3+}, temperature) on the leaching efficiency of metals was observed. Thus, increase in the pH value above 2.0 resulted in the formation of jarosite according to Equation 14.19.

$$3Fe_2(SO_4)_3 + 12H_2O + M_2SO_4 \rightarrow 2M[Fe_3(SO_4)_2(OH)_6] \quad (M = K^+, Na^+, \text{etc.})$$

$$(14.19)$$

The pH value below 1.5 intensified the solution and led to the formation of unfilterable gels due to the dissolution of silicates in sulphuric acid as in Equations 14.20 through 14.22.

$$M_2SiO_4 + 2H_2SO_4 \rightarrow 2MSO_4 + H_2SiO_4 \quad (M = Fe, Zn, Al, \text{etc.}) \quad (14.20)$$

$$H_4SiO_4 \rightarrow H_2SiO_3 + H_2O \quad (14.21)$$

$$H_2SiO_3 \rightarrow SiO_2 + H_2O \quad (14.22)$$

The presence of a high concentration of Fe^{3+} ions was found to inhibit the leaching of Zn and Fe, but enhance the leaching of Cu and Ni as in Equations 14.23 through 14.27.

$$Cu^0 + 2Fe^{3+} \rightarrow Cu^{2+} + 2Fe^{2+} \qquad (14.23)$$

$$Cu_9S_5 + 8Fe^{3+} \rightarrow 5CuS + 8Fe^{2+} \qquad (14.24)$$

$$Cu_5FeS_4 + 4Fe^{3+} \rightarrow 2CuS + CuFeS_2 + 2Cu^{2+} + 4Fe^{2+} \qquad (14.25)$$

$$Fe_3O_4 + 8H^+ \rightarrow Fe^{2+} + 2Fe^{3+} + 4H_2O \qquad (14.26)$$

$$ZnFe_2O_4 + 8H^+ \rightarrow Zn^{2+} + 2Fe^{3+} + 4H_2 \qquad (14.27)$$

Recently, Carranza et al. (2009) reported the application of the BRISA process in which metal from the copper converter slag (9.0% Cu) was leached chemically by ferric sulphate (ferric leaching stage) coupled with the biological generation (bio-oxidation stage) of the leaching reagent by *Acidithiobacillus ferrooxidans*. The regenerated ferric iron is reused in the leaching step running in close circuit, resulting in the dissolution of 93% Cu in 4 h at 60°C, pH 1.67 and 20 g/L pulp density with the fine-sized particles (D80–47.03) in the presence of 11.5 g/L ferric sulphate.

14.3.3 Processing of Metallurgical Sludge

Some R&D work on bioprocessing of theses sludges has been summarised in Table 14.3.

With hardly any significant utilisation of red mud, which is generated to the tune of 66 million tons every year globally at the rate of 1.0–1.5 ton/ton of alumina production (Wu et al., 2008), bio-leaching of a component seems to be a daunting task. Francis and Dodge (1990) treated the solid waste generated during co-precipitation of metal ions from the effluents with iron as goethite (FeOOH) using anaerobic bacteria such as *Clostridium* sp. The dissolution of 55% Ni, 48% Cd, 41% Zn, 59% Fe and 3.2% Cr in shake flask confirmed the instability of such dumps.

Among a few industrial problems, the bio-leaching of copper locked in waste rock dump (1.2 million ton Cu) in China is mentioned. With the possibility of leaching 30% Cu per year by *Acidithiobacillus ferrooxidans* from the dump, an effort was recently made by Wu et al. (2008) to convert the dump into an *in situ* bio-leaching operation. Copper leaching of 44% on bench scale was obtained.

A study by Ostrega et al. (2009) on the high mobilisation of toxic metals (80%–100%) from the stored metallurgy wastes in Poland during a six-stage extraction with autotrophic and heterotrophic bacteria suggests the sequential bio-leaching of the metals from the material as an alternative before it is safely dumped.

14.3.4 Processing of Spent Catalyst/Lithium Ion Batteries

Table 14.3 presents some R&D work on the bioprocessing of spent catalyst/lithium ion batteries (LIBs). The use of LIBs has rapidly increased worldwide due to their wide use as electrochemical power sources in mobile telephones, personal computers, video cameras and other modern-life appliances, including the electric-powered automobiles in the future (Xu et al., 2008). Spent hydro-processing catalysts are generated from oil refineries, which are a valuable source rich in metals such as Al, V, Mo, Ni and Fe. Spent petroleum catalysts constitute a significant amount of the solid waste generated by chemical and petrochemical industries. The amount of spent catalysts generated is 150.000–170.000 tons per year (Dufrense, 2007). Most of the metals used as catalysts are in an oxide form, and in some cases, they are reduced to active metals to catalyse the appropriate reactions needed for the process. The catalysts normally become inactive after multiple and extensive use due to the deposition of organic materials, such as benzene, toluene and other impurities, and are considered hazardous material by the U.S. Environmental Protection Agency (EPA) because of the difficulty of disposal due to the presence of Ni, Mo, V, Co, Pt and Pd (Xu et al., 2008).

Bio-leaching methods for the extraction of cobalt and lithium from spent LIBs containing $LiCoO_2$ were explored, employing acidophilic chemolithotrophic *Acidithiobacillus ferrooxidans*, which uses elemental sulphur and ferrous iron as energy sources, respectively, for the production of sulphuric acid and ferric iron as leaching agents (Mishra et al., 2008). The application of such a bio-leaching method in the treatment of LIBs yielded favourable results. The bacteria were able to grow in a medium containing elemental sulphur and iron as energy sources. The sulphuric acid produced was able to leach the metals from LIBs. Cobalt was leached faster than lithium. The Fe(II) in the medium used in the leaching experiments helps in the growth of *Acidithiobacillus ferrooxidans*. The higher the Fe(II) concentration, the slower the metal dissolution. The Fe(III) ion formed during the course of the investigation was found to precipitate with the metals in the leach residues. Higher solid/liquid ratios prevented bacterial activity in the process as higher metal concentrations are toxic to the organism (Mishra et al., 2008).

Bio-leaching with autotrophic *Acidithiobacillus ferrooxidans* and *Acidithiobacillus thiooxidans* involves Fe(III) as the oxidant and protons to dissolve the metals. Biotic leaching by *Aspergillus niger* and *Penicillium simplicissimum* involves organic acids produced by these fungi as the leaching agents. To study abiotic leaching, reagent-grade organic acids such as citric, oxalic and gluconic acids were used to compare with the results from heterotrophic leaching. Reagent-grade sulphuric acid and ferric chloride were used as leaching agents for a comparison with autotrophic leaching. Results show that the surface area of a fresh catalyst was 23% higher than the spent catalyst, but during the bio-leaching of the spent catalyst by *Acidithiobacillus thiooxidans* and *Acidithiobacillus ferrooxidans*, the surface area increased by 17% and 33%, respectively.

Toxicity characteristic leaching procedure (TCLP) tests were conducted on the residues obtained from bio-leaching and chemical leaching to determine if the residue obeyed the toxic limit set by U.S. EPA. The tests conducted on the spent catalyst (feed material before bio-leaching/chemical leaching) showed that the Ni content exceeded environmental limits set by U.S. EPA, but was found to be within the toxic limit after bio-leaching. Autotrophic bio-leaching conducted with *Acidithiobacillus thiooxidans* was found to be promising with the highest recovery of 29.3% Al, 64.5% Mo and 99.8% Ni. More importantly, bacterial leaching was much faster (15–20 days) compared with fungal leaching (45–50 days). Compared with abiotic controls, bio-leaching was better in metal extraction from the spent catalyst.

An attempt to mobilise Al, Co, Mo and Ni from the spent catalyst obtained from an Iranian oil refinery in batch bio-leaching using *Acidithiobacillus ferrooxidans* showed that the addition of a Fe(II) source into the growth medium at pH 1.8–2.0 resulted in a maximum extraction of 63% Al, 96% Co, 84% Mo and 99% Ni in 30 days. On the other hand, bio-leaching carried out using *Acidithiobacillus thiooxidans* with the addition of a sulphur source to the growth medium at pH 3.9–4.4 resulted in a maximum extraction of 2.4% Al, 83% Co, 95% Mo and 16% Ni in 30 days (Gholami et al., 2011).

14.3.5 Processing of Printed Circuit Boards, Liquid Crystal Displays

The rapid advancement in the technology of electronic devices with lucrative benefits has resulted in a decrease in the life span of most electrical and electronic equipments, including mobile phones, computers, televisions and their printed circuit boards (PCBs). End-of-life equipments/materials of electrical–electronic devices are collectively known as e-waste (WEEE). In the electronic wastes, PCBs have quite a diverse composition, containing metals (10%–27% Cu, 8%–38% Fe, 2%–19% Al, 0.3%–2% Ni, 1%–3% Pb, 200–3000 ppm Ag, 20–500 ppm Au, 10–200 ppm Pd, etc.), ceramics and polymers. The disposal of electronic wastes together with municipal wastes causes major environmental problems in a landfill. On the other hand, electronic wastes can be regarded as a potential secondary resource with respect to the content of the base and precious metal values present in them. However, the economic potential of electronic wastes as a secondary resource is yet to be exploited (Cui and Zhang, 2008).

Though, the biotechnology has been proved to be one of the most promising technologies in the metallurgical sector as mentioned earlier, limited research have been reported for the bio-leaching of WEEE, which are summarised in Table 14.3. Most of these studies were conducted with wild cultures of *Acidithiobacillus thiooxidans* and a consortium of *Acidithiobacillus thiooxidans* with wild cultures and consortium of *Acidithiobacillus thiooxidans* and with fungi (*Aspergillus niger, Penicillium simplicissimum*). Only a few integrated approaches were tried for bioprocessing of PCBs.

The use of fungi (*Aspergillus niger* and *Penicillium simplicissimum*) was reported by Brandl and co-workers (2001). The microbial growth was inhibited in a one-step process with the scrap concentration of above 10 g/L in the medium. However, almost complete dissolution of the metals such as Cu, Pb, Sn and Zn was noticed in a two-step leaching process using a commercial gluconic acid solution.

Ilyas et al. (2007) used a mixed culture containing a moderately thermophilic strain such as *Sulfobacillus thermosulfidooxidans* and an unidentified acidophilic heterotroph (code A1 TSB) isolated from the local environment to leach metals from e-waste. At a scrap (PCBs) concentration of 10 g/L, the metal-adapted mixed consortium leached more than 89% Cu, 81% Ni, 79% Al and 83% Zn in 18 days in the presence of 10 g/L S. The secreted carboxylic acids such as citric, oxalic and succinic acids from the heterotroph was able to increase the leaching rate in this case, which could be due to the synergistic effect of the heterotroph on the growth of *Sulfobacillus thermosulfidooxidans*. Ilyas et al. (2010) further reported a high recovery of metals in a column leaching of the spent PCBs using the metal-adapted culture containing an identified heterotrophic bacterium, *Thermoplasma acidiphilum*, as a synergist along with *Sulfobacillus thermosulfidooxidans*. The acid pre-leach for 27 days followed by bio-leaching for 280 days dissolved 86% Cu, 74% Ni, 80% Zn and 64% Al.

For the recovery of precious metals from leach solutions, standard processes, such as cementation and precipitation, and the use of adsorption methods, including those of several microbial species (bacteria, fungi, algae, proteins, alfalfa), have been explored (Cui and Zhang, 2008). Mention is made of the bioseparation of Au(III), Pd(II) and Cu(II) from the nitrate/aqua regia leach liquor of waste PCBs in a three-step process (Creamer et al., 2006). The biomass of *Desulfovibrio desulfuricans* was able to recover gold selectively as powder with hydrogen, bubbling for 30 min in the first step from a solution at pH 1.5 and then settling for 24 h, followed by recovery of palladium with the supplementation of pre-treated biomass containing bio-Pd0 and hydrogen sparging for 30 min plus settling over 24 h. Copper was recovered by the precipitation method using biogas (ammonia) produced from the culture of *Escherichia coli* followed by centrifuging.

14.4 Perspectives on Integrated Bio-Leaching of Secondary Resources

The integrated bio-hydrometallurgical approach is gaining importance among the extractive metallurgy techniques as a promising technology with economic and environmental benefits for enhanced recovery of base and precious metals. Although there are a variety of well-known, potentially important microbes that participate in bio-leaching and bio-oxidation of sulphide or oxide minerals/material, there will be more potentially useful microbes discovered as studies of commercial bio-oxidation/bio-leaching facilities continue. These discoveries will offer opportunities to learn more about the microorganisms that inhabit these environments, and to control the conditions to better meet the needs of the microbes. Currently, the transfer of fundamental knowledge, regarding bioprocessing of secondary resources to practical and commercial applications of biohydrometallurgy by industry, has challenges at all stages of the process from laboratory demonstration, to small column-scale testing and through further commercial plant designing and operations. There are some key points that should be considered for process development by the industries:

- For the sake of sustainable mining practices and environmental protection, there is a strong need for both developed and developing countries to come forward and adopt emerging integrated bio-leaching approaches, for example, hybrid chemical–biological leaching, and hybrid oxidative–reductive leaching for diversified material/secondary resources. Some of these materials, with heterogeneous or complex chemical compositions, are the spent catalyst, metallurgical sludge, smelter dusts, converter slags, electronic scraps and so on.
- The development process requires time, particularly for secondary resources from process inception to process implementation.
- Along with efficient coupled leaching processes, the most effective microbe process has to fit within engineering design capabilities. The potentially efficient microbe for laboratory testing may or may not demonstrate the same performance in a large-scale commercial facility.
- Knowledge of engineering key elements and understanding of the physical, chemical and microbiological sciences are required for proper implementation and successful practical application of these integrated approaches.
- Economics and metal prices influence the engineering design. The most cost-effective process design may not always be the design most optimal for operating a biological process.
- For the process to succeed, it must be championed by management from inception to acceptance, to commercial application and to continued implementation.

Acknowledgement

This work was supported by the Korea Institute of Geoscience and Mineral Resources (KIGAM).

References

Amiri, F., Mousavi, S. M., Yaghmaei, S. 2011. Enhancement of bioleaching of a spent Ni/Mo hydroprocessing catalyst by *Penicillium simplicissimum*. *Sep Purif Technol* 80: 566–576.

Arwidsson, Z., Allard, B. 2009. Remediation of metal-contaminated soil by organic metabolites from fungi II—Metal redistribution. *Water Air Soil Poll* 207: 5–18.

Aston, J. E., Apel, W. A., Lee, B. D., Peyton, B. M. 2010. Effects of cell condition, pH, and temperature on lead, zinc, and copper sorption to *Acidithiobacillus caldus* strain BC13. *J Hazard Mater* 184: 34–41.

Bakhtiari, F., Atashi, H., Zivdar, M., Bagheri, S. A. S. 2008. Continuous copper recovery from a smelter's dust in stirred tank reactors. *Int J Miner Process* 86: 50–57.

Bosecker, K. 1997. Bioleaching: Metal solubilization by microorganisms. *FEMS Microbiol Rev* 20: 591–604.

Bosshard, P. B., Bachofen, R., Brandl, H. 1996. Metal leaching of fly ash from municipal waste incineration by *Aspergillus niger*. *Environ Sci Technol* 30: 3066–3070.

Brandl, H., Faramarzi, M. A. 2006. Microbe-metal-interactions for the biotechnological treatment of metal-containing solid waste. *China Particuology* 4: 93–97.

Brandl, H., Bosshard, R., Wegmann, M. 2001. Computer-munching microbes: Metal leaching from electronic scrap by bacteria and fungi. *Hydrometallurgy* 59: 319–326.

Brierley, J. A. 1978. Thermophilic iron-oxidizing bacteria found in copper leaching dumps. *Appl Environ Microbiol* 36: 523–525.

Brierley, J. A. 1997. Heap leaching of gold bearing deposits, theory and operational description. In: *Biomining: Theory, Microbes and Industrial Processes*, ed. D. Rawlings, 103–115. New York: Springer.

Brierley, J. A., Brierley, C. L. 1999. Present and future commercial applications of bio-hydrometallurgy. In: *Biohydrometallurgy and the Environment toward the Mining of the 21st Century*, eds. R. Amils and A. Ballester, 81–89. Amsterdam: Elsevier.

Carranza, F., Romero, R., Mazuelos, A., Iglesias, N., Forcat, O. 2009. Biorecovery of copper from converter slag: Slags characterization and exploratory ferric leaching tests. *Hydrometallurgy* 97: 39–45.

Cerruti, C., Curutchet, G., Donati E. 1998. Bio-dissolution of spent nickel-cadmium batteries using *Thiobasillus ferrooxidans*. *J Biotechnol* 62: 209–219.

Cheikh, M., Magnin, J. P., Gondrexon, N., Willisn, J., Hassen, A. 2010. Zinc and lead leaching from contaminated industrial waste sludges using coupled processes. *Environ Technol* 31: 1577–1585.

Cheng, Y., Guo, Z., Liu, X., Yin, H., Qiu, G., Pan, F., Liu, H. 2009. The bioleaching feasibility of Pb/Zn smelting slag and community characteristics of indigenous moderate-thermophilic bacteria. *Bioresour Technol* 100: 2737–2740.

Chi, T. D., Lee, J.-c., Pandey, B. D., Jeong, J., Yoo, K.-k., Huyung, T. H. 2010. Bacterial cyanide generation in presence of metal ions (Na^{1+}, Mg^{2+}, Fe^{2+}, Pb^{2+}) and gold bioleaching from waste PCBs. *J Chem Eng Japan* 44: 692–700. doi:10.1252/jcej.10we232.

Creamer, N. J., Baxter-Plant, V. S., Henderson, M., Macaskie, L. E. 2006. Palladium and gold recovery from precious metal solutions and electronic scrap leachates by *D. desulfuricans*. *Biotechnol Lett* 28: 1475–1484.

Cui, J., Zhang, L. 2008. Metallurgical recovery of metals from electronic waste: A review. *J Hazard Mater* 158: 228–256.

Dobbin, P. S., Powell, A. K., McEwan, A. G., Richardson, D. J. 1995. The influence of chelating agents upon the dissimilatory reduction of Fe(III) by *Shewanella putrefaciens*. *BioMetals* 8(2): 163–173.

Dufrense, P. 2007. Hydroprocessing catalyst regeneration and recycling. *Appl Catal A Gen* 322: 67–75.

Ehrlich, H. L. 1997. Microbes and metals. *Appl Microbiol Biotechnol* 48: 687–692.

Ehrlich, H. L., Newman, D. K. 1996. *Geomicrobiology*. New York: Dekker, CRC Press, Taylor & Francis group LLC.

Elliott, H. A., Shastri, N. L. 1999. Extractive decontamination of metal-polluted soils using oxalate. *Water Air Soil Poll* 110: 335–346.

Faramarzi, M. A., Stagars, M., Pensini, E., Krebs, W., Brandl, H. 2004. Metal solubilization from metal-containing solid materials by cynogenic *Chromobacterium violaceum*. *J Biotechnol* 113: 321–326.

Francis, A. J., Dodge, C. J. 1990. Anaerobic microbial remobilization of toxic metals co-precipitated with iron oxide. *Environ Sci Technol* 24: 373–378.

Gholami, R. M., Borghei, S. M., Mousavi, S. M. 2010. Heavy metals recovery from spent catalyst using *Acidithiobacillus ferrooxidans* and *Acidithiobacillus thiooxidans*. In *Proceedings of International Conference on Chemistry and Chemical Engineering*: 331–335.

Gholami, R. M., Borghei, S. M., Mousavi, S. M. 2011. Bacterial leaching of a spent Mo–Co–Ni refinery catalyst using *Acidithiobacillus ferrooxidans* and *Acidithiobacillus thiooxidans*. *Hydrometallurgy* 106: 26–31.

Guo, Z., Zhang, L., Cheng, Y., Xiao, X., Pan, F., Jiang, K. 2010. Effects of pH, pulp density and particle size on solubilization of metals from Pb/Zn smelting slag using indigenous moderate thermophilic bacteria. *Hydrometallurgy* 104: 25–31.

Hallberg, K., Grail, B. M., du Plessis, C. A., Johnson, D. B. 2011. Reductive dissolution of ferric iron minerals: A new approach for bioprocessing nickel laterites. *Miner Eng* 24: 620–624.

He, Q. X., Huang, X. C., Chen, Z. L. 2011. Influence of organic acids, complexing agents and heavy metals on the bioleaching on the iron from kaolin using Fe(III)-reducing bacteria. *Appl Clay Sci* 51: 478–483.

Hong, K. J., Tokunaga, S., Kajiuchi, T. 2000. Extraction of heavy metals from msw incineration fly ashes by chelating agents. *J Hazard Mater* 75: 57–73.

Ilyas, S., Anwar, M. A., Niazi, S. B., Ghauri, A. 2007. Bioleaching of metals from electronic scrap by moderately *thermophilic acidophilic* bacteria. *Hydrometallurgy* 88: 180–188.

Ilyas, S., Lee, J.-c., Shin, D., Kim, B.-S. 2013. Biohydrometallurgical processing of nonferrous metals from copper smelter slag. *Adv Mat Res* 825: 250–253.

Ilyas, S., Ruan, Chi, Bhatti, H. N., Ghauri, M. A., Anwar, M. A. 2010. Column bioleaching of metals from electronic Scrap. *Hydrometallurgy* 101: 135–140.

Islam, M., Ting, Y. P. 2009. Fungal bioleaching of spent hydroprocessing catalyst: Effect of decoking and particle size. *Adv Mat Res* 17–73: 665–668.

Kaksonen, A. H., Lavonen, L., Kuusenaho, M., Kolli, A., Närhi, H., Vestola, E., Puhakka, J. A., Tuovinen, O. H. 2011. Bioleaching and recovery of metals from final slag waste of the copper smelting industry. *Miner Eng* 24: 1113–1121.

Kim, D. J., Mishra, D., Park, K. H., Ahn, J. G., Ralph, D. E. 2008. Bioleaching of metals from spent petroleum catalyst by acidophilic bacteria in presence of pyrite. *Mater Trans* 49: 2383–2388.

Kohr, W. J. 1995. Method for rendering refractory sulphide ores more susceptible to bio-oxidation. US Patent No. 5431717.

Lee, J-c., Pandey, B. D. 2012. Bioprocessing of solid wastes and secondary resources for metal extraction—A review. *Waste Manage* 32: 3–18.

Lovley, D. R., Phillips, E. J. P. 1988. Novel mode of microbial energy metabolism: Organic carbon oxidation coupled to dissimilatory reduction if iron or manganese. *Appl Environ Microbiol* 54: 1472–1480.

Macaskie, L. E., Dean, A. C. R., Cheetham, A. K., Jakeman, R. J. B., Skar-nulis, A. J. 1987. Cadmium accumulation by a *Citrobacter* sp.: The chemical nature of the accumulated metal precipitate and its location on the bacterial cells. *J Gen Microbiol* 133: 539–544.

Markosyan, G. E. 1972. A new iron-oxidizing bacterium-*Leptospirillum ferrooxidans* nov. gen. nov. sp. *Biol J Armenia* 25: 26–29.

Massianaie, M., Oliazadeh, M., Bagheri, A. S. 2006. Biological copper extraction from melting furnace dust of Sarcheshmeh copper mine. *Int J Miner Process* 81: 58–62.

Mishra, D., Kim, D. J., Ralph, D. E., Ahn, J. G., Rhee, Y. H. 2007. Bioleaching of vanadium rich spent refinery catalysts using sulfur oxidizing lithotrophs. *Hydrometallurgy* 88: 202–209.

Mishra, D., Kim, D. J., Ralph, D. E., Ahn, J. G., Rhee, Y. H. 2008. Bioleaching of metals from spent lithium ion secondary batteries by using *Acidithiobacillus ferrooxidans*. *Waste Manage* 28: 333–338.

Morin, D., Lips, A., Pinches, T., Huisman, J., Frias, C., Norberg, A., Forssberg, E. 2006. BioMinE—Integrated project for the development of biotechnology for metal-bearing materials in Europe. *Hydrometallurgy* 83: 69–76.

Mulligan, C. N., Kamali, M. 2003. Bioleaching of copper and other metals from low grade oxidized mining ores by *Aspergillus niger*. *J Chem Technol Biot* 78: 497–503.

Norris, P. R., Burton, N. P., Foulis, N. A. 2000. Acidophiles in bioreactor mineral processing. *Extremophiles* 4: 71–76.

Ofori-Sarpong, G., Osseo-Asare, K., Tien, M. 2011. Fungal pretreatment of sulfides in refractory gold ores. *Miner Eng* 24: 499–504.

Oliazadeh, M., Massianaie, M., Bagheri, A. S., Shahverdi, A. R. 2006. Recovery of copper from melting furnaces dust by microorganisms. *Miner Eng* 19: 209–210.

Ostrega, B. K., Palddyna, J., Kowalski, J., Jedynak, L., Golimowski, J. 2009. Fractionation study in bioleached metallurgy wastes using six-step sequential extraction. *J Hazard Mater* 167: 128–135.

Rawlings, D. E. 2002. Heavy metal mining using microbes. *Annu Rev Microbiol* 56: 65–91.

Sand, W., Gehrke, T., Hallmann, R., Schippers, A. 1995. Sulfur chemistry, biofilm, and the (in)direct attack mechanism—A critical evaluation of bacterial leaching. *Appl Microbiol Biotechnol* 43: 961–966.

Sand, W., Rohde, K., Sobotke, B., Zenneck, C. 1992. Evaluation of *Leptospirillum ferrooxidans* for leaching. *Appl Environ Microbiol* 58: 85–92.

Santhiya, D., Ting, Y. P. 2005. Bioleaching of spent refinery catalyst using *Aspergillus niger* with high yield oxalic acid. *J Biotechnol* 116: 171–184.

Satroutdinov, A. D., Dedyukhina, E. G., Chistyakova, T. I., Witschel, M., Minkevich, I. G., Eroshin, V. K., Egil, T. 2000. Degradation of metal-EDTA complexes by resting cells of the bacterial strain DSM 9103. *Environ Sci Technol* 34: 1715–1720.

Schinner, F., Burgstaller, W. 1989. Extraction of zinc from industrial waste by *Penicillium* sp. *Appl Environ Microbiol* 55: 1153–1156.

Schippers, A., Sand, W. 1999. Bacterial leaching of metal sulfides proceeds by two indirect mechanisms via thiosulfate or via polysulfides and sulfur. *Appl Environ Microbiol* 65: 319–321.

Sohnle, P. G., Hahn, B. L., Karmarkar, R. 2001. Effect of metals on *Candida albicans* growth in the presence of chemical chelators and human abscess fluid. *J Lab Clin Med* 137: 284–289.

Temple, K. L., Colmer, A. R. 1951. The autotrophic oxidation of iron by a new bacterium *Thiobacillus ferrooxidans*. *J Bacteriol* 62: 605–611.

Vestola, E. A., Kusenaho, M. A., Narhi, H. M. 2010. Acid bioleaching of solid waste materials from copper, steel and recycling industries. *Hydrometallurgy* 103: 74–79.

Vaughan, D. J., Craig, J. R. 1978. *Mineral Chemistry of Metal Sulfides*. Cambridge: Cambridge University Press.

Wackett, L. P., Orme-Johnson, W. H., Walsh, C. T. 1989. Transition metal enzymes in bacterial metabolism. In: *Metal Ions and Bacteria*, eds. T. J. Beveridge and R. J. Doyle, 165–206. New York: Wiley.

Waksman, S. A., Joffe, J. S. 1922. Microorganisms concerned in the oxidation of sulfur in the soil. II. *Thiobacillusthiooxidans*, a new sulfur-oxidizing organism isolated from the soil. *J Bacteriol* 7: 239–256.

Wang, J., Bai, J., Xu, J., Liang, B. 2009. Bioleaching of metals from printed wire boards by *Acidithiobacillus ferrooxidans* and *Acidithiobacillus thiooxidans* and their mixed culture. *J Hazard Mater* 172: 1100–1105.

Wasay, S. A., Barrington, S. F., Tokunaga, S. 1998. Remediation of soils polluted by heavy metals using salts of organic acids and chelating agents. *Environ Technol* 19: 369–379.

Watling, H. R. 2006. The bioleaching of sulphide minerals with emphasis on copper sulphides – A review. *Hydrometallurgy* 84: 81–108.

Willscher, S., Bosecker, K. 2003. Studies on the leaching behaviour of heterotrophic microorganisms isolated from an alkaline slag dump. *Hydrometallurgy* 71: 257–264.

Wu, F., Sun, Y., Xin, B. Li, L. 2008. Bioleaching of nickel and cobalt from spent nickel metal hydride batteries using *Acidithiobacillus ferrooxidans* spp. *J Beijing Inst Technol* 17: 489–494.

Xin, B., Jiang, W., Aslam, H., Zhang, K., Liu, C., Wang, R. Wang, Y. 2012. Bioleaching of zinc and manganese from spent Zn-Mn batteries and mechanism exploration. *Bioresour Technol* 106: 147–153.

Xu, J., Thomas, H. R., Francis, R. W., Lum, K. R., Wang, J., Liang, B. 2008. A review of processes and technologies for the recycling of lithium-ion secondary batteries. *J Power Sources* 177: 512–527.

Yang, T., Xu, Z., Wen, J., Yang, L. 2009. Factors influencing bioleaching copper from waste printed circuit boards by *Acidithiobacillus ferrooxidans*. *Hydrometallurgy* 97: 29–32.

Zhu, N., Zhang, L., Li, C., Cai, C. 2003. Recycling of spent nickel-cadmium batteries based on bioleaching process. *Waste Manage* 23: 703–708.

Zwicker, N., Theobald, U., Zahner, H., Fiedler, H. P. 1997. Optimization of fermentation conditions for the production of ethylene-diamine-disuccinic acid by *Amycolatopsis orientalis*. *J Ind Microbiol Biot* 19: 280–285.

15

Microbiologically Influenced Corrosion and Its Impact on Metals and Other Materials

Tingyue Gu, Dake Xu, Peiyu Zhang, Yingchao Li
and Amy L. Lindenberger

CONTENTS

15.1 Introduction

Microbiologically influenced corrosion (MIC), or biocorrosion, is blamed for causing serious problems in many fields, such as the oil and gas industry, water treatment systems and sewer systems. The integrity, safety and reliability of the pipeline and equipment are severely affected by MIC. In 2006, MIC was a primary suspect in the leakage of Alaska oil pipelines. The leak caused a major production shutdown that resulted in an oil price spike (Jacobson 2007). MIC may account for 20% of all corrosion losses (Flemming 1996).

Sulphate-reducing bacteria (SRB) are the more common microbes involved in MIC. They may be introduced in a system via multiple ways, including secondary oil recovery and hydrotest. SRB oxidise organic carbons to harvest electrons. The electrons are used for sulphate reduction that causes biogenic hydrogen sulphide release (Thauer et al. 2007). Hydrogen sulphide gas is not only toxic to living organisms but is also a cause in reservoir souring and a corrosion threat. SRB tend to live in a biofilm community, which provides protection against harmful environmental factors such as pH swings and biocides. Biofilm is a mass transfer barrier, which may lead to a local shortage of organic carbon at the bottom of the biofilm. Electrogenic microbes such as SRB living as "bottom feeders" on a steel surface may switch from organic carbon to elemental iron as the electron donor for the reduction of an oxidant such as sulphate under biocatalysis, leading to pitting corrosion (Xu and Gu 2011). Besides SRB, electrogenic nitrate-reducing bacteria (NRB) and methanogens may also cause MIC (Uchiyama et al. 2010; Xu et al. 2013) because they are capable of utilising electrons from steel for nitrate and CO_2 reduction, respectively. Gu and Xu (2013) lumped these electrogenic microbes as XRB, that is, X-reducing bugs in which X represents oxidants (electron acceptors), including sulphate, nitrate/nitrite, CO_2 and so on.

Numerous theories have been proposed to demonstrate the mechanisms of MIC (Videla and Herrera 2005). The cathodic depolarisation theory (CDT) was first introduced to explain MIC by von Wolzogen Kühr and van der Vlugt (1934). CDT interprets how MIC occurs from the electrochemical aspect. However, CDT cannot explain MIC caused by hydrogenase-negative SRB and other non-sulphate-reducing bacteria. Gu et al. (2009) presented a new MIC theory called the biocatalytic cathodic sulphate reduction (BCSR) theory. It provides a bioenergetics explanation of MIC. In BCSR, iron dissolution (oxidation) and sulphate reduction in the cytoplasm of cells are two key steps of MIC. The cell potential of the resultant redox reaction is +230 mV, which yields a negative Gibbs free energy change. It means this process is thermodynamically favourable. SRB harvest energy to support their metabolism from this process. SRB biofilms serve as biocatalysts. They lower the high activation energy of sulphate reduction (Gu et al. 2009). Analogously, the theory that explains MIC caused by NRB is called biocatalytic cathodic nitrate reduction (BCNR) (Gu and Xu 2013; Xu et al. 2013). NRB reduce nitrate to N_2 or NH_4^+, with a cell potential of +1196 mV and +805 mV, respectively, when combined with iron oxidation.

MIC has been classified into three categories to distinguish the various mechanisms (Gu 2012a; Gu and Xu 2013). Type I MIC is caused by electrogenic bacteria. Electrogenic bacteria, which can actively form pili for electron transfer and energy distribution, perform respiration metabolism (Chang et al. 2006). They use carbon steel or other non-noble metals as electron donors intentionally because the reduction potentials of the ions in these metallic materials are sufficiently negative to form thermodynamically favourable redox reactions when coupled with the reduction of an oxidant such as sulphate and nitrate. Bacteria utilise electrons released from elemental metal oxidation and reduce the oxidant intracellularly.

Type II MIC is defined as the corrosion caused by the metabolites secreted by microbes. These metabolites are oxidants such as volatile fatty acids. The

MIC caused by acid-producing bacteria (APB) belongs to this category (Gu and Xu 2013). Copper MIC caused by SRB is also Type II MIC. Type III MIC is caused by microbial species that secrete enzymes or other corrosive chemicals, which degrade non-metallic materials containing organic carbon as one of the components.

MIC detection and mitigation are essential for pipeline security. Planktonic cells can be easily detected. However, biofilms are the main culprit in MIC and they are more elusive. The detection and enumeration of sessile cells are more critical. Microbiology tests using culture media may be used for detection. Molecular biology methods such as quantitative polymerase chain reaction (qPCR), confocal laser scanning microscopy (CLSM) and denaturing gradient gel electrophoresis (DGGE) are also applied to identify sessile cells. The main methods used to mitigate MIC in the field are pigging and biocide application (Videla 2002). Pigging can remove most of the established biofilm by physical scratching. Using biocide is a chemical way to reduce the bacteria population. Tetrakis hydroxymethyl phosphonium sulphate (THPS) and glutaraldehyde are two widely used biodegradable biocides in the oil and gas industry. Biocide can be applied during the pigging process. Some pipelines, especially older pipelines, are unpiggable due to their designs. They include pipes with small or multiple diameters, bends and connections, and so on (Tiratsoo 2013). Because of the various defense mechanisms adopted by biofilms, sessile cells in biofilms require 10 times the dosage of biocide for planktonic cells (Mah and O'Toole 2001). Owing to the fact that a large amount of biocide is needed, the cost of biocide treatment is high. Environmental hazards are another concern for the increasing biocide usage in the field. Raad et al. (2003, 2007) introduced ethylenediaminetetraacetic acid (EDTA) as a biocide enhancer to strengthen the effect of biocides while lowering their concentrations. Recently, Kolodkin-Gal et al. (2010) reported that D-tyrosine, D-methionine, D-tryptophan and D-leucine mixed with equal concentration as low as 10 nM (overall) could trigger the disassembly of *Bacillus subtilis* biofilm. Xu et al. (2012b) demonstrated that an equimolar mixture of these four D-amino acids could promote the biocidal effect of THPS. A synergistic THPS and D-tyrosine or D-methionine combination was found to effectively prevent the *Desulfovibrio vulgaris* biofilm formation and remove the established biofilm (Xu et al. 2012a, 2013).

15.2 MIC Mechanisms and Classifications

The essence of anaerobic MIC of elemental iron (Fe^0) is the following Fe^0 oxidation reaction:

$$\text{Anodic: Fe} \rightarrow Fe^{2+} + 2e^- \text{ (iron oxidation)} \quad -E^{\circ\prime} = +447 \text{ mV} \qquad (15.1)$$

TABLE 15.1

Reduction Potentials at pH 7, 25°C, 1 M Solutes (1 Bar Gases) Except H⁺

Redox Couple	n	$E^{o\prime}$ (mV)
CO_2 + acetate/pyruvate	2	−660
Fe^{2+}/Fe^0	2	−447
CO_2 + acetate/lactate	4	−430
CO_2/formate	2	−432
$2H^+/H_2$	2	−414
$2CO_2$/acetate	8	−290
2acetate/butyrate	4	−290
S^0/H_2S	2	−270
$6CO_2$/hexane	8	−250
CO_2/CH_4	8	−244
SO_4^{2-}/HS^-	8	−217
Fumarate/succinate	2	+33
NO_2^-/NH_3	6	+330
NO_3^-/NH_3	8	+360
NO_3^-/NO_2^-	2	+430
$2NO_3^-/N_2$	10	+760
$O_2/2H_2O$	4	+818

Source: From Thauer, R.K., E. Stackebrandt, and W.A. Hamilton. 2007. In *Sulphate-Reducing Bacteria: Environmental and Engineered Systems*, eds. L.L. Barton and W.A. Hamilton, 1–37. 1st ed. Cambridge: University Press.

To keep the oxidation reaction going and to maintain electroneutrality, the electrons released by Reaction 15.1 must be absorbed by an electron acceptor. The reduction potential for Fe^{2+}/Fe^0 is −447 mV, in which $E^{o\prime}$ stands for the reduction potential (also known as redox potential) at 25°C, pH 7, 1 M solutes (1 bar gases) except H⁺ with the standard hydrogen electrode (SHE) as the reference. Because Reaction 15.1 is written as an oxidation reaction, $E^{o\prime} = -447$ mV is expressed as $-E^{o\prime} = +447$ mV to indicate the reaction direction. The prime in $E^{o\prime}$ and other related symbols in bioelectrochemistry denotes pH 7. Table 15.1 shows some $E^{o\prime}$ values used in this work.

15.2.1 BCSR Theory

Sulphate reduction occurs in the cytoplasm of an SRB cell via the adenosine phosphosulphate (adenylylsulphate) (APS) pathway requiring ATP sulphurylase, APS reductase and bisulphite reductase enzymes (Thauer et al. 2007):

$$SO_4^{2-} \rightarrow APS \rightarrow HSO_3^- (\text{bisulphite}) \rightarrow HS^- (\text{bisulphide}) \qquad (15.2)$$

The overall sulphate reduction reaction can be expressed as

$$SO_4^{2-} + 9H^+ + 8e^- \rightarrow HS^- + 4H_2O \qquad E^{o'} = -217\,mV \qquad (15.3)$$

When sulphate reduction is coupled with Fe^0 oxidation, the cell potential for the redox reaction is $\Delta E^{o'} = +230$ mV (calculated from 447 mV − 217 mV) under the conditions defined for E^o. This positive cell potential corresponds to $\Delta G^{o'} = -178$ kJ/mol sulphate based on the following equation:

$$\Delta G^{o'} = -nF\Delta E^{o'} \qquad (15.4)$$

where n is the number of electrons involved in the redox reaction (8 in this case) and F is the Faraday constant (96,485 C/mol). The negative Gibbs free energy change ($\Delta G^{o'}$) value indicates that the redox reaction coupling Fe^0 oxidation with sulphate reduction is thermodynamically favoured (i.e. energy is produced) under the conditions defined for E^o. If actual temperature, concentration and pressure values deviate from these conditions, $E^{o'}$ values will shift based on the Nernst equation. Thus, $\Delta E^{o'}$ and $\Delta G^{o'}$ will change. However, $\Delta G^{o'}$ in this case will still be negative because $\Delta G^{o'} = -178$ kJ/mol is far from borderline negative.

Fe^0 is a fuel molecule. It has been used as the sole energy source in bioenergetics studies not intended for MIC investigations in the past (Biswas and Bose 2005; Ghafari et al. 2008; Ginner et al. 2004). The "burning" of Fe^0 with sulphate as the oxidant produces energy that benefits SRB metabolism. In fact, Fe^0 is slightly more energetic than lactate under the conditions defined for E^o because Fe^{2+}/Fe^0 has a slightly more negative $E^{o'}$ value of −447 mV (Table 15.1) compared with −430 mV for CO_2 + acetate/lactate. Favourable thermodynamics does not mean that corrosion will happen at an appreciable speed if kinetics is retarded. Sulphate reduction has a high activation energy. It will proceed only with SRB biocatalysis involving multiple enzymes in the SRB cytoplasm. Thus, Gu et al. (2009) proposed the BCSR theory by treating Fe^0 oxidation as the anodic reaction and biocatalysed sulphate reduction as the cathodic reaction. The word "cathodic" is used in BCSR merely to indicate that the reaction is a reduction reaction for the purpose of mechanistic modelling (Xu and Gu 2011). There is no physical cathode at the site where sulphate reduction occurs, that is, the SRB cytoplasm.

15.2.2 Electrogenesis in MIC

Organic carbon molecules such as lactate and acetate are soluble. They can diffuse into the SRB cytoplasm to be oxidised. They donate electrons in the oxidation process in the cytoplasm where they are used for sulphate reduction. This is not the case for Fe^0 as an electron donor. The iron (or steel) matrix is insoluble and thus Fe^0 cannot enter the cytoplasm. Fe^0 oxidation occurs extracellularly without catalysis. In SRB MIC that uses sulphate as the terminal electron acceptor, the electrons released by Fe^0 oxidation must be transported across the SRB cell wall to the cytoplasm. This kind of electron transfer is impossible from an iron surface to a planktonic cell because electrons do not "swim" in water. Only sessile cells in a biofilm can do it. Microbes that are capable of cross-cell wall electron transfer are known as electrogens.

The sessile cells in a biofilm can transfer electrons using either direct electron transfer (DET) or mediated electron transfer (MET). In DET, cell membrane-bound proteins such as c-type cytochromes are involved. The cells are either directly in contact with a metal surface or via conductive nanowires (pili) as shown in Figure 15.1. The proteins transfer extracellular electrons from outside the cells to the cytoplasm. This is opposite to the electron transfer direction in the bioanode of a microbial fuel cell (MFC). For cells that are unable to perform DET, redox-active electron mediators (also known as shuttles or carriers) are used by electrogens. These chemicals absorb and discharge electrons when they move between two locations. They are recycled for repeated uses. For hydrogenase-positive SRB, hydrogen (H_2) is used as an electron carrier. Other common electron carriers include riboflavin, nicotinamide adenine dinucleotide (NAD) and flavin adenine dinucleotide (FAD). Gu and Xu (2013) tested riboflavin and FAD to promote (accelerate) MIC by *D. vulgaris* against carbon steel. They found that both of them accelerated weight loss and pit depth growth without changing planktonic and sessile cell densities. Gu and Xu (2010) called these electron mediators MIC promoters (i.e. MIC accelerators). Gu (2011) filed a provisional patent on the use and detection of electron mediators in MIC tests and MIC forensics in 2011.

The classical CDT first proposed by von Wolzogen Kühr and van der Vlugt (1934) describes the use of hydrogenase in the "depolarising" of a cathode to push the iron oxidation reaction forward by removing the adsorbed hydrogen atoms on a cathode. This theory is applicable to hydrogenase-positive SRB. In fact, it is a case of using H_2 as the electron carrier for cross-cell wall electron transfer. In bioelectrochemistry, the use of H_2 as the electron carrier has been well documented, albeit in non-corrosion research such as MFCs (Gu 2012a). Recently, Venzlaff et al. (2013) presented electrochemical data showing direct electron transfer from a steel surface to SRB.

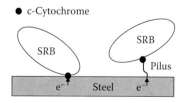

FIGURE 15.1
Schematic illustration of DET across SRB cell walls.

Most cells are not electrogenic because they do not need to transport electrons across cell walls. Typical electron donors such as organic carbons are usually soluble and can diffuse into the cytoplasm. Some cells can become electrogenic when the need arises. Sherar et al. (2011) discovered that pili were formed on the iron surface only in the absence of organic carbon during the culturing of SRB cells isolated from an oil well. It was likely that the SRB cells formed pili to transfer electrons when they switched from organic carbon to Fe^0 as an electron donor. Some SRB isolated from carbon steel pipelines were initially found to be hydrogenase-positive, but they would change to hydrogenase-negative after incubation in an enriched medium (Bryant et al. 1991). It was likely that soluble organic carbon in the culture medium served as electron donor. They readily diffuse to the cytoplasm of the SRB cells without the need for H_2 as the electron carrier. The formation of pili and the production of hydrogenase enzymes consume resources. The SRB cells produce them only when they are needed for electron transfer.

Without the help of pili, direct cell wall to metal surface contact is required for DET. This means only a monolayer of sessile cells in a biofilm are directly involved in harvesting electrons from Fe^0 oxidation. This limits the severity of MIC. With the help of pili networking among cells and between cells and the steel surface, several layers of sessile cells may be involved directly in harvesting electrons from Fe^0 oxidation. Thus, MIC can become more severe. Similarly, electron mediators can allow more than a monolayer of sessile cells to be active in extracellular electron transfer, leading to accelerated MIC (Gu and Xu 2013). Type I MIC's key process is about the transport of extracellular electrons for use in the reduction of an oxidant (e.g. sulphate) in the cytoplasm of a cell. Figure 15.2 illustrates electron transfer in MIC by SRB.

The "motive" for SRB attacks on steel was not fully explained until Xu and Gu (2011) designed a starvation experiment based on their understanding of the bioenergetics of MIC. Their theoretical analysis of Type I SRB MIC suggests that SRB cells starved of organic carbon should be more aggressive because they switch from organic carbon to Fe^0 as the electron donor for energy production. To prove this, they first grew *D. vulgaris* biofilms on C1018 carbon steel coupons in ATCC 1249 medium (full-strength medium) to maturity. Then, the coupons were retrieved and dropped into new anaerobic vials containing modified ATCC 1249 media, including full-strength medium (i.e. full-strength ATCC 1249 medium), full-strength medium minus 90% carbon source (mild starvation), full-strength medium minus 99% carbon source (severe starvation) and full-strength medium minus 100% carbon source (extreme starvation). The vials were incubated for another 7 days at 37°C. The experimental data clearly demonstrated that when removing 90% and 99% of the organic carbon in the culture media, the weight losses increased significantly. Severe organic carbon starvation produced the largest normalised weight loss, while mild organic carbon starvation yielded the deepest pit depth. They suggested that deeper pits required advancement of the SRB

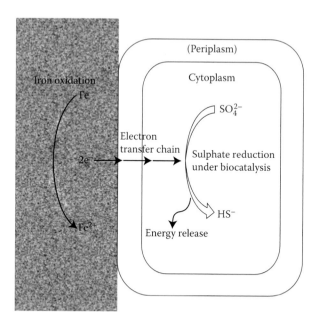

FIGURE 15.2
Electron transport in MIC by SRB. (Periplasm applies only to Gram-negative SRB.)

biofilm towards the newly created pit bottom. This means that the biofilm needed some external organic carbon to grow a little because scavenging the dead cells and exopolymers in the biofilm was insufficient. These data fully supported the theoretical prediction that SRB switched from organic carbon to Fe^0 in order to obtain energy. The data also corroborated with the electron transfer theory in Type I MIC, which requires direct biofilm attachment to the pit bottom for the pit to grow (Figure 15.2).

15.2.3 BCNR Theory

Nitrate injection has been widely used in reservoir souring mitigation to suppress biogenic H_2S production downhole by SRB (Gieg et al. 2011). It promotes respiration by NRB to suppress sulphate respiration by SRB. However, not many people realise that nitrate reduction by NRB can also cause MIC. This means that if nitrate is not completely utilised downhole, it may cause MIC in pipelines. Some SRB can utilise nitrate as well. MIC by NRB can be explained by a theory called BCNR theory that is completely analogous to BCSR. NRB can cause Type I MIC, which means that only electrogenic NRB can do this. Figure 15.3 illustrates the role of electron transport in MIC by NRB.

As shown in the two reactions below, nitrate reduction to either nitrogen or ammonia both have rather large positive $E^{o'}$ values, which means that nitrate

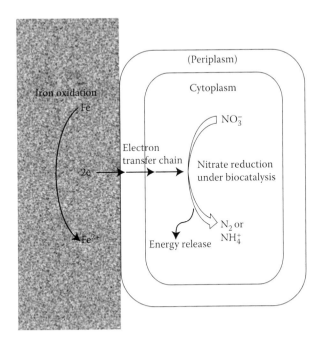

FIGURE 15.3
Electron transport in MIC by NRB. (Periplasm applies only to Gram-negative NRB.)

is a more potent oxidant than sulphate. It leads to far more negative $\Delta G^{o\prime}$ values. This means the nitrate attack on Fe^0 is thermodynamically favoured strongly.

$$2NO_3^- + 10e^- + 12H^+ \rightarrow N_2 + 6H_2O \quad (E^{o\prime} = +760\,mV) \quad (15.5)$$

$$NO_3^- + 8e^- + 9H^+ \rightarrow NH_3 + 3H_2O \quad (E^{o\prime} = +360\,mV) \quad (15.6)$$

With biocatalysis by electrogenic NRB for nitrate reduction, Type I MIC occurs. Xu et al. (2013) found that *Bacillus licheniformis* was more corrosive than typical SRB when it was grown as an NRB. It caused a maximum pit depth of 14.5 μm in 1 week (Figure 15.4).

15.2.4 Type II MIC by Microbes That Secrete Corrosive Metabolites

MIC by SRB and NRB belong to Type I MIC that requires electrogenic microbes at the bottom of the biofilm. Type II MIC does not require electrogenic microbes because it is caused by microbes such as APB that secrete corrosive metabolites such as organic acids. Proton (H^+) is an oxidant. At a

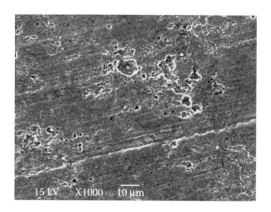

FIGURE 15.4
SEM images of pits on a C1018 coupon caused by *B. licheniformis* biofilm after 7 days of incubation. (Reprinted from *Materials and Corrosion* 63, Xu, D., Y. Li, and T. Gu., D-methionine as a
biofilm dispersal signalling molecule enhanced tetrakis hydroxymethyl phosphonium sulfate
mitigation of *Desulfovibrio vulgaris* biofilm and biocorrosion pitting. 1–9, Copyright 2013, with
permission from Elsevier.)

sufficiently low pH, the proton causes corrosion because it has a sufficiently
high reduction potential that will make the oxidation of Fe^0 coupled with
its reduction thermodynamically favourable. The following Nernst equation can be written to calculate the reduction potential of H^+ at non-standard
conditions:

$$E(2H^+/H_2) = -\frac{2.303RT}{F} \cdot pH - \frac{RT}{2F} \cdot \ln\left(\frac{p_{H_2}}{p_o}\right) \quad (15.7)$$

where R is the universal gas constant, T the temperature (in K) and p_{H_2} the
partial pressure of H_2 (in bar). In Equation 15.7, p_o equals to 1 bar. Plugging
in the conditions defined for $E^{o'}$, the Nernst equation yields $E^{o'}(2H^+/$
$H_2) = -414$ mV, which differs from SHE's 0 mV because $E^{o'}$ requires pH 7,
while SHE requires pH 0 (i.e. 1 M H^+).

As shown below, both proton reduction and free (i.e. undissociated) acetic
acid reduction produce hydrogen gas.

$$2H^+ + 2e^- \rightarrow H_2 \text{ (proton reduction)} \quad (15.8)$$

$$2HAc + 2e^- \rightarrow 2Ac^- + H_2 \text{ (free acetic acid reduction)} \quad (15.9)$$

More recent electrochemical analysis of acetic acid corrosion at the Ohio
University Corrosion Institute suggests that it may be more accurate to
treat HAc as a reservoir of H^+ rather than viewing it as capable of direct

oxidation (Gu and Xu 2013). Oxidants such as H^+ can be directly oxidised without biocatalysis unlike sulphate and nitrate. Thus, acid corrosion happens abiotically in conventional chemical corrosion. Biofilms in Type II corrosion merely provide a locally more concentrated source of oxidants, and planktonic cells may also participate by secreting corrosive metabolites as well, albeit at lower concentrations because sessile cells tend to have a volumetric cell number density that is two orders of magnitude higher (Gu and Xu 2013). Gu (2012b) points out that Type I MIC is caused by microbes that perform respiration while Type II MIC is caused by microbes that grow fermentatively. The former needs external electron acceptors such as sulphate and nitrate, while the latter create their own electron acceptor. Fermentative microbes tend to produce organic acids (e.g. fatty acids), alcohols and so on during their metabolism. APB are typical examples.

Although H_2S is corrosive, Gu and Xu (2013) argued that SRB attack on Fe^0 is primarily Type I MIC due to cross-cell wall electron transport, rather than Type II MIC caused by secreted H_2S. Their view is supported by the fact that starved *D. vulgaris* cells are more aggressive. Interestingly, elemental copper (Cu^0) corrosion by SRB falls under Type II MIC (Gu and Xu 2013). The $E^{o\prime}$ values for Cu^+/Cu^0 and Cu^{2+}/Cu^0 are +520 mV and +340 mV, respectively. This means Cu^0 oxidation coupled with sulphate reduction will have a negative cell potential of –737 mV with Cu^+ as the oxidation product, or –557 mV with Cu^{2+} as the oxidation product. These very negative values lead to very positive $\Delta G^{o\prime}$ values, suggesting that the redox reaction of copper oxidation coupled with sulphate reduction is thermodynamically unfavourable. Thus, this kind of corrosion cannot happen.

Puigdomenech and Taxén (2000) found that the following reaction is strongly favoured thermodynamically:

$$2Cu_{(crystal)} + S^{2-} + 2H^+ \rightarrow Cu_2S_{(crystal)} + H_{2(g)} \tag{15.10}$$

Thus, SRB corrosion of Cu^0 should belong to Type II MIC because it is caused by secreted metabolites. Electron transfer happens outside the cells, rather than in the cytoplasm of SRB cells in this case.

15.2.5 Type III MIC by Microbes That Degrade Organic Materials

The MIC caused by secreted enzymes that degrade organic matter such as polyurethane and other polymers, as well as plasticisers, for the objective of obtaining utilisable organic carbons for the microbes is termed Type III MIC by Gu and Xu (2013). This kind of MIC is better known as biodegradation of organic materials. There have been documented cases of Type III MIC attacks on plastic structures and electric insulation (Gu 2003). Degradation of wire insulation by microbes in older airplanes threatens aging military and civilian aircraft.

15.3 Examples and Analyses of MIC against Metals and Other Materials

There are numerous reported cases of MIC against metals in many different industries in many parts of the world. MIC failures have been reported for many fields, including the oil and gas industry, fire protection systems, copper potable water systems and heat exchangers, wastewater treatment facilities, the pulp and paper industry, and nuclear and hydroelectric power plants. MIC can also be found in many other industries not mentioned in this chapter.

One particular area where there are a large number of reported cases of MIC of metals is in the oil and gas industry. In the oil and gas industry, MIC has been reported more frequently in recent years. One of the main reasons for this is the use of water (often seawater) injection in enhanced oil recovery operations to increase reservoir pressure. The practice brings in nutrients and microbes. One serious example of this is the 2006 oil pipeline failure at the Prudhoe Bay oil field in Alaska (Jacobson 2007). It was discovered that a 0.25″ by 0.5″ hole had formed in the bottom of an oil transit pipeline causing 200,000 gallons of crude oil to leak from the pipeline before it was discovered. This led to the shutdown of the production in order to replace both the leaking pipeline and another parallel pipeline. Upon further investigation, it was believed that MIC was very likely to blame for the failure that had huge financial repercussions. Another serious example of pipeline failure largely attributed to MIC was the rupture of a 30″ natural gas pipeline near Carlsbad, New Mexico, on August 19, 2000 (NTSB 2003). The natural gas that was released as a result of the rupture caught fire and burned for nearly an hour. The gas killed 12 people who happened to be camping under the bridge that supported the pipeline; it also caused damage to two steel suspension bridges that were nearby. The cost was almost $1 billion in damages to the El Paso Natural Gas Company, which operated the pipeline. An additional example of early pipeline failure attributed partly to MIC was first observed in 1997 in a well fluid pipeline off the coast of India that transported well fluid from the well platform to the process platform (Samant et al. 1999). This pipeline failed due to the formation of three pinhole leaks at the bottom of the pipeline only 2.5 years after the pipeline was commissioned. While attempting repairs to the pipeline, several additional leaks were found. It was believed that the infrequent use of biocides and pigging were to blame for the failure. In 1991, MIC was blamed for initial pinhole leaks observed in a new oil and water gathering system in the Lost Hills Field in the San Joaquin Valley, California, after being in commission for only 18 months (DuBose, Fortnum, and Strickland 1996). Additional pinhole leaks were discovered in 1993. Eventually, various pieces of the pipeline had to be replaced costing nearly $800,000.

Another area with many reported cases of MIC is in fire protection systems. Most fire protection systems utilise untreated water that is left stagnant for long periods of time. This creates an environment for bacterial growth and biofilm formation. These microbes cause MIC, which results in pinhole leaks especially at the weld seams and along the bottom of the pipeline. They can also cause biofouling, which can hinder the water from flowing through the pipe in an emergency. These pinhole leaks could easily cause water damage of very expensive equipment or documents. There are also many instances in which the leaking pipelines were replaced by installation companies without any research into the cause of the failure or any suggestions for future prevention. Modifications to a fire protection system such as water recirculation and treatment can help limit and regulate MIC (McReynolds 1998). According to Pope and Pope (2000), MIC can be found in both steel and copper piping of fire protection systems. MIC can sometimes cause failures of these systems within months of initial installation. The buildup of biofilm and other corrosion products in the sprinkler heads can also cause them to fail. One specific example of MIC in fire protection systems was reported at the Nevada Test Site's Device Assembly Facility (DAF) (Edgemon et al. 2010). This facility was initially built in the mid-1980s for underground nuclear weapons testing, but is now used for other purposes. The facility is mostly underground and was built with a fire protection system. The fire protection system was installed with primarily carbon steel coal tar enamel-coated (CTE) pipes, which were welded together. Welding the coated pipes destroyed the integrity of the coating in those areas. This made for an easy place for corrosion, including MIC, to take place. The addition of these defects to the CTE coating also made further delamination of the coating much easier, which only increased the areas susceptible to corrosion. Wall thickness losses ranging from 20% to 80% and tubercles were observed at the welded joints.

Copper is a widely used material in potable water systems and heat exchangers. Copper as a material is relatively corrosion resistant and considered toxic to many microbes in nature (Kikuchi et al. 1999). Despite this, MIC of copper, especially in the form of localised corrosion, has been reported in many parts of the world, especially in low flow and stagnant conditions. One such example of this was the failure of a domestic hot water system installed above the ceiling of a retirement home that failed after only 12 years (Labuda 2003). A portion of the copper pipe had to be removed due to several leaks. Upon examination of the interior of the pipe, a uniform alternating corrosion was observed with average of wall thickness losses alternating between about 28% and 80%. Several through-wall perforations were also found. Another reported case was the repeated failure of the copper heat exchanger coils that networked throughout the campus of a southwestern industrial park built in the early 1980s in the United States (Robinette 2011). The replacement of the leaky coils failed to alleviate the problem, as many of the new replacements failed in as little as 60 days. The situation worsened after one company threatened to leave the industrial park after they observed a coil leak above their critical data processing computers. Further investigation into the matter revealed that MIC was primarily responsible for the pitting corrosion observed in the leaking coils. The coils had to be coated with lepidocrocite (FeOOH) in order to alleviate the problem.

Another area with MIC problems is in wastewater treatment facilities. The MIC problems can occur in many areas of wastewater treatment facilities, especially areas with stagnant conditions and lots of bacteria present, including stainless steels. One example of MIC in this area was the formation a pinhole leak in the rotating disc in a sedimentation tank, which stopped the shaft from rotating (Sreekumari et al. 2004).

MIC has also caused problems in the pulp and paper industry. While problems due to MIC can occur in various areas of the paper machine if the conditions are right, many of the failures are observed in the splash zone at the wet end of the paper machine (Carpén et al. 2001). These components are made of primarily stainless steel. The white waters used in the papermaking process contain many nutrients at perfect temperatures for microbial growth, which create an excellent environment for MIC to occur.

Nuclear power plants have also experienced many failures due to MIC, with some plants spending upwards of $100 million on the problem (Angell 2002). The two main areas that experience MIC in nuclear power plants are heat exchangers and fire protection systems. These systems are vital to the safety of nuclear power plants, and their failure could force a plant to shut down operations. Failure of nickel alloy heat exchanger tubes were reported in the late 1980s at several Ontario Power Generation CANDU power plants less than a year after installation. These power plants use heavy water (D_2O) on the tube side of the heat exchangers and raw water on the shell side. These conditions created an ideal environment for microbes to thrive, especially in the shell side of the heat exchanger, which in addition to being difficult to clean, also reduced fluid velocity.

MIC can also cause problems at hydroelectric power plants. Multiple cases have been reported of MIC of stainless-steel components in hydroelectric power plants (Linhardt and Nichtawitz 2003). An example of this was the pitting corrosion observed after only 18 months of service on the turbine blades and discharge rings of the hydroelectric power plant on the Maas River in the Netherlands. Similar problems were also observed at the hydroelectric power plant on the Mun River in Thailand less than a year after being commissioned.

Besides the severe influence of MIC in industrial fields, the effect of microbes on the corrosion of medical implants has also been investigated. The metal implant surface is continually exposed to the biological environment of the human body, which contains water, organic compounds, microorganisms, enzymes, chlorides, amino acids and so on (Hansen 2008). Several strains of microbes, such as aerobic iron-oxidising bacteria (IOB) and SRB, have been found in the surrounding physiological environment in the human body (Boopathy et al. 2002; Manivasagam et al. 2010; Mylonaki et al. 2006). These bacteria form biofilms, which can cause the biocorrosion of implants and associated infections (Gino et al. 2010; Vianna et al. 2008). Chang et al. (2003) has shown the corrosion of dental metallic materials made of titanium alloys and stainless steel in the presence of *Streptococcus* mutants and the metabolic products, such as lactic acid, carbonic acid and glucan-binding proteins. Moreover, another case from Maruthamuthu et al. (2005) indicated that increasing corrosion currents in nickel, titanium and stainless-steel orthodontic wires were observed in the presence of bacteria such as APB and SRB collected from human saliva. These studies on biocorrosion and electrochemical behaviour of titanium alloys and stainless-steel orthodontic wires in the presence of bacteria reveal that the interactions between the metal implant surface with the microbial cells and their metabolic products have important impacts on implant corrosion propagation and damage (Costerton et al. 2005).

Examples of MIC can be found in a wide range of industries in many parts of the world. MIC is responsible for the degradation of many different metals, including carbon steel, stainless-steel, copper and nickel alloys. Many MIC failures have been reported well before any other corrosion problems were expected, causing a great deal of unanticipated financial losses. MIC even threatens non-metallic materials such as polymers such as polyurethane. A notable case was the polyurethane electrical insulation penetration by filamentous fungi at the Zurich airport in the 1960s. The airport electrical system had to be renovated causing many flight cancellations (Flemming 1998; Pommer and Lorenz 1985). This case belongs to Type III MIC, also known as biodegradation.

15.4 MIC Detection and Mitigation

MIC is mitigated primarily using the old-fashioned "spray and scrub" strategy. Biocides and scrubbing are used to remove microbes. In pipelines, pigs can be deployed to scrub the internal pipeline surface and to spray biocides. In the old days, these robotic devices tended to make a noise that resembled the oink noise made by pigs. Some older pipelines were not designed for pigging due to their tight turns and other pig-unfriendly structures. Because a field system is not sterile, microbes always bounce back after a treatment. Repeated treatments are needed. In the oil and gas industry, the decision for treatment is a big one because downtime, chemicals and labour are costly. Reliable detection is critical to the decision making of whether treatment is required or not. Although various electrical resistance (ER) probes and linear polarisation resistance (LPR) probes have been tried for online biofilm detection, their reliability needs improvement. Recently, Ohio University filed a provisional patent on the utilisation of biofilm electrogenicity for the detection of electrogenic biofilms (Gu 2012c).

15.4.1 Detection of Planktonic and Sessile Microbes in MIC Forensics

It is relatively easy to assay planktonic cells by analysing the microbes in a liquid sample. However, biofilms are the major culprits in MIC. Planktonic cells indicate the possible presence of biofilms in a particular system. Planktonic cell counts are not directly correlated with sessile cell counts. In a pipeline with fluid flow, planktonic cells are relatively evenly distributed while biofilms may be opportunistic and sporadically distributed. This makes the precise detection of biofilms very difficult.

Anaerobic microbes involved in MIC tend to be rather small. For example, *D. vulgaris* is a rod-shaped microbe with a length of roughly 2 μm. They are barely visible under a light microscope at 400× magnification. Fortunately, living *D. vulgaris* cells are motile and this makes counting them under a microscope much easier because artefacts such as small particles are stationary. Using a haemocytometer, cell counts as low as 50,000 cells/mL can be counted. If cell counts are below this, it would be inaccurate because there are not enough cells to be seen on a haemocytometer. To count the cells, one must grow them to increase the number and then back calculate. The so-called most probable number (MPN) enumeration method can be used. It typically uses a liquid culture medium to culture a sample for a period of time. When the cell population increases, they can be enumerated by comparing with a set of standards. For example, SRB cells are cultured with Fe^{2+}-containing medium. The medium will turn black when a sufficient amount of HS^- is secreted. It makes S^{2-} available to form FeS that has a black colour. The time for an inoculated medium to turn black corresponds to the original cell count in the sample used in inoculation. Sessile cells are enumerated the same way, except that they have to be dislodged from a biofilm first. Sometimes, an ultrasound burst lasting 15–30 s can be used to remove sessile cells from a coupon surface used in MIC testing. The brief exposure to ultrasound is not sufficient to kill the sessile cells. Harvested sessile cells are suspended in a liquid and then used as the inoculum for MPN enumeration. The original cell count in the harvested sessile cell sample is used to calculate the sessile cell density (cells/cm²) based on the coupon surface. For cells that are easy to grow on a solid medium in a Petri dish, CFU (colony-forming units) can be used to quantify the cell counts. SRB and other anaerobic cells grow very slowly on a solid medium and they require an anaerobic incubator. Thus, a liquid medium is often used in an anaerobic vial for enumeration. Instead of using the ATCC 1249 liquid medium for MPN, Xu et al. (2012a,b) found that their *D. vulgaris* is more easily enumerated using the Sani-Check SRB test kit (Sani-Check® Product #100, Biosan Laboratories, Warren, MI) that comes with a brush-like dipstick in a solid SRB medium (Figure 15.5). The dipstick is used to dip into a sessile cell suspension before being inserted into the vial

FIGURE 15.5
Sani-Check SRB test kit. (Reproduced with permission from Biosan Laboratories, Inc., Warren, MI, USA.)

for anaerobic incubation at 37°C. No anaerobic chamber is needed for incubation because the solid medium in the sealed vial keeps the oxygen out. The Fe^{2+}-containing solid medium turns black (the FeS film colour) during SRB growth. The time required for the black colour to appear correlates with SRB cell counts (cells/mL) based on the vendor's calibration.

CLSM is a widely used technique in biological science research, which can provide a high-resolution optical image at a selected depth. CLSM is becoming a popular tool for the investigation of biofilm structures due to its non-invasive nature (Mueller et al. 2006). Using a suitable stain, CLSM can even distinguish living and dead cells. Cell density can be quantified with the help of computer software. In laboratory investigations, CLSM can be an effective tool to quantify biocide efficacy. However, when the surviving cell density is too low after a biocide treatment, not enough cells may be visible under the optical microscope. An MPN method is required to increase the cell density.

In MIC forensics, anaerobic biofilms are difficult to preserve once the specimen is exposed to air. Some biofilm cells are not culturable in a lab culture medium. This makes biofilm identification difficult. Luckily, molecular biology techniques do not require living cells because only their molecular markers are assayed. DGGE and fluorescence *in situ* hybridisation (FISH) techniques are used to evaluate the diversity and similarity of the dominant microbes in various microbiological communities collected from the field (Friedrich et al. 1999; Teng et al. 2008). It is helpful to identify these dominant microbial species (Carpén et al. 2012; Teng et al. 2008) when making treatment decisions. The FISH technique employs fluorescent probes that bind to chromosome sections that exhibit a high degree of DNA sequence complementarity. It reveals the presence of specific DNA sequences coded in the probes for the identification of specific microbial species. The DGGE technique can be used to analyse DNA and RNA samples extracted from biofilms. The genetic fingerprints obtained from DGGE are used with the basic local alignment search tool (BLAST). The phylogenetic diversity of corrosive biofilms has been characterised by some researchers using this approach in MIC research (Zhang and Fang 2001).

For biofilm consortia and cultures that have very low concentrations, real-time qPCR can be used. It uses primers to match species-specific sections in DNA molecules extracted from the microbes to fingerprint specific microbial species (Powell et al. 2006). By using a standard curve, the amount of DNA for a microbial species can be correlated with its cell count. Because of its high sensitivity and accuracy, real-time qPCR is widely used to monitor and enumerate microorganisms (Lutterbach et al. 2011; Mitchell et al. 2012). Another popular method of cell enumeration is adenosine triphosphate (ATP) assay. After ATP is dyed with fluorescence, the light emission, which is proportional to the amount of ATP, is measured by ATP photometer. ATP is found in living cells. Thus, ATP assay could be used to quantify the viable microorganisms (Yu et al. 2010). This method, however, is not sensitive enough for low cell counts. Caution must be exercised when using this kind of molecular biology method in biocide efficacy studies because newly killed cells may still have the biomarker molecules (e.g. DNA fragments) intact.

It should be pointed out that the most abundant microbes in a biofilm may not be the MIC culprits as discussed in Section 15.2 of this chapter. In a mixed-culture field biofilm consortium, only one or at most a few bottom layers of microbes are directly involved in Type I MIC. Thus, sampling of the bottom-layer microbes is critical. The existing common mistake of identifying the dominant species of microbes as MIC culprits shows a disregard of MIC mechanisms. The ultimate proof in MIC forensics is the actual corrosion test using the field microbes in a laboratory. After all, a field system is never sterile and you will always find certain microbes. For example, it is nothing unusual to find evidence of the existence of a sulphate reducer in an anaerobic environment when the liquid contains sulphate. Thus, microbiological, molecular biology assay results must be carefully interpreted together with any corrosion test results. It is unfortunate to miss the evidence of MIC, but it can also be very costly to start treatment unnecessarily.

15.4.2 Biocide Mitigation of Biofilms and MIC

Many biocides and biostats (also known as antimicrobials) are used to mitigate MIC and biofouling. They include glutaraldehyde, THPS, bronopol, 2-bromo-2-nitro-1,3-propanediol, tributyl tetradecyl phosphonium chloride, alkyldimethylbenzylammonium chloride, dimethyl benzyl ammonium chloride, chlorine dioxides, calcium hypochlorite, potassium hypochlorite, sodium hypochlorite, dibromonitriloproprionamide (dibromonitrilopropionamide), methylene bisthiocyanate (methylene bis thiocyanate) and 2-(thiocyano-methylthio) benzothiazole (thiocyanomethylthio benzothiazole). Among them, glutaraldehyde and THPS are the top two choices in many large-scale applications, especially in the oil and gas industry because of their broad-spectrum efficacy and good environmental profiles. Both are read-ily biodegradable. THPS is officially a green chemical according to the U.S. Environmental Protection Agency (US EPA 1997). The working mechanism of THPS is less understood than glutaraldehyde, which is a cross-linking agent that reacts with amino and sulphydryl groups in proteins and nucleic acids of microorganisms. Lee et al. (2010) explained that THPS can kill SRB through interfering their energy metabolism. Glutaraldehyde is more effec-tive at higher pH (e.g. pH 8), while THPS prefers slightly acidic pH. Zhao et al. (2009) modelled the degradation of THPS at different temperatures and pH. They found that THPS degrades much faster under alkaline pH. Concentrated THPS is quite acidic. Prolonged exposure causes unacceptable corrosion. A biocide cocktail is often used by mixing a biocide with a sur-factant, a corrosion inhibitor and perhaps a scale remover. A surfactant can help a biocide distribute to a surface (e.g. inner pipe wall) where a biofilm is present, rather than to the body of the pipeline fluid.

Microbes develop resistance after prolonged use of the same biocide. The biocide selectively promotes resistant microbes by killing off susceptible microbes. As a consequence, biocide dosages escalate. In the meantime, environmental regulations in various countries, especially the European community, are becoming more restrictive. Xu et al. (2012c) argued that it is very unlikely that a new blockbuster biocide comparable to glutaraldehyde or THPS will be available on the market any time soon. Thus, they advocate the use of biocide enhancers such as chelators and D-amino acids that can reduce biocide dosages while achieving better biofilm and MIC mitigation outcomes.

EDTA has been used to significantly enhance the eradication of biofilms on catheter surfaces when combined with antibiotics (Raad et al. 2003). This technology was patented by Raad et al. (patent publication numbers: US 6165484 A, US 6509319 B1, US 20110201692 A1). The concern for EDTA being applied in industry is its slow biodegradability and potential accumulation in freshwater systems. Ethylenediaminedisuccinic acid (EDDS) replaced EDTA due to its biodegradability. It was reported by Wen et al. (2010) that the combination of EDDS at a concentration of 2000 (w/w) ppm EDDS with either THPS or glutaraldehyde at (an active) concentration of 30 ppm (w/w) showed a strong inhibition effect against planktonic cells of *D. vulgaris* ATCC 7757 and *Desulfovibrio alaskensis* ATCC 14563 compared with biocide treat-ment alone. EDDS also helped glutaraldehyde in both the prevention of SRB biofilm establishment and the removal of established SRB biofilm test on a C1018 carbon steel surface (Wen et al. 2009).

In oil and gas field applications, methanol has been widely used as a hydrate inhibitor and winterising agent. Wen et al. (2012) and Xu et al. (2012c) found that with the addition of 10% (v/v) methanol, the efficacy of the binary combination of 30 ppm glutaraldehyde and 1000 ppm EDDS was considerably enhanced in the treatment of planktonic SRB growth, prevention of SRB biofilm establishment and mitigation of souring and MIC caused by SRB.

A most recent development is the use of naturally occurring D-amino acids as biocide enhancers. D-Amino acids are far more prevalent than previously thought. They are found in microbes, plants and even humans, playing various biological functions (Xu et al. 2012a). Experimental data reported by Kolodkin-Gal et al. (2010) demonstrated that a low-concentration mixture of D-amino acids (D-tyrosine, D-leucine, D-tryptophan and D-methionine) triggered the disassembly of bacterial biofilms of *B. subtilis*, *Staphylococcus aureus* and *Pseudomonas aeruginosa*. They suggested that these D-amino acids might have substituted the D-alanine terminus in the peptidoglycan molecules in bacterial cell walls. The D-alanine terminus may be a signal molecule for biofilm dispersal. However, when Xu et al. tested D-tyrosine (2012a) and

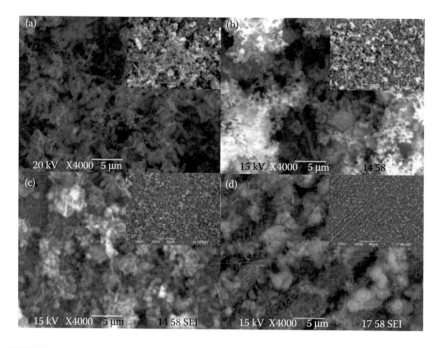

FIGURE 15.6

SEM examination of coupons (initially covered with mature SRB biofilms) after 1-h treatment in a Petri dish containing: (a) a solution of $MgSO_4$ and $(NH_4)_2Fe(SO_4)_2$ at the same concentration as in the ATCC 1249 medium (control), (b) 50 ppm THPS, (c) 100 ppm D-tyrosine, (d) 50 ppm THPS + 1 ppm D-tyrosine, respectively. Scale bars for the small inserted images are 50 μm. (With kind permission from Springer Science+Business Media, *World Journal of Microbiology & Biotechnology*, A synergistic D-tyrosine and tetrakis hydroxymethyl phosphonium sulfate biocide combination for the mitigation of an SRB biofilm, 28(10), 2012a, 3067–3074, Xu, D., Y. Li, and T. Gu.)

TABLE 15.2

Sessile Cell Counts on Coupons (Initially Covered with Mature SRB Biofilms) after Undergoing 1-Hour Treatment

Treatment	After 1-Hour Treatment (cells cm^{-2})
No treatment (control)	$\geq 10^6$
100 ppm D-tyrosine	$\geq 10^5$
50 ppm THPS	$\geq 10^4$
50 ppm THPS + 1 ppm D-tyrosine	<10
100 ppm THPS	<10

Source: From Xu, D., Y. Li, and T. Gu. 2012a. *World Journal of Microbiology & Biotechnology* 28(10): 3067–3074.

D-methionine (2013) individually at much higher concentrations to disperse *D. vulgaris* biofilms on carbon steel surfaces, the results were disappointing. They argued that the *D. vulgaris* biofilm is much more recalcitrant than the biofilms tested by Kolodkin-Gal et al. (2010). They added a 50 ppm (w/w) THPS biocide stress and found that it "convinced" *D. vulgaris* biofilms to disperse. Figure 15.6 shows that 100 ppm D-tyrosine alone did not disperse *D. vulgaris* biofilm while 1 ppm D-tyrosine enhanced the efficacy of 50 ppm THPS greatly by eradicating the mature *D. vulgaris* biofilm (initially grown in the full-strength culture medium prior to biocide treatment) successfully after a 1-h treatment. Their sessile cell count data indicated that a 50 ppm THPS + 1 ppm D-tyrosine combination and 100 ppm THPS alone both achieved a 5-log reduction (Table 15.2). They found that D-methionine at 100 ppm was also an excellent biocide enhancer for 50 ppm THPS in biofilm prevention and eradication, as well as mitigation of MIC pitting by *D. vulgaris*.

15.5 Summary and Perspectives

This chapter discussed the impact of MIC on various metals. Type I MIC mechanisms were dissected in detail because the common SRB attacks on Fe^0 belongs to this category. Bioenergetics and electron transfer theories in bioelectrochemistry were employed to explain the mechanisms. Various MIC examples were given. This chapter also reviewed some common methods to assay planktonic and sessile cell identities and quantities. A new trend in biocide technology using novel biocide enhancers was presented. MIC is becoming more and more important due to reasons such as an aging infrastructure and more frequent practices of water injection in the oil and gas industry. Recent advances in MIC mechanisms and biocide enhancement technologies will contribute to better MIC mitigation. However, there are still many unresolved issues in MIC. More research investment is required.

References

Angell, P. 2002. Use of microbial kinetics to control MIC in the nuclear industry. *Paper presented at NACE Corrosion/2002*, Paper No. 02475, Denver, CO, April 7–11.

Biswas, S., and P. Bose. 2005. Zero-valent iron-assisted autotrophic denitrification. *Journal of Environmental Engineering* 131(8): 1212–1220.

Boopathy, R., M. Robichaux, D. LaFont, and M. Howell. 2002. Activity of sulfate-reducing bacteria in human periodontal pocket. *Canadian Journal of Microbiology* 48(12): 1099–1103.

Bryant, R.D., W. Jansen, J. Boivin, E.J. Laishley, and J.W. Costerton. 1991. Effect of hydrogenase and mixed sulfate-reducing bacterial populations on the corrosion of steel. *Applied and Environmental Microbiology* 57(10): 2804–2809.

Carpén, L., J. Maukonen, and S. Salo. 2012. Accelerated corrosion of carbon steel and zinc in oxygen-free groundwater—Due to the microbiological activity? *Paper presented at NACE Corrosion/2012*, Paper No. C2012-0001397, Salt Lake City, UT, March 11–15.

Carpén, L., T. Hakkarainen, L. Raaska, K. Kujanpää, P. Uutela, K. Mattila, and M. Salkinoja-Salonen. 2001. Simulation of MIC at splash zone areas of the paper industry. *Paper presented at NACE Corrosion/2001*, Paper No. 01245, Houston, TX, March 11–16.

Chang, I.S., H. Moon, O. Bretschger, J.K. Jang, H.I. Park, K.H. Nealson, and B.H. Kim. 2006. Electrochemically active bacteria (EAB) and mediator-less microbial fuel cells. *Journal of Microbiology and Biotechnology* 16(2): 163–177.

Chang, J.-C., Y. Oshida, R.L. Gregory, C.J. Andres, T.M. Barco, and D.T. Brown. 2003. Electrochemical study on microbiology-related corrosion of metallic dental materials. *Bio-Medical Materials and Engineering* 13(3): 281–295.

Costerton, J.W., L. Montanaro, and C.R. Arciola. 2005. Biofilm in implant infections: Its production and regulation. *The International Journal of Artificial Organs* 28(11): 1062–1068.

DuBose, B.W., R.T. Fortnum, and N. Strickland. 1996. A case history of microbiologically influenced corrosion in the Lost Hills Oilfield, Kern County, California. *Paper presented at NACE Corrosion/1996*, Paper No. 96297, Denver, CO, March 24–29.

Edgemon, G.L., G.E.C. Bell, and D.G.L. Baker. 2010. Corrosion of fire protection system lines at the Nevada Test Site's Device Assembly Facility. *Paper presented at NACE Corrosion/2010*, Paper No. 10217, San Antonio, TX, March 14–18.

Flemming, H.-C. 1996. Biofouling and microbiologically influenced corrosion (MIC)—an economic and technical overview. In *Microbial Deterioration of Materials*, eds. E. Heitz, W. Sand, and H.-C. Flemming, 5–14. Heidelberg: Springer.

Flemming, H.-C. 1998. Relevance of biofilms for the biodeterioration of surfaces of polymeric materials. *Polymer Degradation and Stability* 59(1): 309–315.

Friedrich, A.B., H. Merkert, T. Fendert, J. Hacker, P. Proksch, and U. Hentschel. 1999. Microbial diversity in the marine sponge *Aplysina cavernicola* (Formerly *Verongia cavernicola*) analyzed by fluorescence *in situ* hybridization (FISH). *Marine Biology* 134(3): 461–470.

Ghafari, S., M. Hasan, and M.K. Aroua. 2008. Bio-electrochemical removal of nitrate from water and wastewater—A review. *Bioresource Technology* 99(10): 3965–3974.

Gieg, L., T. Jack, and J. Foght. 2011. Biological souring and mitigation in oil reservoirs. *Applied Microbiology and Biotechnology* 92(2): 263–282.

Ginner, J.L., P.J. Alvarez, S.L. Smith, and M.M. Scherer. 2004. Nitrate and nitrite reduction by Fe⁰: Influence of mass transport, temperature, and denitrifying microbes. *Environmental Engineering Science* 21(2): 219–229.

Gino, E., J. Starosvetsky, E. Kurzbaum, R. Armon. 2010. Combined chemical-biological treatment for prevention/rehabilitation of clogged wells by an iron-oxidizing bacterium. *Environmental Science & Technology* 44(8): 3123–3129.

Gu, J.D. 2003. Microbiological deterioration and degradation of synthetic polymeric materials: Recent research advances. *International Biodeterioration & Biodegradation* 52(2): 69–91.

Gu, T. 2011. Methods and Compositions for Applications Related to Microbiologically Influenced Corrosion. Patent Cooperation Treaty Application No.: PCT/US2011/028673.

Gu, T. 2012a. New understandings of biocorrosion mechanisms and their classifications. *Journal of Microbial & Biochemical Technology* 4(4): iii–vi. doi: 10.4172/1948-5948.1000e107.

Gu, T. 2012b. Can acid producing bacteria be responsible for very fast MIC pitting? *Paper presented at NACE Corrosion/2012, Paper No. C2012-0001214*, Salt Lake City, UT, March 11–15.

Gu, T. 2012c. Methods and Devices for the Detection of Biofilms: Patent Cooperation Treaty Application No.: PCT/US2012/052417.

Gu, T., and D. Xu. 2010. Demystifying MIC mechanisms. *Paper presented at NACE Corrosion/2010*, Paper No. 10213, San Antonio, TX, March 14–18.

Gu, T., and D. Xu. 2013. Why are some microbes corrosive and some not? *Paper presented at NACE Corrosion/2013*, Paper No. C2013-0002336, Orlando, FL, March 17–21.

Gu, T., K. Zhao, and S. Nesic. 2009. A practical mechanistic model for MIC based on a Biocatalytic Cathodic Sulfate Reduction (BCSR) theory. *Paper presented at NACE Corrosion/2009*, Paper No. 09390, Atlanta, GA, March 22–26.

Hansen, D.C. 2008. Metal corrosion in the human body: The ultimate bio-corrosion scenario. *The Electrochemical Society Interface* 17(2): 31–34.

Jacobson, G.A. 2007. Corrosion at Prudhoe Bay: A lesson on the line. *Materials Performance* 46(8): 26–34.

Kikuchi, Y., M. Ozawa, T. Sakane, T. Kanamaru, and K. Tohmoto. 1999. Microbiologically influenced corrosion of copper in ground water. *Paper presented at NACE Corrosion/1999*, Paper No. 170, San Antonio, TX, April 25–30.

Kolodkin-Gal, I., D. Romero, S. Cao, J. Clardy, R. Kolter, and R. Losick. 2010. D-amino acids trigger biofilm disassembly. *Science* 328(5978): 627–629.

Labuda, E.M. 2003. Microbiologically induced corrosion of copper piping systems—failure analysis. *Paper presented at NACE Corrosion/2003*, Paper No. 03569, San Diego, CA, March 16–20.

Lee, M.-H.P., S.M. Caffrey, J.K. Voordouw, and G. Voordouw. 2010. Effects of biocides on gene expression in the sulfate-reducing bacterium *Desulfovibrio vulgaris* Hildenborough. *Applied Microbiology and Biotechnology* 87(3): 1109–1118.

Linhardt, P., and A. Nichtawitz. 2003. MIC in hydroelectric powerplants. *Paper presented at NACE Corrosion/2003*, Paper No. 03564, San Diego, CA, March 16–20.

Lutterbach, M.T.S., L.S. Contador, M.M. Galväo, V. de Oliveira, M. Araújo, and G. de S. Pimenta. 2011. Real-time PCR applied to quantification of SRB in bottom water from fuel tanks. *Paper presented at NACE Corrosion/2011*, Paper No. 11233, Houston, TX, March 13–17.

Mah, T.-F.C., and G.A. O'Toole. 2001. Mechanisms of biofilm resistance to antimicrobial agents. *Trends in Microbiology* 9(1): 34–39.

Manivasagam, G, D. Dhinasekaran, and A. Rajamanickam. 2010. Biomedical implants: corrosion and its prevention: A review. *Recent Patents on Corrosion Science* 2, 40–54.

Maruthamuthu, S., A. Rajasekar, S. Sathiyanarayanan, N. Muthukumar, and N. Palaniswamy. 2005. Electrochemical behaviour of microbes on orthodontic wires. *Current Science* 89(6): 988–996.

McReynolds, G. 1998. Prevention of microbiological influenced corrosion in fire protection systems at a semiconductor manufacturing facility. *Paper presented at NACE Corrosion/1998*, Paper No. 286, San Diego, CA, March 22–27.

Mitchell, A., T. Liengen, H. Anfindsen, and S. Molid. 2012. Experience of molecular monitoring techniques in upstream oil and gas operations. *Paper presented at NACE Corrosion/2012*, Paper No. C2012-0001756, Salt Lake City, UT, March 11–15.

Mueller, L.N., J.F. de Brouwer, J.S. Almeida, L.J. Stal, and J.B. Xavier. 2006. Analysis of a marine phototrophic biofilm by confocal laser scanning microscopy using the new image quantification software PHLIP. *BMC Ecology* 6: 1. doi: 10.1186/1472-6785-6-1.

Mylonaki, M., N.B. Rayment, D.S. Rampton, B.N. Hudspith, and J. Brostoff. 2006. Molecular characterization of rectal mucosa-associated bacterial flora in inflammatory bowel disease. *Inflammatory Bowel Diseases* 11(5): 481–487.

National Transportation Safety Board. 2003. *Natural Gas Pipeline Rupture and Fire Near Carlsbad, New Mexico, August 19, 2000.* Pipeline Accident Report NTSB/PAR-03/01. Washington, D.C.

Pommer, E.H., and G. Lorenz. 1985. The behaviour of polyester and polyether polyurethanes towards microorganisms. In *Biodeterioration and Biodegradation of Plastics and Polymers: Proceedings of the Autumn Meeting of the Biodeterioration Society, Held at Cranfield Institute of Technology on 12th to 13th September 1985*, ed. K.J. Seal, 76–86. Issue 1 of Occasional Publication. Kew: Biodeterioration Society.

Pope, D.H., and R. Pope. 2000. Microbiologically influenced corrosion in fire protection sprinkler systems. *Paper presented at NACE Corrosion/2000*, Paper No. 00401, Orlando, FL, March 26–31.

Powell, S.M., S.H. Ferguson, J.P. Bowman, and I. Snape. 2006. Using real-time PCR to assess changes in the hydrocarbon-degrading microbial community in Antarctic soil during bioremediation. *Microbial Ecology* 52(3): 523–532.

Puigdomenech, I., and C. Taxén. 2000. *Thermodynamic Data for Copper Implications for the Corrosion of Copper under Repository Conditions.* Technical Report TR-00-13. Stockholm: Svensk Kärnbränslehantering AB/Swedish Nuclear Fuel and Waste Management.

Raad, I., I. Chatzinikolaou, G. Chaiban, H. Hanna, R. Hachem, T. Dvorak, G. Cook, and W. Costerton. 2003. *In vitro* and *ex vivo* activities of minocycline and EDTA against microorganisms embedded in biofilm on catheter surfaces. *Antimicrobial Agents and Chemotherapy* 47(11): 3580–3585.

Raad, I., H. Hanna, T. Dvorak, G. Chaiban, and R. Hachem. 2007. Optimal antimicrobial catheter lock solution, using different combinations of minocycline, EDTA, and 25-percent ethanol, rapidly eradicates organisms embedded in biofilm. *Antimicrobial Agents and Chemotherapy* 51(1): 78–83.

Robinette, D.J. 2011. A novel adaptation of 1950 vintage technology to prevent pitting of copper heat exchanger tubes in a hydronic system infected by corrosive bacteria. *Paper presented at NACE Corrosion/2011*, Paper No. 11224, Houston, TX, March 13–17.

Samant, A.K., V.K. Sharma, S. Thomas, P.F. Anto, and S.K. Singh. 1999. Investigation of premature failure of a well fluid pipeline in an Indian offshore installation. In *Advances in Corrosion Control and Materials in Oil and Gas Production—Papers from EUROCORR '97 and EUROCORR '98*, eds. L.M. Smith and P.S. Jackman, 180–187. London: IOM Communications.

Sherar, B.W.A., I.M. Power, P.G. Keech, S. Mitlin, G. Southam, and D.W. Shoesmith. 2011. Characterizing the effect of carbon steel exposure in sulfide containing solutions to microbially induced corrosion. *Corrosion Science* 53(3): 955–960.

Sreekumari, K.R., K.R. Akamatsu, T.R. Imamichi, K.R. Hirotani, and Y.R. Kikuchi. 2004. Microbiologically influenced corrosion failure of AISI type 304 stainless steel in a wastewater treatment system. *Paper presented at NACE Corrosion/2004*, Paper No. 04600, New Orleans, LA, March 28-April 1.

Teng, F., Y.T. Guan, and W.P. Zhu. 2008. Effect of biofilm on cast iron pipe corrosion in drinking water distribution system: Corrosion scales characterization and microbial community structure investigation. *Corrosion Science* 50(10): 2816–2823.

Thauer, R.K., E. Stackebrandt, and W.A. Hamilton. 2007. Energy metabolism phylogenetic diversity of sulphate-reducing bacteria. In *Sulphate-Reducing Bacteria: Environmental and Engineered Systems*, eds. L.L. Barton and W.A. Hamilton, 1–37. 1st ed. Cambridge: University Press.

Tiratsoo, J. 2013. *The Ultimate Guide to Unpiggable Pipelines*. Melbourne, Victoria, Australia: Pipelines International.

Uchiyama, T., K. Ito, K. Mori, H. Tsurumaru, and S. Harayama. 2010. Iron-corroding methanogen isolated from a crude-oil storage tank. *Applied and Environmental Microbiology* 76(6): 1783–1788.

US EPA. 1997. THPS Biocides: A New Class of Antimicrobial Chemistry (1997 Designing Greener Chemicals Award). http://www2.epa.gov/green-chemistry/1997-designing-greener-chemicals-award.

Venzlaff, H., D. Enning, J. Srinivasan, K.J. Mayrhofer, A.W. Hassel, F. Widdel, and M. Stratmann. 2013. Accelerated cathodic reaction in microbial corrosion of iron due to direct electron uptake by sulphate-reducing bacteria. *Corrosion Science* 66: 88–96.

Vianna, M.E., S. Holtgraewe, I. Syfart, G. Conrads, and H.P. Horz. 2008. Quantitative analysis of three hydrogenotrophic microbial groups, methanogenic archaea, sulfate-reducing bacteria, and acetogenic bacteria, within plaque biofilms associated with human periodontal disease. *Journal of Bacteriology* 190(10): 3779–3785.

Videla, H.A. 2002. Prevention and control of biocorrosion. *International Biodeterioration & Biodegradation* 49(4): 259–270.

Videla, H.A., and L.K. Herrera. 2005. Microbiologically influenced corrosion: Looking to the future. *International Microbiology* 8(3): 169–180.

Wen, J., D. Xu, T. Gu, and I. Raad. 2012. A green triple biocide cocktail consisting of a biocide, EDDS and methanol for the mitigation of planktonic and sessile sulfate-reducing bacteria. *World Journal of Microbiology and Biotechnology* 28(2): 431–435.

Wen, J., K. Zhao, T. Gu, and I.I. Raad. 2009. A green biocide enhancer for the treatment of sulfate-reducing bacteria (SRB) biofilms on carbon steel surfaces using glutaraldehyde. *International Biodeterioration & Biodegradation* 63(8): 1102–1106.

Wen, J., K. Zhao, T. Gu, and I.I. Raad. 2010. Chelators enhanced biocide inhibition of planktonic sulfate-reducing bacterial growth. *World Journal of Microbiology and Biotechnology* 26(6): 1053–1057.

Von Wolzogen Kühr, C.A.H., and L.S. van der Vlugt. 1934. The graphitization of cast iron as an electrochemical process in anaerobic soils. *Water* 18(16): 147–165.

Xu, D., and T. Gu. 2011. Bioenergetics explains when and why more severe MIC pitting by SRB can occur. *Paper presented at NACE Corrosion/2011*, Paper No. 11426, Houston, TX, March 13–17.

Xu, D., Y. Li, and T. Gu. 2012a. A synergistic D-tyrosine and tetrakis hydroxymethyl phosphonium sulfate biocide combination for the mitigation of an SRB biofilm. *World Journal of Microbiology & Biotechnology* 28(10): 3067–3074.

Xu, D., Y. Li, and T. Gu. 2013. D-methionine as a biofilm dispersal signaling molecule enhanced tetrakis hydroxymethyl phosphonium sulfate mitigation of *Desulfovibrio vulgaris* biofilm and biocorrosion pitting. *Materials and Corrosion* 63: 1–9. doi: 10.1002/maco.201206894.

Xu, D., Y. Li, F. Song, and T. Gu. 2013. Laboratory investigation of microbiologically influenced corrosion of C1018 carbon steel by nitrate reducing bacterium *Bacillus Licheniformis*. *Corrosion Science* 77: 385–390.

Xu, D., J. Wen, W. Fu, T. Gu, and I.I. Raad. 2012b. D-amino acids for the enhancement of a binary biocide cocktail consisting of THPS and EDDS against an SRB biofilm. *World Journal of Microbiology and Biotechnology* 28(4): 1641–1646.

Xu, D., J. Wen, T. Gu, and I. Raad. 2012c. Biocide cocktail consisting of glutaraldehyde, ethylene diamine disuccinate (EDDS), and methanol for the mitigation of souring and biocorrosion. *Corrosion* 68(11): 994–1002.

Yu, P., J. Dillon, and T. Henry. 2010. Identification and control of microbiologically influenced corrosion in a power plant. *Paper presented at NACE Corrosion/2010*, Paper No. 10211, San Antonio, TX, March 14–18.

Zhang, T., and H. Fang. 2001. Phylogenetic diversity of a SRB-rich marine biofilm. *Applied Microbiology and Biotechnology* 57(3): 437–440.

Zhao, K., J. Wen, T. Gu, A. Kopliku, and I. Gruz. 2009. Mechanistic modeling of anaerobic THPS degradation under alkaline condition in the presence of mild steel. *Materials Performance* 48: 62–66.

16

Biosorption of Heavy Metals

Gjergj Dodbiba, Josiane Ponou and Toyohisa Fujita

CONTENTS

16.1 Biosorption: Principles and Methodology

Rapid industrialisation has led to an increase of the amount of heavy metals discharged into the environment. Generally speaking, removal of heavy metals from wastewater is usually achieved by physico-chemical processes, including chemical precipitation, ion exchange, electrochemical treatment, membrane technologies and adsorption to name a few (Kapoor and Viraraghavan 1995; Matheickal and Yu 1996, 1997, 1999; Ahluwalia and Goyal 2007). Each of these methods has its advantages and disadvantages. Chemical precipitation and electrochemical treatments are not always effective, especially when metal ion concentration in an aqueous solution is lower than 50 mg/L (Das et al. 2008). Moreover, such treatments produce large amounts of sludge to be treated with great difficulties. Ion exchange and membrane technologies are complex and expensive (Alluri et al. 2007; Farooq et al. 2010).

Adsorption, on the other hand, is generally recognised as an effective method for the removal of heavy metals from wastewater. Nevertheless, the high cost of activated carbon, which is usually the adsorbent of choice, limits its use in adsorption.

A search for a cost-effective and easily available adsorbent has therefore led to the investigation of biosorption, which is a subcategory of adsorption that involves a solid phase (i.e., the adsorbent or biosorbent, usually a biological material) and a liquid phase (solvent, usually contaminated water) that contain a dissolved species (adsorbate, i.e. a metal ion) to be sorbed (Alluri et al. 2007). In other words, biosorption is a process that utilises inexpensive dead biomass to adsorb heavy metals (Kratochvil and Volesky 1998). It is a process of rapid and reversible binding of ions from aqueous solutions onto functional groups that are present on the surface of biomass (Kapoor and Viraraghavan 1995; Davis et al. 2003; Alluri et al. 2007; Michalak et al. 2013). It is worth noting that biosorption offers several advantages over conventional treatment methods, including cost-effectiveness, efficiency, minimisation of chemical/biological sludge, requirement of additional nutrients, and regeneration of biosorbent with the possibility of metal recovery (Sag 2001).

16.1.1 Sources of Biomass

Materials of agricultural and biological origin, along with industrial by-products, are potential adsorbents (Kapoor and Viraraghavan 1995; Wang and Chen 2006) that are prepared from the naturally abundant or waste biomass of algae, moss, fungi or bacteria (Volesky 1990). The biomass can be cheaply and easily prepared in relatively large quantities, for the biosorption of metals. Biomass can be generated from activated sludge or fermentation wastes (Kapoor and Viraraghavan 1995). Agricultural materials also play an important role as they are widely and easily produced (Farooq et al. 2010). Agricultural products such as wool, rice, straw, coconut husks, peat moss, exhausted coffee (Dakiky et al. 2002), waste tea (Ahluwalia and Goyal 2005), walnut skin, coconut fibre, cork biomass (Chubar et al. 2003), seeds of *Ocimum basilicum* (Melo and D'Souza 2004), defatted rice bran, rice hulls, soya bean hulls and cotton seed hulls (Tarley and Arruda 2004), wheat bran, hardwood (*Dalbergia sissoo*) sawdust, pea pod, cotton and mustard seed cakes (Saeid et al. 2012, 2013) are also proven to be good biomass sources.

A large number of biomass types have already been tested for their metal-binding capability under various conditions, but only those with sufficiently high metal-binding capacity and selectivity are suitable for use in a full-scale biosorption process. The first major challenge is to select the most effective types of biomass from a very extensive list of available and inexpensive biomaterials (Sag 2001). Seaweeds, moulds, fungi, yeasts, bacteria, and crab shells to name a few have been tested for biosorption of, for example, Pb^{2+},

Cu^{2+}, Cd^{2+} and Zn^{2+} ions with very encouraging results (Vieira and Volesky 2000; Davis et al. 2003; Alluri et al. 2007; Das et al. 2008).

16.1.2 Evaluation of Equilibria

Determining metal uptake by an adsorbent or biosorbent is always required in order to express its quality. Generally, the amount of metal ion sorbed at equilibrium (q, mg/g) and the percent removal (R, %) are usually reported. The amount of metal ion sorbed at equilibrium, q, is calculated by using the following equation:

$$q = (C_0 - C) \cdot \frac{V}{M} \, (mg/g) \qquad (16.1)$$

whereas the percent removal is given as

$$R = \frac{C_0 - C}{C_0} \times 100 \, (\%) \qquad (16.2)$$

where C_0 and C are aqueous concentration of adsorbate (in mg/L) before and after adsorption, respectively), V is the volume of the sample (in L), and M is the mass of the dry adsorbent (in g). Between the two, q is considered a better parameter to express and compare the capacities of different biomass types.

The equilibrium relationship between q and the adsorbate concentration C in solution is called an adsorption isotherm. In order to collect the data to construct the isotherm, different amounts of the adsorbent are added in several samples containing an element of known concentration and mechanically shaken for a certain period of time. After filtering, the remaining concentration of ion in the solution is measured. By knowing the initial and the remaining concentrations of ions, q is calculated (Equation 16.1). The data can then be analysed generally using the Langmuir (Langmuir 1916, 1917), Freundlich (Freundlich 1906) or Redlich–Peterson (Redlich and Peterson 1959) models to name a few.

The Langmuir model was originally formulated for gas adsorption on homogeneous surfaces (Langmuir 1916). The model is characterised by linear adsorption at low surface coverage, which becomes non-linear as the adsorption sites approach saturation. In other words, the Langmuir model (Equation 16.3) assumes that the surface of the adsorbent is homogeneous and solute uptake occurs by monolayer adsorption, since the adsorption energy is uniform for each site:

$$q = q_m \frac{K_L C}{1 + K_L C} \qquad (16.3)$$

where q_m is the maximum capacity of the adsorbent for the adsorbate (mass adsorbate/mass adsorbent, mg/g), and K_L (L/mg) is a measure of the affinity of the adsorbate for the adsorbent.

The linearised form of the Langmuir isotherm is given by Equation 16.4 (Langmuir 1916). The maximum capacity q_m and K_L are then determined directly from the intercept and the slope of the straight line of the linearised form of the isotherm.

$$\frac{1}{q} = \frac{1}{q_m K_L}\left(\frac{1}{C}\right) + \frac{1}{q_m} \tag{16.4}$$

Next, $1/q$ is plotted versus $1/C$ and the data have then been extrapolated using the least-squared linear regression in order to calculate the maximum capacity of the adsorbent (q_m).

The Freundlich isotherm (Freundlich 1906), on the other hand, is the earliest known relationship, which assumes that the surface sites of the adsorbent have different binding energies. The Freundlich expression is an empirical equation describing adsorption onto heterogeneous surfaces and is expressed as

$$q = K_F C^{1/n} \tag{16.5}$$

where K_F (mg/g), which indicates the adsorption capacity, and n are both empirical constants. The value of n indicates a favourable adsorption for $1 < n < 10$. The linearised form of Freundlich isotherm is given as

$$\ln q = \ln K_F + \frac{1}{n}\ln C \tag{16.6}$$

Both constants K_F and n are then determined directly from the intercept and the slope of the straight line of the linearised form of the isotherm (Equation 16.6).

In addition, the Redlich–Peterson isotherm model (Redlich and Peterson 1959) includes features of both Langmuir and Freundlich isotherms. It is expressed as

$$q = \frac{K_R \cdot C}{1 + a \cdot C^g} \tag{16.7}$$

where K_R (L/g) and a are the isotherm constants, whereas g is an exponent, which varies from 0 to 1. All three constants are usually evaluated using the generalised reduced gradient (GRG2) non-linear optimisation code.

Regarding the adsorption kinetic, on the other hand, many models have been reported (Yiacoumi and Tien 1995). The pseudo-second-order kinetic model (Ho and McKay 1999) and the intraparticle diffusion model (Crank 1956) are among the most widely applied. The pseudo-second-order equation assumes that the adsorption behaviour is controlled by a second-order reaction. The equation is expressed as follows:

$$\frac{dq_t}{dt} = k(q - q_t)^2 \tag{16.8}$$

where q_t is the amount of metal ion sorbed at time t (mg/g) and k is the rate constant of the pseudo-second-order kinetic model of adsorption (g/mg · min). Integrating Equation 16.8 for the boundary conditions $(t = 0)$ to $(t = t)$ and $(q_t = 0)$ to $(q_t = q_t)$ gives

$$\frac{1}{(q_e - q_t)} = \frac{1}{q_e} + kt \tag{16.9}$$

The linearised form of Equation 16.9 is given by

$$\frac{t}{q_t} = \frac{1}{kq_e^2} + \frac{1}{q_e}t \tag{16.10}$$

If pseudo-second-order kinetic is applicable, the plot of t/q_t versus t can yield a straight line, from which q_e and k can then be determined from the slope and the intercept of the line. In addition, Equation 16.9 can also be rearranged to obtain

$$q_t = \frac{q_e^2 kt}{(1 + q_e kt)} \tag{16.11}$$

The diffusion models, on the other hand, are particularly important where ion exchange or ionic bonding is not as prevalent as in chemisorption processes. According to the theory proposed by Weber and Morris (1963), a functional relationship common to most treatments of intraparticle diffusion is that uptake q_t varies almost with the square root of time $t^{0.5}$. The linear form of the intraparticle diffusion equation is given as

$$q_t = k_i t^{0.5} \tag{16.12}$$

where k_i is the intraparticle diffusion rate constant (mg/g · min$^{0.5}$). If only the intraparticle diffusion takes place, the plot of adsorption capacity q_t at time t versus $t^{0.5}$ would yield a straight line. The k_i value is then calculated from the slope of the straight line. The plot might have more than one 'linear segments' or 'portions'. The initial steep 'segment' is attributed to the diffusion of the adsorbate through the boundary layer diffusion of solute molecules or the external surface of the adsorbent. The second 'segment' describes the gradual adsorption stage, whereas the third 'segment' of the plot is attributed to the final equilibrium stage where intraparticle diffusion starts to slow down due to the extremely low concentration of the adsorbate in the solution.

16.2 Current Trends in Biosorption

A successful biosorption process starts with selecting the type of biomass. In addition, pretreatment and immobilisation are usually done to increase the efficiency of the metal uptake. The adsorbed metal is removed by a desorption process and the biosorbent can be reused for further treatments (Alluri et al. 2007).

It is worth noting that the assessment of the metal-binding capacity of some types of biomass has gained momentum since 1995, when Volesky and Holan (1995) summarised most of the more consistent attempts to identify metal-sorbing biomass types in the microbial world. In addition, there are several other reviews (Kapoor and Viraraghavan 1995; Volesky and Holan 1995; Kratochvil and Volesky 1998; Sag 2001; Davis et al. 2003; Gadd 2009; Michalak et al. 2013) that have summarised various aspects of biosorption and provide some very useful information.

The results of these studies vary widely because of the absence of uniform methodology, that is, results have been reported in different units and in many different ways, often making quantitative comparison not an easy task. However, a general conclusion can be drawn that there are potent biosorbent materials among easily available biomass types from all three groups: algae, fungi and bacteria (Volesky and Holan 1995).

16.2.1 Heavy Metals

Heavy metals are usually classified into three categories: toxic metals (such as Hg, Cr, Pb, Zn, Cu, Ni, Cd, As, Co, Sn, etc.), precious metals (such as Pd, Pt, Ag, Au, Ru, etc.) and radionuclides (such as U, Th, Ra, Am, etc.) (Donati 2009). Broadly speaking, heavy metals occur in immobilised form in sediments and as ores in nature. However, owing to various human activities such as ore mining and industrial processes, the natural biogeochemical cycles are disrupted, causing increased deposition of heavy metals in terrestrial and aquatic environment. Release of these pollutants without proper treatment poses a significant threat to both environment and public health, as they are both non-biodegradable and toxic (Volesky 1990; Volesky and Holan 1995; Wang and Chen 2006; Alluri et al. 2007).

Heavy metal removal by biosorption has been extensively investigated during the last several decades. Several researchers have reviewed different aspects of heavy metal biosorption (Volesky 1990; Fujita et al. 1992; Kapoor and Viraraghavan 1995, 1997, 1998; Volesky and Holan 1995; Kratochvil and Volesky 1998; Seki and Suzuki 1998; Seki et al. 2000; Davis et al. 2003; Ahluwalia and Goyal 2007; Ponou et al. 2011, 2014; Paudyal et al. 2012), and identified some potential biomaterials with high metal-binding capacity, including marine algae (e.g. *Sargassum natans*), bacteria (e.g. *Bacillus subtilis*), fungi (e.g. *Rhizopus arrhizus*), yeast (e.g. *Saccharomyces cerevisiae*) and waste microbial biomass from fermentation and food industry (Das et al. 2008). Sag (2001), for instance, compared biosorption of heavy metal ions on various free and immobilised fungal cells in different reactor systems and found out that biosorption preference for metals decreases in the following order: Cd > Co > Cr > Au = Cu > Fe > Ni > Th > U > Pb > Hg > Zn.

16.2.2 Rare Earths

Several other researchers, on the other hand, have investigated the toxicity of rare earth elements (REE) (Hebert 1907; Graca et al. 1962, 1964; Bruce et al. 1963; Zhang et al. 2000; Filipi et al. 2007), which are often used in industry for the production of glass additives, fluorescent materials, catalysts, ceramics, lighters, superconductors, magnets or condensers (Andres et al. 2003). As early as 1939, Botti studied the adsorption on activated carbon of cerium (Ce), a member of the rare earth group, and showed that adsorption of the rare earths on organic complexes does occur (Vickery 1946). In 1991, Bayer (1991) showed that lanthanum, europium and terbium were accumulated during growth, in the space between the inner and outer membranes of the cell envelope (periplasmic space) of *Escherichia coli*. Later, Diatloff et al. (1995a–c), for instance, showed that REE lanthanum and cerium could have a negative effect on the root elongation of corn. In 2003, Andres (2003) summarised several published lanthanide biosorption experiments, and reported that the sorption capacities ranged between 2 and 1100 µmol/g depending on the microorganism being used and the experimental conditions. Inoue and Alam (2013) prepared two different types of adsorption gels from biomass wastes. The first gel was produced from persimmon peel rich, and was employed for the removal of uranium (U(VI)) and thorium (Th(IV)) from rare earths, whereas the second gel was prepared from chitosan, a basic polysaccharide, produced from shells of crustaceans (i.e., wastes generated in marine product industry), and was used for the adsorption of rare earths. Generally, the experiments took place at the pH of hydroxide precipitation, for instance, 7.8 for lanthanum (La), 6.8 for europium (Eu) and 6.30 for lutetium (Lu) (Andres et al. 2003).

16.3 Biosorption of Cr Using Pineapple Leaves: A Case Study

It is known that many microorganisms can remarkably concentrate heavy metal ions or economically recover metals from dilute streams. If the organisms are bacteria or unicellular algae, they may not be easily harvested and are, thus, difficult to handle after heavy metal accumulation. One potential solution to this problem has been the development of porous polymeric beads containing immobilised microbial material. The primary problem of this technique is the loss of active organism sites in preparation of the beads.

There has been steady progress in studying the biosorption of heavy metals, resulting in the identification of some biomass types that show a very promising uptake of metallic ions. The focus is on several fungal strains generated as a by-product of fermentation processes and several species of abundant brown marine algae (Volesky and Holan 1995).

Wang and Chen (2006) reviewed the biosorption of heavy metals by *S. cerevisiae* and compared metal-binding capacity for various heavy metals under different conditions. They assessed the isotherm equilibrium as well as kinetic models and reported that lead and uranium could be removed from dilute solutions more effectively in comparison with other metals.

Kuyucak and Volesky (1988) reported that algal biomass of *S. natans* and *Ascophyllum nodosum* outperformed ion exchange resins in sequestering gold and cobalt, respectively, from solutions. They indicated that non-living biomass of *S. cerevisiae* and *R. arrhizus* exhibited higher metal uptake capacity than the living biomass for the uptake of copper, zinc, cadmium and uranium. The solution pH affected the metal uptake capacity of the biomass whereas the equilibrium biosorption isotherms were independent of the initial concentration of the metal in the solution.

Mukhopadhyay et al. (2011) compared the reported literature for the feasibility of *Aspergillus niger* for efficient removal of toxic trace metals from industrial wastewater. The sorption and regeneration process data are compared and evaluated using batch and column experimentation.

Generally speaking, the important factors that influence the biosorption processes are (i) pH, (ii) biomass concentration in solution, (iii) initial concentration of metals and (iv) temperature (Mosbah and Sahmoune 2013). Among all the factors, pH seems to be the most important factor during the process of biosorption (Friis and Myerskeith 1986; Galun et al. 1987; Das 2012). Biomass concentration strongly affects the specific metal uptake. Lower values of biomass concentration causes interference between the binding sites (Fourest and Roux 1992), attributing the responsibility of the specific uptake decrease due to metal concentration shortage in the solution.

Among the heavy metals, chromium is widely distributed in the Earth's crust and exits in oxidation states of +2 to +6, where Cr(VI) is the most toxic form with carcinogenic and mutagenic effects. However, drinking-water intake can be the toxicity factor when the total chromium content exceeds

25 µg/L. Thus, the World Health Organization has established 0.005 mg/L as a maximum allowable concentration of Cr(VI) in drinking water (World Health 1996). In view of the toxic nature of Cr(VI) in wastewater, the treatment process is proving to be indispensable. In the present work, in order to carry out the structure of pineapple leaves with heating, the material is carbonised at 623 or 723 K and used as adsorbents as well as the dried one for removing Cr(VI) anions from aqueous solution. Parametric factors such as pH, contact time and solution temperature are investigated to point out the mechanism of the system.

16.3.1 Synthesis of Adsorbent

Pineapple leaves were washed thoroughly with distilled water and then dried at 353 K for at least 48 h to remove the adherent moisture. After drying, the material was heated in the electric furnace, for 1 h under N_2 atmosphere (gas flow: 15 L/min) to avoid any oxidation during the carbonisation and then desiccated until it was completely cooled. The sample was then sieved to obtain a fraction of fine particles all passing 0.6 mm.

16.3.2 Batch Adsorption Studies

Except where noted, a stock solution containing 1000 mg/L of $K_2Cr_2O_7$ was diluted with deionised water to adjust the concentration of Cr(VI) at 5 mg/L. The pH of the solution was adjusted using 1 mol/L of HCl or NaOH aqueous solution. The adsorbent was then mixed with Cr(VI) solution, resulting in a concentration of 10 g/L. The mixture was then shaken in a bath at 293 K and 196 rpm. The sample was then filtered using a 0.2-µm filter and the filtrate was then analysed for Cr(VI). Next, the amount of Cr(VI) ions adsorbed q (mg/g) was calculated by using Equation 16.1, whereas the percent removal R was calculated using Equation 16.2. All the investigations were done at least three times to avoid any discrepancy in experimental data and the relative standard deviation was estimated at $1 \pm 0.5\%$.

16.3.3 Equipment

Thermogravimetric/differential thermal analysis (Thermo plus EVO TG 8120/Rigaku) was first carried out under air atmosphere. The carbonisation of pineapple leaves was carried out by using an electric furnace (HPM-0G/AS ONE). The specific surface area and pore diameter were measured using pore size analyser (NOVA 2200 e/TFII) by an adsorption–desorption process with N_2. The zeta potential was figured out by electrophoresis analysis (LS-2000E/Photal OTSUKA Electronic). The reciprocal shaking bath (FTB-01/AS ONE) was used for the batch experiment process. The pH of the solution was measured with a pH meter (HM-25R/TDA DK) previously calibrated. The Cr(VI) content in treated solution was analysed using inductively coupled plasma–optimal emission spectrometer (ICP-OES) (PerkinElmer/ Optima 5300). In order to investigate the structure of the adsorbents and find out the functional group involved in the adsorption process, a Fourier transform infrared spectrometer (FT-IR 4100/DR410/ATR PRO 450-S) was also used.

16.3.4 Quantitative and Structure Analysis of the Adsorbent

The recorded thermograms of raw pineapple leaves and the FT-IR spectra of pineapple leaves carbonised at 623 or 723 K as well as the ones dried at 353 K are shown in Figure 16.1. The aim of the analysis is to find any possible change in the structure of the material by heating as well as quantifying the weight loss. In the thermalgravimetric analysis (TGA) thermogram, around 383 K, the sample loses 65% of its initial weight resulting of the moisture content; however, the remaining weight after 823 K was 2% representation of the ash content of the material. Although, in the differential thermal analysis (DTA) thermogram, the heat was absorbed by the materials and two main peaks were observed. The first peak was around 353 K.

In order to investigate the mechanism governing the process at 353 K, raw pineapple leaves were heated at 353 K for more than 48 h and were subject to FT-IR analysis (Figure 16.2). The spectrum of 353 K dried pineapple leaves indicates the dominant peaks at wave numbers 1043.54, 2916.59 and 3350.60 cm^{-1} for SiO, unsaturated alkene and surface O–H stretching,

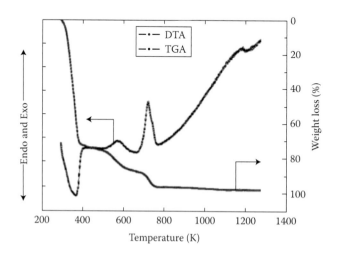

FIGURE 16.1

Thermograms for raw pineapple leaves. (*Note:* No carbonisation of the material was carried out prior to the analysis; heating rate: 283 K/min, reactive gas: air.) (Reprinted from *Chemical Engineering Journal* 172(2–3), Ponou, J. et al., Sorption of Cr(VI) anions in aqueous solution using carbonized or dried pineapple leaves, 906–13, Copyright 2011, with permission from Elsevier.)

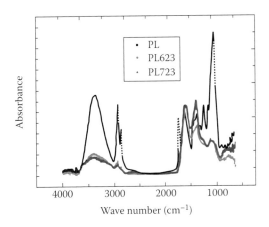

FIGURE 16.2
(**See colour insert**.) FT-IR spectra of pineapple leaves carbonised at 723 K (PL723) or 623 K (PL623) and dried pineapple leaves (PL). (Reprinted from *Chemical Engineering Journal* 172(2–3), Ponou, J. et al., Sorption of Cr(VI) anions in aqueous solution using carbonized or dried pineapple leaves, 906–13, Copyright 2011, with permission from Elsevier.)

respectively. Moreover, 353 K dried pineapple leaves are mainly composed of cellulose, lingnocellulose-lignin and amorphous silicate.

The second peak observed between 623 and 723 K is the reason for setting the carbonisation temperature. The dominant peaks of 623 or 723 K carbonised pineapple leaves are found at wave numbers 1392.96 and 1613.64 cm^{-1} that represent carboxylate anion and unsaturated alkene stretching, respectively. Further analysis on FT-IR shows that carbonised materials are mainly composed of humic acid and fumarodinitrile.

The different properties of the synthesised adsorbents are listed in Table 16.1. The specific surface area and average pore diameter increased with an increase in the temperature of carbonisation, suggesting that heating enhances the available adsorption site.

16.3.5 Eh–pH Diagram

Eh–pH diagram shows the thermodynamic stability areas of chromium species in an aqueous solution. The upper and lower stability limits of water are shown in the diagrams with dotted lines. Figure 16.3 was obtained by selected Cr as the main element and K, H and O as others at 293 K. Chromium molality was set at 1.7×10^{-5} mol/L, similar to chromium concentration in the experimental solution and the pressure sets at 1 Pa. The diagram indicates all the possible species involved in the reaction. It can be seen that the pH not only influences the properties of the sorbent surface but also affects metal speciation in the solution.

TABLE 16.1

Physical and Chemical Characteristics of Dried Pineapple Leaves (PL),
623 K Carbonised Pineapple Leaves (PL623) and 723 K Carbonised Pineapple
Leaves (PL723)

Adsorbents	Specific Surface Area (m²/g)	Average Pore Diameter (nm)	Moisture Content (%)	Ash Content (%)	Point of Zero Charge
PL723	374.90	3.66	1	16	2.1
PL623	44.08	0.93	4	13	2.2
PL	6.84	0.29	5	6	2.8

Source: Reprinted from *Chemical Engineering Journal* 172(2–3), Ponou, J. et al., Sorption of Cr(VI) anions in aqueous solution using carbonized or dried pineapple leaves, 906–13, Copyright 2011, with permission from Elsevier.

16.3.6 Adsorption Kinetics

Figure 16.4 depicts the intraparticle diffusion (Equation 16.12) plots for the adsorption of Cr(VI) on the selected adsorbents. The related plots corresponding to the adsorbents figured out three linear portions of segment that indicated three phenomena governing the adsorption process. The first section of the plot indicates the so-called film diffusion where the adsorbate moves from

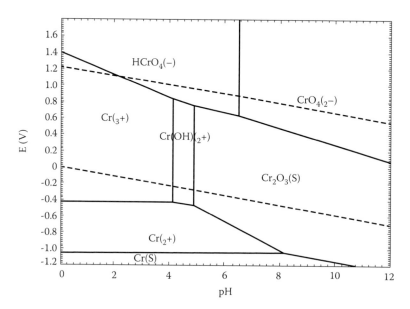

FIGURE 16.3

Eh–pH diagram of the system Cr, K and H_2O (temperature: 293 K; Cr concentration: 1.7×10^{-5} mol/L). (Reprinted from *Chemical Engineering Journal* 172(2–3), Ponou, J. et al., Sorption of Cr(VI) anions in aqueous solution using carbonized or dried pineapple leaves, 906–13, Copyright 2011, with permission from Elsevier.)

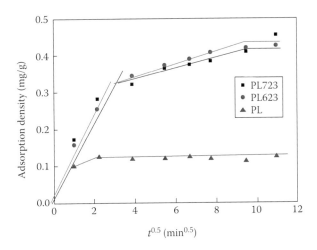

FIGURE 16.4
Intraparticle diffusion plot of removal of Cr(VI) by pineapple leaves carbonised at 723 K (PL723) or 623 K (PL623) and dried pineapple leaves (PL) (experimental conditions: initial pH = 2; initial concentration: [Cr(VI)] = 5 mg/L; adsorbent dosage: 10 g/L; temperature: 263 K). (Reprinted from *Chemical Engineering Journal* 172(2–3), Ponou, J. et al., Sorption of Cr(VI) anions in aqueous solution using carbonized or dried pineapple leaves, 906–13, Copyright 2011, with permission from Elsevier.)

the liquid phase to the adsorbent surface. The second linear segment corresponds to intraparticle diffusion. Adsorption on an activated group occurs in the third linear step. The intraparticle diffusion model also indicated that the system is governed by both film diffusion and intraparticle diffusion.

Figure 16.5 shows the plots of t/q_t versus t of the selected adsorbents at the experimental conditions. The results suggested that the adsorption of Cr(VI) onto selected adsorbents follows the pseudo-second-order kinetic model. It was found that the percent removal of 90.1% and the maximum adsorption capacity of 18.77 mg/g were achieved when pineapple leaves carbonised at 723 K were used as adsorbent. Consistent with an endothermic reaction, the mechanism was found to be chemisorption and the adsorption was well described by the pseudo-second-order kinetic model.

16.4 Future Prospects and Directions

Biosorption no doubt presents an economic alternative for today's mining, mineral, and wastewater treatment industries at a time when high-grade mineral resources are being depleted, energy costs are increasing and adverse environmental effects are becoming more apparent as a result of conventional technologies (Farooq et al. 2010). The major advantages of biosorption over

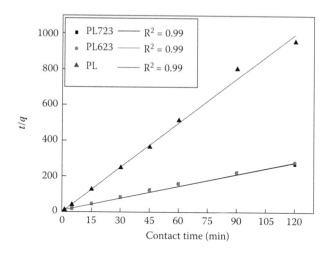

FIGURE 16.5

Pseudo-second-order kinetic model for adsorption of Cr(VI) by using pineapple leaves carbonised at 723 K (PL723) or 623 (PL623) and dried pineapple leaves (PL); (experimental conditions: initial pH = 2; initial concentration: [Cr(VI)] = 5 mg/L; adsorbent dosage: 10 g/L; temperature: 293 K). (Reprinted from *Chemical Engineering Journal* 172(2–3), Ponou, J. et al., Sorption of Cr(VI) anions in aqueous solution using carbonized or dried pineapple leaves, 906–13, Copyright 2011, with permission from Elsevier.)

conventional treatment methods include low cost, high efficiency, minimisation of chemical and biological sludge, regeneration of biosorbent and possibility of metal recovery (Kratochvil and Volesky 1998). The disadvantages of biosorption are (i) early saturation, that is, when metal interactive sites are occupied, metal desorption is necessary prior to further use; (ii) the potential for biological process improvement (e.g. through genetic engineering of cells) is limited because cells are not metabolising and (iii) there is no potential for biologically altering the metal valency state (Ahluwalia and Goyal 2007).

The scope of biosorption will broaden in the coming years. The use of biomass as an adsorbent for heavy metal pollution control can generate revenue for industries presently wasting the biomass and at the same time ease the burden of disposal costs associated with the waste biomass produced (Kapoor and Viraraghavan 1995). Undoubtedly, a range of new biosorbents will be introduced that will be better able to adsorb valuable or toxic elements. This would be of considerable value to plant operators by assisting them in improving the recovery of the elements.

References

Ahluwalia, S. S. and D. Goyal. Removal of heavy metals by waste tea leaves from aqueous solution. *Engineering in Life Sciences* 5(2), Apr 2005: 158–62.

Ahluwalia, S. S. and D. Goyal. Microbial and plant derived biomass for removal of heavy metals from wastewater. *Bioresource Technology* 98(12), Sep 2007: 2243–57.

Alluri, H. K., S. R. Ronda, V. S. Settalluri, J. Singh, S. V. Bondili, and P. Venkateshwar. Biosorption: An eco-friendly alternative for heavy metal removal. *African Journal of Biotechnology* 6(25), Dec 28 2007: 2924–31.

Andres, Y., A. C. Texier, and P. Le Cloirec. Rare earth elements removal by microbial biosorption: A review. *Environmental Technology* 24(11), Nov 2003: 1367–75.

Bayer, M. E. and M. H. Bayer. Lanthanide accumulation in the periplasmic space of *Escherichia coli*-B. *Journal of Bacteriology* 173(1), Jan 1991: 141–49.

Bruce, D. W., B. E. Hietbrink, and K. P. Dubois. Acute mammalian toxicity of rare earth nitrates and oxides. *Toxicology and Applied Pharmacology* 5(6), 1963: 750.

Chubar, N., J. R. Carvalho, and M. J. N. Correia. Cork biomass as biosorbent for Cu(II), Zn(II) and Ni(II). *Colloids and Surfaces A—Physicochemical and Engineering Aspects* 230(1–3), Dec 10 2003: 57–65.

Crank, J. *The Mathematics of Diffusion*. Oxford: Clarendon Press, 1956.

Dakiky, M., M. Khamis, A. Manassra, and M. Mer'eb. Selective adsorption of chromium(VI) in industrial wastewater using low-cost abundantly available adsorbents. *Advances in Environmental Research* 6(4), Oct 2002: 533–40.

Das, N. Remediation of radionuclide pollutants through biosorption—An overview. *Clean-Soil Air Water* 40(1), Jan 2012: 16–23.

Das, N., R. Vimala, and P. Karthika. Biosorption of heavy metals—An overview. *Indian Journal of Biotechnology* 7(2), Apr 2008: 159–69.

Davis, T. A., B. Volesky, and A. Mucci. A review of the biochemistry of heavy metal biosorption by brown algae. *Water Research* 37(18), Nov 2003: 4311–30.

Diatloff, E., F. W. Smith, and C. J. Asher. Rare-earth elements and plant-growth. 1. Effects of lanthanum and cerium on root elongation of corn and mungbean. *Journal of Plant Nutrition* 18(10), 1995a: 1963–76.

Diatloff, E., F. W. Smith, and C. J. Asher. Rare-earth elements and plant-growth. 2. Responses of corn and mungbean to low concentrations of lanthanum in dilute, continuously flowing nutrient solutions. *Journal of Plant Nutrition* 18(10), 1995b: 1977–89.

Diatloff, E., F. W. Smith, and C. J. Asher. Rare-earth elements and plant-growth. 3. Responses of corn and mungbean to low concentrations of cerium in dilute, continuously flowing nutrient solutions. *Journal of Plant Nutrition* 18(10), 1995c: 1991–2003.

Donati, E. R. *Biohydrometallurgy: A Meeting Point between Microbial Ecology, Metal Recovery Processes and Environmental Remediation (IBS 2009): Selected, Peer Reviewed Papers from the 18th International Conference IBS*, 13–17 September 2009, Bariloche, Argentina. Stafa-Zurich; UK: Trans Tech Pub., 2009.

Farooq, U., J. A. Kozinski, M. A. Khan, and M. Athar. Biosorption of heavy metal ions using wheat based biosorbents—A review of the recent literature. *Bioresource Technology* 101(14), Jul 2010: 5043–53.

Filipi, R., K. Nesmerak, M. Rucki, Z. Roth, I. Hanzlikova, and M. Tichy. Acute toxicity of rare earth elements and their compounds. *Chemicke Listy* 101(10), 2007: 793–98. [In Czech].

Fourest, E. and J. C. Roux. Heavy-metal biosorption by fungal mycelial by-products—Mechanisms and influence of pH. *Applied Microbiology and Biotechnology* 37(3), Jun 1992: 399–403.

Freundlich, H. M. F. Adsorption in solids. *Zeitschrift fuer Physikalische Chemie, Stoechiometrie und Verwandtschaftslehre* 57, 1906: 385–470.

Friis, N. and P. Myerskeith. Biosorption of uranium and lead by *Streptomyces longwoodensis*. *Biotechnology and Bioengineering* 28(1), Jan 1986: 21–28.

Fujita, T., E. Kuzuno, and M. Mamiya. Adsorption of metal ions by river algae. [In Japanese.] *Journal of the Mining and Materials Processing Institute of Japan* 108(2), 1992: 123–128.

Gadd, G. M. Biosorption: Critical review of scientific rationale, environmental importance and significance for pollution treatment. *Journal of Chemical Technology & Biotechnology* 84(1), 2009: 13–28.

Galun, M., E. Galun, B. Z. Siegel, P. Keller, H. Lehr, and S. M. Siegel. Removal of metal ions from aqueous solutions by *Penicillium* biomass: Kinetic and uptake parameters. *Water, Air, and Soil Pollution* 33(3–4), 1987: 359–71.

Graca, J. G., J. B. Feavel, and F. C. Davison. Comparative toxicity of stable rare-earth compounds. *Archives of Environmental Health* 8(4), 1964: 555.

Graca, J. G., J. B. Feavel, and F. C. Davison. Comparative toxicity of stable rare-earth compounds. 2. Effect of citrate and edetate complexing on acute toxicity in mice and guinea-pigs. *Archives of Environmental Health* 5(5), 1962: 437.

Hebert, A. The toxicity relative to slat of chromium, aluminium and magnesium, compared with the analogous properties of rare earth. *Comptes Rendus Hebdomadaires Des Seances De L Academie Des Sciences* 145, 1907: 337–40. [In French.]

Ho, Y. S. and G. McKay. Pseudo-second order model for sorption processes. *Process Biochemistry* 34(5), Jul 1999: 451–65.

Inoue, K. and S. Alam. Refining and mutual separation of rare earths using biomass wastes. *Jom* 65(10), 2013: 1341–1347.

Kapoor, A. and T. Viraraghavan. Biosorption of heavy metals on *Aspergillus niger*: Effect of pretreatment. *Bioresource Technology* 63(2), Feb 1998: 109–13.

Kapoor, A. and T. Viraraghavan. Fungal biosorption—An alternative treatment option for heavy metal bearing wastewaters: A review. *Bioresource Technology* 53(3), 1995: 195–206.

Kapoor, A. and T. Viraraghavan. Heavy metal biosorption sites in *Aspergillus niger*. *Bioresource Technology* 61(3), Sep 1997: 221–27.

Kratochvil, D. and B. Volesky. Advances in the biosorption of heavy metals. *Trends in Biotechnology* 16(7), Jul 1998: 291–300.

Kuyucak, N. and B. Volesky. Biosorbents for recovery of metals from industrial solutions. *Biotechnology Letters* 10(2), 1988: 137–42.

Langmuir, I. The constitution and fundamental properties of solids and liquids. Part I: Solids. *Journal of the American Chemical Society* 38, Jul–Dec 1916: 2221–95.

Langmuir, I. The constitution and fundamental properties of solids and liquids. Part II: Liquids. *Journal of the American Chemical Society* 39, Jul 1917: 1848–906.

Matheickal, J. T. and Q. M. Yu. Biosorption of lead from aqueous solutions by marine algae *Ecklonia radiata*. *Water Science and Technology* 34(9), 1996: 1–7.

Matheickal, J. T. and Q. Yu. Biosorption of lead(II) from aqueous solutions by *Phellinus badius*. *Minerals Engineering* 10(9), Sep 1997: 947–57.

Matheickal, J. T. and Q. M. Yu. Biosorption of lead(II) and copper(II) from aqueous solutions by pre-treated biomass of Australian marine algae. *Bioresource Technology* 69(3), Sep 1999: 223–29.

Melo, J. S. and S. F. D'Souza. Removal of chromium by mucilaginous seeds of *Ocimum basilicum*. *Bioresource Technology* 92(2), Apr 2004: 151–55.

Michalak, I., K. Chojnacka, and A. Witek-Krowiak. State of the art for the biosorption process—A review. *Applied Biochemistry and Biotechnology* 170(6), Jul 2013: 1389–416.

Mosbah, R. and M. N. Sahmoune. Biosorption of heavy metals by *Streptomyces* species—An overview. *Central European Journal of Chemistry* 11(9), Sep 2013: 1412–22.

Mukhopadhyay, M., S. B. Noronha, and G. K. Suraishkumar. A review on experimental studies of biosorption of heavy metals by *Aspergillus niger*. *Canadian Journal of Chemical Engineering* 89(4), Aug 2011: 889–900.

Paudyal, H., B. Pangeni, K. N. Ghimire, K. Inoue, K. Ohto, H. Kawakita, and S. Alam. Adsorption behavior of orange waste gel for some rare earth ions and its application to the removal of fluoride from water. *Chemical Engineering Journal* 195, 2012: 289–96.

Ponou, J., J. Kim, L. P. Wang, G. Dodbiba, and T. Fujita. Sorption of Cr(VI) anions in aqueous solution using carbonized or dried pineapple leaves. *Chemical Engineering Journal* 172(2–3), 2011: 906–13.

Ponou, J., L. P. Wang, G. Dodbiba, K. Okaya, T. Fujita, K. Mitsuhashi, T. Atarashi, G. Satoh, and N. Noda. Recovery of rare earth elements from aqueous solution obtained from Vietnamese clay minerals using dried and carbonized parachlorella. *Journal of Environmental Chemical Engineering* 2, 2014: 1070–81.

Redlich, O. and D. L. Peterson. A useful adsorption isotherm. *Journal of Physical Chemistry* 63(6), 1959: 1024–24.

Saeid, A., K. Chojnacka, and G. Balkowski. Two-phase exponential model for describing kinetics of biosorption of Cr(III) ions by microalgae *Spirulina maxima*. *Chemical Engineering Journal* 197, Jul 2012: 49–55.

Saeid, A., K. Chojnacka, M. Korczynski, D. Korniewicz, and Z. Dobrzanski. Biomass of *Spirulina maxima* enriched by biosorption process as a new feed supplement for swine. *Journal of Applied Phycology* 25(2), Apr 2013: 667–75.

Sag, Y. Biosorption of heavy metals by fungal biomass and modeling of fungal biosorption: A review. *Separation and Purification Methods* 30(1), 2001: 1–48.

Seki, H. and A. Suzuki. Biosorption of heavy metal ions to Brown Algae, Macrocystis pyrifera, Kjellmaniella crassiforia, and Undaria pinnatifida. *Journal of Colloid and Interface Science* 206(1), 1998: 297–301.

Seki, H., A. Suzuki, and Y. Iburi. Biosorption of heavy metal ions to a marine microalga, Heterosigma akashiwo (Hada) Hada. *Journal of Colloid and Interface Science* 229(1), 2000: 196–98.

Tarley, C. R. T. and M. A. Z. Arruda. Biosorption of heavy metals using rice milling by-products. Characterizations and application for removal of metals from aqueous effluents. *Chemosphere* 54(7), Feb 2004: 987–95.

Vickery, R. C. Adsorption on carbon of rare earth organic complexes. *Nature* 158(4018), 1946: 623–24.

Vieira, R. H. S. F. and B. Volesky. Biosorption: A solution to pollution. *International Microbiology* 3, 2000: 17–24.

Volesky, B. and Z. R. Holan. Biosorption of heavy-metals. *Biotechnology Progress* 11(3), May–Jun 1995: 235–50.

Volesky, B. *Biosorption of Heavy Metals*. Boca Raton, FL: CRC Press, 1990.

Wang, J. L. and C. Chen. Biosorption of heavy metals by *Saccharomyces cerevisiae*: A review. *Biotechnology Advances* 24(5), Sep-Oct 2006: 427–51.

Weber, W. J. and J. C. Morris. Kinetics of adsorption on carbon from solution. *Journal of the Sanitary Engineering Division* 89(2), 1963: 31–60.

World Health Organization. *Guidelines for Drinking-Water Quality*. 4th ed. (ISBN:978 92 4 154815 1): World Health Organization, 2011. (Available from internet: http://www.who.int/water_sanitation_health/publications/2011/dwq_guidelines/en/).

Yiacoumi, S. and C. Tien. *Kinetics of Metal Ion Adsorption from Aqueous Solutions: Models, Algorithms, and Applications*. Boston, London: Kluwer Academic Publishers, 1995.

Zhang, H., J. Feng, W. F. Zhu, C. Q. Liu, S. Q. Xu, P. P. Shao, D. S. Wu, W. J. Yang, and J. H. Gu. Chronic toxicity of rare-earth elements on human beings—Implications of blood biochemical indices in REE-high regions, South Jiangxi. *Biological Trace Element Research* 73(1), Jan 2000: 1–17.

17

Biosynthesis of Nanomaterials

Dhanasekar Naresh Niranjan, Jayakumar Pathma, Raman Gurusamy and Natarajan Sakthivel

CONTENTS

17.1 Introduction

The greener and biologically inspired approaches in material science, nanoscience and bionanoscience are becoming increasingly popular due to the cost-effectiveness and eco-friendliness (Anastas 2012, Anastas and Horvath, 2012; Oxana et al. 2013). The prefix *nano* derived from the Greek word *nanos* meaning 'dwarf' has gained tremendous interest in the field of physics, chemistry, material science and other biomedical sciences. The foremost concept of nanotechnology was understood from the famous lecture of Richard Feynman at the American Institute of Technology in 1959. Nanoparticles are widely classified into two types, namely, organic and inorganic. Inorganic nanoparticles include iron; noble metals such as silver, gold and platinum; and semiconductor nanoparticles such as TiO_2 and ZnO_2, while organic nanoparticles include carbon nanoparticles. Generally, two different modes, namely, bottom-up and top-down approaches, are employed for the synthesis of nanoparticles. Although the top-down technique involves the breakdown of larger-sized materials into nanosized compounds, the bottom-up method is generally employed for the chemical and biological routes for the nanoparticle synthesis, which usually involves the union of one or more small molecules into molecular structures, resulting in nanometre range. Nevertheless, the synthesis of nanoparticles with a diverse range of chemical compositions, sizes, shapes and controlled monodispersity is one of the challenging aspects in the field of nanotechnology since the physical, chemical, optical and electronic properties of the nanoscopic materials depend on the size and shape of the nanoparticle (Eustis and El-Sayed 2006).

Although physical and chemical procedures such as ultraviolet (UV) irradiation, microwave treatment, aerosol treatment, laser ablation, ultrasonic fields and photochemical reduction form the primary route of synthesis of nanoparticles, their high expense and the release of toxic and hazardous by-products restrict their use in the field of biomedical sciences. Hence, researchers in the field of material sciences and bionanosciences turned their attention towards biological approaches. Moreover, biological methods are safe, cost-effective, sustainable and eco-friendly which results in better control over size, production, shape and crystallinity. Various biophysical and optical techniques such as UV-Vis spectroscopy, x-ray diffraction (XRD), Fourier transform infrared spectroscopy (FT-IR), scanning electron microscopy (SEM) and transmission electron microscopy (TEM) are commonly employed to characterise both inorganic and organic nanoparticles. UV-Vis spectroscopy forms the basic and primary mode of the detection of nanoparticles, whose wavelength lies between 300 and 800 nm (Feldheim and Foss 2002). The absorbance measurements in the wavelength ranges of 400–450 nm and 500–550 nm are typical for silver and gold nanoparticles, respectively (Huang and Yang 2004; Shankar et al. 2004). SEM and TEM give possible information regarding the size and shape of the nanoparticles produced (Schaffer et al. 2009). FT-IR spectroscopy helps in identifying the possible functional groups present on the surface of the nanoparticle and the crystalline nature of the material was confirmed from XRD (Chithrani et al. 2006). This chapter describes the recent developments made towards the biological approaches on the synthesis of nanomaterials.

17.2 Bionanomaterials

17.2.1 Silver

17.2.1.1 Biosynthesis of Nanomaterials Using Fungi

The generation of large biomass, easy handling, bioavailability, high metal tolerance and accumulation, mineral solubilising activity and less time consumption make fungi extremely superior over other microbes. Moreover, fungi are fastidious organisms and can be cultivated with ease. In addition, the fungal biomass can withstand flow pressure, agitation and other conditions in bioreactors or other chambers compared to plant materials and bacteria. In recent years, fungi such as *Epicoccum nigrum* (Sheikhloo et al. 2011), *Alternaria alternata* (Sarkar et al. 2012), *Penicillium* sp. (Du et al. 2011), *Penicillium rugulosum* (Mishra et al. 2012), *Penicillium purpurogenum* NPMF (Nayak et al. 2011), *Phoma macrostoma* (Sheikhloo and Salouti 2012), *Rhizopus oryzae* (Das 2012), *Rhizopus stolonifer* (Binupriya et al. 2010b), *Trichoderma asperellum* (Mukherjee et al. 2008), *Fusarium solani* (Ingle et al. 2009) and *Cylindrocladium floridanum* (Narayanan and Sakthivel 2011) have been reported for the synthesis of various organic and inorganic nanoparticles (Table 17.1).

17.2.1.1.1 Intracellular Mode of Synthesis of Silver Nanoparticles

An enormous amount of reports are dedicated to green and eco-friendly synthesis of silver nanoparticles. Silver nanoparticles have been synthesised by both intracellular and extracellular modes of reduction by microbes. Nevertheless, the reduction of the size could be attributed to the nucleation of the particles inside the organism. The fungus *C. floridanum* was able to selectively accumulate silver on the surface of the mycelium when incubated with silver nitrate (Narayanan and Sakthivel 2011). Interestingly, the mycelia deposited >205 mg of silver when compared to the fungus *Phoma* sp. 3.2883, which deposited only 13.4 mg/g of the dry body mass. In addition, the size of the spherical nanoparticle was also in the range of 5–55 nm. On the other hand, the detoxificating fungus *Phoma* sp. 3.2883 yielded a particle size of 71.06 nm (Chen et al. 2003). In a similar manner, *Aspergillus flavus* has also resulted in the production of silver nanoparticles on incubation with silver ions for a period of 3 days. The fungus exhibited yeast- and mould-like morphologies on treatment with silver nitrate. However, the spectroscopic analysis shows the presence of three absorption bands corresponding to 420, 220 and 280 nm, respectively. The peak at 220 nm signifies the presence of amides and the band at 280 nm identifies tryptophan/tyrosine residues, which are the major ingredients responsible for preventing the flocculation of the synthesised nanoparticles. The diffraction study shows the presence of chitin microfibrils in the fungal matrix. TEM analysis shows the presence of spherical nanoparticles on the surface of the mycelia with an

TABLE 17.1

Biosynthesis of Nanoparticles Using Fungi

Fungi	Nanoparticle	Mode/Shape	Size	Reference
Fusarium oxysporum	Barium titanate	Extracellular/quasi-spherical	4–5 nm	Bansal et al. (2006)
F. oxysporum	Zirconia	Extracellular/quasi-spherical	3–11 nm	Bansal et al. (2004)
F. oxysporum	Silica	Extracellular/quasi-spherical	5–15 nm	Bansal et al. (2005)
F. oxysporum	Titania	Extracellular/spherical	6–13 nm	Bansal et al. (2005)
Fusarium sp. (LB-1)	Gold	Intracellular	35 nm	Gupta et al. (2011)
Neurospora crassa	Silver	Intracellular/spherical	11 nm	Castro-Longoria et al. (2011)
N. crassa	Gold	Intracellular/spherical	32 nm	Castro-Longoria et al. (2011)
N. crassa	Silver–gold	Intracellular/spherical	–	Castro-Longoria et al. (2011)
Rhizopus oryzae	Gold	Intracellular	15 nm	Das et al. (2012)
Penicillium rugulosum	Gold	Extracellular	30 nm	Mishra et al. (2012)
Penicillium (SD-10)	Gold	Intracellular	47 nm	Gupta et al. (2011)
P. brevicompactum	Gold	Extracellular/spherical	25–60 nm	Mishra et al. (2011)
Epicoccum nigrum	Gold	Both intracellular and extracellular/spherical and rod	5–50 nm	Sheikhloo et al. (2011)
Alternaria alternata	Silver	Extracellular/spherical	20–60 nm	Gajbhiye et al. (2009)
A. alternata	Gold	Spherical, triangular and hexagonal	12 ± 5 nm	Sarkar et al. (2011)

Organism	Metal	Location/Shape	Size	Reference
Phoma macrostoma	Gold	Intracellular/spherical, triangle and rod	100–200	Gajbhiye et al. (2009)
Aspergillus terreus	Silver	Extracellular/spherical	1–20 nm	Li et al. (2012)
Aspergillus (SD-9)	Gold	Intracellular	45 nm	Gupta et al. (2011)
A. fumigates	Silver	Extracellular/spherical and triangular	5–60 nm	Bhainsa and D'Souza (2006)
A. clavatus	Gold	Intracellular/nanotriangles	20–35 nm	Verma et al. (2011)
A. ochraceus	Silver	Intracellular/ND	<20 nm	Vijayakumar and Prasad (2009)
A. oryzae var. *viridis*	Silver	Extracellular/spherical	5–50 nm	Binupriya et al. (2010a)
Bipolaris nodulosa	Silver	Extracellular/spherical, hexahedral and semi-pentagonal	10–60 nm	Saha et al. (2010)
F. solani (USM-3799)	Silver	Extracellular/spherical	5–35 nm	Ingle et al. (2009)
F. semitectum	Silver	Extracellular/spherical	10–60 nm	Basavaraja et al. (2008)
Sclerotium rolfsii	Gold	Extracellular/spherical, triangular, hexagonal, decahedral and rod	25 nm	Narayanan et al. (2011)
Trichoderma asperellum	Silver	Extracellular/ND	13–18 nm	Mukherjee et al. (2008)
T. reesei	Silver	Extracellular/ND	5–50 nm	Vahabi et al. (2011)
Verticillium sp.	Silver	Intracellular/ND	25 ± 12 nm	Mukherjee et al. (2011)
Verticillium sp.	Gold	Intracellular/spherical, triangular and hexagonal	20 ± 8 nm	Mukherjee et al. (2001)

average size of 8.92 ± 1.61 nm (Vigneshwaran et al. 2007). Filamentous fungi are of much interest since it results in the production of highly monodispersed stable nanoparticles. In this instance, Castro-Longoria et al. (2011) reported the use of filamentous non-pathogenic fungus *Neurospora crassa* for the production of Ag, Au and Ag–Au (bimetallic) nanoparticles. TEM analysis shows the presence of quasi-spherical-shaped nanoparticles with a size range of 3–50 nm, with a mean diameter of 11 nm for silver. The size of the formed gold nanoparticles was also spherical with a mean diameter of 32 nm. However, the bimetallic nature of the nanoparticle varied as the ratio of Au:Ag varies. The particle size ranges between 3 and 90 nm for Au:Ag 70:30; 3 and 110 nm for Au:Ag 50:50 and 4 and 45 nm for Au:Ag 30:70. In addition to TEM, confocal microscopic images further show the deposition of the nanoparticles on the surface of the mycelium. Similarly, other forms of filamentous fungi such as *Cladosporium cladosporioides* (Balaji et al. 2009), *Aspergillus* (Binupriya et al. 2010a), *Penicillium* sp. (Basavaraja et al. 2008) and *Fusarium* sp. (Syed and Ahmed 2012) have also been employed for the synthesis of metal nanoparticles.

Although the filamentous fungi find potent roles in the synthesis of metal nan, nevertheless, they have been considered a great threat to plants and humans and hence, it makes the handling and disposal of the fungal biomass a major inconvenience in industrial applications. There is a need for developing a safer and eco-friendly approach of testing non-pathogenic fungi for the successful synthesis and capping of metal nanoparticles.

17.2.1.1.2 Extracellular Mode of Synthesis of Silver Nanoparticles

Fungi are considered to be promising candidates as efficient secretors of extracellular enzymes and proteins, thus making large-scale production much easier. Nanoparticles are easy to find when the fungal enzymes or other components of fungal cell wall are secreted in the medium. The extracellular process has led to the production of reduced-sized nanoparticles and it is usually devoid of cellular components from the cell. The secretion of enzymes and proteins into the surrounding medium serves as the capping agent for the nanoparticle. The extracellular method of producing nanoparticles has wider applications since it is devoid of unusual cellular matrix components, higher yield and better viability. Binupriya et al. (2010a) showed the bioreduction of silver ions to silver nanoparticles using the live and dead cell filtrates of the aflatoxin fungus *A. oryzae* var. *viridis*. For the first time, dead cell filtrate was utilised for the production of silver colloids. Spherical- and ellipsoidal-shaped nanoparticles with a size range between 5 and 50 nm were observed for dead and live cell extracts, respectively. Furthermore, total organic carbon content analysis showed that the dead cell-free extracts release the enzymes/proteins twofold to the surrounding medium when compared to live cell-free extracts (Binupriya et al. 2010a). Another fungus *Fusarium semitectum* also yielded spherical nanoparticles with a size range of 10–60 nm. FT-IR results confirm the presence of amide I band and amide II band, which arises due to the carbonyl and amide stretch of the proteins, respectively (Basavaraja et al. 2008). Similarly, the fungus *Aspergillus tamarii* incubated with silver nitrate under dark results in the production of monodispersed spherical nanoparticles with a size range of 25–50 nm (Kumar et al. 2012a). However, *C. cladosporioides* resulted in the production of polydispersed spherical nanoparticles with the size ranging from 10 to 100 nm (Balaji et al. 2009).

17.2.1.2 Bacteria-Mediated Synthesis of Silver Nanoparticles

Wei et al. (2012) reported the use of *Bacillus amyloliquefaciens* for the synthesis of silver nanoparticles under solar power. When the cell-free extract of *B. amyloliquefaciens* was incubated with an aqueous solution of silver nitrate under the influence of sun rays, it resulted in the change of colour from dark yellow to brown within 1 min. However, further increases in the reaction time resulted in the formation of an orange red colour exhibiting a strong plasmon band at 423 nm. Interestingly, the silver nanoparticles synthesised at a higher solar intensity of about 30,000 and 40,000 lx were stable for more than 2 months. Electron microscopy study showed the formation of circular and triangular morphologies with a size range of 4.8–23.7 nm. FT-IR analysis shows the presence of carboxyl and hydroxyl groups of amino acids such as Asp, Glu and Tyr that are responsible for the reduction of Ag^{2+} to Ag^0 as well as stabilisation of the silver nanoparticles. Nevertheless, these silver nanoparticles exhibited strong antibacterial activity against *Escherichia coli* and *B. subtilis* (Wei et al. 2012). Production of anisotrophic silver nanoparticles was achieved by controlling the growth kinetics pattern of the silver-resistant psychrophilic bacterium, *Morganella psychotolerans*. Interestingly, for the first time, voltammetric analysis was performed on the bacterium for the production of silver nanoparticles. When the cell-free extract of the bacterium was exposed to an aqueous solution of silver nitrate in the absence of light, it resulted in the production of muddy green colour exhibiting a strong plasmon band in the range of 350–530 nm. At 5°C and 15°C, spherical and some nanoplate-like morphologies were observed with a size range of 2–5 and 100–150 nm, respectively. In addition to spherical nanoparticles, triangular- and hexagonal-shaped nanoparticles were seen at a higher temperature (25°C). Polymerase chain reaction confirms the presence of silver binding gene (silE) present on the periplasmic region of the bacterium (Ramanathan et al. 2011). Captivatingly, thermophilic bacterium resulted in the production of both silver and gold nanoparticles on incubation of the cell-free extract of *Geobacillus stearothermophilus* with the corresponding metal salts. Monodispersed spherical silver nanoparticles with a size range of 5–8 nm were obtained from TEM. However, gold nanoparticles were polydispersive in nature with an average dimension of 5–35 nm (Fayaz et al. 2011).

17.2.1.3 *Plant Extracts-Mediated Synthesis of Silver Nanoparticles*

Plants form another ideal source for the synthesis of metal nanoparticles. A reduced size with different morphologies of metal nanoparticles has been reported using plant extracts. Huang et al. (2011) has reported that the size and shape of the AgNPs varies with an increase in the concentration of silver salt. TEM analysis revealed that 1 and 3 mM AgNO$_3$ have yielded spherical-shaped morphologies with a size range of 50–100 nm. In addition to spheroidal shape, plate-like nanoparticles were also reported with a size of about >100 nm in the case of 5 mM AgNO$_3$ solution (Huang et al. 2011). Table 17.2 lists the production of metal nanoparticles from plant extracts. However, Kumar et al. (2012) have reported the bi-functional role (reducing and capping agent) of *Terminalia chebula* extract for the synthesis of silver nanoparticles (Kumar et al. 2013). The extract of *T. chebula* contains a high level of polyphenolic materials, specifically ellagic acid, which favours the reduction of Ag^{2+} ions to Ag^0. The reducing action of gallic acid and the stabilising potential of glucose makes the synthesis much more rapid, that is, the total reaction time was <25 min when compared to the nanomaterials synthesised using the extract of *Acalypha indica*, which takes a total reaction period of >4 h. HR-TEM analysis revealed the presence of anisotrophic nature with spherical, triangular and pentagonal morphologies with a size <100 nm (Krishnaraj et al. 2010).

Similarly, the leaves of pepper also resulted in the formation of AgNPs. The reaction between the pepper extract and silver nitrate yielded spherical nanoparticles with a size range of 5–60 nm (Mallikarjuna et al. 2012). With the fruit extract of the common vegetative plant *Annona squamosa*, silver nanoparticles were produced under two different temperatures, namely, 25°C and 60°C. The intensity of the silver nanoparticle solution was less under ambient condition; on the other hand, the intensity tends to increase with a rise in temperature. The nanoparticles were spherical in shape with a size range of 20–60 nm and an average diameter of 35 ± 2 nm. Gas chromatography–mass spectrometry (GC–MS) analysis reveals C=O and CHO as the major constituents present in the extract, which aid in the reduction of silver ions to Ag^0. Nevertheless, OH groups prevent the agglomeration of the nanoparticles in aqueous solution (Kumar et al. 2012b). *Rosa rugosa*, commonly known as Ramanas rose, is well known for its hepatic activity and also its role in reversing cancerous growth. The synthesis of silver nanoparticles takes place through the addition of 60 mL of 1 mM AgNO$_3$ with 2.5 mL of the leaf extract. TEM images signify the presence of spherical nanoparticles with an average size between 30 and 60 nm (Dubey et al. 2010). Ponarulselvam et al. (2012) demonstrated the antihelminthic activity of the silver nanoparticles synthesised using the extract of *Catharanthus roseus*. The extract obtained by boiling the leaves in 100 mL distilled water was used for the synthesis of silver nanoparticles. A broad surface plasmon resonance (SPR) was observed at around 410 nm indicating the polydispersive nature of the nanoparticles.

TABLE 17.2

Biosynthesis of Nanoparticles Using Plant Extracts

Biomaterial	Nanoparticle	Mode of Synthesis	Shape	Size	Reference
Black tea (leaf)	Silver	Extracellular	Spherical	20 nm	Uddin et al. (2012)
Cacumen platycladi (leaf)	Silver	Extracellular	Sphere/plate like	50–100 nm	Huang et al. (2011)
Terminalia chebula (fruit)	Palladium, iron	Extracellular	Spherical, pentagonal, triangular	<100 nm	Kumar et al. (2013)
Acalypha indica (leaf)	Silver	Extracellular	Spherical	20–30 nm	Krishnaraj et al. (2010)
Catharanthus roseus (leaf)	Silver	Extracellular	Spherical	35–55 nm	Ponarulselvam et al. (2012)
Iresine herbstii (leaf)	Silver	Extracellular	–	44–64 nm	Dipankar et al. (2012)
Trachyspermum ammi and *Papaver somniferum* (seed)	Silver	Extracellular	Triangular, spherical	87 nm, 3.2 nm and 7.6 µm	Vijayaraghavan et al. (2012)
Tribulus terrestris (fruit)	Silver	Extracellular	Spherical	16–28 nm	Gopinath et al. (2012)
Artemisia nilagirica (leaf)	Silver	Extracellular	Spherical, square, hexagonal, triangular	10–45 nm, 70–90 nm, 10–25 nm, 45–60 nm	Vijayakumar et al. (2013)
Abelmoschus esculentus (seed)	Gold	Extracellular	Spherical	45–75 nm	Jayaseelan et al. (2013)
Euphorbia prostrata (leaf)	Silver	Extracellular	Rod	25–80 nm	Abduz Zahir et al. (2012)

Spherical nanoparticles with a size range of about 35–55 nm were obtained (Ponarulselvam et al. 2012).

For the first time, rod-shaped silver nanoparticles were obtained with the extract of *Coscinium fenestratum*. The UV-Vis spectroscopic studies showed the presence of two SPR bands at 420 and 650 nm, respectively. The band at 420 nm confirms the nanoparticle synthesis, while the peak at 650 nm is specific for rods. The presence of berberines and terpenoids in the extract is responsible for the reduction of Ag^{2+} ions and the presence of –COOH on the surface of the nanoparticles protects them from getting agglomerated (Jacob et al. 2012). Contrastingly, Dipankar and Murugan (2012) used only 5 g of fresh leaves of the forest plant *Iresine herbstii* for the synthesis of silver nanoparticles. 10 mL of the plant extract was challenged with 90 mL of 1 mM silver nitrate solution to produce silver nanoparticles, which were confirmed through a colour change from dark yellow to brownish gray. FT-IR analysis indicates the presence of phenols and terpenoids (having amine, ketone, aldehyde and carboxylic acid), which are involved in capping the silver nanoparticles. These spherical nanoparticles with a size range of about 44–64 nm exhibited good antibacterial activity with a least concentration of 50 μg/mL against several bacterial species such as *Staphylococcus aureus*, *Enterococcus faecalis*, *Pseudomonas aeruginosa* and *E. coli*, while the growth of *Klebsiella pneumonia* was inhibited at a minimum concentration of 100 μg/mL. Besides antibacterial potential, these silver nanoparticles also exhibited good cytotoxic and antioxidant activity (Dipankar and Murugan 2012).

Melia azedarach, known for its therapeutic properties for centuries, have also resulted in the development of silver nanoparticles. The synthesis of nanoparticles was carried out at four different temperatures, namely, 30°C, 60°C, 90°C and 95°C. However, the intensity was high at 95°C, which favours the formation of spherical- and cubical-shaped nanoparticles with a size of about 78 nm (Sukirtha et al. 2012). Similar to *M. azedarach*, another thermostable plant, *Morinda citrifolia* also produced a dark brown colour at 90°C confirming the silver nanoparticle synthesis. These flavonoid-stabilised nanoparticles were spherical in nature with a size range of 10–60 nm with an average size of 29 nm. The crude silver nanoparticles exhibited good antibacterial potential when compared to the purified one (SathishKumar et al. 2012).

17.2.2 Gold

Gold (Au) nanoparticles can also be synthesised using a wide range of greener and eco-friendly methods. The history of gold starts from early 2500 BC, when the Chinese started using gold for therapeutics. However, the effect on mycobacterium marked the beginning of the medicinal value of gold. Apart from tubercle bacilli, gold nanoparticles and its corresponding derivatives are used in rheumatic diseases such as juvenile arthritis, palindromic rheumatism and discoid lupus erythematosus (Daniel and Astruc 2004; Felson et al. 1990; Shaw 1999).

17.2.2.1 Biosynthesis of Gold Nanoparticles Using Fungi

17.2.2.1.1 Intracellular Production of Gold Nanoparticles Using Fungi

Gold nanoparticles can also be produced by both intracellular and extracellular methods. The intracellular synthesis of gold nanoparticles by the fungus *P. macrostoma* isolated from a copper mine was studied. TEM illustrated the production of triangular, spherical and rod-like morphologies with a size range of 100–200 nm (Sheikhloo and Salouti 2012). The enzymes secreted by the fungus *Penicillium* sp. were believed to act as a reducing agent to synthesise intra- and extracellular gold nanoparticles in an aqueous solution of gold chloride (Du et al. 2011). Spherical nanoparticles were observed in the case of intra- and extracellular production with an average diameter of 50 and 45 nm, respectively. However, nanoparticles were accumulated on the surface of the mycelium for intracellular production. Similarly, another eucaryotic organism, namely, the *E. nigrum* was employed for both extracellular and intracellular accumulation of gold nanoparticles. The mycelial mesh helps in the fabrication of 50 nm size range of spherical- and rod-shaped nanoparticles. These nanoparticles were clearly visible on the surface of the cell wall and cytoplasm, which is identified from TEM studies (Sheikhloo et al. 2011). Das et al. (2012) reported the accumulation of gold nanoparticles with a size of <15 nm on the surface of the mycelium of *R. oryzae* when auric acid was added. Sodium dodecyl sulphate–polyacrylamide gel electrophoresis (SDS-PAGE) showed the presence of 42- and 45-kDa proteins responsible for nanoparticle formation, and the presence of an 80-kDa protein attributed to the capping agent.

17.2.2.1.2 Extracellular Production of Gold Nanoparticles Using Fungi

Sarkar et al. (2012) reported the use of plant pathogenic fungus *A. alternata* for the bioproduction of gold nanoparticles. TEM analysis shows the presence of a wide range of nanoparticles such as spherical, triangular and hexagonal morphologies with an average size of 2–30 nm (Sarkar et al. 2012). Similarly, when the cell-free extract of the fungus *A. clavatus* was challenged with auric acid, it reduced Au^{3+} to Au^0. UV-Vis spectroscopic study shows that there is a steady increase in the absorbance after 48 h. However, there occurs a steady shift in the plasmon resonance from 540 to 555 nm after 72 h. This change in the plasmon resonance depends on several factors such as dielectric constant of the medium, particle size and the interparticle distance (Link and El-Sayed 2003; Verma et al. 2011). TEM analysis showed that the purified gold nanoparticles were triangular in shape with an average size range from 20 to 35 nm. In addition, low-resolution TEM images show the presence of both sharp and sniped angled nanotriangles. The proteins/enzymes secreted by the cells to the surrounding medium were found to be responsible for the nanoparticle synthesis with different morphology. For the first time, Rehman et al. (2011) reported the formation of gold nanowires using a combination of tea extract and the fungal spores of the filamentous fungus *A. niger*. The catechins and other polyphenols such as flavonoids and terpenoids played a vital role in reducing gold chloride solution to gold nanoparticles. The formed gold nanoparticles that were in a size range of 10–25 nm were then used to accumulate the fungal hyphae resulting in the formation of gold wires after heat treatment at 350°C, 450°C and 550°C (Rehman et al. 2011).

17.2.2.2 Bacteria-Mediated Synthesis of Gold Nanoparticles

Suresh et al. (2011) reported the use of the γ-proteobacterium *Shewanella oneidensis* for the production of discrete spherical-shaped gold nanoparticles. The change of pale yellow to deep purple indicates the formation of gold nanoparticles, whose plasmon peak was centred at 520 nm. The peptides and other proteins secreted by the bacterium to the surrounding medium favour the reduction and stabilisation of gold nanoparticles. TEM analysis shows the presence of spherical nanoparticles with a size range of 2–50 nm (Suresh et al. 2011). Similarly, intracellular production of gold nanoparticles was achieved by incubating the biomass of the bacterium *Bacillus licheniformis* with $HAuCl_4$. XRD study confirms the presence of crystalline nature of the gold nanoparticles. TEM analysis shows the presence of polydisperse nanocubes with a range of 10–100 nm in size (Kalishwaralal et al. 2009). Arunkumar et al. (2013) reported the dual-mode synthesis of gold nanoparticles using the bacterium *Micrococcus luteus*. The extracellular enzyme α-amylase and the cell wall teichuronic acid from the bacterium *M. luteus* were responsible for the reduction of Au^{3+} to Au^0. TEM studies show the presence of monodisperse spherical nanoparticles with an average size of 6 and 50 nm for α-amylase and teichuronic acid, respectively (Arunkumar et al. 2013). Nevertheless, another eco-friendly approach was adopted by Radhika Rajasree and Suman (2012) for the synthesis of gold nanoparticles using the cell-free extract of the Gram-negative soil bacterium *P. fluorescens*. TEM study shows the presence of spherical-shaped nanoparticles with a range of 20–80 nm in size. Proteins present in the extract were responsible for the reduction and stabilisation of the gold nanoparticles (Radhika Rajasree and Suman 2012).

17.2.2.3 Plant Extracts-Mediated Synthesis of Gold Nanoparticles

The use of plant extracts in the synthesis of metal nanoparticles has gained tremendous interest due to the non-toxic nature of the plant extracts and other plant components. Plant-mediated synthesis of nanoparticles takes place by simple mixing of the required concentration of metal salt with the aqueous extract of plant under ambient conditions. However, the size and shape of the nanoparticle depends on various factors such as the concentration of plant extract and the metallic salt, pH, temperature and time of the reaction process. Ali et al. (2011) reported the synthesis of gold and silver nanoparticles within a few minutes using the leaf extract of a common pepper plant known as *Mentha piperita*. Spherical morphology with a size of 90 nm was observed from SEM. FT-IR studies shows the presence of phytochemicals such as flavonoids, menthol, steroids, polysaccharides and other proteins, which act as capping and reducing agents for the synthesised gold and silver nanoparticles. In addition, silver and gold nanoparticles were effective in inhibiting the growth of both *S. aureus* and *E. coli* (Ali et al. 2011). On the other hand, gold nanoparticles showed antibacterial activity against *S. aureus* only. The crushed, boiled and filtered leaves of coriander were used to synthesise gold nanoparticles of various shapes such as spherical, triangular, including truncated triangle, and decahedral with an average size of 20.65 ± 7.09 nm. The C–N and N–H stretching of the amides helps in capping the gold nanoparticles (Badri Narayanan and Sakthivel 2008). The fruit extract of *Prunus domestica*, commonly known as the plum fruit, was employed for the first time by Dauthal and Mukhopadhyay (2012) for the synthesis of gold nanoparticles. The boiled and filtered extract of the fruit resulted in the production of spherical and a few irregular triangular-shaped structures having an average diameter of 20 ± 6 nm (Dauthal and Mukhopadhyay 2012).

17.2.3 Platinum

Platinum nanoparticles have also gained much importance in the field of biology due to mimicking the components involved in electron transport chain and diseases related to oxidative stress (Hikosaka et al. 2008). Syed and Ahmad (2012) reported the use of the fungus *Fusarium* sp. for the extracellular synthesis of protein-capped platinum nanoparticles, which were spherical in shape and with a size range of 5–30 nm. However, the fungus, *F. oxysporum* f. sp. *lycopersici* resulted in the production of platinum nanoparticles of various shapes such as circles, squares, rectangles, hexagons and pentagons with a size range of 10–100 nm by both intercellular and extracellular processes (Riddin et al. 2006). Surprisingly, Konishi et al. (2007) reported the use of bacteria for the synthesis of biogenic platinum nanoparticles. The metal-ion-reducing bacterium *Shewanella algae* on treatment with H_2PtCl_6 was able to deposit discrete platinum nanoparticles of 5 nm on the periplasmic surface under resting conditions at neutral pH, on the supplementation of lactate as an electron donor (Konishi et al. 2007). Apart from reduction, the bacterium does not help in adsorption of the platinum like that of the sulphate-reducing bacterium *Desulfovibrio desulfuricans* (Yong et al. 2002). Later, *Desulfovibrio desulfuricans* was also responsible for the reduction of Pt(IV) to Pt(0). Two different hydrogenase enzymes were responsible for the step-wise degradation to platinum nanoparticles. Initially, O_2-sensitive cytoplasmic hydrogenase favours the reduction of two electrons from platinum (IV) to platinum (II). Later, another two-electron bioreduction takes place, converting Pt(II) to Pt(0), which involves an O_2-tolerant or protected periplasmic hydrogenase (Riddin et al. 2009).

In addition to microbes, plant extracts have also been exploited for the synthesis of platinum nanoparticles. Song et al. (2010) reported the non-enzymatic reduction of platinum ions with the leaves of *Dipyros kaki*. HR-TEM analysis reveals the presence of spheres and plates with an average size of 2–20 nm. FT-IR analysis shows the presence of terpenoids bearing amines, ketones, reducing sugars, aldehydes and carboxylic acids, which help in reducing and stabilizing the platinum nanoparticles (Song et al. 2010). In addition, an increase in the concentration of the leaf extract from 5% to 10% with further increase in the reaction temperature from 25°C to 95°C resulted in 100% reduction of Pt(IV) to Pt(0). Captivatingly, natural sweet honey also acts as a possible reducing agent in the production of platinum nanoparticles. Increase in the reaction period from 10 h to 20 h resulted in the production of nanowires with a maximum of 5–15 nm. On the other hand, monodispersed, spherical morphologies with a much more reduced size of 2.2 nm were observed when the reaction completed within 2 h. Moreover, these bio-inspired platinum nanoparticles find a strong role in the production of quinone dyes (Santhanalakshmi et al. 2007; Venu et al. 2008).

17.2.4 Zinc Oxide

Inorganic nanoparticles, especially zinc oxide nanoparticles, have gained considerable attention in the field of drug delivery and food packaging due to their unique features such as biocompatibility, controlled release and specific targeting of the cell. Jayaseelan et al. (2012) reported the production of spherical-shaped zinc oxide nanoparticles with an average size of 57.72 nm exhibiting a strong and sharp band at 374 nm using the culture supernatant of the bacterium *Aeromonas hydrophila*. Interestingly, for the first time, the produced zinc oxide nanoparticles showed maximum inhibition of 22 and 19 mm against *P. aeruginosa* and *A. flavus* with a minimum concentration of 25 µg/mL of zinc oxide nanoparticle solution (Jayaseelan et al. 2012). On the other hand, two different materials, namely, the leaf extract and the gel obtained from the *Aloe vera* plant, were utilised for the synthesis of polydispersed zinc oxide nanoparticles with an average size of 35 and 45 nm, respectively. However, spherical-shaped nanoparticles with a larger size of 60 nm were obtained under chemical conditions. A complete (100%) reduction of metal oxide to nanoparticles takes place within 6 h; however, only 50% conversion takes place under a chemical process. Thus, the plant extract- and gel-produced zinc nanoparticles are of smaller size when compared to nanoparticles produced through the chemical process (Sangeetha et al. 2011).

17.2.5 Calcium Carbonate

Calcium carbonate nanoparticles were of special interest because of their industrial importance in the manufacturing of paints, plastics and paper and in dental fields. The biological synthesis of calcium carbonate nanocrystals was first reported using the fungus *F. oxysporum*. Interestingly, Ahmed et al. (2004) found the aggregation of circular calcium carbonate nanocrystals with an increase in the perimeter and the central area as the reaction period increases from day 1 to day 3. Nevertheless, at higher magnification, circular quasi-bilayer structures were formed. Energy dispersive x-ray analysis (EDAX) analysis shows the presence of strong signals such as Ca and C at the centre and N and S at the periphery of the nanoparticle. Moreover, this also suggests that proteins released from F. oxysporum might be responsible for the reduction of calcium carbonate to calcium crystallites. On the other hand, when the fungus *Trichothecium* sp. was incubated with aqueous solution of calcium carbonate, it led to the production of calcium carbonate nanocrystals with various shapes ranging from flat opened to smooth plate-like morphologies (Ahmad et al. 2004). Zhang et al. (2011b) reported the use of the houseplant *Epipremnum aureum* under various pH for the synthesis of highly stable calcium carbonate nanoparticles. SEM analysis indicates the presence of pancake morphology-like nanoparticles with an average diameter of 22 µm and a thickness of 3 µm. In addition, cubic- and rod-shaped structures were obtained in the absence of *E. aureum* leaves. However, a change of rhombohedral smooth surface crystals to rhombohedrals with truncated corners was obtained when the initial pH was increased from 5.7 to 7.7. Further increases in the pH to 9.7 led to the production of monodispersed disc-like particles with an average diameter and thickness of 6 and 3 µm, respectively. Besides pH, $CaCO_3$ concentration has also led to the change in the shape and symmetry of calcium carbonate nanoparticles, varying from rhombohedral with a convex symmetry to pseudo-dodecahedral and further modification to bread-like morphologies at 10, 50 and 100 mM, respectively (Zhang et al. 2011b). Apart from the use of microbes, Guo et al. (2013) showed the potential role of amino acids in the synthesis of $CaCO_3$ nanocrystals with various morphologies. SEM images show the transformation of irregular-shaped nanocrystals to cubic forms on increasing the concentration of L-valine from 1 to 50 mM. However, arginine–$CaCO_3$ nanoparticles were ellipsoidal in shape with an average diameter of 25 µm. Nevertheless, the formation of microspheres takes place only when the concentration of arginine was increased to 50 mM. Interestingly, irregular morphologies of larger-sized microspheres and hollow spheres with an average size of 34 µm were obtained in the case of 50 mM aqueous solution of L-serine (Guo et al. 2013).

17.2.6 Copper and Its Oxides

Gunalan et al. (2007) demonstrated the production of copper oxide nanoparticles using the extract of *Aloe barbadensis* from UV-Vis spectroscopic studies, showing the presence of two SPR bands at 265 and 285 nm, respectively. However, an increase in the concentration of leaf extract from 0% to 50% resulted in the shifting of the SPR from the region of lower wavelengths to the region of higher wavelengths. TEM images show the presence of monodispersed spherical and quasi-linear-shaped nanoparticles with an average size ranging from 15 to 30 nm. FT-IR analysis indicates the presence of phenolic compounds and terpenoids, which were found to cap the copper oxide nanoparticles. On the other hand, the proteins present in the extract help in covering the nanoparticle, thus preventing the agglomeration and contributing to the stability of the nanoparticle (Gunalan et al. 2007). Captivatingly, for the first time, microwave thermal treatment was adopted for the rapid reduction of Cu^{2+} to Cu^0. When the extract of *Terminalia arjuna* bark obtained through microwave irradiation was mixed with copper nitrate, it resulted in the change of colour from pale yellow to dark brown within 8 min, indicating the rapid formation of copper oxide nanoparticles. SEM analysis signifies the presence of uniform monodispersed well-separated spherical-shaped nanoparticles with a size range of 20–30 nm. In addition to FT-IR, the ^{13}C NMR spectrum confirms the presence of amides, reducing sugars, aromatic benzenes and other organic compounds, which play a vital role in the production and stabilisation of the copper nanoparticle. In addition, the copper nanoparticles dried under ambient conditions displayed better antibacterial and antifungal activity than the nanoparticles dried at 70°C. Nevertheless, these nanoparticles showed maximum inhibition against *E. coli* and *S. aureus* with least inhibition against *P. aeruginosa* and *S. typhi*, respectively (Yallapa et al. 2013).

17.2.7 Iron

The iron nanoparticle has gained tremendous interest in wastewater treatment (Xu et al. 2012) and other biomedical applications such as tissue repair, drug delivery, magnetic resonance imaging (MRI), hyperthermia and cellular therapy (Gupta and Gupta 2005). Machado et al. (2013) reported the fact that the size, shape and agglomeration behaviour of the nanoparticles depends on the ratio of leaf extract and the solvent employed for the synthesis of iron oxide nanoparticles. Out of 26 different fruit species, the leaf extracts of pomegranates, mulberry and cherry resulted in the production of spherical-shaped nanoparticles with an average size of 10–30 nm, while the remaining leaf extract of the fruits resulted in a larger size with much flocculation (Machado et al. 2013). Similarly, the caffeine and polyphenols present in the tea leaves helps in the complexation of iron salts, thus contributing to the reduction of $Fe(NO_3)_3$ to Fe. However, the oxidised form of these polyphenols binds to the surface of the nanoparticles, thus preventing the agglomeration behaviour of the nanoparticle. Spherical nanoparticles, nearly monodispersed with an average size of 50 nm, were observed by TEM (Nadagouda et al. 2010). In addition to palladium nanoparticles, *T. chebula* extract also produced chain-like iron oxide nanoparticles with a size range of <80 nm. UV-Vis spectroscopic studies show the presence of a single narrow peak at 527 nm, whose aqueous solution of the nanoparticle exhibits a dark brown colour (Kumar et al. 2013). Apart from the plant extracts, sulphate-reducing bacteria such as

Magnetospirillium (Farina et al. 1990) and *Actinobacter* sp. (Bharde et al. 2008) also resulted in the production of spherical-shaped iron sulphide nanoparticles, which are typically in the range of 20 nm.

17.2.8 Other Metals

Zhang et al. (2011a) demonstrated the use of *P. alcaliphila* for the extracellular synthesis of selenium nanoparticles. When the bacterium was incubated aerobically with or without polyvinylpyrrolidone (PVP) in an aqueous solution of sodium selenite pentahydrate, it resulted in the reduction of SeO_3^{2-} to Se^0 in 6 h, and a change of colour from grey to red. FT-IR analysis shows the presence of proteins on the surface of the selenium nanoparticles acting as ion transporters. FESEM studies show the transformation of m-Se nanospheres to t-Se nanorods with an average size of 20 ± 5 nm. The oxidation and reduction enzymes secreted from the bacterial cell wall contribute to the initial reduction of SeO_3^{2-} to form Se nuclei. This Se nuclei growth pattern follows the Ostwald ripening mechanism. Later, through Gibbs–Thomson law, these small Se nanoparticles fuse to form larger nanoparticles (Gates et al. 2002; Wang et al. 2010b; Zhang et al. 2011). Similarly, the culture supernatant of the bacterium *B. subtilis* challenged with sodium selenite resulted in the gradual change of colour from light yellow to light red. Further incubation after 48 h resulted in the transformation to dark red, indicating the formation of m-Se spherical nanoparticles with an average size of 50–150 nm. These nanoparticles on prolonged reaction increase in perimeter to form nanoparticles 500 nm in size (Wang et al. 2010a).

Inorganic sulphide nanoparticles such as CdS, PbS, ZnS, CdSe and PbSe have gained potential interest due to a wide range of applications in the field of photoelectric diodes, solar cells, photocatalysis, IR-photodetector and other biosensors (David and Michael 2006; Ionov et al. 2006; Nag et al. 2008). Prasad and Jha (2010) reported for the first time the use of both procaryotes and eucaryotes for the synthesis of cadmium sulphide nanoparticles. *Lactobacillus* sp. and *Saccharomyces cerevisiae* resulted in the deposition of orange-yellow clusters of CdS nanoparticles on incubation of the aqueous solution of cadmium chloride with cell-free supernatant. UV-Vis spectroscopic analysis shows the presence of bands at 393 and 369 nm for *Lactobacillus* sp. and yeast, respectively (Prasad and Jha 2010). The high electronegative and oxidoreductive potential of *Lactobacillus* sp. and other nutritive and environmental conditions such as increasing/decreasing the partial pressure of gaseous hydrogen plays a critical role in synthesising spherical nanoparticles of 2.62 nm size. Similarly, cytosolic oxidoreductive enzymes and quinones favour yeast for the production of nanoparticles of much smaller size (2.02 nm) when compared to *Lactobacillus*. Although the plant pathogenic fungus *F. oxysporum* was employed for the synthesis of almost all the organic and inorganic nanoparticles, strontium carbonate is of peculiar interest due to its immense role in the field of ceramics. SDS-PAGE shows that the proteins of molecular weights 33 and 50 kDa present in the outer membrane of the fungus are responsible for the biotransformation of Sr^{2+} ions to strontium carbonate crystals. In addition, a sharp peak at 260 nm further illustrates the presence of proteins bound to the surface of $SrCO_3$ nanocrystals. SEM analysis indicates the presence of needle-like morphologies of $SrCO_3$ nanocrystals with a size range of 10–50 nm. However, the increase in the size of the $SrCO_3$ needles to 100 nm was due to the removal of the proteins on treatment with NaOCl. This further confirms the presence of proteins in designing needle-like morphologies (Rautaray et al. 2004). Velayutham et al. (2012) reported the potential role of irregular-shaped TiO_2 nanoparticles synthesised using the leaf extract of *C. roseus* against the hematophagous fly *Hippobosca maculata* and the sheep louse *Bovicola ovis*. TEM studies reveal the presence of TiO_2 nanoparticles with a size ranging from 25 to 100 nm (Velayutham et al. 2012). Similarly, chemical synthesis of Ni nanoparticles also exhibited antiparasitic activity against the wide range of parasites such as the larvae of cattle ticks *Rhipicephalus (Boophilus) microplus* and *Hyalomma anatolicum anatolicum*, *Anopheles subpictus*, *Culex quinquefasciatus* and *Culex gelidus* (Rajakumar et al. 2013).

17.3 Conclusion

Chemical and physical methods of synthesis of metallic nanoparticles are expensive and in addition pose a threat to environmental safety due to the hazardous nature of the chemicals agents used. This has led to the urge to explore and implement cost-effective and environmentally friendly biological methods of nanoparticle synthesis. Hence, over the past few years, bacteria, fungi, algae and plants were employed for nanoparticle synthesis, which finds a wide range of applications in cosmetics, medicine, electronics and agriculture. Further research on biological synthesis of non-toxic nanomaterials would pave the way to identify non-pathogenic biological systems for nanoparticle synthesis on a commercial scale.

References

Abduz Zahir, A., and Abdul Rahuman, A. 2012. Evaluation of different extracts and synthesised silver nanoparticles from leaves of *Euphorbia prostrata* against *Haemaphysalis bispinosa* and *Hippobosca maculate*. *Vet Parasitol* 187:511–520.
Ahmad, A., Rautaray, D., and Sastry, M. 2004. Biogenic calcium carbonate: Calcite crystals of variable morphology by the reaction of aqueous Ca^{2+} ions with fungi. *Adv Funct Nanomater* 11:1075–1080.

Ali, D. M., Thajuddin, N., Jeganathan, K., and Gunasekaran, M. 2011. Plant extract mediated synthesis of silver and gold nanoparticles and its antibacterial activity against clinically isolated pathogens. *Colloids Surf B: Biointerfaces* 85:360–365.

Anastas, P. T. (ed.). 2012. *Handbook of Green Chemistry—Green Processes.* (Volume 3, 3-book set), John Wiley & Sons, New Jersey, USA.

Anastas, P. T., and Horvath, I. T. 2012. *Green Chemistry for a Sustainable Future.* (1st edn), Wiley 2, Hoboken, NJ, USA.

Arunkumar, P., Thanalakshmi, M., Kumar, P., and Premkumar, K. 2013. Micrococcus mediated dual mode synthesis of gold nanoparticles: Involvement of extracellular α-amylase and cell wall teichuronic acid. *Colloids Surf B: Biointerfaces* 103:517–522.

Badri Narayanan, K., and Sakthivel, N. 2008. Coriander leaf mediated biosynthesis of gold nanoparticles. *Mater Lett* 62:4588–4590.

Balaji, D. S., Basavaraja, S., Deshpande, R., Mahesh, D. B., Prabhakar, B. K., and Venkataraman, A. 2009. Extracellular biosynthesis of functionalized silver nanoparticles by strains of *Cladosporium cladosporioides* fungus. *Colloids Surf B: Biointerfaces* 68:88–92.

Bansal, V., Poddar, P., Ahmad, A., Sastry, M. 2006. Room temperature biosynthesis of ferroelectric barium titanate nanoparticles. *J Am Chem Soc* 128:11958–11963.

Bansal, V., Rautaray, D., Ahmad, A., Sastry, M. 2004. Biosynthesis of Zirconia nanoparticles using the fungus *Fusarium oxysporum*. *J Mater Chem* 14:3303–3305.

Bansal, V., Rautaray, D., Bharde, A., Ahire, K., Sanyal, A., Ahmad, A., Sastry, M. 2005. Fungus mediated biosynthesis of silica and titania nanoparticles. *J Mater Chem* 15:2583–2589.

Basavaraja, S., Balaji, S. D., Lagashetty, A., Rajasab, A. H., and Venkataraman, A. 2008. Extracellular biosynthesis of silver nanoparticles using the fungus *Fusarium semitectum*. *Mater Res Bull* 43:1164–1170.

Bhainsa, K. C., and D'Souza, S. F. 2006. Extracellular biosynthesis of silver nanoparticles using the fungus *Aspergillus fumigatus*. *Colloids Surf B: Biointerfaces* 47:160–164.

Bharde, A. A., Parikh, R. Y., Baidakova, M., Jouen, S., Hannoyer, B., and Enoki, T. 2008. Bacteria-mediated precursor-dependent biosynthesis of superparamagnetic iron oxide and iron sulfide nanoparticles. *Langmuir* 24:5787–5794.

Binupriya, A. R., Sathishkumar, M., and Yun, S-I. 2010a. Myco-crystallization of silver ions to nanosized particles by live and dead cell filtrates of *Aspergillus oryzae* var. *viridis* and its bactericidal activity toward *Staphylococcus aureus* KCCM 12256. *Ind Eng Chem Res* 49:852–858.

Binupriya, A. R., Sathishkumar, M., and Yun, S-I. 2010b. Biocrystallization of silver and gold ions by inactive cell filtrate of *Rhizopus stolonifer*. *Colloids Surf B: Biointerfaces* 79:531–534.

Castro-Longoria, E., Vilchis-Nestor, A. R., and Avalos-Borja, M. 2011. Biosynthesis of silver, gold and bimetallic nanoparticles using the filamentous fungus *Neurospora crassa*. *Colloids Surf B: Biointerfaces* 83:42–48.

Chen, J. C., Lin, Z. H., and Ma, X. X. 2003. Evidence of the production of silver nanoparticles via pretreatment of *Phoma* sp.3.2883 with silver nitrate. *Lett Appl Microbiol* 37:105–108.

Chithrani, B. D., Ghazani, A. A., and Chan, W. C. W. 2006. Determining the size and shape dependence of gold nanoparticle uptake into mammalian cells. *Nano Lett* 6:662–8.

Daniel, M. C., and Astruc, D. 2004. Gold nanoparticles: Assembly, supramolecular chemistry, quantum-size-related properties, and applications toward biology, catalysis, and nanotechnology. *Chem Rev* 104:293–346.

Das, S. K., Liang, J., Schmidt, M., Laffir, F., and Marsili, E. 2012. Biomineralization mechanism of gold by zygomycete fungi *Rhizopus oryzae*. *ACS Nano* 6:6165–6173.

Dauthal, P., and Mukhopadhyay, M. 2012. *Prunus domestica* fruit extract-mediated synthesis of gold nanoparticles and its catalytic activity for 4-nitrophenol reduction. *Ind Eng Chem Res* 51:13014 – 13020.

David, C., and Michael, C. 2006. Design and numerical modeling of normal-oriented quantum wire infrared photodetector array. *Infrared Phys Technol* 48:227–234.

Dipankar, C., and Murugan, S. 2012. The green synthesis, characterization and evaluation of the biological activities of silver nanoparticles synthesized from *Iresine herbstii* leaf aqueous extracts. *Colloids Surf B: Biointerfaces* 98:112–119.

Du, L., Xia, L., and Feng, J-X. 2011. Rapid extra-/intracellular biosynthesis of gold nanoparticles by the fungus *Penicillium* sp. *J Nanopart Res* 13:921–930.

Dubey, S. P., Lahtinen, M., and Sillanpaa, M. 2010. Green synthesis and characterizations of silver and gold nanoparticles using leaf extract of *Rosa rugosa*. *Colloids Surf A Physicochem Eng Aspects* 364:34–41.

Eustis, S., and El-Sayed, M. A. 2006. Why gold nanoparticles are more precious than pretty gold: Noble metal surface plasmon resonance and its enhancement of the radiative and nonradiative properties of nanocrystals of different shapes. *Chem Soc Rev* 35:209–217.

Farina M., Esquivel, D. M. S., and de Barros, H. G. P. L. 1990. Magnetic iron-sulphur crystals from a magnetotactic microorganism. *Nature* 343:256–258.

Fayaz, A. M., Girilal, M., Rahman, M., Venkatesan, R., and Kalaichelvan, P. T. 2011. Biosynthesis of silver and gold nanoparticles using thermophilic bacterium *Geobacillus stearothermophilus*. *Process Biochem* 46:1958–1962.

Feldheim, D. L., Foss, C. A. 2002. *Metal Nanoparticles: Synthesis, Characterization, and Applications*. Marcel Dekker, New York.

Felson, D. T., Anderson, J. J., and Meenan, R. F. 1990. The comparative efficacy and toxicity of second-line drugs in rheumatoid arthritis. Results of two metaanalyses. *Arthritis Rheum* 33:1449–1461.

Gajbhiye, M., Kesharwani, J., Ingle, A., Gade, A., and Rai, M. 2009. Fungus-mediated synthesis of silver nanoparticles and their activity against pathogenic fungi in combination with fluconazole. *Nanomed Nanotechnol Biol Med* 5:382–386.

Gates, B., Mayers, B., Cattle, B., and Xia, Y. 2002. Synthesis and characterization of uniform nanowires of trigonal selenium. *Adv Funct Mater* 12:219–227.

Gopinath, V., Mubarak Ali, D., Priyadarshini, S., Meera Priyadharsshini, N., Thajuddin, N., and Velusamy, P. 2012. Biosynthesis of silver nanoparticles from *Tribulus terrestris* and its antimicrobial activity: A novel biological approach. *Colloids Surf B: Biointerfaces* 96:69–74.

Gunalan, S., Sivaraj, R., and Venckatesh, R. 2007. *Aloe barbadensis* Miller mediated green synthesis of mono-disperse copper oxide nanoparticles: Optical properties. *Spectrochim Acta A Mol Biomol Spectrosc* 97:1140–1144.

Guo, Y., Wang, F., Zhang, J., et al. 2013. Biomimetic synthesis of calcium carbonate with different morphologies under the direction of different amino acids. *Res Chem Intermed* 39:2407–2415.

Gupta, S., Devi, S., and Singh, K. 2011. Biosynthesis and characterization of Au-nanostructures by metal tolerant fungi. *J Basic Microbiol* 51:601–606.

Gupta, A. J., and Gupta, M. 2005. Synthesis and surface engineering of iron oxide nanoparticles for biomedical applications. *Biomaterials* 26:3995–4021.

Hikosaka, K., Kim, J., Kajita, M., Kanayama, A., and Miyamoto, Y. 2008. Platinum nanoparticles have an activity similar to mitochondrial NADH:ubiquinone oxidoreductase. *Colloids Surf B: Biointerfaces* 66:195–200.

Huang, H., and Yang, X. 2004. Synthesis of polysaccharide-stabilized gold and silver nanoparticles: A green method. *Carbohydrate Res* 339:2627–31.

Huang, J., Zhan, G., Zheng, B., et al. 2011. Biogenic silver nanoparticles by *Cacumen Platycladi* extract: Synthesis, formation mechanism, and antibacterial activity. *Ind Eng Chem Res* 50:9095–9106.

Ingle, A., Rai, M., Gade, A., and Bawaskar, M. 2009. *Fusarium solani*: A novel biological agent for the extracellular synthesis of silver nanoparticles. *J Nanopart Res* 11:2079–2085.

Ionov, L., Sapra, S., Synytska, A., Rogach, A. L., Stamm, M., and Diez, S. 2006. Fast and spatially resolved environmental probing using stimuli-responsive polymer layers and fluorescent nanocrystals. *Adv Mater* 18:1453.

Jacob, S. J., Mohammed, H., Murali, K., and Kamarudeen, M. 2012. Synthesis of silver nanorods using *Coscinium fenestratum* extracts and its cytotoxic activity against Hep-2 cell line. *Colloids Surf B: Biointerfaces* 98:7–11.

Jayaseelan, C., Abdul Rahuman, A., Vishnu Kirthi, A., et al. 2012. Novel microbial route to synthesize ZnO nanoparticles using *Aeromonas hydrophila* and their activity against pathogenic bacteria and fungi. *Spectrochim Acta* A 90:78–84.

Jayaseelan, C., Ramkumar, R., Abdul Rahuman, A., and Perumal, P. 2013. Green synthesis of gold nanoparticles using seed aqueous extract of *Abelmoschus esculentus* and its antifungal activity. *Ind Crops Prod* 45:423–429.

Kalishwaralal, K., Deepak, V., Ram Kumar Pandian, S., and Gurunathan, S. 2009. Biological synthesis of gold nanocubes from *Bacillus licheniformis*. *Bioresour Technol* 100:5356–5358.

Konishi, Y., Ohno, K., Saitoh, N., et al. 2007. Bioreductive deposition of platinum nanoparticles on the bacterium *Shewanella algae*. *J Biotechnol* 128:648–653.

Krishnaraj, C., Jagan, E. G., and Rajasekar, S. 2010. Synthesis of silver nanoparticles using *Acalypha indica* leaf extracts and its antibacterial activity against water borne pathogens. *Colloids Surf B: Biointerfaces* 76:50–56.

Kumar, K. M., Mandal, B. K., Kumar, K. S., Reddy, P. S., and Sreedhar, B. 2013. Biobased green method to synthesise palladium and iron nanoparticles using *Terminalia chebula* aqueous extract. *Spectrochim Acta A Mol Biomol Spectrosc* 102:128–133

Kumar, R. R., Priyadharsani, K. P., and Thamaraiselvi, K. 2012a. Mycogenic synthesis of silver nanoparticles by the Japanese environmental isolate *Aspergillus tamarii*. *J Nanopart Res* 14:860.

Kumar, R., Roopan, S. M., Prabhakarn, A., Khanna, V. G., and Chakroborty, S. 2012b. Agricultural waste *Annona squamosa* peel extract: Biosynthesis of silver nanoparticles. *Spectrochim Acta A* 90:173–176.

Li, G, He, D, Qian, Y, et al. 2012. Fungus-mediated green synthesis of silver nanoparticles using Aspergillus terreus. *Int J Mol Sci* 13:466–476.

Link, S., and El-Sayed, M. A. 2003. Optical properties and ultrafast dynamics of metallic nanocrystals. *Annu Rev Phys Chem* 54:331.

Machado, S., Pinto, L., Grosso, J. P., Nouws, H. P. A., Albergaria, J. T., and Delerue-Matos, C. 2013. Green production of zero-valent iron nanoparticles using tree leaf extracts. *Sci Total Environ* 445:1–8.

Mallikarjuna, K., John Sushma, N., Narasimha, G., Manoj, L., and Deva Prasad Raju, L. 2012. Phytochemical fabrication and characterization of silver nanoparticles by using *Pepper* leaf broth. *Arabian J Chem.* In Press.

Mishra, A, Tripathy, S. K., Wahab, R., et al. 2011. Microbial synthesis of gold nanoparticles using the fungus Penicillium brevicompactum and their cytotoxic effects against mouse mayo blast cancer C_2C_{12} cells. *Appl Microbiol Biotechnol* 92:617–630.

Mishra, A., Tripathy, S. K., and Yun, S-I. 2012. Fungus mediated synthesis of gold nanoparticles and their conjugation with genomic DNA isolated from *Escherichia coli* and *Staphylococcus aureus. Process Biochem* 47:701–711.

Mukherjee, P, Ahmad, A, Mandal, D., et al. 2011. Fungus-mediated synthesis of silver nanoparticles and their immobilization in the mycelial matrix: A novel biological approach to nanoparticle synthesis. *Nano Lett* 1:515–519.

Mukherjee, P., Ahmad, A., Mandal, D., et al. 2001. Bioreduction of $AuCl_4^-$ ions by the fungus, *Verticillium* sp. and surface trapping of the gold nanoparticles formed. *Angew Chem Int Ed* 40:3585–3588.

Mukherjee, P., Roy, M., Mandal, B. P., et al. 2008. Green synthesis of highly stabilized nanocrystalline silver particles by a non-pathogenic and agriculturally important fungus *T. asperellum. Nanotechnology* 19:075103.

Nadagouda, M. N., Castle, A. B., Murdock, R. C., Hussain, S. M., and Varma, R. S. 2010. *in vitro* biocompatibility of nanoscale zerovalent iron particles (NZVI) synthesized using tea polyphenols. *Green Chem* 12:114–122.

Nag, A., Sapra, S., Sengupta, S., et al. 2008. Luminescence in Mn-doped CdS nanocrystals. *Bull Mater Sci* 31:561.

Narayanan, K. B., and Sakthivel, N. 2011. Facile green synthesis of gold nanostructures by NADPH-dependent enzyme from the extract of *Sclerotium rolfsii. Colloids Surf A Physicochem Eng Aspects* 380:156–161.

Narayanan, K. B., and Sakthivel, N. 2011. Heterogeneous catalytic reduction of anthropogenic pollutant, 4-nitrophenol by silver-bionanocomposite using *Cylindrocladium floridanum. Bioresour Technol* 102:10737–10740.

Nayak, R. R., Pradhan, N., Behera, D., et al. 2011. Green synthesis of silver nanoparticle by *Penicillium purpurogenum* NPMF: The process and optimization. *J Nanopart Res* 13:3129–3137.

Oxana, V., Kharissova, H. V., Rasika, D., Boris, I., and Perez, B. O. 2013. The greener synthesis of nanoparticles. *Trends Biotechnol* 31:240–248.

Ponarulselvam, S., Panneerselvam, C., Murugan, K., Aarthi, N., Kalimuthu, K., and Thangamani, S. 2012. Synthesis of silver nanoparticles using leaves of *Catharanthus roseus* Linn. G. Don and their antiplasmodial activities. *Asian Pac J Trop Biomed* 2:574–580.

Prasad, K., and Jha, A. K. 2010. Biosynthesis of CdS nanoparticles: An improved green and rapid procedure. *J Colloid Interface Sci* 342:68–72.

Radhika Rajasree, S. R., and Suman, T. Y. 2012. Extracellular biosynthesis of gold nanoparticles using a gram negative bacterium *Pseudomonas fluorescens. Asian Pac J of Trop Dis* 2:S795–S799.

Rajakumar, G., Rahuman, A. A., Velayutham, K., et al. 2013. Novel and simple approach using synthesized nickel nanoparticles to control blood-sucking parasites. *Vet Parasitol* 191:332–339.

Ramanathan, R., O'Mullane, A. P., Parikh, R. Y., Smooker, P. M., Bhargava, S. K., and Bansal, V. 2011. Bacterial kinetics-controlled shape-directed biosynthesis of silver nanoplates using *Morganella psychrotolerans. Langmuir* 27:714–719.

Rautaray, D., Sanyal, A., Adyanthaya, S. D., Ahmad, A., and Sastry, M. 2004. Biological synthesis of strontium carbonate crystals using the fungus *Fusarium oxysporum*. *Langmuir* 20:6827–6833.

Rehman, A., Majeed, M. I., Ihsan, A., et al. 2011. Living fungal hyphae-templated porous gold microwires using nanoparticles as building blocks. *J Nanopart Res* 13:6747–6754.

Riddin, T. L., Gericke, M., and Whiteley, C. G. 2006. Analysis of the inter- and extracellular formation of platinum nanoparticles by *Fusarium oxysporum* f. sp. *lycopersici* using response surface methodology. *Nanotechnology* 17:3482–3489.

Riddin, T. L., Govender, Y., Gericke, M., and Whiteley, C. G. 2009. Two different hydrogenase enzymes from sulphate-reducing bacteria are responsible for the bioreductive mechanism of platinum into nanoparticles. *Enzyme Microbiol Technol* 45:267–273.

Saha, S., Sarkar, J., Chattopadhyay, D., Patra, S., Chakraborty, A., and Acharya, K. 2010. Production of silver nanoparticles by a phytopathogenic fungus *Bipolaris nodulosa* and its antimicrobial activity. *Digest J Nanomater Biostruct* 5:887–895.

Sangeetha, G., Rajeshwari, S., and Venckatesh, R. 2011. Green synthesis of zinc oxide nanoparticles by aloe barbadensis miller leaf extract: Structure and optical properties. *Mater Res Bull* 46:2560–2566.

Santhanalakshmi, J., Kasthuri, J., and Rajendiran, N. 2007. Studies on the platinum and ruthenium nanoparticles catalysed reaction of aniline with 4-aminoantipyrine in aqueous and microheterogeneous media. *J Mol Catalysis A Chem* 265:283–291.

Sarkar, J., Chattopadhyay, D., Patra, S., et al. 2011. *Alternaria alternata* mediated synthesis of protein capped silver nanoparticles and their genotoxic activity. *Digest J Nanomater Biostruct* 6:563–573.

Sarkar, J., Ray, S., Chattopadhyay, D., Laskar, A., and Acharya, K. 2012. Mycogenesis of gold nanoparticles using a phytopathogen *Alternaria alternata*. *Bioprocess Biosyst Eng* 35:637–643.

Sathishkumar, G., Gobinath, C., Karpagam, K., Hemamalini, V., Premkumar, K., and Sivaramakrishnan, S. 2012. Phyto-synthesis of silver nanoscale particles using *Morinda citrifolia* L. and its inhibitory activity against human pathogens. *Colloids Surf B: Biointerfaces* 95:235–240.

Schaffer, B., Hohenester, U., Trugler, A., and Hofer, F. 2009. High-resolution surface plasmon imaging of gold nanoparticles by energy-filtered transmission electron microscopy. *Phys Rev B* 79. (4) 041401.ISSN 1098–0121.

Shankar, S. S., Rai, A., Ahmad, A., and Sastry, M. 2004. Rapid synthesis of Au, Ag, and bimetallic Au core–Ag shell nanoparticles using Neem (*Azadirachta indica*) leaf broth. *J Colloid Interface Sci* 275:496–502.

Shaw, I. C. 1999. Gold-based therapeutic agents. *Chem Rev* 99:2589–2600.

Sheikhloo, Z., and Salouti, M. 2012. Intracellular biosynthesis of gold nanoparticles by Fungus *Phoma macrostoma*. *Synthesis Reactivity Inorg Metal Org Nano Metal Chem* 42:65–67.

Sheikhloo, Z., Salouti, M., and Katiraee, F. 2011. Biological synthesis of gold nanoparticles by fungus *Epicoccum nigrum*. *J Clust Sci* 22:661–665.

Song, Y. J., Eun-Yeong, K., and Beom Soo, K. 2010. Biological synthesis of platinum nanoparticles using *Diopyros kaki* leaf extract. *Bioprocess Biosyst Eng* 33:159–164.

Sukirtha, R., Priyanka, K. M., Antony, J. J., et al. 2012. Cytotoxic effect of Green synthesized silver nanoparticles using *Melia azedarach* against *in vitro* HeLa cell lines and lymphoma mice model. *Process Biochem* 47:273–279.

Suresh, A. K., Pelletier, D. A., Wang, W., et al. 2011. Biofabrication of discrete spherical gold nanoparticles using the metal-reducing bacterium *Shewanella oneidensis*. *Acta Biomater* 7:2148–2152.

Syed, A., and Ahmad, A. 2012. Extracellular biosynthesis of platinum nanoparticles using the fungus *Fusarium oxysporum*. *Colloids Surf B: Biointerfaces* 97:27–31.

Uddin, M. J., Chaudhuri, B., Pramanik, K., Middya, T. R., and Chaudhuri, B. 2012. Black tea leaf extract derived Ag nanoparticle-PVA composite film: Structural and dielectric properties. *Mater Sci Eng* 177:1741–1747.

Vahabi, K., Ali Mansoori, G., AND Karimi, S. 2011. Biosynthesis of silver nanoparticles by fungus Trichoderma Reesei (A route for large-scale production of AgNPs). *Insciences J* 1:65–79.

Velayutham, K., Rahuman, A. A., Rajakumar, G., et al. 2012. Evaluation of *Catharanthus roseus* leaf extract-mediated biosynthesis of titanium dioxide nanoparticles against *Hippobosca maculate* and *Bovicola ovis*. *Parasitol Res* 111:2329–2337.

Venu, R., Ramulu, T. S., Anandakumar, S., Rani, V. S., and Kim, C. G. 2008. Bio-directed synthesis of platinum nanoparticles using aqueous honey solutions and their catalytic applications. *Colloids Surf B: Biointerfaces* 66:195–200.

Verma, V. C., Singh, S. K., Solanki, R., and Prakash, S. 2011. Biofabrication of anisotropic gold nanotriangles using extract of endophytic *Aspergillus clavatus* as a dual functional reductant and stabilizer. *Nanoscale Res Lett* 6:16–22.

Vigneshwaran, V., Ashtaputre, N. M., Varadarajan, P. V., Nachane, R. P., Paralikar, K. M., and Balasubramanya, R. H. 2007. Biological synthesis of silver nanoparticles using the fungus *Aspergillus flavus*. *Mat Lett* 61:1413–1418.

Vijayakumar, P. S., and Prasad, B. L. V. 2009. Intracellular biogenic silver nanoparticles for the generation of carbon supported antiviral and sustained bactericidal agents. *Langmuir* 25:11741–11747.

Vijayakumar, M., Priya, K., Nancy, F. T., Noorlidah, A., and Ahmed, A. B. A. 2013. Biosynthesis, characterisation and anti-bacterial effect of plant-mediated silver nanoparticles using *Artemisia nilagirica*. *Ind Crops Prod* 41:235–240.

Vijayaraghavan, K., Kamala Nalini, S. P., Udaya Prakash, N., and Madhankumar, D. 2012. One step green synthesis of silver nano/microparticles using extracts of *Trachyspermum ammi* and *Papaver somniferum*. *Colloids Surf B: Biointerfaces* 94:114–117.

Wang, L., Yang, B., and Zhang, J. 2010a. Extracellular biosynthesis and transformation of selenium nanoparticles and application in H_2O_2 biosensor. *Colloids Surf B: Biointerfaces* 80:94–102.

Wang, M. C. P., Zhang, X., Majidi, E., Nedelec, K., and Gates, B. D. 2010b. Electrokinetic assembly of selenium and silver nanowires into macroscopic fibers. *ACS Nano* 4:2607–2614.

Wei, X., Luo, M., Li, W., et al. 2012. Synthesis of silver nanoparticles by solar irradiation of cell-free *Bacillus amyloliquefaciens* extracts and $AgNsO_3$. 2012. *Bioresour Technol* 103:273–278.

Xu, P., Guang Ming, Z., Dan Lian, H., and Chong Ling, F. 2012. Use of iron oxide nanomaterials in wastewater treatment: A review. *Sci Total Environ* 424:1–10.

Yallapa, S., Manjanna, J., and Sindhe, M. A. 2013. Microwave assisted rapid synthesis and biological evaluation of stable copper nanoparticles using *T. arjuna* bark extract. *Spectrochim Acta A Mol Biomol Spectrosc* 110:108–115.

Yong, P., Rowson, N. A., Farr, J. P. G., Harris, I. R., and Macaskie, L. E. 2002. Bioaccumulation of palladium by *Desulfovibrio desulfuricans*. *J Chem Technol Biotechnol* 77:593–601.

Zhang, W., Chen, Z., Liu, H., Zhang, L., Gao, P., and Li, D. 2011a. Biosynthesis and structural characteristics of selenium nanoparticles by *Pseudomonas alcaliphila*. *Colloids Surf B: Biointerfaces* 88:196–201.

Zhang, X., Huang, F., Shen, Y., Xie, A., and Lin, Z. 2011b. Biomimetic growth of $CaCO_3$ pancakes on the leaves of *Epipremnum aureum*. *Russian J Phys Chem A* 85:2187–2191.

18

Microbial Remediation of Acid Mine Drainage

Laura G. Leff, Suchismita Ghosh, G. Patricia Johnston
and Alescia Roberto

CONTENTS

18.1 Introduction

Acid mine drainage (AMD) is common worldwide and represents a major threat to water quality and ecosystem health (Sasowsky et al., 2000; Kimball et al., 2002; Neculita et al., 2007). Low pH along with leaching of heavy metals under AMD conditions deteriorates water quality and has adverse effects on aquatic life (Jennings et al., 2008).

AMD is caused by mining of various minerals, which results in exposure of sulphide-bearing materials to oxygen and water. The impact of AMD is compounded by leaching of metals and is inherently biogeochemical in nature. On the one hand, some microorganisms of specific types thrive in AMD-impacted environments and exacerbate the situation, while others can be used to remediate AMD-impacted systems. In this chapter, we discuss the biogeochemistry of AMD and the unique role microorganisms play in AMD, remediation of AMD and approaches used for bioremediation.

18.2 Biogeochemistry and Microbiology of AMD Environments

AMD occurs when metal sulphide minerals, such as iron sulphides (e.g. pyrite [FeS_2], chalcopyrite [$CuFeS_2$], pyrrhotite [FeS], arsenopyrite [FeAsS] and pentlandite [FeNiS]), are exposed to water and oxygen. Spontaneous oxidation of these and other sulphide minerals result in the generation of metal-laden, acidic solutions.

Although a wide range of different physicochemical properties contribute to AMD formation (e.g. temperature and solution chemistry), dissolution rates of sulphide minerals and net acidity of resulting effluent is determined by the composition of parent rock material and its grain size (Baker and Banfield, 2003). In most acid mine sites, pyrite is the predominant mineral (Skousen et al., 2000).

The oxidation of pyrite is an electrochemical process that occurs via abiotic and/or biotic processes (Equation 18.1). The main oxidants of pyrite are oxygen (O_2) and ferric iron (Fe^{3+}), but because oxygen is a less effective oxidant than Fe^{3+}, dissolution of pyrite is primarily driven by initial oxidation of ferrous iron (Fe^{2+}) by O_2 (Equation 18.2), resulting in a net gain of Fe^{3+} to drive further sulphide oxidation (Equation 18.3; Baker and Banfield, 2003). However, oxidation of Fe^{2+} by O_2 is slow under acidic conditions (pH $\leq$ 4).

$$FeS_2 + 3.5O_2 + H_2O \rightarrow Fe^{2+} + 2SO_4^{2-} + 2H^+ \tag{18.1}$$

$$14Fe^{2+} + 3.5O_2 + 14H^+ \rightarrow 14Fe^{3+} + 7H_2O \tag{18.2}$$

$$FeS_2 + 14Fe^{3+} + 8H_2O \rightarrow 15Fe^{2+} + 2SO_4^{2-} + 16H^+ \tag{18.3}$$

Typically, aerobic iron- and sulphur-oxidising bacteria and archaea accelerate pyrite oxidation by several orders of magnitude via regeneration of Fe^{3+} through metabolic processes (Equation 18.3; Singer and Stumm, 1970; Johnson and Hallberg, 2003). Edwards et al. (2000) suggest that ~75% of AMD occurrences are derived from biological oxidation of pyrite.

Oxidation of sulphides via bacterial–mineral interactions occurs by three mechanisms: indirect leaching, contact ('direct') leaching or symbiotic leaching (reviewed by Tributsch, 2001). In brief, indirect leaching is oxidation of Fe^{2+} to Fe^{3+} in solution by free-living microorganisms. In contrast, contact leaching occurs as a result of sulphur extraction and/or electrochemical dissolution of pyrite facilitated at and/or within the extracellular polymeric layer of chemoautotrophic biofilms. In the case of sulphur extraction, oxidation of sulphur occurs via a disproportionation pathway, where tetrathionate, a sulphur intermediate, is oxidised to colloidal elemental sulphur (S^0) (Equation 18.4; Norlund et al., 2009). S^0 aggregates act as intermediate energy storage compounds, which may be used by co-occurring heterotrophic microbes during oxidation of organic molecules (Equation 18.5; Rojas-Chapana et al., 1996).

$$S_4O_6^{2-} + H_2O \rightarrow S_2O_3^{2-} + SO_4^{2-} + S_{colloidal}^0 + 2H^+ \tag{18.4}$$

$$HCOOH + S^0 \rightarrow CO_2 + H_2S \tag{18.5}$$

Electrochemical dissolution occurs as a result of sulphide decomposition via electron extraction by Fe^{3+} ions accumulated in extracellular polymeric layer (EPL) (Tributsch, 2001). Symbiotic leaching occurs during utilisation of soluble and particulate sulphides, which are liberated during contact leaching.

Microbes inhabiting AMD-impacted sites represent metabolically and physiologically diverse communities composed of bacteria (e.g. proteobacteria, nitrospira, firmicutes and acidobacteria), archaea (e.g. *Thermoplasma* spp.) and eucarya (ciliates, flagellates, algae and amoeba). Their phylogeny and physiology are explained in more detail elsewhere (e.g. Hallberg, 2010).

Microbial community composition, which is influenced by local mineralogy and variations in oxygen content, pH, temperature, ionic strength, reducing conditions, metal composition and concentrations, and the physical location, varies among AMD sites (reviewed in Johnson and Hallberg, 2003; Hallberg, 2010). For instance, 'acid streamers' of the Cae Coch pyrite mine in Wales (pH 2–3) are dominated by iron-oxidising bacteria (*Acidithiobacillus* spp. and '*Ferrovum myxofaciens*'), whereas the warmer (35–45°C) and more acidic (pH 0–2) streams at the Richmond mine in Iron Mountain, California, are dominated by both iron-oxidising bacteria *Leptospirillum* spp. and archaea '*Ferroplasma acidarmanus*'. In comparison, pyrite mines with sulphur species as their main energy source, like the Frasassi Cave Complex in Italy, are dominated by sulphur oxidisers *Acidithiobacillus thiooxidans*, *Sulfobacillus* and a bacterium related to *Acidimicrobium* (Jones et al., 2011).

To thrive in AMD environments, microbial communities form consortial arrangements, where co-occurrence of specific species is based on coordination of synergistic metabolic interactions (Norlund et al., 2009). For example, synergistic relationships occur in aggregates of the chemoautotrophic iron and sulphur oxidiser, *Acidithiobacillus ferrooxidans*, and the autotrophic sulphur oxidiser and facultative heterotrophic elemental sulphur reducer, *Acidiphilium* sp. (Norlund et al., 2009). *A. ferrooxidans* forms an external layer that surrounds *Acidiphilium* sp., allowing partitioning of sulphur metabolism: disproportionation of tetrathionate to colloidal S^0 in the exterior layer, and the coupling of oxidation of organic molecules with the reduction of S^0 to sulphide in the interior (Equations 18.4 and 18.5; Norlund et al., 2009). Partitioning of specific elemental metabolism, through the formation of geochemical niches, reduces competition for nutrients (Headd and Summers, 2013).

In summary, because AMD-impacted ecosystems host a plethora of micro-organisms with varying physiological capabilities, further characterisation of functional consortia could advance remediation strategies for AMD. As described below, remediation alters microbial community composition and use of microbes in biomediation has advantages, under some circumstances, over abiotic strategies.

18.3 Remediation Strategies for AMD

The long-term and expensive liability of AMD has led to the development of an array of techniques for control and/or abatement. One such technique, 'source control', has been utilised as a preventative measure for controlling the formation of AMD, such as excluding oxygen and/or water by flooding and sealing abandoned deep mines. Another approach involves underwater storage of potentially acid-producing disposed mine tailings, thus minimising contact between minerals and dissolved oxygen (Li et al., 1997). Use of a layer of sediment or organic material over the tailings improves effectiveness by limiting oxygen and preventing resuspension. However, 'source control' measures are not always possible given the inherent variability among mines and their environmental conditions, and the practical difficulties involved in the prevention of AMD formation. Thus, abiotic and/or biotic remediation strategies are used to minimise the impact of AMD on receiving streams and rivers.

Two principal goals of remediation are to reduce metal loading (e.g. iron, aluminium and manganese) and to raise the pH. To achieve these goals, remediation strategies utilise physical, chemical or biological processes either singly or in combination. Often, these mitigation strategies involve a range of processes, including pH control, oxidation/reduction, adsorption/absorption, complexation, flocculation/settling, chelation, electrochemistry, biological mediation, sedimentation and ion exchange and crystallisation. However, in most cases, neutralisation (pH control) is the preferred method for treating AMD. Selection of an appropriate remediation system depends on economic constraints and environmental conditions of the impacted sites (i.e. flow rates and acid loads).

Remediation of AMD can be divided into either abiotic or biotic treatment strategies, and in turn each of these strategies can be further subdivided as either active or passive treatment systems (Figure 18.1; Johnson and Hallberg, 2005a,b). Active abiotic systems involve continuous application of

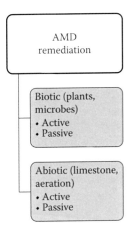

FIGURE 18.1
AMD remediation categories (for examples of each type and details of the differences, see Johnson and Hallberg, 2005a).

alkaline materials to neutralise acidic influent, whereas passive systems use natural or constructed systems to accelerate the rate at which acidic influent is neutralised. In comparison, biotic treatment strategies utilise biological organisms for neutralisation of acidic influent and immobilisation of metal contaminants.

18.3.1 Abiotic Remediation

In both active and passive abiotic systems, mitigation of AMD is primarily achieved by the neutralisation of acidic water. An increase in pH accelerates oxidation of metal ions (e.g. Fe^{2+} to Fe^{3+} and Mn^{2+} to Mn^{3+}/Mn^{4+}) present in the solution, which results in the precipitation of these ions as hydroxides and carbonates (Stumm and Morgan, 1996). For example, iron, one of the dominant metals in AMD, oxidises to ferric iron, which is subsequently reduced to ferric hydroxide. Dissolution of these hydroxides results in the formation of an iron-rich sludge, which, depending on mine water chemistry, may co-precipitate other metals, such as manganese.

Active treatment systems improve the water quality of AMD sites via input of chemical reagents and/or energy (Younger et al., 2002). These treatment strategies involve continuous dosing with alkaline chemicals, such as NaOH, $Ca(OH)_2$, CaO, Na_2CO_3 or NH_3 (Skousen et al., 1990) and incorporation of oxygen to accelerate metal ion oxidation and precipitation (Younger et al., 2002).

Selection of an active treatment system and chemical reagent(s) used in remediation is dependent upon several factors, including environmental parameters of the impacted site (e.g. pH, redox, total suspended solid concentration and metal concentrations), flow rate and geographical location. In addition, the selection of remediation projects are often determined by the extent of mining activity, environmental regulations of local, regional and federal governments, and availability of electrical power to the impaired site.

Generally, active treatment systems are most appropriate at active mine sites, where changes in influent chemistry are frequent and land availability is limited (Younger et al., 2002). To aid in the selection and cost-effective design of treatment strategies, computer programs, such as AMDTreat (Means et al., 2003), are often utilised. Through the use of computer-aided design modules, active systems have become very reliable in terms of effective neutralisation of AMD effluent acidity and reducing effects on receiving waters. Active systems often employ multi-step addition of metal-removing reagents along with neutralising agents (Aube and Payant, 1997). Other conventional strategies, such as ion exchange, solvent extraction, activated carbon adsorption, cementation, reverse osmosis and evaporation, have been used; however, they are not suitable to handle large volumes of AMD, especially when pre-treatment is required (Gaydardjiev et al., 1996).

Although active remediation strategies are very effective in raising the influent pH of AMD sites, a large variation in alkalinity generation and effective removal of metals remains an issue (Faulkner and Skousen, 1993), likely the result of chemical variability of mine waters (Hedin et al., 1994). Additionally, these technologies are not cost-effective or sustainable, as AMD may continue for decades after a mines' decommissioning; disposal of sludge generated is expensive and contributes to long-term land and water pollution (Matlock et al., 2002; Kurniawan et al., 2006).

Abiotic, passive treatment systems require large tracks of land and are most well suited for closed mine sites. The most common treatment methods include open limestone channels, open limestone drains, limestone-leaching beds, slag-leaching beds, diversion wells, anoxic limestone drains (Hedin and Watzlaf, 1994), successive alkalinity producing systems, also referred to as vertical flow wetlands, and reduction and alkalinity-producing systems (Zipper and Jage, 2001). With these strategies, the pH of mine drainage is raised either by direct contact of acidic water with alkaline rock, like limestone, or via the mixing of acid mine effluent with alkaline water. Often, passive treatment systems use some form of aggregated-carbonate rock with or without the addition of organic matter, and are designed to accommodate slow reaction rates. Because of the slower reaction kinetics of neutralising AMD acidity in these systems, residence time and the dissolution rate of alkaline material (usually limestone) are a determining factor in the design.

The stable chemistry and flow rates of AMD make passive treatment more economical than active (Skousen and Ziemkiewicz, 2005). However, abiotic passive treatment systems have drawbacks. In AMD with high concentrations of ferric iron or aluminium, the buildup of hydroxide precipitates decrease anoxic limestone drain permeability leading to remediation failure.

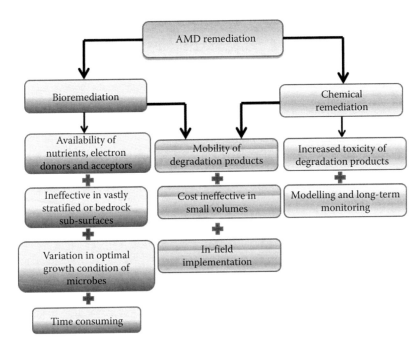

FIGURE 18.2
Summary of important limitations on remediation.

Because of the limitations of abiotic remediation methods, biotic AMD treatments provide some advantages over other forms of remediation strategies, including low capital and labour cost, sustainability, and minimal detrimental effects, such as introduction of high concentrations or organic and inorganic chemicals, on the environment (Figure 18.2).

18.3.2 Biotic Remediation

Biotic remediation utilises biological means (microbial metabolism, phytoremediation, etc.) to ameliorate AMD by increasing pH, and removal of metals via reduction of sulphate to insoluble sulphide. Phytoremediation is carried out in constructed wetlands that have relatively low maintenance costs. However, improvement of AMD water quality by construction of wetlands with macrophytes (*Typha* and *Phragmites*) has not been very effective in treating AMD (Johnson and Hallberg, 2002). Therefore, bioremediation of AMD relies primarily on the ability of microorganisms to generate alkalinity and immobilise metals, thus reversing reactions leading to AMD. The abundance of both ferric iron and sulphate in AMD makes reduction of these two species of prime importance in AMD-impacted waters.

Microbial bioremediation strategies utilise the ability of microbes to generate alkalinity and/or immobilise metal species. Microbial reductive processes such as denitrification, methanogenesis, sulphate reduction and iron and manganese reduction are responsible for increased alkalinity. However, factors such as chemical composition, availability of electron donors/acceptors, temperature and pH determines the extent to which microbes can contribute to neutralisation of AMD. Microbes are capable of utilising many combinations of electron donors and electron acceptors to drive their metabolism. Such metabolic diversity allows them to survive, adapt and, indeed, thrive in harsh environments impacted by AMD. Bioremediation applies these basic principles to select suitable combinations of microbes that can succeed at available electron donor/acceptor/contaminant concentrations and in turn remediate AMD.

Active biological treatment systems, such as sulphidogenic bioreactors (Boonstra et al., 1999; Johnson, 2000), depend on the production of H_2S by sulphate-reducing bacteria (SRB) to remove metal sulphides and generate alkalinity. To optimise H_2S production, bioreactors are designed to protect SRB that are sensitive to acidity, by pre-treatment with alkaline chemicals. SRB belonging to groups *Desulfovibrio*, *Desulfomicrobium*, *Desulfobacter* and *Desulfotomaculum* (Luptakova and Kusnierova, 2005; Martins et al., 2008) utilised in these systems convert sulphate to sulphide. Additionally, these organisms in the presence of organic carbon generate bicarbonate under anaerobic conditions (Bechard et al., 1994; Kuyucak and St-Germain, 1994). Carbon compounds (e.g. lactic acid and acetate) used by SRB are generally products of microbial fermentation that occur under anaerobic conditions.

The chemical basis of SRB remediation involves sulphate reduction coupled with organic matter oxidation:

$$2CH_2O + SO_4^{2-} + H^+ \rightarrow H_2S + 2HCO_3^- \tag{18.6}$$

Bicarbonate released in the process results in alkalinity. This first step of sulphate reduction produces H_2S, which then forms insoluble metal (Me) complexes that are precipitated as sulphides:

$$H_2S + Me^{2+} \rightarrow MeS + 2H^+ \tag{18.7}$$

One example of passive bioremediation is constructed water bodies (wetlands) that are designed with an aerobic top layer and anaerobic bottom layer (Goldsborough and Robinson, 1996). The aerobic layer promotes metal oxidation and hydrolysis, causing precipitation and physical retention of Fe, Al and Mn hydroxides, while the anaerobic bottom layer allows reducing reactions, such as sulphate reduction. The reduction of metals occurs by dissimilatory metal reduction, where microbes utilise metals as the terminal electron acceptor for anaerobic respiration (Ramasamy and Kamaludeen, 2006).

Under anaerobic conditions, heterotrophic bacteria, including *Pseudomonas* sp., *Clostridium* sp. and *Desulfovibro* sp., use iron and manganese as electron acceptors, resulting in the reduction of these species to metal sulphides that precipitate.

These systems focus on plant–microbe-mediated interactions for the removal of metals and acidity from AMD (e.g. Greger and Landberg, 1999; Mendez and Maier, 2008). Plants used in this process are fast-growing, high biomass-producing, deep-rooted and capable of tolerating and accumulating high metal concentrations. Plant root exudates stimulate the survival and action of bacteria, and its root system assists in the spread of bacteria through the soil.

Another approach to bioremediation is through the use of bioreactors. In compost bioreactors, key reactions occur under anaerobic conditions and alkalinity, and biogenic sulphides are generated by microbially mediated reactions. Indigenous iron and SRB play an important role in compost bioreactors (Vile and Wieder, 1993). Filtration of suspended and colloidal materials and adsorption of metals by the organic matrix improves AMD quality in compost bioreactors. The reducing and alkalinity-producing system (Kepler and McCleary, 1994) is a variation of this (Younger et al., 2002). In this system, removal of oxygen and reduction of iron and sulphate occurs as AMD seeps downwards through a layer of compost and pass through a limestone gravel bed. Finally, water is drained into a sedimentation pond to precipitate and retain iron hydroxides.

Permeable reactive barriers (PRBs) used to mitigate AMD are based on the same principles as compost bioreactors (Benner et al., 1997). PRBs are constructed by digging a trench or pit in the flow path of contaminated water followed by filling with reactive material (such as mixture of organic solids and limestone gravel: Younger et al., 2002).

In iron-oxidation bioreactors, *Acidithiobacillus ferrooxidans* (an obligate acidophile capable of oxidising a variety of reduced inorganic sulphur compounds) is immobilised onto a solid matrix (packed bed reactors also referred to as biological contactors [Long et al., 2003]). There is considerable biodiversity of ferrous iron-oxidising procaryotes, so other species may be employed (Hallberg and Johnson, 2001). Biosulfide and Thiopaq are two technologies that use off-line sulphidogenic bioreactors. The Biosulfide system is composed of two independently operating components, a biological and a chemical (Rowley et al., 1997). The chemical circuit receives untreated AMD where it is exposed to hydrogen sulphide generated in the biological circuit. The semi-treated AMD then enters the biological circuit that harbours SRB. Effectiveness can be improved by external alkali addition (inorganic) over that produced by SRB. Unlike the Biosulfide process, the Thiopaq system utilises two distinct microbial processes, that is, precipitation of metal sulphides due to conversion of sulphate to sulphide by SRB and use of sulphide-oxidising bacteria to convert hydrogen sulphide to elemental sulphur.

18.4 Remediation Impacts on Microbial Communities

Remediation of AMD impacts the extant microbial community by directly affecting their composition, structure and function. The microbial community structure in AMD systems is shaped by environmental factors (e.g. pH, temperature and metal concentrations [Bond et al., 2000; Baker and Banfield, 2003; Zhang et al., 2007]). Microbes employ a variety of mechanisms to resist and cope with toxic metals, including metal oxidation, metal reduction, methylation, demethylation, enzymatic reduction, metal–organic complexion, metal ligand degradation, intracellular and extracellular metal sequestration, metal efflux pumps, exclusion by permeability barrier, and production of chelators.

Changes in water chemistry resulting from remediation affect the survival of existing microflora. For example, Ghosh et al. (2012) observed that alteration in pH and metal ion concentration over time played a major role in driving the shifts in the bacterial community structure at a site remediated via steel slag leach beds and open limestone channel mitigating systems. Specifically, in post-remediation, there was an overall increase in bacterial abundance, and the community structure was similar to an unimpacted reference stream. Decreased bacterial abundance prior to remediation could be linked to the restricted microbial habitats in AMD (Baker and Banfield, 2003).

18.5 Role of Extant Microbial Communities in Remediation Success

A lack of understanding of the role of extant microbial communities may limit the success of remediation. Basic metabolic groups in AMD environments include lithoautotrophs, that oxidise Fe^{2+} and S^- released by pyrite dissolution, organo-heterotrophs that derive carbon from lithoautotrophs, lithohet-erotrophic iron and sulphur oxidisers, and anaerobes capable of coupling the oxidation of sulphur or organic carbon to Fe^{3+} reduction. Intermediate sulphur and sulphoxy oxidation reaction products also act as electron donors. Sulphate and Fe^{3+} act as electron acceptors mostly under micro-aerophilic and anoxic conditions. Photosynthesis is an important source of energy and fixed carbon in areas of AMD exposed to sunlight (Hiraishi et al., 2000).

Microbial communities of AMD environments contribute to the maintenance of acidity, which results in further deterioration of environmental conditions (Baker and Banfield, 2003). Studies conducted over a broad range of geographical regions have revealed characteristic bacterial communities found only in these environments (Druschel et al., 2004). Bacteria in the genera *Thiobacillus*, *Sulfolobus* and *Thiomicrospira* are common inhabitants of acidic environments, and tend to be important players in AMD maintenance.

These organisms derive energy from the oxidation of sulphide minerals resulting in the production of sulphuric acid.

Post-remediation in microbial community composition may aid AMD mitigation due to increased metabolic diversity. The success of remediation depends on the knowledge of factors that contribute to the deterioration of that environment and an understanding of the role played by microbes.

18.6 Lab- and Small-Scale Experiments

Bacterial metabolism, specifically that of SRB, can be of significant value in the removal of metals from AMD water (Dvorak et al., 1992). Because these organisms promote the precipitation of metals, SRB have been used in AMD bioremediation, and laboratory studies have been conducted to help develop bioremediation strategies. For example, the edible oil substrate (EOS®) process involves utilisation of emulsified 'edible' plant oils, such as soya oil, for groundwater bioremediation (Lindow and Borden, 2004, 2005). The EOS process is a cost-effective process due to slow degradation and hydrogen release. The efficacy of such low-cost effective alternatives to enhance AMD bioremediation has been evaluated in an AMD microcosms study by Lindow and Borden (2004). EOS stimulated the growth of iron oxidisers and SRB.

In addition, Gilbert et al. (2005) demonstrated that columns inoculated with SRB had high initial removal efficiency of heavy metals from AMD water. They also found that municipal compost used as a carbon source did not sustain the growth of SRB; however, metal removal was mostly achieved through precipitation of hydroxides and hydrated carbonates due to increased pH and sorption. Martins et al. (2008) described a two-stage AMD bioremediation strategy using a calcite tailing column and an anaerobic flow reactor and sludge as inoculum for SRB.

Overall, these lab studies suggest that SRB are promising candidates in AMD bioremediation. Use of SRB provides simpler, less intrusive and cost-effective treatment strategies for AMD. These studies also looked at alternative low-cost, readily available, nutrient sources and suitable electron donors or electron acceptors to encourage proliferation of the SRB used in remediation.

18.7 Case Studies of Remediation

The major constraints for field studies on AMD bioremediation are that each site is unique in its intrinsic parameters (such as metal contaminants and microbial community) and physical parameters (pH, metal concentration, hydrology, etc.) (Figure 18.2). Such practical considerations lead to the choice of materials and design that requires proper optimisation for the success of bioremediation. Also, technologies for these impacted sites must be low-cost, simple to implement and require little or no ongoing maintenance. Case studies are described below highlighting some variations in approaches and outcomes.

In Pennsylvania, the Blacklick Creek Watershed contains many surface coal mines and refuse dumps that contribute to a high acid load to the streams (Black Creek Watershed Association, 2006). Based on a pilot system, the Black Creek Watershed Association (BCWA) built a sulphate-reducing bioreactor in 2002 (Acid Mine Drainage Clearinghouse, 2004). The reactor generates alkalinity, reduces sulphate and decreases iron and aluminium concentrations. The resulting effluent has met water quality standards, with pH of 6.6 and about 99% removal of aluminium, iron and nickel. However, the reactor failed to decrease manganese concentrations.

The Wheal Jane mine located in the Carnon Valley, Cornwall, UK, is a closed metal mining site. AMD was released into the Fal Estuary and Carnon River (Hallberg and Johnson, 2003). Extreme acidity, high concentrations of iron, zinc, cadmium and arsenic and yellow-orange precipitate led to the need for remediation. A passive treatment plant was built in 1994 to treat AMD before it reached the Carnon River (Whitehead and Prior, 2005). The treatment plant had multi-cell treatment systems that utilised one of three pretreatment methods to raise pH: lime dosing to pH 5.0, an anoxic limestone drain or a lime-free system without pre-treatment. Mine water was allowed to flow through an anaerobic cell for sulphate reduction by SRB. Increase in pH was observed in all three bioreactors. However, acidic, oxygenated water entering the bioreactors created suboptimal conditions for SRB growth as they failed to tolerate the low pH leading to the poor performance of the bioreactor (Johnson and Hallberg, 2005a,b).

The Lilly/Orphan Boy Mine located south of Elliston, Montana, is an abandoned lead mine (MWTP, 2004). An *in situ* bioreactor was constructed in 1994 with a permeable organic substrate for sulphate reduction by SRB. The bioreactor was monitored from 1994 to 2005, and is still in place. The bioreactor was effective at raising the pH to near neutral and decreasing most metal concentrations such as iron by 65% and magnesium by 76% and almost complete removal of aluminium, arsenic, cadmium, copper and zinc (MWTP, 2004). The results suggest that the construction of an *in situ* bioreactor involves minimal cost, maintenance, and space requirements, and may be suitable for remote, difficult-to-access, abandoned mine sites.

18.8 Microbial Metabolism and AMD Bioremediation

Bioremediation of AMD relies on capabilities of microbial communities to mitigate acidity via a suite of diverse metabolic processes (Johnson and Hallberg,

2005a,b). For example, denitrification, ammonification, methanogenesis, and sulphate, iron and manganese reduction are involved in the alkalinisation of AMD, whereas microbial-mediated mobilisation of metals occurs via the dissimilatory reduction of sulphate to sulphide and biological oxidation of ferric iron to ferrous iron as mentioned in the earlier sections. The production of alkalinity through microbial action can be accomplished through different physiological processes; for example, ammonification and methanogenesis use hydrogen ions to generate alkalinity (Ibeanusi et al., 2012).

Denitrification is an alternative approach to reduce the acidity of AMD systems through microbial nitrate reduction to nitrogen gas (Baeseman et al., 2005). Although denitrification is a predominantly heterotrophic process and requires organic carbon, which may be limited in AMD-impacted environments, facultative heterotrophic bacteria may carry out denitrification under carbon-limiting conditions. Autotrophic denitrification has also been described, in which acetate, thiosulphate, ferrous iron and pyrite (all common compounds present in AMD) are used as electron donors (Baeseman et al., 2005). For example, *Paracoccus ferrooxidans* strain BDN-1, a facultative autotroph, isolated from a denitrification bioreactor has been shown to use thiosulphate and thiocyanate as inorganic electron donors, and nitrate, nitrite, nitrous oxide as well as oxygen as electron acceptors (Kumaraswamy et al., 2006).

More recently, genes involved in the nitrogen cycle have been isolated from AMD environments (Xie et al., 2011). A total of 35 *nif* genes (for nitrogen fixation), 78 genes (*ureC* and *gdh*) for ammonification and 70 genes (*narG*, *nirS*, *nirK*, *narB*, *norB*, *nosZ*) for denitrification were detected in AMD environments. Most of the sequences were related to uncultured organisms.

Iron and manganese are both important under AMD conditions (Johnson and Hallberg, 2005a,b). Iron reduction is ubiquitous under anaerobic conditions and is altered by dissimilatory iron-reducing bacteria, including *Clostridium*, *Bacteroides* and *Geobacter* species (Wang et al., 2009). Increased alkalinity is achieved by the reduction of iron compounds, as indicated in Equation 18.8 (reviewed by Johnson and Hallberg, 2005a,b):

$$Fe(OH)_3 + 3H^+ + e^- \rightarrow Fe^{2+} + 3H_2O \qquad (18.8)$$

Iron-oxidising, heterotrophic, acidophilic bacteria such as *Ferromicrobium acidophilus* and *Acidomicrobium ferrooxidans* are capable of reducing iron and using ferric iron as an electron acceptor under low redox potentials (Bridge and Johnson, 1998; Bond et al., 2000). In AMD environments, both iron- and sulphur-oxidising bacteria depend on mineral sulphides as their energy source (Natarajan et al., 2006). More recently, Sanchez-Andrea et al. (2012) reported that members of the genera *Acidiphilium* and *Acidithiobacillus* are capable of iron reduction under anoxic conditions.

Fe(III) reducers tend to use simpler molecules as organic electron donors than SRB, and under steady-state conditions, they compete with SRB for carbon sources. However, when organic electron donors are in excess, iron and sulphate reducers can survive without competition (Barlett et al., 2012 and references therein). SRB can oxidise ferrous iron at pH > 5 under anaerobic, aerobic or partially anaerobic conditions (Gadd, 2010). For instance, *Acidothiobacillus ferrooxidans, Leptospirillum ferrooxidans, Sulfolobus* spp., *Acidianus brierley* and *Sulfobacillus thermosulfidooxidans* oxidise ferrous iron enzymatically. On the other hand, *Gallionella ferrugine* and *Leptothrix* spp. oxidise ferrous iron under partially reduced conditions (Ehrlich and Newman, 2009).

Sulphur metabolism is perhaps the most important process in AMD environments, and genes encoding dissimilatory sulphate reduction (*dsrA* and *dsrB*) have been found at AMD sites. Dissimilatory sulphite reductase (dsr) is a multi-subunit enzyme essential to sulphate-reducing metabolism since it catalyses the final steps in sulphate and sulphite reduction (Giloteaux et al., 2010 and references therein). Thus, examination of these genes has been very useful as an alternative approach to infer physiology and diversity from environmental samples. For instance, Xie et al. (2011) reported 81 *dsrA* genes from uncultured and cultured SRB in three copper mines in China. Moreau et al. (2010) recovered *dsrAB* sequences from sites exposed to AMD for over 100 years and found that *Desulfovibrio* species dominated the community; other sequences corresponded to *Desulfosarcina, Desulfococcus, Desulfobulbus* and *Desulfosporosinus*.

Manganese-reducing and manganese-oxidising bacteria play an important role in the manganese cycle since manganese oxides accumulates in sediments. In anoxic sediments associated with AMD, manganese and iron hydroxides can be used as terminal electron acceptors for microbial respiration (Lovley, 2000). More specifically, manganese-reducing bacteria can mobilise oxidised or fixed manganese into the aqueous solution (Gadd, 2010). Manganese-reducing bacteria are taxonomically diverse and can reduce oxidised manganese under oxic or anoxic conditions. However, most manganese reducers can reduce Mn(IV) indirectly without enzymatic action by producing metabolic products such as hydrogen sulphide, pyruvates and ferrous iron (Ehrlich and Newman, 2009).

18.9 Use of SRB in Bioremediation

SRB have been used extensively for AMD abatement because of their capacity to reduce sulphate to sulphide and increase pH (carbon metabolism of SRB; Garcia et al., 2001). SRB use sulphite, sulphate, thiosulphate or sulphur as terminal electron acceptors for respiration (dissimilatory sulphate reduction) while oxidising organic carbon as in Equation 18.9 (Baumgartner et al., 2006):

$$SO_4^{2-} + 2[CH_2O] + OH^- \rightarrow HS^- + 2HCO_3^- + 2H_2O \qquad (18.9)$$

or alternatively for disproportionation of the sulphur compound in their anaerobic energy metabolism (Jorgensen and Bak, 1991 and references therein). SRB are key organisms involved in the formation of sulphide minerals, especially pyrite (Natarajan et al., 2006) and are a very diverse group of anaerobic chemoorganotrophic bacteria, including members of the *Desulfovibrio, Desulfomicrobium, Desulfobacter, Desulfosarcina, Desulfotomaculum, Thermodesulfobacterium* and *Archaeoglobus* genera (Luptakova and Kusnierova, 2005). More recently, other microorganisms, including members of the genera *Desulfosporosinus, Syntrophobacter* and *Desulfurella* have been retrieved from acidic environments (Sanchez-Andrea et al., 2012).

There is contradicting information about the type of carbon sources that SRB can utilise. Several authors have reported that SRB prefer simple organic compounds as carbon sources (e.g. Cohen, 2006), while numerous laboratory and field studies have shown successful SRB activity when complex organic compounds were used (e.g. Zagury et al., 2006). Wood chips, leaf compost, poultry manure, mushroom compost and so on have been tested as potential organic carbon sources for SRB (Cocos et al., 2002). SRB preferentially degrade some organic acids and not others (Kuever et al., 2001). For instance, *Desulforhabdus amnigenus* can use lactate, acetate and propionate as a carbon source (Kleikemper et al., 2002). In general, organic and cellulosic materials (Gibert et al., 2004), fatty acids (Beaulieu et al., 2000) and fermentative products (Johnson and Hallberg, 2005a,b) are among the suitable substrates.

One of the challenges in the biological treatment of AMD is the selection of a suitable organic substrate to allow for efficient and economic remediation. Because AMD contains low concentrations of dissolved organic carbon (<10 mg/L) (Zagury et al., 2006), effectiveness of bioreactors using SRB relies on choosing the most appropriate organic electron donor or mixture to stimulate microbial metabolism (Coetser et al., 2006).

To optimise SRB success, a variety of assays to determine the degradability of organic carbon sources have been established (Johnson and Hallberg, 2005a,b). Chemical characterisation and analyses of carbon sources before implementation of laboratory and field studies have proved somewhat useful for choosing the right carbon source for SRB stimulation. Among these analyses are the determination of dissolved organic carbon, total nitrogen, total carbon, total inorganic carbon, carbon/nitrogen ratios, percentage of protein content, cellulose content and easily available substance assays (in Zagury et al., 2006 and references therein).

Other conditions influence SRB success; SRB require a pH of 5–8 to grow, although some success has been achieved under low pH values (<5). For instance, Kolmert and Johnson (2001) reported SRB in acid mine water with the pH lower than 3, and Tsukamoto et al. (2004) indicated SRB growth at pH 2.5. Nevertheless, most bioreactor experiments acclimate their systems to pH near neutrality. Anaerobiosis is another factor that will determine the success of bioremediation. SRB need anaerobic environments to grow and thrive. In the majority of laboratory experiments, this is achieved by ensuring nitrogen flow into the columns/cylinders and is monitored by measuring the redox potential. SRB need a reduced environment with a redox potential lower than −100 mV to grow; however, in most bench-scale and bioreactor experiments, SRB tend to grow under positive mV values (Zaluski et al., 2003). This can be attributed to the fact that some SRB are not obligate anaerobes. Sulphate reducers have such a metabolic flexibility and adopt different lifestyles either as sulphidogenic, acetogenic or hydrogenic (Plugge et al., 2011). These metabolic capabilities give them the advantage of survival under changed environmental conditions. Temperature, although important, is not too critical for SRB growth since SRB can tolerate temperature between −5°C and 75°C (Postgate, 1984). In fact, most bioreactor experiments are not affected by changes in temperature (Reisinger et al., 2000).

The treatment of AMD by SRB has been widely investigated by using batch, column and bioreactor experiments in laboratory settings. Most laboratory studies have been effective in increasing pH and removing sulphate and metals, and used either column reactors or bioreactors. To achieve reliable results, some specific conditions must be met to ensure SRB growth: pH near neutrality, adequate physical substrate, carbon source, anaerobic environment, and SRB sources are needed for bioreactor studies. The physical substrate selection is critical since bacteria are naturally found attached to soils and sediments (Lyew and Sheppard, 1997). Crushed stone, crushed limestone, quartz sand, gravel and ordinary sand have been used as effective substrates for SRB growth (e.g. Christensen et al., 1996, Costa and Duarte, 2005) and in some instances, lime is added initially to the bioreactor to bring neutrality to the system and stimulate SRB growth.

Other limitations are related to the physical setting of the bioreactors. SRB require a solid matrix to establish and survive; when this is not supplied, sulphate reduction rates are usually lower. Hydraulic retention time (HRT) in the bioreactor is critical to allow enough time for SRB metabolism. In general, shorter HRT slows SRB activity, alkalinisation and metal precipitation (Zagury et al., 2006), while longer HRT results in the depletion of carbon sources.

Additionally, many heavy metals are toxic to SRB because they can deactivate enzymes by reacting with their functional group (reviewed by Utgikar et al., 2002). Specifically, dissolved sulphide ions can have both negative effects on SRB by reducing their metabolic activity. Metal sulphides act as physical barriers, preventing access of sulphate and organic matter to enzymes for metabolism.

Although the majority of bench-scale laboratory experiments using SRB are successful, most experiments cannot be replicated in the field. In fact, most *in situ* studies of AMD using SRB have been limited to measuring the activity and diversity of SRB and monitoring if/how SRB were active (Kleikemper et al., 2002). A number of factors at bench scale or in the field determine the success or failure of bioremediation using SRB; however, it is important to note that most limitations have been assessed only under laboratory experiments. For experimental settings, polyacrylate cylinders and glass columns are the most common design to provide anaerobiosis (La et al., 2003). Sources of SRB are usually supplemented as inocula of known microbial consortia, but in the majority of studies acid mine water or soils/sediments from acid mine environments are used to introduce SRB into the reactor (e.g. La et al., 2003).

18.10 Conclusions

AMD is a common problem around the world and microbes play an explicit role in its maintenance and remediation. Both biotic and abiotic systems for remediation have been used and, in particular, SRB play an important and well-studied role in bioremediation. However, most of our understanding of SRB's role in AMD remediation is based on laboratory studies and the extension of these findings to the field is problematic. Bioremediation success is impacted by variations in environmental conditions among AMD-impacted sites; thus, there is no single solution for remediation.

References

Acid Mine Drainage Clearinghouse. 2004. *6th Annual Pennsylvania Statewide Conference on Abandoned Mine Reclamation: Yellow Creek Treatment Site.* http://www.amrclearinghouse.org/SpecialEvents/2004AMRConf/Yellow%20Creek.html, accessed on November 2013.

Aube, B.C. and S. Payant. 1997. The Geco process: A new high density sludge treatment for acid mine drainage. *Proceedings of the Fourth International Conference on Acid Rock Drainage*, May 30–June 6, 1997, Vancouver, BC, vol. I, p. 165–80.

Baeseman, J., R. Smith and J. Silverstein. 2005. Denitrification potential in stream sediments impacted by acid mine drainage: Effects of pH, various electron donors, and iron. *Microbial Ecology* 51: 232–241.

Baker, B.J. and J.F. Banfield. 2003. Microbial communities in acid mine drainage. *FEMS Microbiology Ecology* 44:139–152.

Barlett M., K. Zhuang, R. Mahadevan and D. Lovley. 2012. Integrative analysis of *Geobacter* spp. and sulfate-reducing bacteria during uranium bioremediation. *Biogeosciences* 9:1033–1040.

Baumgartner, L., R. Reid, C. Dupraz, A. Decho, D. Buckley et al. 2006. Sulfate reducing bacteria in microbial mats: Changing paradigms, new discoveries. *Sedimentary Geology* 185:131–145.

Beaulieu, S., G.J. Zagury, L. Deschênes and R. Samson. 2000. Bioactivation and bioaugmentation of a passive reactor for acid mine drainage treatment. In: *Environmental Issues and Management of Waste in Energy and Mineral Production*, ed. R.K. Singhal and A.K. Mehrotra, AA Balkema, Rotterdam, the Netherlands, pp. 533–537.

Bechard, G., H. Yamazaki, D. Gould and P. Bedard. 1994. Use of cellulosic substrates for the microbial treatment of acid mine drainage. *Journal of Environmental Quality* 23:111–116.

Benner S.G., D.W. Blowes and C.J. Ptacek. 1997. A full-scale porous reactive wall for prevention of acid mine drainage. *Ground Water Monitoring and Remediation* 17:99–107.

Blacklick Creek Watershed Association, Inc. 2006. http://bcwa-inc.org/index.html, accessed on November 15, 2013.

Bond P., S. Smriga and J. Banfield. 2000. Phylogeny of microorganisms populating a thick, subaerial, predominantly lithotrophic biofilm at an extreme acid mine drainage site. *Applied and Environmental Microbiology* 66:3842–3849.

Bond, P.L., G.K. Druschel and J.F. Banfield. 2000. Comparison of acid mine drainage microbial communities in physically and geochemically distinct ecosystems. *Applied and Environmental Microbiology* 66:4962–4971.

Boonstra J., R. van Lier, G. Janssen, H. Dijkman and C.J.N. Buisman. 1999. Biological treatment of acid mine drainage. In: *Biohydrometallurgy and the Environment toward the Mining of the 21st Century*, ed. R. Amils and A. Ballester, vol. 9B. Amsterdam: Elsevier, pp. 559–567.

Bridge, T. and D. Johnson. 1998. Reduction of soluble iron and reductive dissolution of ferric iron-containing minerals by moderately thermophilic iron-oxidizing bacteria. *Applied and Environmental Microbiology* 64:2181–2186.

Christensen B., M. Laake and T. Lien. 1996. Treatment of acid mine water by sulfate-reducing bacteria: Results from a bench scale experiment. *Water Research* 30:1617–1624.

Cocos I., G. Zagury, B. Clément and R. Samson. 2002. Multiple factor design for reactive mixture selection for use in reactive walls in mine drainage treatment. *Water Research* 32:167–177.

Coetser S., W. Pulles, R. Heath and T. Cloete. 2006. Chemical characterization of organic electron donors for sulfate reduction for potential use in acid mine drainage treatment. *Biodegradation* 17:169–179.

Cohen, R. 2006. Use of microbes for cost reduction of metal removal from metals and mining industry waste streams. *Journal of Cleaner Production* 14:1146–1157.

Costa M. and J. Duarte. 2005. Bioremediation of acid mine drainage using acidic soil and organic wastes for promoting sulfate reducing bacteria activity on a column reactor. *Water, Air, Soil and Pollution* 165:325–345.

Druschel, G.K., B.J. Baker, T.M. Gihring and J.F. Banfield. 2004. Acid mine drainage biogeochemistry at Iron Mountain, California. *Geochemical Transactions* 5(2):13–32.

Dvorak, D.H., R.S. Hedin, H.M. Edenborn and P.E. McIntire. 1992. Treatment of metal-contaminated water using bacterial sulfate reduction: Results from pilot-scale reactors. *Biotechnology and Bioengineering* 40:609–616.

Edwards, K.J., P.L. Bond and J.F. Banfield. 2000. Characteristics of attachment and growth of Thiobacillus caldus on sulphide minerals: A chemotactic response to sulphur minerals? *Environmental Microbiology* 2:324–332.

Ehrlich, H.L. and D.K. Newman. 2009. *Geomicrobiology*, 5th edn. Boca Raton, FL: CRC Press/Taylor & Francis.

Faulkner, B.B. and J.G. Skousen. 1993. Monitoring of passive treatment systems: An update. *Proceedings of the 14th West Virginia Surface Mine Drainage Task Force Symposium*. Morgantown, WV: West Virginia Publication Services.

Gadd, G. 2010. Metals, minerals and microbes: Geomicrobiology and bioremediation. *Microbiology* 156:609–643.

Gaydardjiev, S., M. Hadjihristova and R. Tichy. 1996. Opportunities for using two low-cost methods for treatment of metal-bearing aqueous streams. *Minerals Engineering* 9:947–964.

Garcia C., D. Moreno, A. Ballester, M. Blazquez and F. González. 2001. Bioremediation of an industrial acid mine water by metal-tolerant sulfate-reducing bacteria. *Minerals Engineering* 14:997–1008.

Ghosh, S., M. Moitra, C.J. Woolverton and L.G. Leff. 2012. Effects of remediation on the bacterial community of an acid mine drainage impacted stream. *Canadian Journal of Microbiology* 58:1316–1326.

Gilbert, O., J. de Pablo, J.L. Cortina and C. Ayora. 2005. Municipal compost-based mixture for acid mine drainage bioremediation: Metal retention mechanisms. *Applied Geochemistry* 20:1648–1657.

Gibert, O., J. Pablo, J. Cortina and C. Ayora. 2004. Chemical characterization of natural organic substrates for biological mitigation of acid mine drainage. *Water Research* 38:4186–4196.

Giloteaux, L., M. Goñi-Urriza and R. Duran. 2010. Nested PCR and new primers for analysis of sulfate reducing bacteria in low-cell biomass environments. *Applied and Environmental Microbiology* 76:2856–2865.

Goldsborough, L.G. and G.G.C. Robinson. 1996. Pattern in wetlands. In: *Algal Ecology: Freshwater Benthic Ecosystems*, eds. R.J. Stevenson, M.L. Bothwell and R.L. Lowe, Academic Press. San Diego, CA. pp. 77–117.

Greger, M. and T. Landberg. 1999. Use of willow in phytoextraction. *International Journal of Phytoremediation* 1:115–123.

Hallberg, K.B. and D.B. Johnson. 2001. Biodiversity of acidophilic micro- organisms. *Advances in Applied Microbiology* 49:37–84.

Hallberg, K.B. and D.B. Johnson. 2003. Passive mine water treatment at the former Wheal Jane Tin Mine, Cornwall: Important biogeochemical and microbiological lessons. *Land Contamination and Reclamation* 11:213–220.

Hallberg, K.B. 2010. New perspectives in acid mine drainage microbiology. *Hydrometallurgy* 104:448–453.

Headd, B.E. and A. Summers. 2013. Evidence for niche partitioning revealed by the distribution of sulfur oxidation genes collected from areas of a terrestrial sulfidic spring with differing geochemical conditions. *Applied and Environmental Microbiology* 79:1171–1182.

Hedin, R.S. and G.R. Watzlaf. 1994. The effects of anoxic limestone drains on mine water chemistry. p. 185–194. *Proceedings, Third International Conference on the Abatement of Acidic Drainage*, Pittsburgh, PA.

Hedin, R.S., G.R. Watzlaf and R.W. Nairn. 1994. Passive treatment of acid mine drainage with limestone. *Journal of Environmental Quality* 23:1338–1345.

Hiraishi, A., Y. Matsuzawa, T. Kanbe and N. Wakao. 2000. *Acidisphaera rubrifaciens* gen. nov., sp. nov., an aerobic bacteriochlorophyll-containing bacterium isolated from acidic environments. *International Journal of Systematic and Evolutionary Microbiology* 50:1539–1546.

Ibeanusi V., E. Jackson, J. Coffen and Y. Jeliani. 2012. Assessing bioremediation of acid mine drainage in coal mining sites using a predictive neural network-based decision support system (NNDSS). *Bioremediation and Biodegradation* 3:148.

Jennings, S.R., D.R. Neuman and P.S. Blicker. 2008. *Acid Mine Drainage and Effects on Fish Health and Ecology: A Review*. Bozeman, MT: Reclamation Research Group Publication.

Johnson, D.B. 2000. Biological removal of sulfurous compounds from inorganic wastewaters. In: *Environmental Technologies to Treat Sulfur Pollution: Principles and Engineering*, ed. P. Lens and Pol L. Hulshoff, London: 7th International Association on Water Quality, pp. 175–206.

Johnson D.B. and K.B. Hallberg. 2002. Pitfalls of passive mine water treatment. *Reviews in Environmental Science and Biotechnology* 1:335–343.

Johnson, D.B. and K.B. Hallberg. 2003. The microbiology of acidic mine waters. *Research in Microbiology* 154(7):466–473.

Johnson, D.B. and K.B. Hallberg. 2005a. Acid mine drainage remediation options: A review. *Science of the Total Environment* 338(1–2):3–14.

Johnson, D.B. and K.B. Hallberg. 2005b. Biogeochemistry of the compost bioreactor components of a composite acid mine drainage passive remediation system. *Science of the Total Environment* 338(1–2):81–93.

Jones, D.S., H.L. Albrecht, K.S. Dawson, et al. 2011. Community genomic analysis of an extremely acidophilic sulfur-oxidizing biofilm. *The ISME Journal* 6:158–170.

Jorgensen, B. and F. Bak. 1991. Pathways and microbiology of thiosulfate transformations and sulfate reduction in a marine sediment (Kattegat, Denmark). *Applied and Environmental Microbiology* 57:847–856.

Kepler, D.A. and E.C. McCleary. 1994. *Successive Alkalinity-Producing Systems (SAPS) for the Treatment of Acidic Mine Drainage*. Washington, DC: Bureau of Mines Special Publication SP 06B-94, 1, 185–194.

Kimball, B.A., R.L. Runkel, K. Walton-Day and K.E. Bencala. 2002. Assessment of metal loads in watersheds affected by acid mine drainage by using tracer injection and synoptic sampling: Cement Creek, Colorado, USA. *Applied Geochemistry* 1:1183–1207.

Kleikemper, J., M. Schroth, W. Sigler, M. Schmucki, S. Bernasconi and J. Zeyer. 2002. Activity and diversity of sulfate reducing bacteria in a petroleum hydrocarbon contaminated aquifer. *Applied and Environmental Microbiology* 68:1516–1523.

Kuever J., M. Konneke, A. Galushko and O. Drzyzga. 2001. Reclassification of *Desulfobacterium phenolicum* as *Desulfobacula phenolica* com. Nov and description of strain Sax(T) as *Desulfotignum balticum* gen. nov. sp. Nov. *International Journal of Systematic and Evolutionary Microbiology* 51:171–177.

Kumaraswamy, R., K. Sjollema, G. Kuenen, M. Loosdrecht and G. Muyzer. 2006. Nitrate-dependent [Fe (II) EDTA] 2- oxidation by *Paracoccus ferrooxidans* sp. Nov., isolated from denitrifying bioreactor. *Systematic and Applied Microbiology* 29:276–286.

Kurniawan, T.A., W.-H. Lo, and G.Y.S. Chan. 2006. Degradation of recalcitrant compounds from stabilized landfill leachate using a combination of ozone-GAC adsorption treatment. *Journal of Hazardous Materials* B137:433–455.

Kolmert, Å. and D.B. Johnson. 2001. Remediation of acidic waste waters using immobilised, acidophilic sulfate-reducing bacteria. *Journal of Chemical Technology and Biotechnology* 76:836–843.

Kuyucak, N. and P. St-Germain. 1994. *In situ* treatment of acid mine drainage by sulphate reducing bacteria in open pits: Scale-up experiences. In: *International Land Reclamation and Mine Drainage Conference and 3rd International Conference on Abatement of Acid Drainage*. Pittsburgh, PA. pp. 26–29.

La, H., K. Kim, Z. Quan, Y. Cho and S. Lee. 2003. Enhancement of sulfate reduction activity using granular sludge in anaerobic treatment of acid mine drainage. *Biotechnology Letters* 25:503–508.

Li, M.G., B.C. Aube and L.C. St-Arnaud. 1997. Considerations in the use of shallow water covers for decommissioning reactive tailings. *Proceedings of the Fourth International Conference on Acid Rock Drainage*, May 30–June 6, 1997, Vancouver, BC, vol. I, p. 115–30.

Lindow, N.L. and R.C. Borden. 2005. Anaerobic bioremediation of acid mine drainage using emulsified soybean oil. *Mine Water and the Environment* 24:199–208.

Lindow, N.L. and R.C. Borden. 2004. Anaerobic bioremediation of acid mine drainage using EOS. In: *Proceedings of the 2004 National Meeting of the American Society of Mining and Reclamation and the 25th West Virginia Surface Mine Drainage Task Force*. Morgantown, WV. pp. 1192–1204.

Long Z.E., Y.H. Huang, Z.L. Cai, W. Cong and O.Y. Fan. 2003. Biooxidation of ferrous iron by immobilized *Acidithiobacillus ferrooxidans* in poly(vinyl alcohol) cryogel carriers. *Biotechnology Letters* 25:245–9.

Lovley D.R. 2000. Fe(III) and Mn(IV) reduction. In: *Environmental Microbe–Metal Interactions*, ed. D.R. Lovley, Washington, DC: ASM Press, pp. 3–30.

Luptakova, A. and M. Kusnierova. 2005. Bioremediation of acid mine drainage contaminated by SRB. *Hydrometallurgy* 77:97–102.

Lyew, D. and J. Sheppard. 1997. Effects of physical parameters of a gravel bed on the activity of sulphate reducing bacteria in the presence of acid mine drainage. *Journal of Chemistry Technology and Biotechnology* 70:223–230.

Martins, M.S.F., E.S. Santos, R.J.J. Barros and M.C.S. da Silva Costa. 2008. Treatment of acid mine drainage with sulphate-reducing bacteria ising a two-stage bioremediation process. In: *Proceedings of the 10th International Mine Water Association (IMWA) Congress*. Czech Republic, IMWA.

Matlock, M.M., B.S. Howerton and D.A. Atwood. 2002. Chemical precipitation of heavy metals from acid mine drainage. *Water Research* 36:4757–4767.

Means, B., B. McKenzie and T. Hilton. 2003. A computer-based model for estimating mine drainage treatment costs. *Proceedings of the 24th Annual West Virginia Surface Mine Drainage Task Force Symposium*, Morgantown, WV.

Mendez, M.O. and R.M. Maier. 2008. Phytostabilization of mine tailings in arid and semiarid environments—An emerging remediation technology. *Environmental Health Perspectives* 116:278–283.

Mine Waste Technology Program, 2004. Annual Report 2004: Sulfate Reducing Bacteria Demonstration. http://www.epa.gov/ORD/NRMRL/std/mtb/mwt/annual/annual2004/adwt/sulfatereducingbacteri ademo.htm, accessed on November 15, 2013.

Moreau, J., R. Zierenberg and J. Banfield. 2010. Diversity of dissimilatory sulfite reductase genes (DSRAB) in a salt marsh impacted by long-term acid mine drainage. *Applied and Environmental Microbiology* 76:4819–4828.

Natarajan, K., S. Subramanian and J.J. Braun. 2006. Environmental impact of metal mining-biotechnological aspects of water pollution and remediation—An Indian experience. *Journal of Geochemical Exploration* 88:45–48.

Neculita, C.M., G.J. Zagury and B. Bussiere. 2007. Passive treatment of acid mine drainage in bioreactors using sulfate-reducing bacteria: Critical review and research needs. *Journal of Environmental Quality* 36:1–16.

Norlund, K.L.I., G. Southam, T. Tyliszczak et al. 2009. Microbial architecture of environmental sulfur processes: A novel syntrophic sulfur-metabolizing consortia. *Environmental Science & Technology* 43:8781–8786.

Plugge C., W. Zhang, J. Scholten and J. Stams. 2011. Metabolic flexibility of sulfate reducing bacteria. *Frontiers in Microbiology* 2:2–8.

Postgate, J.R. 1984. *The Sulphate Reducing Bacteria*, 2nd ed. Cambridge, UK: Cambridge University Press.

Ramasamy, K. and Kamaludeen S.P.B. 2006. Bioremediation of Metals: Microbial Processes and Techniques; In: *Environmental Bioremediation Technologies*, ed. S.N. Singh and R.D. Tripathi, NY: Springer Publication, pp. 173–187.

Reisinger, R.W., J.J. Gusek, and T.C. Richmond. 2000. Pilot-scale passive treatment test of contaminated waters at the historic Ferris-Haggarty Mine, Wyoming. In: *Proceedings of the 5th International Conference on Acid Rock Drainage*. Denver, CO, pp. 1071–1077.

Rojas-Chapana, J.A., M. Giersig and H. Tributsch. 1996. The path of sulfur during the bio-oxidation of pyrite by *Thiobacillus ferrooxidans*. *Fuel* 75:923–930.

Rowley, M., D.D. Warkentin and V. Sicotte. 1997. Site demonstration of the biosulfide process at the former Britannia mine. In: *Proceedings of the Fourth International Conference on Acid Rock Drainage*. May 30–June 6, 1997, Vancouver, BC, vol. IV, pp. 1531–1548.

Sanchez-Andrea, I., K. Knittel, R. Amman, R. Amils and J. Sanz. 2012. Quantification of Tinto River sediment microbial communities: Importance of sulfate-reducing bacteria and their role in attenuating acid mine drainage. *Applied and Environmental Microbiology* 78:4638–4645.

Sasowsky I.D., A. Foos and C.M. Miller. 2000. Lithic controls on the removal of iron and remediation of acidic mine drainage. *Water Research* 34:2742–2746.

Singer, P.C. and W. Stumm. 1970. Acid mine drainage: The rate determining step. *Science* 167:1121–1123.

Skousen, J. and P. Ziemkiewicz. 2005. Performance of 116 passive treatment systems for acid mine drainage. *Proceedings of the National Meeting of the American Society of Mining and Reclamation, Breckenridge*, CO, June 19–23, 2005, Lexington, KY: Published by ASMR.

Skousen, J.G., A. Sextone and P.F. Ziemkiewicz. 2000. Acid mine drainage control and treatment. Chapter 6. In: *Reclamation of Drastically Disturbed Lands*. American Society of Agronomy, Crop Science Society of America, Soil Science Society of America, 677 S. Segoe Rd., Madison, WI 53711, USA, Agronomy Monograph no. 41.

Skousen, J.K., K. Politan, T. Hilton and A. Meeks. 1990. Acid mine drainage treatment systems: Chemicals and costs. *Green Lands* 20:31–37.

Stumm, W. and J.J. Morgan. 1996. Aquatic chemistry: *Chemical Equilibria and Rates in Natural Waters*. 3rd edn. New York, NY: Wiley-Interscience, p. 470.

Tributsch, H. 2001. Direct versus indirect bioleaching. *Hydrometallurgy* 59:177–185.

Tsukamoto, T.K., H.A. Killion and G.C. Miller. 2004. Column experiments for microbiological treatment of acid mine drainage: Low-temperature, low-pH and matrix investigations. *Water Research* 38:1405–1418.

Utgikar, V., S. Harmon, N. Chaudhary, H. Tabak, R. Govind and J. Haines. 2002. Inhibition of sulfate reducing bacteria by metal sulfide formation in bioremediation of acid mine drainage. *Environmental Toxicology* 17:40–48.

Vile M.A. and R.K. Wieder. 1993. Alkalinity generation by Fe(III) reduction versus sulfate reduction in wetlands constructed for acid mine drainage treatment. *Water, Air and Soil Pollution* 69:425–41.

Wang, X., J. Yang, X. Chen, G. Sun and Y. Zhue. 2009. Phylogenetic diversity of dissimilatory ferric iron reducers in paddy soils of Hunan, South China. *Journal of Soils and Sediments* 9:568–577.

Whitehead, P.G. and H. Prior. 2005. Bioremediation of acid mine drainage; an introduction to the Wheal Jane wetlands project. *Science of the Total Environment* 338:15–21.

Xie, J., Z. He, X. Liu, X. Liu, J. Van Nostrand, Y. Deng, L. Wu, J. Zhou and G. Qiu. 2011. Geochip-based analysis of the functional gene diversity and metabolic potential of microbial communities in acid mine drainage. *Applied and Environmental Microbiology* 77:991–999.

Younger, P.L., S.A. Banwart and R.S. Hedin. 2002. *Mine Water: Hydrology, Pollution, Remediation.* Dordrecht, The Netherlands: Kluwer Academic Publishers.

Zagury G., V. Kulniens and C. Neculita. 2006. Characterization and reactivity assessment of organic substrates for sulphate reducing bacteria in acid mine drainage treatment. *Chemosphere* 64:944–954.

Zaluski, M.H., Trudnowski, J.M., M.A. Harrington-Baker and D.R. Bless. 2003. Postmortem findings on the performance of engineered SRB field-bioreactors for acid mine drainage control. In: *Proceedings of the 6th International Conference on Acid Rock Drainage.* Cairns, QLD, pp. 845–853.

Zhang, D., G. Xu, W. Zhang and S.D. Golding. 2007. High salinity fluid inclusions in the Yinshan polymetallic deposit from the Le-De metallogenic belt in Jiangxi Province, China: Their origin and implications for ore genesis. *Ore Geology Reviews* 31:247–260.

Zipper, C. and C. Jage. 2001. *Passive Treatment of Acid-Mine Drainage with Vertical-Flow Systems.* Virginia Polytechnic Institute and State University, Virginia Cooperative Extension, Publication Number 460-133, 16 p.

19

Microbiology of Arsenic-Contaminated Groundwater

Pinaki Sar, Dhiraj Paul, Angana Sarkar, Rahul Bharadwaj
and Sufia K. Kazy

CONTENTS

19.1 Introduction

Arsenic (As), the notorious chemical element, is ubiquitously present in sediments, soils and aquifers throughout the world (Smith et al., 1992; Haque et al., 2008; Pearcy et al., 2011). Natural enrichment of groundwater with As concentration exceeding the safe level of World Health Organization (WHO) causes a very critical water quality problem in many parts of the world, including southern and eastern Asia, the United States, and countries of the European Union (Nath et al., 2011; Smedley and Kinniburgh, 2013 and references therein). Groundwater concentration of As shows a very large range (<0.5–5000 µg/L) spreading across more than 70 countries (Sharma et al., 2014). Prolonged consumption of As-rich groundwater has very serious environmental implications. Severe health problems associated with As-contaminated drinking water have been documented from around the world, particularly from countries such as Argentina, Bangladesh, Burkina Faso, China, Chile, Cambodia, Hungary, India, Laos, Mexico, Nepal, Romania, Spain, Taiwan, Thailand, the United States, Vietnam and others. A rough estimate based on available literature indicates that more than 100 million people are exposed to >10 µg As/L and >45 million people are exposed to concentrations as high as 50 µg As/L or beyond (Sharma et al., 2014). Among the affected areas, the most severe effect has been observed in southern and eastern Asia, particularly countries such as India and Bangladesh, where more than 30 million people are at risk from high levels (>50 µg/L) of naturally occurring As in groundwater (http://www.who.int/inf-fs/fr). The other affected areas of south and east Asia are the Red River delta and the Mekong delta (>10 million people exposed), and Chindwin–Irrawady, Salween, Brahmaputra, Ganges, Indus and Chenab river basins (Sharma et al., 2014). Considering the health of millions of people worldwide, it is considered to be an issue of critical importance to improve our understanding of the biogeochemical behaviour of As in groundwater flow systems and in particular the role of microorganisms in its mobilisation within the subsurface environment (Pearcy et al., 2011; Sarkar et al., 2013 and references therein).

Mobilisation of As in contaminated niches occurs through a complex set of conditions and biogeochemical processes. These include the interplay of hydro-geo-bio-chemical processes and human interactions (Kar et al., 2010; Smedley and Kinniburgh, 2013). Owing to spatial variations in hydrogeochemical and microbiological properties, the As release mechanism may vary with location. In most of the contaminated aquifers, elevated As concentrations typically result from the mobilisation and transport of this metalloid under natural conditions. Although the release of As due to weathering and the mobilisation of As-bearing minerals or from geothermal sources over millions of years seems to contribute a large part of the contamination; lower amounts are of anthropogenic origin, for example, smelting, mining and agricultural activities (Muller et al., 2007; Pepi et al., 2007). The latter activities may directly add various As compounds into the environment or cause/exacerbate mobilisation (Smedley and Kinniburgh, 2013).

A number of mechanisms, including (i) reductive dissolution of As-rich Fe-oxyhydroxides, (ii) oxidation of As-rich pyrite, (iii) weathering of minerals that contain phosphate/ammonia/iron and so on, have been proposed to explain the subsurface mobilisation of As in groundwater (Islam et al., 2004; Mailloux et al., 2009; Hery et al., 2010; Sarkar et al., 2013). Each mechanism includes both biotic and abiotic components and probably plays a role under certain condition(s). It has been established beyond doubt that microbial mobilisation of As into an aqueous phase is one of the main sources of As in contaminated groundwater across many parts of the world (Drewniak et al., 2012). Inhabiting microorganisms within aquifer sediments and/or groundwater have been shown to affect the As geochemistry by catalysing redox transformations and other reactions that affect the mobility of this metalloid in subsurface environments (Oremland and Stolz, 2005; Bachate et al., 2012). Chemical speciation of As is related to its oxidation states, and in an aqueous environment, this regulates the mobility and toxicity of this metalloid. The predominant form of inorganic As in oxic subsurface systems is arsenate (As^{5+}), which tends to be strongly adsorbed onto the solid phases (such as ferrioxyhydroxide, ferrihydrite, apatite, alumina, etc.). Conversely, arsenite (As^{3+}) is more prevalent in an anoxic environment and gets adsorbed poorly onto fewer such minerals yet is more mobile (and notably more toxic too) than As^{5+} (Oremland and Stolz, 2005). Arsenite is more (100 times) toxic and can inhibit various dehydrogenases; it can bind sulphydryl groups of proteins and dithiols such as glutaredoxin, while arsenate acts as a structural analogue of phosphate and inhibits oxidative phosphorylation by producing unstable arsenylated derivatives (Pepi et al., 2007).

In response to their natural coexistence, microorganisms have evolved dynamic mechanisms for facing the toxicity of As in the environment. In this sense, arsenic speciation and mobility is also affected by microbial metabolism (i.e. reduction, oxidation and methylation as part of their resistance and respiratory processes) that participates in the biogeochemical cycle of the element (Stolz et al., 2006; Muller et al., 2007; Páez-Espino et al., 2009). Redox transformations of As by microorganisms mediated by specific enzymes or respiratory chains facilitate their energy requirements for growth as well as a means of resistance to cope with high arsenic in their environment (Oremland and Stoltz, 2005; Silver and Phung, 2005; Chang et al., 2010). These organisms inhabiting As-rich habitats and undertaking important biogeochemical reactions are taxonomically diverse and metabolically versatile (Oremland and Stolz, 2005; Slyemi and Bonnefoy, 2012). Some bacteria can reduce As^{5+} to As^{3+} during their anaerobic respiration or as a means of As detoxification, while others oxidise As^{3+} to As^{5+} during their chemolithoautotrophic/heterotrophic metabolism. The latter group uses energy and reducing power from As^{3+} oxidation during CO_2 fixation or other anabolic reactions and cell growth under aerobic or anaerobic (nitrate reducing) conditions (Oremland and Stolz, 2005).

The natural abundance of As in subsurface environment often guided the evolution of appropriate genetic systems within the inhabiting microorganisms, facilitating their survival (as a means of detoxification and satisfying metabolic requirements). Among these, As resistance system (*ars*) appears to be widely distributed within the procaryotes. The multiplicity of *ars* mechanisms affects the transformation between soluble and insoluble forms of As, thus regulating their toxicity (Saltikov and Olson, 2002; Jackson and Dugas, 2003; Achour-Rokbani et al., 2010). The presence of dissimilatory As^{5+} reductase (*arr*) system and its possible role in subsurface As mobilisation has also been studied. Although reductive dissolution of Fe/Mn oxides/hydroxides under the aquifer condition with a supply of electrons/carbon from available organic matter has been proposed to explain the release of As in Bangladesh, several other studies do indicate the presence of alternate/additional microbial processes as well.

In the following sections, microbial interaction with As and the composition of bacterial communities present in various contaminated aquifers are presented. Sources of As, its toxicity, speciation, distribution and mobility are described in Section 19.2. A broad overview of groundwater As contamination is presented in Section 19.3, bacterial interaction with As is elaborated in Section 19.4 followed by a detailed description of microbial diversity in As-contaminated aquifers in Section 19.5. Sections 19.6 and 19.7 describe the microbial role in As mobilisation and microbial route towards the mitigation of the As problem, respectively.

19.2 Arsenic in the Environment

19.2.1 Sources of Arsenic

Arsenic is ubiquitously distributed in the atmosphere, pedosphere, hydrosphere and biosphere of our planet (Lièvremont et al., 2009). Naturally existing minerals are the key source of As contamination in the environment. Arsenic has been found to be associated with more than 200 different minerals, containing several metals such as Fe, S, Au, Ag, Cu, Sb, Ni, Co and so on (Huang, 2014). Some of the most frequent As-rich mineral are listed in Table 19.1.

The biogeochemical cycle of As involves several physico-chemical processes, including redox reactions, adsorption and desorption, competitive adsorption (ion exchange), solid-phase precipitation and dissolution in which microbiological processes play a crucial role (Kar et al., 2010). Several factors such as redox potential (Eh), pH, dissolved organic carbon (DOC),

TABLE 19.1

Basic Information Regarding Arsenic

Overview of Arsenic	*Value*
Atomic number	33
Atomic weight	74.9216
Density	5.727 g/cm^3
Melting point	817°C
Boiling point	614°C
Oxidation States	*Chemical Formula*
Arsenate	As^{5+}
Arsenite	As^{3+}
Arsenide	As^{3-}
Elemental arsenic	As^0
Source	*Types*
Anthropogenic sources	Smelter slag, coal combustion, mine tailings, tanning waste, production of pigment for paints and dyes, application of As-based pesticides
Natural source	Hot spring, volcanic rock, groundwater
Abundance (Location/Source)	*Arsenic Concentration (mg L^{-1}/mg kg^{-1})*
Groundwater	0.01–3
Lakes	0.001–1.1
Seawater	0.001–0.006
Geothermal water	0.01–6
Soil	5–1500
Groundwater sediment	5–5000
Sedimentary rock	5–700
Mine drainage	0.1–85
Arsenic-Bearing Minerals	*Composition*
Realgar	AsS
Orpiment	As$_2$S$_3$
Claudetite	As$_2$O$_3$
Arsenolite	As$_2$O$_3$
Arsenopyrite	FeAsS
Loellingite	FeAs$_2$
Scorodite	FeAsO$_4 \cdot 2H_2O$
Pharmacosiderite	Fe$_3$(AsO4)$_2$(OH)$_3 \cdot 5H_2O$
Enargite	Cu$_3$AsS$_4$
Niccolite	NiAs
Annabergite	(Ni,Co)$_3$(AsO$_4$)$_2 \cdot 8H_2O$
Cobaltite	CoAsS
Tennantite	(Cu,Fe)$_{12}$As$_4$S$_{13}$
Conichalcite	CaCu(AsO$_4$)(OH)

Continued

TABLE 19.1 (Continued)

Basic Information Regarding Arsenic

Haematolite	$(Mn,Mg)_4Al(AsO_4)(OH)_8$
Hoernesite	$Mg_3(AsO_4)_2 \cdot 8H_2O$
Beudantite	$PbFe_3(AsO_4)(SO_4)(OH)_6$
Minerals on Which As Can Be Adsorbed	*Composition*
Geothite	$FeOOH$
Siderite	$FeCO_3$
Clinochlore	$(Mg_5Al)(AlSi_3)O_{10}(OH)_8$
Illite	$(K,H_3O)(Al,Mg,Fe)_2(Si,Al)_4O_{10}[(OH)_2,(H_2O)]$
Amphibole	$(Mg,Fe)_7Si_8O_{22}(OH)_2$
Hornblende	$Ca_2(Mg,Fe,Al)_5(Al,Si)_8O_{22}(OH)_2$
Muscovite	$KAl_2(Si_3Al)O_{10}(OH,F)_2$
Glauconite	$(K,Na)(Fe^{3+},Al,Mg)_2(Si,Al)_4O_{10}(OH)_2$

and chemical speciation take a lead part in these processes (Kar et al., 2010; Smedley and David, 2013). Along with natural sources, several anthropogenic activities, including precious metals mining, pharmaceutical manufacturing, wood processing, glassmaking, electronic, chemical weapons making and so on are also responsible for As contamination (Huang, 2014). Agricultural use of arsenic as a fungicide has also contributed to As poisoning. The key origin of atmospheric As contamination is the As-rich fossil fuels, producing highly toxic As oxide (As_2O_3), while mine wastewater and solubilisation of subsurface minerals are responsible for the contamination of aqueous environment (Drewniak and Sklodowska, 2013) (Figure 19.1).

19.2.2 Distribution, Speciation and Mobility of Arsenic in the Environment

Arsenic occurs ubiquitously but at variable concentrations in rocks, unconsolidated sediments and soils around the world due to its association with various minerals (Table 19.1). Change in the physico-chemical behaviour of aquifers and water–sediment interaction (dissociation, association, adsorption, desorption, dissolution, etc.) as well as several other abiotic and biotic processes cause the release of As into the groundwater. Arsenic concentration in groundwater is found to be very wide (<0.05–5000 µg/L) (Sharma et al., 2014). Among the four oxidation states (arsenate [As^{5+}], arsenite [As^{3+}], arsenide [As^{3-}] and elemental arsenic [As^0]), As^{3+} and As^{5+} species are the most abundant in aquatic environments (Kudo et al., 2013). Occurrence, distribution and mobility of these As species are dependent on the complex interplay of both biological processes (oxidation, reduction, methylation) and local physico-chemical conditions (redox potential [Eh] and pH) of the environment. At a low pH (pH $\leq$ 6.9) and oxidising environment, the predominant form is As^{5+} ($H_2AsO_4^-$), while at higher pH levels, $HAsO_4^{2-}$ is the most

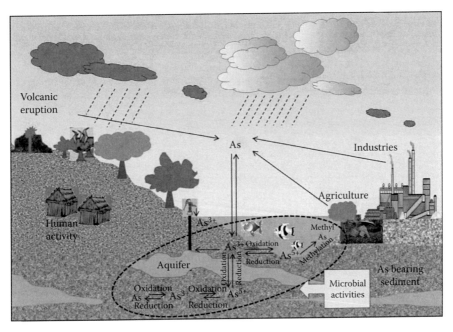

FIGURE 19.1
(**See colour insert.**) Global As biogeochemical cycle in environment. (Modified after Mukhopadhyay, R. et al. *FEMS Microbiology Reviews* 26(3), 2002: 311–325.)

abundant form. Arsenate, being negatively charged, adsorbs easily on the oxidised minerals of iron (Fe), magnesium (Mg), manganese (Mn) and/or sulphide (S), thereby restricting its availability in the aqueous phase (Martin et al., 2001). In most natural water with pH levels ≤9.2 as well as in slightly reductive environments, $As(OH)_3$ is the most abundant form (Páez-Espino et al., 2009). Solubility of As mainly depends on its speciation, which makes it bio-available even if reduction of As^{5+} to As^{3+} results in the mobilisation of this element (Drewniak and Sklodowska, 2013). Elemental As is not common in the environment while organic arsines (monomethyl, dimethyl, and/or trimethyl) are only found in extremely reducing environments (Dopp et al., 2009).

19.2.3 Toxicity of Arsenic

Among the several species of As, the gaseous form, arsine (As_3H_3), is the most toxic followed by arsenite (As^{3+}), arsenoxides (As_2O_3 and As_2O_5), arsenate (As^{5+}) arsonium compounds (R_4As^+) and elemental arsenic (As^0) (Silver and Phung, 2005). Although previously methylated As species were assumed to be less toxic, more recent observations indicate that they are more toxic than their inorganic counterparts (Dopp et al., 2009). The most abundant species of inorganic As (As^{3+} and As^{5+}) in the environment show their toxicity in different modes. After entering the cell, As^{5+} substitutes for phosphate in normal phosphorylation, leading to the production of unstable arsenical by-products such as ADP–As^{5+}, and thereby causing disruption of ATP (adenosine 5′-triphosphate) synthesis (Rosen et al., 2011). These unstable arsenical by-products spontaneously hydrolyse to ADP and As^{5+}, thus preventing ATP formation. Other metabolic processes such as ATP-dependent transport, glycolysis, the pentose phosphate pathway and signal transduction pathways (two-component and phosphorelay systems, chemotaxis, etc.) can also be interrupted likewise. On the other hand, As^{3+} has a strong affinity for sulphydryl groups (Silver and Phung, 2005). Arsenite reacts with cysteine groups as well as glutathione, glutaredoxin and thioredoxin present on active sites of many enzymes, which ultimately control intracellular redox homeostasis, DNA synthesis and repair, protein folding, sulphur metabolism and xenobiotic detoxification. Arsine gas binds to red blood cells, causing haemolysis by damaging membranes (Dopp et al., 2009). Drinking As-contaminated water over a long period of time (more than 5 years) results in chronic poisoning, including several skin diseases (e.g. melanosis, keratosis, carcinoma and skin cancer), gangrene, cancers of the bladder, kidney and lung, and diseases of the blood vessels of the legs and feet, diabetes, high blood pressure and reproductive disorders.

19.3 Groundwater Arsenic Contamination

Arsenic contamination in aquifers supplying drinking water creates a severe problem throughout the world. A number of countries, including Bangladesh, India, Chile, China, Mexico, Taiwan, Argentina, Vietnam, the United States and so on, are facing high levels of As contamination in different aquifers used for drinking as well as agricultural purposes. Besides these countries, As contamination has also been noted in many other countries such as the United Kingdom, Sweden, Ghana, Canada, Spain, Germany, France and so on through mining activities, volcanic eruption and/or anthropogenic activities (Mukherjee et al., 2006). A higher level of As contamination is found in aquifers having diverse environmental conditions, including varied redox states or levels of chemically active species of other elements, organic carbon, pH and so on (Basu et al., 2014). In the following section, a brief description of the global scenario of natural As contamination in groundwater is presented.

19.3.1 Asia

In Asia, several countries from the south (Bangladesh, India, Nepal, Pakistan and Sri Lanka), southeast (Cambodia, Laos and Vietnam) and east (China

TABLE 19.2

Natural Contamination in Groundwater throughout the World

Country/Region	Year of First Discovery	Reference
Argentina	1917	Arguello et al. (1938)
Mexico	1983	Cebrián et al. (1994)
United States	1970	Goldsmith et al. (1972)
Bangladesh	1992	Chakraborti et al. (2008)
Cambodia	2000	World Bank Policy Report (2005)
China (Inner Mongolia, Xinjiang and Shanxi)	1980s	Sun et al. (2001); World Bank Policy Report (2005)
West Bengal, India	1983	Chakraborti et al. (2002)
Bihar, India	2002	Chakraborti et al. (2008)
Uttar Pradesh, India	2003	Chakraborti et al. (2008)
Jharkhand, India	2003	Chakraborti et al. (2008)
Manipur, India	2004	Singh (2004)
Assam, India	2004	Mukherjee et al. (2006)
Chhattisgarh, India	1999	Chakraborti et al. (1999)
Lao PDR	2001	World Bank Policy Report (2005)
Myanmar	2001	World Bank Policy Report (2005)
Nepal	2001	Tandukar et al. (2001); World Bank Policy Report (2005)
Pakistan	2000	Nickson et al. (2005)
Taiwan	1960s	Chiou et al. (2001); Smedley and Kinniburgh (2002)

and Taiwan) are exposed to higher As contamination in the groundwater (Mukherjee et al., 2006; Fendorf et al., 2010) (Table 19.2). The level of As contamination and its major sources as reported in different parts of the affected countries are discussed below.

Groundwater As contamination has been reported from several states of India (e.g. West Bengal, Bihar, Jharkhand, Uttar Pradesh, Assam, Arunachal Pradesh, Tripura and Manipur). Among the affected states, the most severe contamination is noted in West Bengal, where people are being exposed to contaminated water since the 1970s, although it was first reported only in 1983 (Bhattacharya et al., 1997; Chakraborti et al., 2004; Nickson et al., 2007). In West Bengal, nine districts out of 19 (spreading across 3500 villages) are affected by elevated As concentration ($>50\ \mu g/L$). It is estimated that tens of millions of people of the state are under the threat (Mukherjee et al., 2006). The source of groundwater As contamination in West Bengal is geogenic. The alluvial sediment of West Bengal is formed by the sedimentation of the rivers Ganges, Brahmaputra and Meghna, representing the largest fluvio-deltaic basin. Arsenic contamination is mainly noted in shallower aquifers, which were formed during the Holocene age. Arsenic concentration in the solid phase of the aquifer here is nearly consistent and represented mostly by As^{5+}. It is found to be co-precipitated in or co-adsorbed on various Fe- and Mn-rich clastic and authigenic minerals or mineral phases (Fe-oxide/ hydroxidecoated sand, FeOH and Fe OOH and authigenic pyrites). Aqueous As, in contrast, is dominated by As^{3+} and its concentration showed high spatial and depth variations. Many mechanisms have been proposed within which microbe-mediated reductive dissolution plays an important role in As mobilisation within the Bengal delta aquifers (Islam et al., 2004; Gault et al., 2005; Hery et al., 2010). Groundwater As contamination in Bihar was first detected in 2002 (Chakraborti et al., 2009), indicating that about 39% of the 9500 samples analysed contained As above 10 µg/L and that 23% contained As greater than 50 µg/L (Mukherjee et al., 2006). Subsequently, As contamination was reported from the states of Jharkhand (in the year 2003), Uttar Pradesh 2004, Assam and Tripura (2006) and Manipur (2008). The sediment of northern India contains a high amount of clay and organic matter, which may retain and release As in groundwater aquifers (Mukherjee et al., 2006; Nickson et al., 2007).

Naturally occurring As-contaminated groundwater was first detected in Bangladesh in 1993. In Bangladesh, the groundwater As contamination problem is the worst in the world with more than 85 million people reported to be at risk (Mukherjee et al., 2006). Ninety-seven percent of the population in the country uses groundwater for drinking and domestic purposes as surface water is mismanaged. High levels of As in groundwater are causing widespread poisoning in Bangladesh. Different studies have addressed various aspects of the As issue in Bangladesh. Mukherjee et al. (2006) have reported that out of the water samples collected from deep wells of 64 districts, 59 contained >10 µg/L and 50 showed concentrations >50 µg/L. Maximum As contamination is noted in shallower aquifers compared to deeper aquifers. Arsenic contamination in groundwater occurs due to the characteristic nature of aquifers such as high organic content, reducing environment and As-rich minerals, which possibly favours microbial-mediated As mobilisation (Ravenscroft et al., 2005; Dhar et al., 2011).

In Cambodia, groundwater As contamination was first reported in the year 2000 with concentration ranging from 1 to 1340 µg As/L and with an average of 163 µg As/L (Berg et al., 2007; Buschmann et al., 2008). Higher As contamination is observed in the Mekong River, particularly in Prey, Veng, Kratie and Kandal provinces, where As^{3+} is found predominantly in the groundwater due to the reducing nature of the aquifer. In some places such as Kampong Cham Province, groundwater As is present as As^{5+} also. The mechanism of As release in the Mekong River basin of Cambodia is the reductive dissolution of As-rich Fe (oxy) hydroxide (Kocar et al., 2008; Rowland et al., 2008). Arsenic contamination in the groundwater of Taiwan has been observed since 1960. The southwestern part of the country is mostly affected by As contamination (Mukherjee et al., 2006; Al Lawati et al., 2012). Various studies have proposed that As mobilisation in Taiwan occurs by reductive dissolution of As-bearing Fe (oxy) hydroxides and recently it has been observed that the composition and origin of organic matter present in Taiwanese aquifer sediments play an important role in As mobilisation (Al Lawati et al., 2012). Serious effects of As contamination are noted in Taiwan, for example, a large population shows blackfoot disease due to higher levels of As toxicity. Blackfoot disease is mostly noted in Chai-Nan Plain, southwestern Taiwan, where As contamination is very high (Li et al., 2011). Arsenic contamination in the groundwater of China was first reported in the 1980s (World Bank Policy Report, 2005). In several parts of China, including Xinxiang, Shanxi and Inner Mongolia, the As level in drinking water is found to be as high as 4400 µg/L. It is reported that more than 5 million people of China are exposed to As-contaminated drinking water with concentrations >50 µg/L (Sun et al., 2001). New, high-As areas are continuously recognised in parts of the Jilin, Ningxia, Qinghai and Anhui provinces and in the suburbs of Beijing. The presence of higher concentrations of As in the groundwater of these regions is associated with alluvial/deltaic sedimentary deposits and with the prevalence of reducing conditions (Rahman et al., 2009).

Groundwater resources in large alluvial deltas of the Red River in northern Vietnam as well as the Mekong River in southern Vietnam have been used extensively for drinking purpose. It is reported that approximately 10 million people in the Red River delta depend on a tubewell system for irrigation and drinking purposes (Nguyen et al., 2009; Jessen, 2012) and 65% of these wells exceed the50 µg/L. Nguyen et al. (2009) reported that As levels in several wells of the Red River valley exceed 3050 µg/L with an average value of 430 µg/L. In the Mekong delta, more than 1 million people are at risk as more than 40% of the tubewells have higher than 100 µg As/L (Hug et al., 2008; Nguyen et al., 2009; Kim et al., 2011). Besides these countries in Asia, As contamination in the groundwater is also reported from Nepal, Pakistan, Lao PDR, Myanmar and Indonesia (Mukherjee et al., 2006; Rahman et al., 2009).

19.3.2 North and South America

The levels of As contamination in the groundwater in the United States vary due to the combination of climate and geology. It is observed that more than 10% of the wells contain As concentrations above the WHO limit (>10 µg/L). Arsenic contamination is mostly observed in the western part of the country. In the last decade, further investigations on groundwater in New England, Michigan, Minnesota, South Dakota, Oklahoma and Wisconsin states suggests that As concentrations exceeding 10 µg/L are more widespread and common than previously recognised (Welch et al., 2000; Scanlon et al., 2009). The release of As into the groundwater occurs due to the reaction of iron oxide with either natural or anthropogenic (i.e. petroleum products) organic carbon. Iron oxide can also release As to alkaline aquifers of the western states (Scanlon et al., 2009). In Argentina, groundwater As contamination was reported in 1917 (Arguello et al., 1938). More than 1 million people of this country are exposed to this toxic metalloid with concentration that exceeds drinking water standard. The northern part of the country, that is, Chaco-Pampean plain, is the most affected area. It is noted that shallow wells are mostly contaminated by higher levels of As. The As-contaminated aquifers are formed due to the alluvial deposits from the Rio Dulce River (Bundschuh et al., 2004). The sources of As contamination in the drinking water of Chile are rivers originating from Cordillera de los Andes. Higher natural As contamination is noted in the rivers of northern Chile, particularly from the region of Antofagasta (Zaldivar and Wetterstrand, 1978; Cáceres et al., 1992). At the beginning of the 1960s, the first dermatological symptom was noted. The highest exposure period for the major population was 1958–1970. People of this region are exposed up to 860 µg As/L, depending on the rivers used for water supply (Mukherjee et al., 2006). In 1958, a high level of As in the groundwater of Mexico was observed for the first time as the cause of adverse health effects of the inhabitants of Comarca Lagunera (Cebrián et al., 1994). Subsequently, As concentrations above the Mexican drinking water standard have also been detected at other sites, including Zimapan valley, Rio Grande and The Yaqui Valley. The complex geology of Mexico with igneous and sedimentary rocks as well as active tectonic setting predisposes the environment to natural groundwater contamination (Armienta et al., 1997).

19.3.3 Europe

In Europe, higher As concentration in the groundwater is noted in the regions of the Pannonian Basin (Hungary, Romania, Croatia and Serbia), the area of Thessaloniki (Greece), the Kuthahya Plain (Turkey) and the Duero Basin (Spain). In most places, natural groundwater contains As with a broad concentration regime (maximum up to 5000 µg/L). In Turkey, As concentrations as high as 10,000 µg/L are also measured in the groundwater (Etterlin et al., 2008). In Hungary, it is estimated that 0.5 million people use drinking water with >10 µg/L, while in Croatia, 0.2 million people are considered to be affected by the consumption of intoxicated drinking water (Mukherjee et al., 2006; Etterlin et al., 2008). It is noted that in most of the basins such as the Pannonian Basin, the Duero Basin and in the area of Thessaloniki, As is released due to reductive dissolution of Fe (oxy) hydroxides, while at other sites, geothermal or anthropogenic activities are responsible (Etterlin et al., 2008).

19.4 Bacterial Interaction with Arsenic

Bacteria interact with As using an array of mechanisms, including assimilation, methylation, demethylation, oxidation and reduction (Stolz et al., 2010). Figure 19.2 represents several mechanisms of As metabolism in procaryotes. These reactions protect the microorganisms from As toxicity in the prevailing

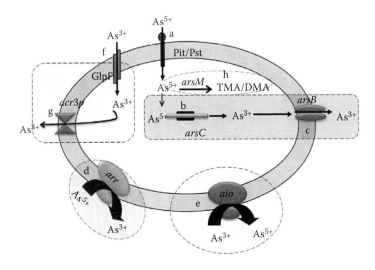

FIGURE 19.2

(**See colour insert.**) Schematic diagram of representing mechanisms of bacteria–As interaction. Arsenate (As^{5+}) enters the cells through the phosphate transporters (a); after entering the cells, As^{5+} is reduced to As^{3+} by cytosolic As^{5+} reductase, *arsC* (b); following reduction, As^{3+} is extruded out from the cell interior through efflux pump *arsB* (c). As^{5+} can be used as the terminal electron acceptor during respiration under anaerobic environment (d). As^{3+} can be oxidised to serve as an electron donor or a mechanism of resistance via periplasmic *aio* system (e). As^{3+} enters the cell through the aquaglyceroporin channel (f) and is directly extruded out from the cells via another As^{3+} transporter *acr3p* (g). Furthermore, inorganic As can also be transformed into organic species by methylation (h).

environment by enhancing resistance mechanisms and using some of these in energy generation. Microorganisms inhibiting contaminated niches often use As as a potential metabolic resource. For example, bacteria can utilise toxic As^{3+} as an electron donor or As^{5+} as a terminal electron acceptor during their anaerobic chemolithotrophic or heterotrophic aerobic and anaerobic metabolisms, respectively (Escudero et al., 2013).

19.4.1 Arsenic Uptake and Accumulation in Bacteria

Bacterial uptake of As^{3+} and As^{5+} occurs adventitiously through essential nutrient transport systems such as phosphate and glycerol transporters (Stolz et al., 2010; Escudero et al., 2013) (Figure 19.2). Arsenate, being a phosphate analogue, is able to use both of the two phosphate transporters—low-affinity Pit (phosphate inorganic transport) system as and the high-affinity Pst (phosphate specific transport) system for entry into the cell (Slaughter et al., 2012). The uptake pathway in *Escherichia coli* uses both Pst and Pit, while in *Saccharomyces cerevisiae*, the Pit system is mostly used (Rosen and Liu, 2009; Cavalca et al., 2013). On the other hand, As^{3+} is taken up by the aquaglyceroporin/glycerolophosphate transporter GlpF due to its structural and functional analogy (Sanders et al., 1997). These aquaglyceroporins are widespread in bacteria and they often allow the uptake of As^{3+} adventitiously by these channels. GlpF homologues have been identified in *E. coli*, *Sinorhizobium meliloti* and *Pseudomonas putida*, and are likely to facilitate As^{3+} transport across the cell membrane in these species (Lu et al., 2003; Yang et al., 2005). Though the accumulation of As inside the cell is common for a plant system, it is not frequent in bacteria (Gulz et al., 2005; Wang et al., 2006; Yang et al., 2012). As accumulation is observed in *Marinomonas communis* and *Pseudomonas* sp., which could accumulate 2.3 and 4 mg As/g dw, respectively (Takeuchi et al., 2007; Joshi et al., 2008).

19.4.2 Basic Detoxification Mechanism

Microorganisms have developed a number of mechanisms to counteract the deleterious effects of As. These strategies include (i) limiting the amount of As entering the cell by increasing the specificity of phosphate uptake pathways (Páez-Espino et al., 2009); (ii) producing less toxic organic compounds by methylation (Bentley and Chasteen, 2002); (iii) volatilisation of organic As species (Qin et al., 2006); and (iv) As^{5+} reduction and extrusion of reduced As^{3+} from the cell (Rosen, 1999). Basic detoxification mechanisms against As are almost similar for both Gram-negative and Gram-positive bacteria, which exploit the resistance system based on the *ars* operon encoded either on the chromosome, or on the plasmid or both (Cavalca et al., 2013). Principal components of this operon are As^{5+} reductase enzyme (ArsC) and As^{3+} transporter (ArsB). Two discrete families of As^{3+} transporters, namely, ArsB and Acr3p have been described in bacteria, which are involved in the extrusion of As^{3+} from the cell (Achour-Rokbani et al., 2007; Giloteaux et al., 2013) (Figure 19.3a). The first one is ArsB permease, which mainly functions as a uniporter using the membrane potential to extrude As^{3+} but it can also be coupled with ArsA and work more proficiently as ATP-driven As^{3+} pump that provides enhanced As^{3+} resistance (Figure 19.2). In addition to As^{3+}, proteins of the ArsB family also actively expel Sb^{3+} oxyanions in a wide range of microorganisms (Meng et al., 2004) (Figure 19.3b). Little is known about the second family of As^{3+} transporter Acr3p or Arr3p. Although members of this family are found in bacteria, archaea and fungi, but have been functionally characterised in fewer bacterial species, including *Ochrobactrum tritici*, *Bacillus subtilis*, *Synechocystis* sp. and *Corynebacterium glutamicum* (Rosen, 2002; López-Maury et al., 2003; Ordóñez et al., 2005; Achour-Rokbani et al., 2007; Cai et al., 2009) (Figure 19.3c). The Acr3p proteins in *B. subtilis* and *S. cerevisiae* have been reported to be involved in specifically transporting As^{3+}, whereas the homologue of Acr3p in *Synechocystis* sp. promotes both As^{3+} and Sb^{3+} extrusion. Based on phylogenetic dissimilarities of Acr3p genes, they can be divided into two subfamilies, namely, Acr3(1)p and Acr3(2)p (Rosen et al., 2011). Prevalence of *arsB* genes is observed in Firmicutes and γ-Proteobacteria, while *acr3* gene is mostly present in Actinobacteria and α-Proteobacteria

(a)

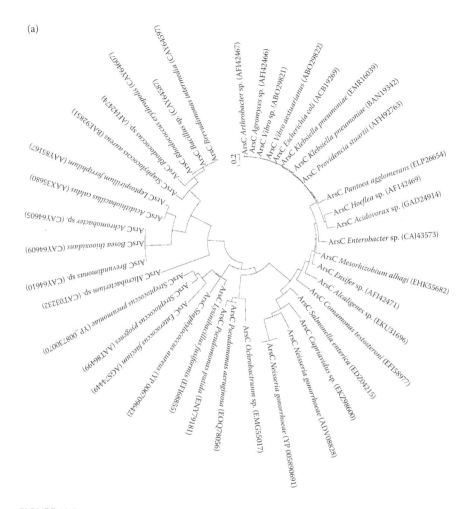

FIGURE 19.3

(a) Phylogenetic tree of deduced amino acid sequences of As^{5+} reductase gene (*arsC*) from different As-resistant bacteria. The scale bar 0.2 indicates 20% nucleotide sequence substitution.

(Achour-Rokbani et al., 2007). Distribution of these transporters among microorganisms suggests that evolution and horizontal gene transfer (HGT) of both the transporters ArsB and Acr3p family might have happened simultaneously in microbial evolution (Mukhopadhyay et al., 2002). Besides the *ars* detoxification system, bacteria can also use respiratory As^{5+} reductase (*arr*) as a mode of detoxification. This enzyme activity is well characterised in several anaerobic and facultative bacteria, including *Chrysiogenes arsenates*, *Bacillus selenitireducens* and *Shewanella* sp. strain ANA-3 (Krafft and Macy, 1998; Afkar et al., 2003; Saltikov et al., 2005; Drewniak and Sklodowska, 2013; Escudero et al., 2013). Some bacteria also employ the As^{3+} oxidation mechanism to produce the less toxic As^{5+}. A special As detoxification mechanism is

(b)

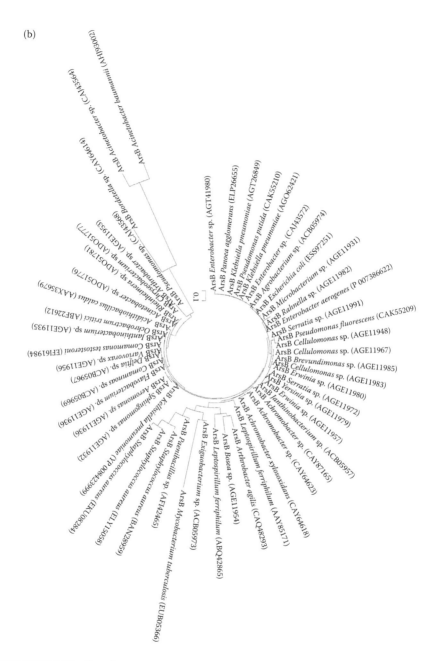

FIGURE 19.3 (Continued)
(b) Phylogenetic tree of deduced amino acid sequences of As^{3+} efflux pump gene (*arsB*) from different As-resistant bacteria. The scale bar 0.1 indicates 10% nucleotide sequence substitution.

(c)

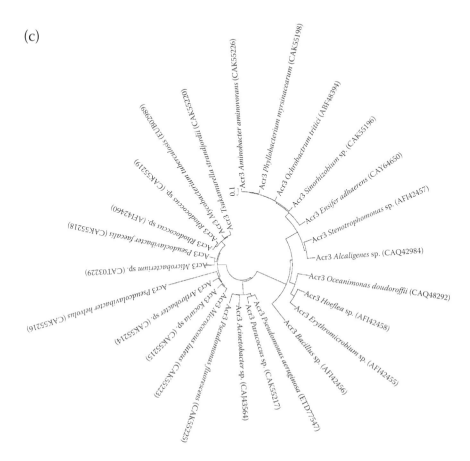

FIGURE 19.3 (Continued)
(c) Phylogenetic tree of deduced amino acid sequences of As^{3+} transporter gene (*acr3*) from different As-resistant bacteria. The scale bar 0.1 indicates 10% nucleotide sequence substitution.

observed in *S. meliloti*, where As^{5+} is taken up via a phosphate transport system and is reduced to As^{3+} by ArsC, which is further extruded from the cell by the downhill movement through the aquaglyceroporin channel (AqpS). Thus, in *S. meliloti*, the combination of AqpS and ArsC efficiently forms a new pathway of As detoxification (Yang et al., 2005). Biomethylation is considered to be another detoxification process where inorganic As is biomethylated to both the volatile species such as monomethylarsine (MMA), dimethylarsine (DMA) and trimethylarsine (TMA) as well as non-volatile species such as methylarsonic acid, dimethylarsinic acid and trimethylarsenic oxide (Chen et al., 2013). However, not all methylated products are always less toxic than inorganic forms of As. Although As methylation by fungi and other eucaryotes has been well studied, little is known for bacterial systems (Bentley and Chasteen, 2002; Yin et al., 2011).

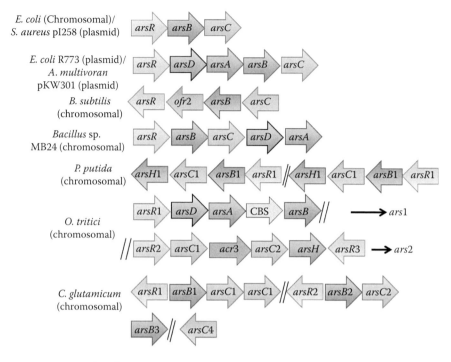

FIGURE 19.3 (Continued)
(See colour insert.) (d) Arrangement of gene clusters in *ars* operon in different bacteria.

19.4.3 Arsenic in Microbial Metabolism

Despite its severe toxicity, As is readily used by a large number of procaryotes for cell growth and metabolism. Microbial As transformation includes four fundamental processes: reduction, oxidation, methylation and demethylation (Figure 19.2).

19.4.3.1 Bacterial Arsenate Reduction

Exhaustive studies have been carried out on the mechanisms underlying microbial reduction of As^{5+} to As^{3+} (Oremland and Slotz, 2005; Páez-Espino et al., 2009; Stolz et al., 2010; Escudero et al., 2013). Bacteria undergo As reduction via two distinct mechanisms: cytosolic and respiratory. The first one is widespread in nature, providing detoxification of the cells via expression of *ars* operon, found either in chromosomal locations or in plasmids of Gram-negative bacteria belonging to the α-, β- and γ-Proteobacteria as well as in Gram-positive Firmicutes and Actinobacteria (Escudero et al., 2013). The mechanism conferring As resistance by *ars* operon has been extensively studied in more than 50 genera, including *Escherichia, Staphylococcus, Corynebacterium, Ochrobactrum, Pseudomonas, Acinetobacter, Exiguobacterium,*

Aeromonas, Vibrio, Psychrobacter, Enterobacter, Bacillus, Pantoea, Halanaerobium and so on (Jackson and Dugas, 2003; Anderson and Cook, 2004; Silver and Phung, 2005; Drewniak et al., 2008; Stolz et al., 2010; Liao et al., 2011; Pepi et al., 2011; Drewniak et al., 2012; Hamamura et al., 2012; Yang et al., 2012; Wu et al., 2013) (Figure 19.3a,b and c). The *ars* genes appear systematically and co-transcribed by a large variety of genomic configurations arranged in three to five member fashions depending upon the bacterial species. The core genes of this system include cytosolic As^{5+} reductase *arsC*, membrane-bound As^{3+} efflux pump *arsB* and transcriptional repressor *arsR* (Escudero et al., 2013) (Figure 19.3d). In this operon, additional members are also reported: *arsA* is an ATPase that provides energy to *arsB* for the extrusion of As^{3+} and Sb^{3+} and *arsD* is an As chaperone for the *arsAB* pump (Lin et al., 2007). The three-member gene set (*arsRBC*) is abundant in the chromosome and/or plasmid of *E. coli, Pseudomonas fluorescens* MSP3 and in the *Staphylococcus* plasmids pI258 and pSX267. The five-member *ars* operon arranged in *arsRDABC* fashion is found frequently in plasmids of *E. coli* R773 and R46 and *Acidophilus multivurum* AIU301 and pKW301 (Cavalca et al., 2013; Escudero et al., 2013). Other genes that have also been found in *ars* operons are *arsH* and rhodanese (thiosulphate transferase). The direct involvement of the latter components in As resistance has not been established well. In *Shigella flexneri, P. putida, P. aeruginosa, S. meliloti* and *Yersinia*, virulence plasmid pYV and *arsH* are also required to confer complete As resistance (Vorontsov et al., 2007; Ye et al., 2007). More recently, new arrangements of *ars* operon with multiple As^{5+} reductase genes have been discovered in a number of bacteria, including *C. glutamicum, Desulfovibrio desulfuricans, O. tritici SCII24T, Shewanella* sp., *Microbacterium* sp. and *Geobacillus kaustophilus* (Ordóñez et al., 2005; Li and Krumholz, 2007; Branco et al., 2008; Murphy and Saltikov, 2009; Achour-Rokbani et al., 2010; Cuebas et al., 2011; Villadangos et al., 2011). Thorough characterisation of ArsC enzyme has been done in *E. coli, B. subtilis* and *Synechrocystis* sp. (Rosen, 1999; Martin et al., 2001; Li et al., 2007). Although different studies reported a common origin for the *arsC* genes, based on the source of reducing power, three unrelated groups of ArsC with common biochemical function but different evolutionary relationship have been identified (Mukhopadhyay et al., 2002; Saltikov and Olson, 2002; Jackson and Dugas, 2003; Cavalca et al., 2010). These three groups are (a) glutaredoxin–glutathione-coupled enzyme associated with plasmids and chromosomes of Gram-negative bacteria as well as both As^{3+}-oxidising and (respiratory) As^{5+}-reducing bacteria, (b) less studied glutaredoxin-dependent As^{5+} reductase found in yeasts and (c) thioredoxin-coupled As^{5+} reductase found in Gram-positive as well as Gram-negative Proteobacteria (Martin et al., 2001; Messens et al., 2004). This divergence may occur mainly due to HGT of *arsC* gene among the microorganisms through convergent evolution (Mukhopadhyay et al., 2002). The phenomenon of HGT of *arsC* genes among diverse bacterial members has been studied within the strains isolated from several As-contaminated habitats (Cai et al., 2009; Villegas-Torres et al., 2011).

The other type of As^{5+} reductase is dissimilatory or respiratory As^{5+} reductase (arrA). This has been described mainly in obligate or facultative anaerobic bacteria affiliated to diverse phylogenetic groups (Figure 19.4a). During this process, bacterial cells gain metabolic energy by "breathing or respiring" As^{5+} as terminal electron acceptor (Saltikov and Newman, 2003). Although thermodynamic considerations suggest that respiratory reduction of As^{5+} can provide enough energy for microbial growth, its distribution among bacteria is limited due to overall toxicity of As (Cavalca et al., 2013). Rapid emergence of phylogenetically diverse As^{5+}-respiring bacteria in a wide range of environmental samples (from freshwater sediments, hypersaline lake waters, hot springs, deep-sea hydrothermal vents and gold mines) as well as in enrichment cultures suggests that they are widespread and metabolically active in nature (Silver and Phung, 2005; Stolz et al., 2010; Kudo et al., 2013). Dissimilatory As^{5+} reduction was first observed in *Geospirillum arsenophilus* MIT-13 strain (Ahmann et al., 1994). Subsequently, many other bacteria with respiratory As^{5+} reductase activity have been reported. These include *Desulfotomaculum auripigmentum* (Newman et al., 1997), *Sulfurospirillum barnessi* and *S. arsenophilum* (Stolz et al., 1999; Malasarn et al., 2004), *Bacillus* spp. (Santini et al., 2004; Wu et al., 2013), *Desulfuroporosinus* sp. (Pérez-Jimenez et al., 2005), *Wollinella succinogenes* (Stolz et al., 2006), *Alkaliphilus metalliredigenes* (Stolz et al., 2006), *Clostridium* sp. (Stolz et al., 2006), *Shewanella* sp. (Malasarn et al., 2008), *Alkaliphilus oremlandii* (Fisher et al., 2008), *Halarsenatibacter silvermanii* (Blum et al., 2009), *Desulfohalophilus alkaliarsenatis* (Sorokin et al., 2012; Blum et al., 2012), *Geobacter* sp. (Ohtsuka et al., 2013) and *Anaeromyxobacter* sp. (Kudo et al., 2013). In spite of its relatively wide distribution, respiratory As^{5+} reductase enzyme has been characterised in fewer cases (Páez-Espino et al., 2009). To date, only the respiratory As^{5+} reductase (Arr) from *C. arsenates, B. selenitireducens* and *Shewanella* sp. strain ANA-3 have been purified and characterised (Krafft and Macy, 1998; Afkar et al., 2003; Malasarn et al., 2008). Arsenate respiratory reductase (Arr) is a heterodimer protein with a catalytic subunit, ArrA, and a smaller electron transfer protein, ArrB. ArrA contains the molybdopterin centre and a [3Fe-4S] cluster while ArrB subunit contains four [4Fe-4S] clusters. Though the core enzyme ArrAB is highly conserved, the *arr* operon differs from organism to organism with respect to a number of genes. *Shewanella* sp. strain ANA has only two genes in the core enzyme (*arrAB*), while *Desulfitobacterium hafniense* has a gene encoding a putative membrane anchoring peptide as well as a multi-component regulatory system (*arrABC* or *arrSKRCAB*) (Malasarn et al., 2008; Kim et al., 2012) (Figure 19.4b). Recently, transcription of As respiration and resistance genes of *Geobacter lovleyi* has been characterised during *in situ* uranium bioremediation (Giloteaux et al., 2013). Though little is known regarding the genetic control of its expression, this enzyme, due to the conserve feature, is considered as a marker for the oxidation state of the environment.

Although these two reduction mechanisms (*ars* operon and *arr* operon mediated) are not directly correlated, in some cases, the *arr* and *ars* operons

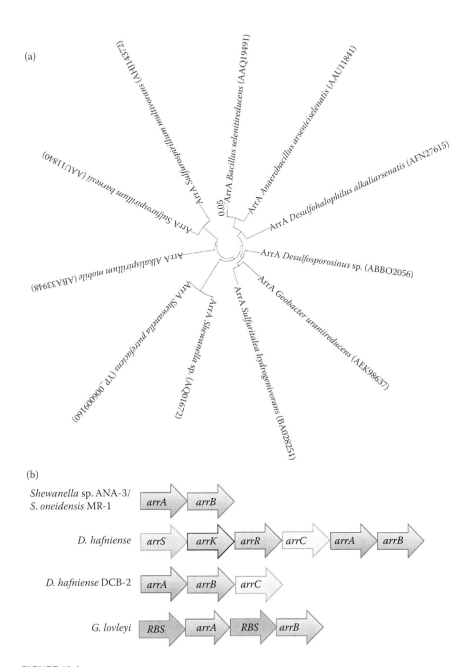

FIGURE 19.4
(**See colour insert.**) (a) Phylogenetic tree of deduced amino acid sequences of dissimilatory As^{5+} reductase gene (*arrA*) from different As-resistant bacteria. The scale bar 0.05 indicates 5% nucleotide sequence substitution. (b) Arrangement of gene clusters in *arr* operon in different bacteria.

lie in close proximity, within the genome of the organism suggesting an 'arsenic metabolism island,' which may be in *cis* or *trans* position (Stolz et al., 2010). It has been well established that these two systems play key roles in the solubilisation of As from sediment bound in subsurface environment, leading to major contamination of aquifers (Islam et al., 2004; Sutton et al., 2009; Zhang et al., 2012).

19.4.3.2 Bacterial Arsenite Oxidation

Arsenite oxidation is widely distributed in metabolically and taxonomically diverse bacteria. To date, more than 50 phylogenetically diverse As^{3+} oxidising strains distributed among 25 genera isolated from various contaminated niches (soils, mine tailings, river sediments, groundwater and geothermal springs) have been described for their ability to oxidise As^{3+} enzymatically (Santini et al., 2000; Oremland et al., 2002; Salmassi et al., 2002; Macur et al., 2004; Santini et al., 2004; Silver and Phung, 2005; Inskeep et al., 2007; Fan et al., 2008; Liao et al., 2011; Hammura et al., 2012; Song et al., 2012; Escudero et al., 2013; Yamamura et al., 2013) (Figure 19.5a). Since the first report of As^{3+}-oxidising bacterium belonging to the genus *Achromobacter* (Green 1919), many As^{3+}-oxidising bacteria distributed among different genera, including *Thermus* (Gihring et al., 2001), *Alcaligenes* (Anderson et al., 2002), *Achromobacter* (Santini et al., 2000; Fan et al., 2008; Cai et al., 2009; Bachate et al., 2012), *Agrobacterium* (Salmassi et al., 2002; Hao et al., 2012), *Herminiimonas* (Muller et al., 2006), *Thiomonas* (Duquesne et al., 2008), *Ochrobactrum* (Branco et al., 2008), *Pseudomonas* (Cai et al., 2009; Campos et al., 2010), *Polaromonas* (Osborne et al., 2010), *Bosea* (Liao et al., 2011), *Ancylobacter* (Andreoni et al., 2012), *Acidovorax* (Huang et al., 2012) and *Bordetella* (Bachate et al., 2012), have been reported. These include both heterotrophic as well as chemolithoautotrophic As^{3+} oxidisers (Oremland and Stolz, 2005), wherein they use the energy and reducing power from As^{3+} oxidation during CO_2 fixation and cell growth under both aerobic (Santini et al., 2000) and anaerobic nitrate-reducing conditions (Oremland et al., 2002; Rhine et al., 2007).

The As^{3+} oxidase enzyme was first isolated and characterised in *Alcaligenes faecalis* by Anderson et al. (1992). The detailed mechanism of As^{3+} oxidation has become clear only after the investigation of the crystal structure of this enzyme from this strain in 2001, revealing its heterodimeric organisation (Ellis et al., 2001). Bacterial As^{3+} oxidase belongs to the dimethyl sulphoxide (DMSO) reductase of the molybdenum family consisting of (i) small subunit with a Rieske [2Fe-2S] cluster, (ii) large subunit harbouring molybdopterin guanosine dinucleotide at the active site and iron-binding [3Fe-4S] cluster (Ellis et al., 2001) (Figure 19.5b). Arsenite oxidase enzyme has also been thoroughly characterised in several species of *Ralstonia*, *Arthrobacter* and *Rhizobium* (Prasad et al., 2009; van Lis et al., 2013; Kalimuthu et al., 2014).

Arsenite oxidase gene encoding two subunits (formerly known as *aoxA* and *aoxB*) has been recognised and characterised for the first time in the

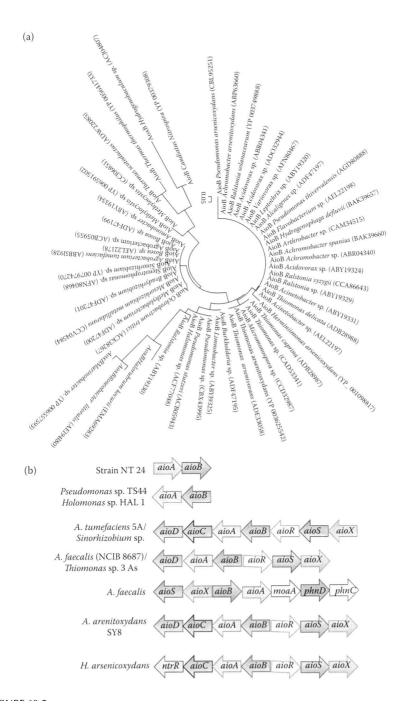

FIGURE 19.5
(**See colour insert.**) (a) Phylogenetic tree of deduced amino acid sequences of As^{3+} oxidase gene (*aioB*) from different As-resistant bacteria. The scale bar 0.05 indicates 5% nucleotide sequence substitution. (b) Arrangement of gene clusters in *aio* operon in different bacteria.

heterotrophic bacteria *Herminiimonas arsenicoxydans* (Weeger et al., 1999). Later, homologues of these genes were identified in a variety of organisms (Silver and Phung, 2005; Halter et al., 2011; Escudero et al., 2013). Different nomenclatures were adopted, such as *aroB* or *asoB* for the small and *aroA* or *asoA* for the large subunit genes in the chemolithoautotrophic As^{3+} oxidiser NT-26 and in *A. faecalis* strain NCBI8687. Recently, Lett et al. (2012) unified the nomenclature of these genes encoding As^{3+} oxidase. The small and large sub-units of the As^{3+} oxidase have now been termed as *aioB* and *aioA*, respectively. In most As^{3+}-oxidising bacteria, the synthesis of this enzyme is generally reg-ulated by As^{3+}. A complex mechanism for the expression of structural genes of As^{3+} oxidase (*aoxAB*) involving quorum sensing as well as a two-compo-nent signal transduction system was described in *Agrobacterium tumefaciens* 5A (Kashyap et al., 2006), while two-component regulatory genes, histidine sensor (*aoxS*) and transcriptional regulator (*aoxR*), located directly upstream of *aoxAB*, were identified as AoxSR system in the heterotrophic bacterium *O. tritici* SCII24 (Branco et al., 2008), *Ralstonia* sp. and *H. arsenicoxydans* (Muller et al., 2007; Koechler et al., 2010) and in the chemolithoautotrophic bacte-rium NT-26 (Sardiwal et al., 2010). A novel As^{3+} oxidase gene, *arxA*, has been identified in the genome sequence of chemolithoautotrophic *Alkalilimnicola ehrlichii* MLHE-1, which couples As^{3+} oxidation to nitrate reduction (Zargar et al., 2010). The identification of the As^{3+} oxidase gene from soil, sediment, and geothermal mats with different chemical characteristics and various levels of As contamination has suggested its wide distribution among the microorganisms (Inskeep et al., 2007; Rhine et al., 2007; Cai et al., 2009; Halter et al., 2011; Heinrich-Salmeron et al., 2011; Hamamura et al., 2012; Sultana et al., 2012; Majumder et al., 2013). The whole-genome sequencing of highly efficient As^{3+} oxidising bacteria *Achromobacter arsenitoxydans* SY8, haloaro-matic acid-degrading bacterium *Achromobacter xylosoxidans* A8, *Pseudomonas stutzeri* TS44, *Agrobacterium tumefaciens* 5A, *Halomonas* sp. and *Acidovorax* sp. has been recently done (Strnad et al., 2011; Li et al., 2012; Hao et al., 2012; Lin et al., 2012; Huang et al., 2012).

19.4.3.3 Methylation/Demethylation of Arsenic

The methylation of As is considered as a basic detoxification process. Arsenic methylation in bacteria is not so frequent though, in eucaryotic cells, it is more familiar (Bentley and Chasteen, 2002). The pathway of methylation was first proposed by Challenger (1951) in the fungus *Scopulariopsis brevicaulis*, involving a series of steps in which the reduction of As^{5+} was followed by oxidative addition of a methyl group (Bentley and Chasteen, 2002). The same pathway has also been suggested for procaryotes, although the formation of arsine is more common in bacteria (Dopp et al., 2009). Volatilisation of arsenicals was first reported in *Methanobacterium bryantii* in the early 1970s (McBride and Wolfe, 1971). Since then, several bacteria (e.g. *Methanobacterium formicium, Clostridium collagenovorans, Disulfovibrio gigas* and *Disulfovibrio vulgaris*) have been shown to volatilise methylated arsenical species (Michalke et al., 2000). Although many of the enzymes involved in this complex process of As methylation remain unknown, a methyl transferase, ArsM, from *Rhodobacter sphaeroides* conferring resistance to As and generating TMA has been identified (Qin et al., 2006). Compared to the methylation process, less is known about the mechanism of demethylation. It was first suggested in *S. brevicaulis* and *Penicillium notatum* by Challenger (1951). Soil bacteria affiliated to *Alcaligenes*, *Pseudomonas*, *Mycobacterium* and *Cyanobacteria* have been reported for demethyling mono- and dimethyl As compounds (Bentley and Chasteen, 2002; Yin et al., 2011; Chen et al., 2013).

19.5 Microbial Diversity in As-Contaminated Aquifers

19.5.1 Bengal Delta Plain

Microbial communities within the As-contaminated aquifers of BDP are explored by analysing groundwater and sediment samples collected from parts of Bangladesh (Sutton et al., 2009; Sultana et al., 2011; Gorra et al., 2012). Few studies have used samples from West Bengal and Assam as well (Gault et al., 2005; Ghosh and Sar, 2013; Sarkar et al., 2013). Characteristics of bacterial populations and/or their potential role in As mobilisation within the subsurface environment are mainly investigated by monitoring the population shifts using laboratory-based microcosm experiments and/or culture-based approaches. Composition of bacterial communities as reported in these studies is presented in Figure 19.6. It is observed that the most abundant populations in different communities are represented by members of Proteobacteria, particularly of class β > γ > α-Proteobacteria along with Bacteroidetes, Actinobacteria, Firmicutes and so on (Figure 19.7). Identification of bacterial groups at lower taxonomic level revealed predominance of a number of genera (viz. *Pseudomonas*, *Rhizobium*, *Methylophilales*, *Burkholderiales* of γ-Proteobacteria, α-Proteobacteria, and β-Proteobacteria) representing the major populations (Figure 19.8).

Members of the genus *Acidovorax* (β-Proteobacteria) and *Pseudomonas* (γ-Proteobacteria) are most frequently detected in the sample from Assam and West Bengal, respectively (Hery et al., 2010; Ghosh and Sar, 2013) (Figure 19.9). Bacterial strains affiliated to *Acidovorax* mostly isolated from Assam showed strong relatedness with previously retrieved As-resistant *Acidovorax* sp. and type strain *Acidovorax delafieldii* (Fan et al., 2008; Cai et al., 2009). Members of the genus *Acidovorax* are known be chemoorganotrophic, nitrate reducing, and capable of mobilising iron in the subsurface environment. Members of this genus are found within the less contaminated sites north of the Brahmaputra River in Assam as well as a predominant

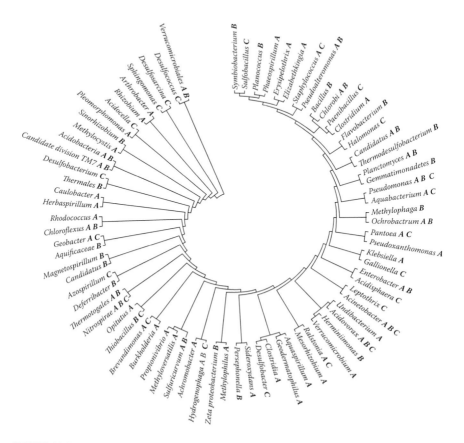

FIGURE 19.6

Phylogenetic tree of different bacterial genera identified from various As-contaminated sites of the world. The tree is constructed using Jukes–Cantor distances and 1000 bootstraps analyses are conducted. Sequences retrieved from Bengal delta aquifers, others countries including Southeast Asia and As-containing acid mine drainage sites are suffixed here as 'A', 'B' and 'C', respectively, in bold italic font.

group in high As-contaminated areas south of the Brahmaputra River and east of the Ganges (Chakdaha district of West Bengal). Huang et al. (2012) have also observed the presence of *Acidovorax* as the predominate group in As-contaminated soil from a gold mine in Daye, Hubei Province, China.

Members of the genus *Pseudomonas* are reported as a predominant group in several As-rich sites of Bangladesh, West Bengal and Assam (Hery et al., 2010; Sultan et al., 2011; Ghosh and Sar, 2013) (Figure 19.8). Bacterial strains affiliated to *Pseudomonas* are found to be closely related to *P. putida* and other *Pseudomonas* strains reported to have (i) siderophore-mediated Fe acquisition ability and (ii) As resistance property (Matthijs et al., 2009; Davolos and Pietrangeli, 2013). Members of this genus can grow aerobically as well as anaerobically using nitrate as the terminal electron acceptor. The average redox potential at different parts of the Bengal delta aquifer shows its

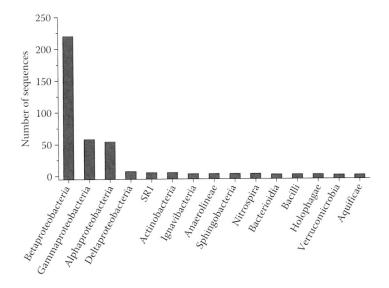

FIGURE 19.7
(**See colour insert.**) Distribution of major phylogenetic groups at phylum level retrieved from different sites of Bengal delta plain.

partially reduced condition, which seems to be favourable for bacterial denitrification coupled to organic matter degradation (Stumm and Morgan, 1995). This correlated very well with a noticeable abundance of denitrifying pseudomonads related to *Pseudomonas* sp. K-50, a bacterium able to denitrify nitrate with concomitant consumption of oxygen or *Pseudomonas mendocina* ZW27, which are capable of even aerobic denitrification (Takaya et al., 2003). The ability of such strains for simultaneous nitrate and oxygen utilisation

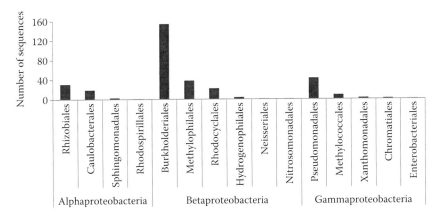

FIGURE 19.8
(**See colour insert.**) Distribution of major phylogenetic groups at genus level retrieved from different sites of Bengal delta plain.

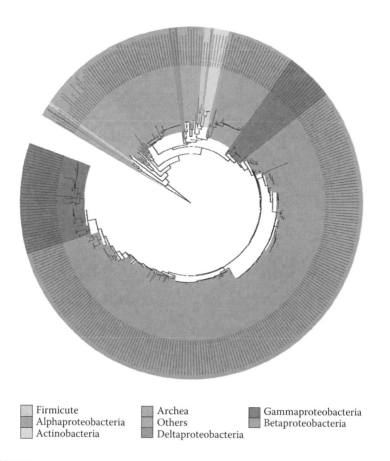

Firmicute

Alphaproteobacteria

Actinobacteria

Archea

Others

Deltaproteobacteria

Gammaproteobacteria

Betaproteobacteria

FIGURE 19.9
(See colour insert.) Phylogenetic tree of different bacterial genera identified from various As-contaminated sites of Bengal delta plain.

suggests their potential role in the formation of reducing/anoxic conditions as well.

Members of the genera *Brevundimonas* (α-Proteobacteria) are abundantly detected from the northern part of the Brahmaputra River and as a minor population from the eastern part of the Ganges (West Bengal), but not from Bangladesh (Ghosh and Sar, 2013; Sarkar et al., 2013) (Figure 19.8). It is observed that *Brevundimonas* isolates from BDP shows lineages with *B. bullata* and *B. nasdae*. Similarity of these isolates is also found with As-resistant *Brevundimonas* spp. reported from As-rich mine environments (Drewniak et al., 2008). Most of the *Brevundimonas* species retrieved from Assam are hypertolerant to both forms of As, two-thirds of which could withstand even higher concentrations of As^{5+}, that is, up to 550 mM and a few could grow at 50 mM of As^{3+}. Members of the genus *Brevundimonas* are capable of utilising different types of organic carbon sources and As^{5+} and SO_4^{2-} as their electron acceptors under anaerobic growth (Ghosh and Sar, 2013). A few *Brevundimonas* spp. isolated from Assam also show their ability for chemo-lithotropic growth in the presence of As^{3+}. Together with high As tolerance, the use of diverse organic carbon during heterotrophic growth, anaerobic metabolism utilising As^{5+} and even chemolithotrophic metabolism possibly indicate the metabolic robustness of this group required for survival in the oligotrophic nutrient-limited environment of the Bengal delta.

Another group of bacteria affiliated to *Rhizobium* and *Agrobacterium* species of α-Proteobacteria is abundantly detected from Assam and West Bengal (Ghosh and Sar, 2013; Sarkar et al., 2013). The presence of the same group is less frequently reported from Bangladesh. Most of the *Rhizobium* strains retrieved from Assam and West Bengal are hypertolerant to both forms of As (As^{3+} and As^{5+}), and many of them also showed superior tolerance to other heavy metals as well. The presence of plasmid is observed within several As- and other metal-resistant strains (Sarkar et al., 2013). Most of the *Rhizobium* strains isolated from the western part of the Bengal basin also showed As^{5+} reductase, siderophore and phosphates activities (Sarkar et al., 2013). *Rhizobium* strains isolated from Assam are capable of utilising different types of organic carbon sources and can utilise alternate electron acceptors under anaerobic growth (Ghosh and Sar, 2013).

Methane-utilising genera *Methyloversatilis* and *Methylophilus* of (β-Proteobacteria) are noted from As-rich sites of both West Bengal and Bangladesh (Hery et al., 2010; Sultana et al., 2011). Facultative methanol-utilising *Methylophilus* species are detected as an abundant group in Bangladesh. It is noted that *Methylophilus* can grow on a limited range of more complex organic compounds (Jenkins et al., 1987), which may be available to the inhabitant microbes in BDP aquifers that contain relatively lower dissolved organic matter (Kar et al., 2010). A related species, *Methylophilus* sp. EHg7, is reported as a heavy-metal- and As^{5+}-resistant bacterium by Marco et al. (2004). Members of *Methyloversatilis* are reported to be capable of utilising methane and multi-carbon compounds.

Several groups of β-Proteobacteria, namely, *Aquamonas*, *Aquabacterium* and *Aquaspirillum* are noted from Bangladesh (Sutton et al., 2009; Sultana et al., 2011). These bacteria are mainly observed in a deep aquifer (Pleistocene aquifers) that is considered to be maintaining oxic to suboxic conditions (Sutton et al., 2009). The presence of *Aquabacterium* and other facultative aerobic organisms, capable of utilising oxygen as well as nitrate as terminal electron acceptor is consistent with the suboxic conditions found in Pleistocene sediments of Bangladesh (Kalmbach et al., 1999; Swartz et al., 2004). Bacterial genera including *Hydrogenophaga*, *Acinetobacter* and *Herbaspirillum*, known for their chemolithoautotrophic mode of metabolism, are reported from both West Bengal and Bangladesh (Sultana et al., 2011; Sarkar et al., 2013; Ghosh and Sar, 2013). Although often isolated as heterotrophs, these organisms can use energy and reducing power from the oxidation of various inorganic elements including $As^{3+}/Fe^{2+}/Mn^{2+}$ during CO_2 fixation or other anabolic reactions and grow under aerobic or anaerobic oligotrophic environments (Sutton et al., 2009; Sarkar et al., 2013). *Acinetobacter* species has been shown to be capable of As oxidation; however, the As resistance genes *ars*R and *ars*H present in these organisms are indicative of a survival strategy rather than a means of chemotropic growth (Fournier et al., 2006; Fan et al., 2008). Members of the genus *Hydrogenophaga* identified in the eastern part of BDP with an As concentration of 332 mg/L have been found to be associated with As oxidising biofilm (Salmassi et al., 2006).

Well-known iron-reducing bacteria *Geobacter* and other members of *Deltaproteobacteria* are reported from the BDP. Their role in As mobilisation in groundwater with respect to BDP are well documented by several investigators where they use this model of iron-reducing bacteria for sediment-based microcosm study. A close lineage of *Geobacter* species detected in BDP with *G. sulfurreducens* capable of releasing As via direct enzymatic reduction of As^{5+} to As^{3+} as well as indirectly through Fe^{3+} reduction is noted. It is reported that *G. sulfurreducens* is unable to conserve energy for growth via the dissimilatory reduction of As^{5+}, although it is able to grow in medium containing fumarate as the terminal electron acceptor in the presence of As^{5+} (Islam et al., 2004, 2005; Gault et al., 2005).

Among the bacterial groups detected as relatively minor populations, *Propionivibrio, Burkholderiales, Herbauspirillum, Sideroxydans, Nitrosomonas, Vogesella, Ideonella, Achromobacter, Ralstonia, Undibacterium* and so on are reported from different As-rich sites of BDP (Gault et al., 2005; Sutton et al., 2009; Hery et al., 2010; Sultana et al., 2011; Ghosh and Sar, 2013; Sarkar et al., 2013) (Figure 19.6). Within these groups, *Achromobacter, Herbaspirillum* and *Burkholderiales* are known for their role in the As biogeochemical cycle (Ghosh and Sar, 2013; Sarkar et al., 2013). *Achromobacter* species isolated from West Bengal show higher As resistance properties (40 mM As^{3+} and 100 mM As^{5+}) and are positive for siderophores, acid phosphatase and arsenite oxidase or arsenate reductase activities (Sarkar et al., 2013).

19.5.2 Microbial Diversity within As-Contaminated Groundwater from Other Parts of Southeast Asia and Other Parts of the World

Besides the BDP, microbial diversity is investigated within As-rich groundwater from other parts of Asia, Europe and America. For the purpose of analysis, we have used the bacterial community composition reported from 15 samples. It is noted that bacterial communities are dominated by aerobic, facultative anaerobic organisms belonging to α-, β- and γ-Proteobacteria, Actinobacteria and Firmicutes. There is an abundance of genera *Thiobacillus, Herminimonas, Acidovorax, Hydrogenophaga, Gallionella* (members of β-Proteobacteria); *Pseudomonas, Methylophaga, Enterobacter, Actinobacter* (members of γ-Proteobacteria); *Symbiobacterium, Planococcus, Bacillus* (members of Firmicutes) along with *Planctomyces, Acidobacteria, Verrucomicrobiales, Acitinobacter, Sulfuricuruum, Thermotogae* and iron-reducing *Geobacter*

(Bruneel et al., 2006; Kinegam et al., 2008; Halter et al., 2011; Al Lawati et al., 2012; Jareonmit et al., 2012; Sheik et al., 2012; Price et al., 2013; Jiang et al., 2014) (Figure 19.6). Most of these organisms show relatedness with highly As-resistant, sulphide/thiosulphate-oxidising, denitrifying and aromatic hydrocarbon-degrading bacteria. In response to the change of groundwater environment, relative abundance of major bacterial groups are varied. Samples with high concentration of As, methane and Fe^{2+} and low concentration of SO_4^{2-} and NO_3^- are dominated by *Acinetobacter, Geobacter, Thermoprotei* and *Methanosaeta*. In contrast, *Pseudomonas* and *Nitrosophaera* are abundant and *Thermoprotei* and methanogens are absent in groundwater with lower levels of As and methane, but higher concentrations of SO_4^{2-} and NO_3^- (Kinegam et al., 2008; Jiang et al., 2014).

19.6 Microbial Role in As Mobilisation

Increased research in the last few years have provided several lines of evidences indicating that indigenous bacteria play a critical role in As mobilisation from alluvial or deltaic sediments of BDP, many sites of southeast Asia, the United States and many other countries. Many mechanisms have been proposed to explain the bacterial role in As release, including reductive dissolution of Fe oxide/oxyhydroxides (Islam et al., 2004; McArthur et al., 2004), oxidation of Fe sulphide (pyrite) (Loeppert et al., 1997) and also mineral bioweathering during nutrient acquisition (Mailloux et al., 2009).

19.6.1 Reducing Environment

With respect to most extensively studied As-rich groundwater (at BDP, Hetao plain and few other), it is observed that these aquifers maintain a moderate–strong reducing condition. Mobilisation of As from subsurface sediment to groundwater under such conditions is generally observed in geologically young (Holocene) fluvial sedimentary deposits associated with modern deltas (Pearcy et al., 2011). The shallow aquifers of the Ganges–Brahmaputra delta region of West Bengal (India) and Bangladesh are the prime examples (World Bank Policy Report, 2005; Kar et al., 2010; Pearcy et al., 2011). In presence of reducing conditions, desorption of As from the mineral sources in aquifers sediment is favoured. Most important, As bearing minerals in such aquifers are metal oxides (especially Fe oxides or Fe [oxy] hydroxide) and sulphides (especially pyrite). A number of reducing mechanisms are proposed, including the reduction of solid-phase As^{5+} to As^{3+}, desorption of As from Fe oxides (As-bearing Fe^{3+} oxides) and reductive dissolution of the oxides themselves (Smedley and Kinniburgh, 2002; Stüben et al., 2003; Islam et al., 2004; World Bank Policy Report, 2005). There are both biotic and abiotic pathways present for the reduction mechanisms. Lovley et al. (1991) compared the reduction of Fe (oxy) hydroxide in aquatic sediments by three bacterial strains (a sediment isolate GS-15, *Clostridium pasteurianum* and *E. coli*) to chemical reduction by a number of different organic compounds. They found that bacterial oxidation of organic matter coupled to the Fe^{3+} reduction in aquatic sediments is faster and more extensive than chemical reduction at neutral pH, indicating that microorganisms are mainly responsible for Fe^{3+} reduction in alluvial or deltaic sediments, which are directly or indirectly involved in As mobilisation. Islam et al. (2004) observed that anaerobic metal-reducing bacteria play a role in the formation of toxic, mobile As^{3+} in sediments from the BDP. Islam et al. (2005) reported that Fe^{3+}-reducing bacteria *Geobacter* and *Geothrix* are capable of controlling As mobilisation by reducing As-bearing Fe^{3+} oxide minerals from the western part of BDP. In subsequent time, several other studies have also demonstrated that anaerobic metal-reducing bacteria can play a key role in As mobilisation from sediments (Gault et al., 2005; Hery et al., 2010). Interestingly, it is highlighted that the processes of Fe and As release are decoupled. The role of reductive mechanisms has been implicated by several studies conducted with samples from Bangladesh (Nickson et al., 1998; Acharyya, 2002; Bose and Sharma, 2002; McArthur et al., 2004; Swartz et al., 2004, Dhar et al., 2011), the United States and Mexico (Welch and Lico, 1998; Welch et al., 2000; Kneebone et al., 2002), New Zealand (Aggett and O'Brien, 1985), Switzerland (Azcue and Nriagu, 1995), southeast Asiatic countries such as Cambodia, Vietnam and Laos (Kocar et al., 2008; Rowland et al., 2008; Kim et al., 2011) and other alluvial sediment-containing countries as well.

19.6.2 Source of Organic Carbon

The possible role of organic carbon available in the subsurface environment of BDP and other As-rich aquifers is observed by several investigators (McArthur et al., 2004) and few from other places (Postma et al., 2007). It is noted that even very low concentration of labile organic matter are able to support a rapid microbial As mobilisation process than organic carbon lean sediment via metal reduction, where generally organic matter acts as carbon/electron donors and metal acts as electron acceptors for microbial growth and metabolism (Islam et al., 2004; Gault et al., 2005; Hery et al., 2010; Dhar et al., 2011). Sedimentary organic matter is derived from peat layers and external sources, which are either of anthropogenic or natural origin, such as surface-derived materials, decaying phytoplankton and other plant or animal material. In particular, the total organic carbon (TOC) content in the sediments of the BDP region is reported to be typically very low (<1%) (Ravenscroft et al., 2001; McArthur et al., 2004; Rowland et al., 2006). It has been hypothesised that both quantity and quality of labile organic matter and electron donor required for driving the As mobilisation reactions are important for microbial action (Hery et al., 2010). Labile organic matter is a mixture of many types of functional groups, including primary metabolites such as fatty acids, carbohydrates, sugars, and amino acids and also contains considerable amounts of inorganic impurities, such as Fe, Mn and Al.

19.6.3 Mineral Dissolution

It is well known that the subsurface environment in As-rich aquifers are nutrient limited and oligotrophic in nature and therefore, indigenous microorganisms often produce different types of organic acids, ligands, polysaccharides and so on to promote the dissolution of minerals for nutrient acquisition. It is known that during nutrient acquisition, microbes access the trace nutrient elements, but along with trace elements they also release toxic elements such as As from the host minerals to the groundwater (Milloux et al., 2009). Among the several modes of bacterial nutrient acquisition, the production of siderophores is probably very well known in mobilising various heavy metals (Gadd, 2004). Siderophores are small, high-affinity Fe-chelating compounds secreted by microorganisms. In addition to Fe, siderophores can also bind to other metals such as Al, Mn, Mg, Cr and so on. According to earlier studies, it is reported that there may be a link between As mobilisation as a by-product and Fe acquisition through siderophores from minerals like Fe (oxy) hydroxides and hydroxyapatite, present in the Bengal basin (Mailloux et al., 2009; Pal and Mukherjee, 2009). Matlakowska et al. (2008) indicated that *Pseudomonas* strains isolated from mining sites produced siderophores, which could promote mineral dissolution and mobilisation of the more toxic As^{3+} species in the environment. The role of siderophore in As mobilisation by *Pseudomonas azotoformans*, *Aspergillus niger*, *Mycobacterium* sp., *Rhodococcus erythropolis* and the well-known siderophore producer *Rhizobium leguminosarum* is also studied earlier (Sriyosachati and Cox, 1986; Vala et al., 2006; Nair et al., 2007). However, the presence of such phenotypes within the bacteria from As-contaminated groundwater of Bengal delta is first reported by Sarkar et al., 2013. They observed that taxonomically distinct bacteria sharing important properties related to As resistance, transformation and nutrient acquisition are abundant in As-rich groundwater. Mineral dissolution may also occur via the by-product of microbial metabolites. Phosphate-limited cells of *Burkholderia fungorum* mobilise As from apatite as a by-product of mineral weathering for nutrient acquisition (Mailloux et al., 2009). Microbial degradation of glucose to gluconic acid by *B. fungorum* in close contact with apatite is a likely cause of As released from the mineral structure into the groundwater (Mailloux et al., 2009). Frey et al. (2010) highlighted the potential of organic acids and HCN-producing microorganisms to dissolve minerals, which may be a probable mechanism to release mineral-associated As as well. Therefore, mineral dissolution by microorganisms during nutrient acquisition, which may directly or indirectly release As and other toxic metals, is important for future research work.

19.7 Microbial Route towards Mitigation of Arsenic Problem

Microbial interaction that leads to change in speciation of As as well as its toxicity and mobility has been considered as an effective tool for developing bioremediation process. Water treatment systems for As removal are very complex. Any effective treatment of As-contaminated water has to target both the toxic As^{3+} and As^{5+} forms. Under oxidising condition, such as those prevailing in surface water, the predominant species is As^{5+}. On the other hand, under mildly reducing conditions, such as those prevailing in most groundwater, As^{3+} is the thermodynamically stable form. Arsenite interacts to smaller extent with most solid surfaces; therefore, it is more difficult to remove by the application of conventional treatment methods, such as coagulation, filtration, ion exchange, lime softening, adsorption on iron oxides or activated alumina and reverse osmosis (Jekel, 1994; Katsoyiannis and Zouboulis, 2004). Optimisation of these processes generally requires a preliminary pre-oxidation of As^{3+} to As^{5+}, which is often performed with chemical oxidants such as chlorine, hypochlorite, ozone permanganate or hydrogen peroxide (Jekel, 1994). These methods have a high cost, are energy consuming, disruptive to the environment and difficult for implementation or maintenance. Since bacteria play an essential role in the geochemical cycle of As, it will be worth it to use them in biosystems to treat As-contaminated groundwater. A number of well-characterised bacteria (in terms of their physiological and metabolic properties) have been used to develop various innovative treatment systems adapted to different environmental situations.

Microbial activity can decrease As concentration through sorption, biomethylation, biomineralisation, complexation and oxidation–reduction processes (Table 19.3). They have evolved the biochemical mechanisms to exploit As either as an electron acceptor for anaerobic respiration, or as an electron donor to support chemoautotrophic fixation of CO_2 into cell carbon. People have adopted these metabolic activities of microorganisms which may be advantageous for bioremediation of this metalloid (Table 19.2). Some of the studies used a two-stage approach of As removal by utilising the As^{3+} oxidising ability of various strains, including *Microbacterium lacticum, H. arsenicoxydans, Ralstonia eutropha, P. putida, B. indicus, Leptothrix* sp. and *Gallionella* sp. with Fe-oxides/prozzolona as adsorbent matrix in fixed-bed bioreactors (Katsoyiannis and Zouboulis, 2004; Battaglia-Brunet et al., 2006; Lièvremont et al., 2009; Shakya et al., 2013). Though As^{3+}-oxidising bacteria are used to remove As from water, the As^{5+}-reducing strains are efficient for cleaning up As-contaminated soil, where remobilisation of strongly adsorbed As^{5+} is required. Reduced As^{3+} is then removed by precipitation or complexation with sulphide. For the remediation of As-contaminated mine tailings, gold mines, soils and so on, several As^{5+}/SO_4^{3-}-reducing bacteria have been employed successfully (Chung et al., 2006; Ghodsi et al., 2011; Gorra et al., 2012).

TABLE 19.3

Microorganism-Based Arsenic Bioremediation from Aqueous Phases

Microbiological Process	Microorganisms	Efficiency	Technology	Reference
Biosorption	*Gallionella ferruginea* and *Leptothrix ochracea*	Up to 95%	Fixed-bed upflow Filtration unit	Katsoyiannis and Zouboulis (2004)
	Ralstonia eutropha	>99%	Batch experiment	Mondal et al. (2008)
	Rhodococcus sp.	77.3 mg/g	Batch experiment	Prasad et al. (2011)
	Azotobacter sp.	>96%	Batch experiment	Gauri et al. (2011)
	Bacillus cereus	32.42 mg/g	Batch experiment	Miyatake and Hayashi (2011)
	Bacillus subtilis	11 times higher than normal cell	Batch experiment	Yang et al. (2012)
Bioaccumulation	*Escherichia coli* (engineered bacteria)	50–60-fold	Batch experiment	Kostal et al. (2004)
	Marinomonas communis	2.3 mg As/g dw	Batch experiment	Takeuchi et al. (2007)
	Pseudomonas sp.	4 mg As/g dw	Batch experiment	Joshi et al. (2008)
	Escherichia coli (engineered bacteria)	>80-fold	Batch experiment	Singh et al. (2010)
	Bacillus sp.	25.4% As^{5+} and 30.4% As^{3+}	Batch experiment	Majumder et al. (2013)
Bioreduction	*Bacillus selenatarsenatis*	As is reduced significantly	Batch experiment	Yamamura et al. (2008)
Bioxidation	CAsO1 bacterial consortium and *Thiomonas arsenivorans*	9-fold	Upflow column reactor	Michel et al. (2007)
	Rhodococcus equi	95%	Packed-bed reactor	Bag et al. (2010)
	Thiomonas arsenivorans	48.2%–99.3%	Packed-bed reactor	Dastidar et al. (2012)
Biomethylation	*Desulfuromonas palmitatis*	90%	Batch experiment	Vaxevanidou et al. (2008)
	Pseudomonas putida	Volatise As efficiently	Batch experiment	Chen et al. (2013)
	Bacillus sp.	13.5%	Batch experiment	Liu et al. (2013)
	Sphingomonas sp.			

Engineered bacteria (mostly *E. coli*) acting as bioadsorbent are considered as an attractive tool for the low-cost and efficient As removal system (Kostal et al., 2004). Modification of As resistance operon using DNA shuffling in a metagenomic library from an industrial effluent treatment plant sludge revealed a novel As^{5+} resistance gene (arsN) encoding a protein similar to acetyl transferases. Overexpression of this novel gene led to higher As resistance in *E. coli* (Chauhan et al., 2009). Recently, the soil bacterium *P. putida*, which is used in the remediation of As-contaminated sites, has actually been engineered to catalyse biomethylation of As (Chen et al., 2013). These investigations draw attention to the possibility of combining both natural and modified pathways for hyper As accumulation, which will lead to efficient removal of As from water.

References

Acharyya, S.K. Arsenic contamination in groundwater affecting major parts of southern West Bengal and parts of western Chhattisgarh: Source and mobilization process. *Current Science* 82(6), 2002: 740–743.

Achour-Rokbani, A., P. Bauda, and P, Billard. Diversity of arsenite transporter genes from arsenic-resistant soil bacteria. *Research in Microbiology* 158(2), 2007: 128–137.

Achour-Rokbani, A., A. Cordi, P. Poupin, P. Bauda, and P. Billard. Characterization of the ars gene cluster from extremely arsenic-resistant *Microbacterium* sp. strain A33. *Applied and Environmental Microbiology* 76(3), 2010: 948–955.

Afkar, E., J. Lisak, C. Saltikov, P. Basu, R. S. Oremland, and J. F. Stolz. The respiratory arsenate reductase from *Bacillus selenitireducens* strain MLS10. *FEMS Microbiology Letters* 226(1), 2003: 107–112.

Aggett, J., and Glennys A. O'Brien. Detailed model for the mobility of arsenic in lacustrine sediments based on measurements in Lake Ohakuri. *Environmental Science & Technology* 19(3), 1985: 231–238.

Ahmann, D., A.L. Roberts, L.R. Krumholz, and F.M. Morel. Microbe grows by reducing arsenic. *Nature* 371, 1994: 750–758.

Al Lawati, W.M., A. Rizoulis et al. Characterisation of organic matter and microbial communities in contrasting arsenic-rich Holocene and arsenic-poor Pleistocene aquifers, Red River Delta, Vietnam. *Applied Geochemistry* 27(1), 2012: 315–325.

Anderson, C.R., and G.M. Cook. Isolation and characterization of arsenate-reducing bacteria from arsenic-contaminated sites in New Zealand. *Current Microbiology* 48(5), 2004: 341–347.

Anderson, G.L., Jeffrey Williams, and R. Hille. The purification and characterization of arsenite oxidase from *Alcaligenes faecalis*, a molybdenum-containing hydroxylase. *Journal of Biological Chemistry* 267(33), 1992: 23674–23682.

Andreoni, V., R. Zanchi, L. Cavalca, A. Corsini, C. Romagnoli, and E. Canzi. Arsenite oxidation in *Ancylobacter dichloromethanicus* As3–1b strain: Detection of genes involved in arsenite oxidation and CO2 fixation. *Current Microbiology* 65(2), 2012: 212–218.

Arguello, R.A., D.D. Cenget, and E.E. Tello. Cancer and endemic arsenism in the Cordoba region. *Rev Argent Dermatosifiorg* 22, 1938: 461–487.

Armienta, M.A., R. Rodriguez, A. Aguayo, N.G. Villaseñor, and O. Cruz. Arsenic contamination of groundwater at Zimapán, Mexiko. *Hydrogeology Journal* 5(2), 1997: 39–46.

Azcue, J.M., and J.O. Nriagu. Impact of abandoned mine tailings on the arsenic concentrations in Moira Lake, Ontario. *Journal of Geochemical Exploration* 52, 1, 1995: 81–89.

Bachate, S.P., R.M. Khapare, and K.M. Kodam. Oxidation of arsenite by two β-proteobacteria isolated from soil. *Applied Microbiology and Biotechnology* 93(5), 2012: 2135–2145.

Bag, P., P. Bhattacharya, and R. Chowdhury. Bio-detoxification of arsenic laden ground water through a packed bed column of a continuous flow reactor using immobilized cells. *Soil and Sediment Contamination* 19(4), 2010: 455–466.

Basu, A., D. Saha, R. Saha, T. Ghosh, and B. Saha. A review on sources, toxicity and remediation technologies for removing arsenic from drinking water. *Research on Chemical Intermediates* 40(2), 2014: 447–485.

Battaglia-Brunet, F., C. Joulian et al. Oxidation of arsenite by *Thiomonas* strains and characterization of *Thiomonas arsenivorans* sp. nov. *Antonie van Leeuwenhoek* 89(1), 2006: 99–108.

Bentley, R., and T.G. Chasteen. Microbial methylation of metalloids: Arsenic, antimony, and bismuth. *Microbiology and Molecular Biology Reviews* 66(2), 2002: 250–271.

Berg, M., C. Stengel et al. Magnitude of arsenic pollution in the Mekong and Red River Deltas— Cambodia and Vietnam. *Science of the Total Environment* 372, 2, 2007: 413–425.

Bhattacharya, P., D. Chatterjee, and G. Jacks. Occurrence of arsenic-contaminated groundwater in alluvial aquifers from delta plains, Eastern India: Options for safe drinking water supply. *International Journal of Water Resources Development* 13(1), 1997: 79–92.

Blum, J.S., S. Han et al. Ecophysiology of *Halarsenatibacter silvermanii* strain SLAS-1T, gen. nov., sp. nov., a facultative chemoautotrophic arsenate respirer from salt-saturated Searles Lake, California. *Applied and Environmental Microbiology* 75(7), 2009: 1950–1960.

Blum, J.S., T.R. Kulp et al. *Desulfohalophilus alkaliarsenatis* gen. nov., sp. nov., an extremely halophilic sulfate-and arsenate-respiring bacterium from Searles Lake, California. *Extremophiles* 16(5), 2012: 727–742.

Bose, P., and A. Sharma. Role of iron in controlling speciation and mobilization of arsenic in subsurface environment. *Water Research* 36(19), 2002: 4916–4926.

Branco, R., A.-P. Chung, and P.V. Morais. Sequencing and expression of two arsenic resistance operons with different functions in the highly arsenic-resistant strain *Ochrobactrum tritici* SCII24T. *BMC Microbiology* 8(1), 2008: 95.

Bruneel, O., R. Duran, C. Casiot, F. Elbaz-Poulichet, and J-C. Personné. Diversity of microorganisms in Fe-As-rich acid mine drainage waters of Carnoules, France. *Applied and Environmental Microbiology* 72(1), 2006: 551–556.

Bundschuh, J., B. Farias, R. Martin, A. Storniolo, P. Bhattacharya, J. Cortes, G. Bonorino, and R. Albouy. Groundwater arsenic in the Chaco-Pampean plain, Argentina: Case study from Robles county, Santiago del Estero province. *Applied Geochemistry* 19(2), 2004: 231–243.

Buschmann, J., M. Berg et al. Contamination of drinking water resources in the Mekong delta floodplains: Arsenic and other trace metals pose serious health risks to population. *Environment International* 34(6), 2008: 756–764.

Cáceres, V.L., D.E. Gruttner, and N.R. Contreras. Water recycling in arid regions: Chilean case. *Ambio* 21(2), 1992: 138–144.

Cai, L., G. Liu, C. Rensing, and G. Wang. Genes involved in arsenic transformation and resistance associated with different levels of arsenic-contaminated soils. *BMC Microbiology* 9(1), 2009: 4–13.

Campos, V.L., C. Valenzuela et al. *Pseudomonas arsenicoxydans* sp nov., an arsenite-oxidizing strain isolated from the Atacama desert. *Systematic and Applied Microbiology* 33(4), 2010: 193–197.

Cavalca, L., A. Corsini, V. Andreoni, and G. Muyzer. Microbial transformations of arsenic: Perspective for biological removal of arsenic from water. *Future Microbiology* 8(1), 2013: 753–768.

Cavalca, L., R. Zanchi et al. Arsenic-resistant bacteria associated with roots of the wild *Cirsium arvense* (L.) plant from an arsenic polluted soil, and screening of potential plant growth-promoting characteristics. *Systematic and Applied Microbiology* 33(3), 2010: 154–164.

Cebrián, M.E., A. Albores, M. Aguilar, and E. Blakely. Chronic arsenic poisoning in the north of Mexico. *Human Toxicology* 2, 1983: 121–133.

Chang, J.S., I.H. Yoon, K.R. Kim, J. An, and K.W. Kim. Arsenic detoxification potential of aox genes in arsenite-oxidizing bacteria isolated from natural and constructed wetlands in the Republic of Korea. *Environ Geochem Health* 32, 2010: 95–105.

Chakraborti, D., B.K. Biswas et al. Arsenic groundwater contamination and sufferings of people in Rajnandgaon district, Madhya Pradesh, India. *Current Science* 77(4), 1999: 502–504.

Chakraborti, D., B. Das et al. Status of groundwater arsenic contamination in the state of West Bengal, India: A 20-year study report. *Molecular Nutrition and Food Research* 53, 2009: 542–551.

Chakraborti, D., J. Singh et al. Groundwater arsenic contamination in Manipur, one of the seven North-Eastern hill states of India: A future danger. *Environmental Geology* 56(2), 2008: 381–390.

Chakraborti, D., M.M. Rahman et al. Arsenic calamity in the Indian subcontinent: What lessons have been learned? *Talanta* 58(1), 2002: 3–22.

Chakraborti, D., M.K. Sengupta et al. Groundwater arsenic contamination and its health effects in the Ganga-Meghna-Brahmaputra plain. *Journal of Environmental Monitoring* 6(1), 2004: 74–83.

Challenger, F. Biological methylation. *Advances in Enzymology and Related Areas of Molecular Biology* 12, 1951: 429–491.

Chauhan, N.S., R. Ranjan, H.J. Purohit, V.C. Kalia, and R. Sharma. Identification of genes conferring arsenic resistance to *Escherichia coli* from an effluent treatment plant sludge metagenomic library. *FEMS Microbiology Ecology* 67(1), 2009: 130–139.

Chen, L.-x., J.-t. Li et al. Shifts in microbial community composition and function in the acidification of a lead/zinc mine tailings. *Environmental Microbiology* 15(9), 2013: 2431–2444.

Chen, J., J. Qin, Y.-G. Zhu, V. de Lorenzo, and B. P. Rosen. Engineering the soil bacterium *Pseudomonas putida* for arsenic methylation. *Applied and Environmental Microbiology* 79(14), 2013: 4493–4495.

Chiou, H.-Y., S.-T. Chiou et al. Incidence of transitional cell carcinoma and arsenic in drinking water: A follow-up study of 8,102 residents in an arseniasis-endemic area in northeastern Taiwan. *American Journal of Epidemiology* 153(5), 2001: 411–418.

Chung, J., X. Li, and B. E. Rittmann. Bio-reduction of arsenate using a hydrogen-based membrane biofilm reactor. *Chemosphere* 65(1), 2006: 24–34.

Cuebas, M., A. Villafane, M. McBride, N. Yee, and E. Bini. Arsenate reduction and expression of multiple chromosomal ars operons in *Geobacillus kaustophilus* A1. *Microbiology* 157(7), 2011: 2004–2011.

Dastidar, A., and Y.-T. Wang. Modeling arsenite oxidation by chemoautotrophic *Thiomonas arsenivorans* strain b6 in a packed-bed bioreactor. *Science of the Total Environment* 432(2), 2012: 113–121.

Davolos, D., and B. Pietrangeli. A molecular study on bacterial resistance to arsenic-toxicity in surface and underground waters of Latium (Italy). *Ecotoxicology and Environmental Safety* 96, 2013: 1–9.

Dhar, R.K., Y. Zheng et al. Microbes enhance mobility of arsenic in Pleistocene aquifer sand from Bangladesh. *Environmental Science and Technology* 45(7), 2011: 2648–2654.

Dopp, E., A.D. Kligerman, and R.A. Diaz-Bone. Organoarsenicals. Uptake, metabolism, and toxicity. *Metal Ions in Life Sciences* 7, 2009: 231–265.

Drewniak, L., A. Styczek, M. Majder-Lopatka, and A. Sklodowska. Bacteria, hyper-tolerant to arsenic in the rocks of an ancient gold mine, and their potential role in dissemination of arsenic pollution. *Environmental Pollution* 156(3), 2008: 1069–1074.

Drewniak, L., N. Maryan, W. Lewandowski, S. Kaczanowski, and A. Sklodowska. The contribution of microbial mats to the arsenic geochemistry of an ancient gold mine. *Environmental Pollution* 162, 2012: 190–201.

Drewniak, L., and A. Sklodowska. Arsenic-transforming microbes and their role in biomining processes. *Environmental Science and Pollution Research* 20(11), 2013: 7728–7739.

Duquesne, K., A. Lieutaud et al. Arsenite oxidation by a chemoautotrophic moderately acidophilic *Thiomonas* sp.: From the strain isolation to the gene study. *Environmental Microbiology* 10(1), 2008: 228–237.

Ellis, P.J., T. Conrads, R. Hille, and P. Kuhn. Crystal structure of the 100 kDa arsenite oxidase from *Alcaligenes faecalis* in two crystal forms at 1.64 Å and 2.03 Å. *Structure* 9(2), 2001: 125–132.

Escudero, L.V, E.O. Casamayor, G. Chong, C. Pedro´s-Alio´, and C. Demergasso. Distribution of microbial arsenic reduction, oxidation and extrusion genes along a wide range of environmental arsenic concentrations. *PLoS ONE*, 8, 2013: e78890.

Etterlin, F., M. Berg, and H. Rowland. Arsenic contamination in European groundwater resources. 2008.

Fan, H., C. Su, Y. Wang, J. Yao, K. Zhao, and G. Wang. Sedimentary arsenite-oxidizing and arsenate-reducing bacteria associated with high arsenic groundwater from Shanyin, Northwestern China. *Journal of Applied Microbiology* 105(2), 2008: 529–539.

Fendorf, S., H.A. Michael, and A. van Geen. Spatial and temporal variations of groundwater arsenic in South and Southeast Asia. *Science* 328(5982), 2010: 1123–1127.

Fisher, E., A.M. Dawson et al. Transformation of inorganic and organic arsenic by *Alkaliphilus oremlandii* sp. nov. Strain OhILAs. *Annals of the New York Academy of Sciences* 1125(1), 2008: 230–241.

Fournier, P.-E., D. Vallenet et al. Comparative genomics of multidrug resistance in *Acinetobacter baumannii*. *PLoS Genetics* 2(1), 2006: e7.

Frey, B., S.R. Rieder et al. Weathering-associated bacteria from the Damma glacier forefield: Physiological capabilities and impact on granite dissolution. *Applied and Environmental Microbiology* 76(14), 2010: 4788–4796.

Gadd, G.M. Microbial influence on metal mobility and application for bioremediation. *Geoderma* 122(2), 2004: 109–119.

Gault, A.G., F.S. Islam, and D.A. Polya. Microcosm depth profiles of arsenic release in a shallow aquifer, West Bengal. *Mineralogical Magazine* 69(5), 2005: 855–863.

Gauri, S.S., S. Archanaa et al. Removal of arsenic from aqueous solution using pottery granules coated with cyst of *Azotobacter* and portland cement: Characterization, kinetics and modeling. *Bioresource Technology* 102(10), 2011: 6308–6312.

Ghodsi, H., M. Hoodaji, A. Tahmourespour, and M.M. Gheisari. Investigation of bioremediation of arsenic by bacteria isolated from contaminated soil. *African Journal of Microbiology Research* 5(32), 2011: 5889–5895.

Ghosh, S., and P. Sar. Identification and characterization of metabolic properties of bacterial populations recovered from arsenic contaminated ground water of North East India (Assam). *Water Research* 47(19), 2013: 6992–7005.

Gihring, T.M., G.K. Druschel, R.B. McCleskey, R. J. Hamers, and J. F. Banfield. Rapid arsenite oxidation by *Thermus aquaticus* and *Thermus thermophilus*: Field and laboratory investigations. *Environmental Science & Technology* 35(19), 2001: 3857–3862.

Giloteaux, L., D.E. Holmes et al. Characterization and transcription of arsenic respiration and resistance genes during *in situ* uranium bioremediation. *The ISME Journal* 7(2), 2013: 370–383.

Goldsmith, J.R., M., Deane, J. Thom, and G. Gentry. Evaluation of health implications of elevated arsenic in well water. *Water Research* 6(10), 1972: 1133–1136.

Gorra, R., Go. Webster et al. Dynamic microbial community associated with iron–arsenic co-precipitation products from a groundwater storage system in Bangladesh. *Microbial Ecology* 64(1), 2012: 171–186.

Green, H.H. Isolation and description of a bacterium causing oxidation of arsenite to arsenate in cattle dipping baths. *Union S. Africa Dept. Agrie. 5th and 6th Repts. Direc. Vet. Research* April 1918. 1919: 595–610.

Gulz, P.A., G. Satish-Kumar, and S. Rainer. Arsenic accumulation of common plants from contaminated soils. *Plant and Soil* 272(1), 2005: 337–347.

Halter, D., A. Cordi et al. Taxonomic and functional prokaryote diversity in mildly arsenic-contaminated sediments. *Research in Microbiology* 162(9), 2011: 877–887.

Hamamura, N., K. Fukushima, and T. Itai. Identification of antimony-and arsenic-oxidizing bacteria associated with antimony mine tailing.*Microbes and Environments/JSME* 28(2), 2012: 257–263.

Hao, X., L. Yanbing et al. Genome sequence of the arsenite-oxidizing strain *Agrobacterium tumefaciens* 5A. *Journal of Bacteriology* 194(4), 2012: 903–903.

Haque, S., J. Ji, and K.H. Johannesson. Evaluating mobilization and transport of arsenic in sediments and groundwaters of Aquia aquifer, Maryland, USA. *Journal of Contaminant Hydrology* 99(1), 2008: 68–84.

Heinrich-Salmeron, A., C. Audrey et al. Unsuspected diversity of arsenite-oxidizing bacteria as revealed by widespread distribution of the aoxB gene in prokaryotes. *Applied and Environmental Microbiology* 77(13), 2011: 4685–4692.

Hery, M., B.E. Van Dongen et al. Arsenic release and attenuation in low organic carbon aquifer sediments from West Bengal. *Geobiology* 8(2), 2010: 155–168.

Huang, J.-H. Impact of microorganisms on arsenic biogeochemistry: A review. *Water, Air, & Soil Pollution* 225(2), 2014: 1–25.

Huang, Y., H. Li et al. Genome sequence of the facultative anaerobic arsenite-oxidizing and nitrate-reducing bacterium *Acidovorax* sp. strain NO1. *Journal of Bacteriology* 194(6), 2012: 1635–1636.

Hug, S. J., O.X. Leupin, and M. Berg. Bangladesh and vietnam: Different groundwater compositions require different approaches to arsenic mitigation. *Environmental Science & Technology* 42(17), 2008: 6318–6323.

Inskeep, W.P., R.E. Macur, N. Hamamura, T.P. Warelow, S.A. Ward, and J.M. Santini. Detection, diversity and expression of aerobic bacterial arsenite oxidase genes. *Environmental Microbiology* 9(4), 2007: 934–943.

Islam, F.S., C. Boothman, A.G. Gault, D.A. Pola, and J.R. Lloyd, Potential role of the Fe (III)-reducing bacteria *Geobacter* and *Geothrix* in controlling arsenic solubility in Bengal delta sediments. *Mineralogical Magazine* 69(5), 2005: 865–875.

Islam, F.S., A.G. Gault et al. Role of metal-reducing bacteria in arsenic release from Bengal delta sediments. *Nature* 430(6995), 2004: 68–71.

Jackson, C.R., and S.L. Dugas. Phylogenetic analysis of bacterial and archaeal arsC gene sequences suggests an ancient, common origin for arsenate reductase. *BMC Evolutionary Biology* 3(1), 2003: 18–28.

Jareonmit, P., M. Mehta, M.J. Sadowsky, and K. Sajjaphan. Phylogenetic and phenotypic analyses of arsenic-reducing bacteria isolated from an old tin mine area in Thailand. *World Journal of Microbiology and Biotechnology* 28(5), 2012: 2287–2292.

Jekel, M.R. Removal of arsenic in drinking water treatment. In: *Arsenic in the Environment*. Nriagu, J.O., (Eds.), New York, Wiley-Interscience, 1994: 119–130.

Jenkins, O., D. Byrom, and D. Jones. Methylophilus: A new genus of methanol-utilizing bacteria. *International Journal of Systematic Bacteriology* 37(4), 1987: 446–448.

Jessen, S., D. Postma et al. Surface complexation modeling of groundwater arsenic mobility: Results of a forced gradient experiment in a Red River flood plain aquifer, Vietnam. *Geochimica et Cosmochimica Acta* 98, 2012: 186–201.

Jiang, Z., P. Li, Y. Wang, B. Li, Y. Deng, and Y. Wang. Vertical distribution of bacterial populations associated with arsenic mobilization in aquifer sediments from the Hetao plain, Inner Mongolia. *Environmental Earth Sciences* 71(1), 2014: 311–318.

Joshi, D.N., J.S. Patel, S.J.S. Flora, and K. Kalia. Arsenic accumulation by *Pseudomonas stutzeri* and its response to some thiol chelators. *Environmental Health and Preventive Medicine* 13(5), 2008: 257–263.

Kalimuthu, P., M.D. Heath, J.M. Santini, U. Kappler, and P.V. Bernhardt. Electrochemically driven catalysis of *Rhizobium* sp. NT-26 arsenite oxidase with its native electron acceptor cytochrome552. *Biochimica et Biophysica Acta (BBA)—Bioenergetics* 1837(1), 2014: 112–120.

Kalmbach, S., W. Manz, J. Wecke, and U. Szewzyk. *Aquabacterium* gen. nov., with description of *Aquabacterium citratiphilum* sp. nov., *Aquabacterium parvum* sp. nov. and *Aquabacterium commune* sp. nov., three *in situ* dominant bacterial species from the Berlin drinking water system. *International Journal of Systematic Bacteriology* 49(2), 1999: 769–777.

Kar, S., J.P. Maity et al. Arsenic-enriched aquifers: Occurrences and mobilization of arsenic in groundwater of Ganges Delta Plain, Barasat, West Bengal, India. *Applied Geochemistry* 25(12), 2010: 1805–1814.

Kashyap, D.R., L.M. Botero, W.L. Franck, D.J. Hassett, and T.R. McDermott. Complex regulation of arsenite oxidation in *Agrobacterium tumefaciens*. *Journal of Bacteriology* 188(3), 2006: 1081–1088.

Katsoyiannis, I.A., and A.I. Zouboulis. Application of biological processes for the removal of arsenic from groundwaters. *Water Research* 38(1), 2004: 17–26.

Kim, K.-W., P. Chanpiwat, H.T. Hanh, K. Phan, and S. Sthiannopkao. Arsenic geochemistry of groundwater in Southeast Asia. *Frontiers of Medicine* 5(4), 2011: 420–433.

Kinegam, S., T. Yingprasertchai et al. Isolation and characterization of arsenite-oxidizing bacteria from arsenic-contaminated soils in Thailand. *World Journal of Microbiology and Biotechnology* 24(12), 2008: 3091–3096.

Kneebone, P.E., P.A. O'Day, N. Jones, and J.G. Hering. Deposition and fate of arsenic in iron-and arsenic-enriched reservoir sediments. *Environmental Science & Technology* 36(3), 2002: 381–386.

Kocar, B.D., M.L. Polizzotto et al. Integrated biogeochemical and hydrologic processes driving arsenic release from shallow sediments to groundwaters of the Mekong delta. *Applied Geochemistry* 23, 11, 2008: 3059–3071.

Koechler, S., J. Cleiss-Arnold et al. Multiple controls affect arsenite oxidase gene expression in *Herminiimonas arsenicoxydans*. *BMC Microbiology* 10(1), 2010: 53–61.

Kostal, J., R. Yang, C.H. Wu, A. Mulchandani, and W. Chen. Enhanced arsenic accumulation in engineered bacterial cells expressing ArsR. *Applied and Environmental Microbiology* 70(8), 2004: 4582–4587.

Krafft, T., and J.M. Macy. Purification and characterization of the respiratory arsenate reductase of *Chrysiogenes arsenatis*. *European Journal of Biochemistry* 255(3), 1998: 647–653.

Kudo, K., N. Yamaguchi et al. Release of arsenic from soil by a novel dissimilatory arsenate-reducing bacterium, *Anaeromyxobacter* sp. strain PSR-1. *Applied and Environmental Microbiology* 79(15), 2013: 4635–4642.

Lett, M.-C., D. Muller, D. Lièvremont, S. Silver, and J. Santini. Unified nomenclature for genes involved in prokaryotic aerobic arsenite oxidation. *Journal of Bacteriology* 194(2), 2012: 207–208.

Li, X., and L. R. Krumholz. Regulation of arsenate resistance in *Desulfovibrio desulfuricans* G20 by an arsRBCC operon and an arsC gene. *Journal of Bacteriology* 189(0), 2007: 3705–3711.

Li, X., Y. Hu et al. Genome sequence of the highly efficient arsenite-oxidizing bacterium *Achromobacter arsenitoxydans* SY8. *Journal of Bacteriology* 194(5), 2012: 1243–1244.

Li, Z., H. Hong et al. Characterization on arsenic sorption and mobility of the sediments of Chia-Nan Plain, where Blackfoot disease occurred. *Environmental Earth Sciences* 64(3), 2011: 823–831.

Liao, V.H.-C., Y.-J. Chu et al. Arsenite-oxidizing and arsenate-reducing bacteria associated with arsenic-rich groundwater in Taiwan. *Journal of Contaminant Hydrology* 123(1), 2011: 20–29.

Lièvremont, D., P.N. Bertin et al. Arsenic in contaminated waters: Biogeochemical cycle, microbial metabolism and biotreatment processes. *Biochimie* 91(10), 2009: 1229–1237.

Lin, Y., H. Fan et al. Draft genome sequence of *Halomonas* sp. strain HAL1, a moderately halophilic arsenite-oxidizing bacterium isolated from gold-mine soil. *Journal of Bacteriology* 194(1), 2012: 199–200.

Lin, Y.-F., J. Yang et al. ArsD: an As (III) metallochaperone for the ArsAB As (III)-translocating ATPase. *Journal of Bioenergetics and Biomembranes* 39(5–6), 2007: 453–458.

Liu, S., Y. Hou, and G. Sun. Synergistic degradation of pyrene and volatilization of arsenic by cocultures of bacteria and a fungus. *Frontiers of Environmental Science & Engineering* 7(2), 2013: 191–199.

Loeppert, A.J., K. Raven et al. Arsenate and arsenite retention and release in oxide and sulfide dominated systems. Texas Water Resources Institute. Available electronically from http://hdl.handle.net/1969, 1, 1997: 6154.

López-Maury, L., F.J. Florencio et al. Arsenic sensing and resistance system in the cyanobacterium *Synechocystis* sp. strain PCC 6803. *Journal of Bacteriology* 185(18), 2003: 5363–5371.

Lovley, D.R., E.J.P. Phillips et al. Enzymic versus nonenzymic mechanisms for iron (III) reduction in aquatic sediments. *Environmental Science & Technology* 25(6), 1991: 1062–1067.

Lu, D., P. Grayson, and K. Schulten. Glycerol conductance and physical asymmetry of the *Escherichia coli* glycerol facilitator GlpF. *Biophysical Journal* 85(5) 2003: 2977–2987.

Mailloux, B.J., E. Alexandrova et al. Microbial mineral weathering for nutrient acquisition releases arsenic. *Applied and Environmental Microbiology* 75(8), 2009: 2558–2565.

Majumder, A., K. Bhattacharyya et al. Arsenic-tolerant, arsenite-oxidising bacterial strains in the contaminated soils of West Bengal, India. *Science of the Total Environment* 463, 2013: 1006–1014.

Malasarn, D., C.W. Saltikov et al. arrA is a reliable marker for As (V) respiration. *Science* 306(5695), 2004: 455–455.

Malasarn, D., J.R. Keeffe et al. Characterization of the arsenate respiratory reductase from *Shewanella* sp. strain ANA-3. *Journal of Bacteriology* 190(1), 2008: 135–142.

Marco, P., C.C. Pacheco et al. Novel pollutant-resistant methylotrophic bacteria for use in bioremediation. *FEMS Microbiology Letters* 234(1), 2004: 75–80.

Martin, P., S. DeMel et al. Insights into the structure, solvation, and mechanism of ArsC arsenate reductase, a novel arsenic detoxification enzyme. *Structure* 9(11), 2001: 1071–1081.

Matlakowska, R., L. Drewniak et al. Arsenic-hypertolerant pseudomonads isolated from ancient gold and copper-bearing black shale deposits. *Geomicrobiology Journal* 25(7–8), 2008: 357–362.

Matthijs, S., G. Laus et al. Siderophore-mediated iron acquisition in the entomopathogenic bacterium *Pseudomonas entomophila* L48 and its close relative *Pseudomonas putida* KT2440. *Biometals* 22(6), 2009: 951–964.

McArthur, J.M., D.M. Banerjee et al. Natural organic matter in sedimentary basins and its relation to arsenic in anoxic ground water: The example of West Bengal and its worldwide implications. *Applied Geochemistry* 19(8), 2004: 1255–1293.

McBride, B.C., and R.S. Wolfe. Biosynthesis of dimethylarsine by *Methanobacterium*. *Biochemistry* 10(23), 1971: 4312–4317.

Meng, Y.-L., Z. Liu et al. As (III) and Sb (III) uptake by GlpF and efflux by ArsB in *Escherichia coli. Journal of Biological Chemistry* 279(18), 2004: 18334–18341.

Messens, J., I. Van Molle et al. How thioredoxin can reduce a buried disulphide bond. *Journal of Molecular Biology* 339(3), 2004: 527–537.

Michalke, K., E.B. Wickenheiser et al. Production of volatile derivatives of metal (loid) s by microflora involved in anaerobic digestion of sewage sludge. *Applied and Environmental Microbiology* 66(7), 2000: 2791–2796.

Michel, C., M. Jean et al. Biofilms of As (III)-oxidising bacteria: Formation and activity studies for bioremediation process development. *Applied Microbiology and Biotechnology* 77(2), 2007: 457–467.

Miyatake, M., and S. Hayashi. Characteristics of arsenic removal by *Bacillus cereus* strain W2. *Resources Processing* 58(3), 2011: 101–107.

Mondal, P., C.B. Majumder, and B. Mohanty. Treatment of arsenic contaminated water in a laboratory scale up-flow bio-column reactor. *Journal of Hazardous Materials* 153(1), 2008: 136–145.

Mukherjee, A., M.K. Sengupta et al. Arsenic contamination in groundwater: A global perspective with emphasis on the Asian scenario. *Journal of Health, Population and Nutrition* 24(2), 2006: 142–163.

Mukhopadhyay, R., B.P. Rosen et al. Microbial arsenic: from geocycles to genes and enzymes. *FEMS Microbiology Reviews* 26(3), 2002: 311–325.

Muller, D., D.D. Simeonova et al. *Herminiimonas arsenicoxydans* sp. nov., a metalloresistant bacterium. *International Journal of Systematic and Evolutionary Microbiology* 56(8), 2006: 1765–1769.

Muller, D., C. Médigue et al. A tale of two oxidation states: Bacterial colonization of arsenic-rich environments. *PLoS Genetics* 3(4), 2007: e53.

Murphy, J.N., and C.W. Saltikov. The ArsR repressor mediates arsenite-dependent regulation of arsenate respiration and detoxification operons of *Shewanella* sp. strain ANA-3. *Journal of Bacteriology* 191(21), 2009: 6722–6731.

Nair, A., A.A. Juwarkar et al. Production and characterization of siderophores and its application in arsenic removal from contaminated soil. *Water, Air, and Soil Pollution* 180(1–4), 2007: 199–212.

Nath, B., J.P. Maity et al. Geochemical characterization of arsenic-affected alluvial aquifers of the Bengal Delta (West Bengal and Bangladesh) and Chianan Plains (SW Taiwan): Implications for human health. *Applied Geochemistry* 26(5), 2011: 705–713.

Newman, D.K., T.J. Beveridge et al. Precipitation of arsenic trisulfide by *Desulfotomaculum auripigmentum. Applied and Environmental Microbiology* 63(5), 1997: 2022–2028.

Nguyen, V.A., S. Bang et al. Contamination of groundwater and risk assessment for arsenic exposure in Ha Nam province, Vietnam. *Environment International* 35(3), 2009: 466–472.

Nickson, J.M.M., B. Shrestha et al. Arsenic and other drinking water quality issues, Muzaffargarh District, Pakistan. *Applied Geochemistry* 20(1), 2005: 55–68.

Nickson, R., J. McArthur et al. Arsenic poisoning of Bangladesh groundwater. *Nature* 395(6700), 1998: 338–338.

Nickson, R., C.S., P. Mitra et al. Current knowledge on the distribution of arsenic in groundwater in five states of India. *Journal of Environmental Science and Health Part A* 42(12), 2007: 1707–1718.

Ohtsuka, T., N. Yamaguchi et al. Arsenic dissolution from Japanese paddy soil by a dissimilatory arsenate-reducing bacterium *Geobacter* sp. OR-1. *Environmental Science and Technology* 47, 2013: 6263–6271.

Ordóñez, E., M. Letek et al. Analysis of genes involved in arsenic resistance in *Corynebacterium glutamicum* ATCC 13032. *Applied and Environmental Microbiology* 71(10), 2005: 6206–6215.

Oremland, R.S., S.E. Hoeft et al. Anaerobic oxidation of arsenite in Mono Lake water and by a facultative, arsenite-oxidizing chemoautotroph, strain MLHE-1. *Applied and Environmental Microbiology* 68(10), 2002: 4795–4802.

Oremland, R.S., and J.F. Stolz. Arsenic, microbes and contaminated aquifers. *Trends in Microbiology* 13(2), 2005: 45–49.

Osborne, T.H., H.E. Jamieson, K.A. Hudson-Edwards, D.K. Nordstrom, S.R. Walker, S.A. Ward, and J.M. Santini. Microbial oxidation of arsenite in a subarctic environment: Diversity of arsenite oxidase genes and identification of a psychrotolerant arsenite oxidiser. *BMC Microbiology* 10(1), 2010: 205.

Páez-Espino, D., J. Tamames et al. Microbial responses to environmental arsenic. *Biometals* 22(1), 2009: 117–130.

Pal, T., and P.K. Mukherjee. Study of subsurface geology in locating arsenic-free groundwater in Bengal delta, West Bengal, India. *Environmental Geology* 56(6), 2009: 1211–1225.

Pearcy, C.A., D.A. Chevis et al. Evidence of microbially mediated arsenic mobilization from sediments of the Aquia aquifer, Maryland, USA. *Applied Geochemistry* 26(4), 2011: 575–586.

Pepi, M., M. Volterrani et al. Arsenic-resistant bacteria isolated from contaminated sediments of the Orbetello Lagoon, Italy, and their characterization. *Journal of Applied Microbiology* 103(6), 2007: 2299–2308.

Pepi, M., G. Protano et al. Arsenic-resistant *Pseudomonas* spp. and *Bacillus* sp. bacterial strains reducing As (V) to As (III), isolated from Alps soils, Italy. *Folia Microbiologica* 56(1), 2011: 29–35.

Pérez-Jiménez, J.R, C. DeFraia et al. Arsenate respiratory reductase gene (arrA) for *Desulfosporosinus* sp. strain Y5. *Biochemical and Biophysical Research Communications* 338(2), 2005: 825–829.

Postma, D., F. Larsen et al. Arsenic in groundwater of the Red River floodplain, Vietnam: Controlling geochemical processes and reactive transport modeling. *Geochimica et Cosmochimica Acta* 71(21), 2007: 5054–5071.

Prasad, K.S., V. Subramanian et al. Purification and characterization of arsenite oxidase from *Arthrobacter* sp. *Biometals* 22(5), 2009: 711–721.

Prasad, K.S., P. Srivastava et al. Biosorption of As (III) ion on *Rhodococcus* sp. WB-12: Biomass characterization and kinetic studies. *Separation Science and Technology* 46(16), 2011: 2517–2525.

Price, R.E, R. Lesniewski et al. Archaeal and bacterial diversity in an arsenic-rich shallow-sea hydrothermal system undergoing phase separation. *Frontiers in Microbiology* 4, 2013.

Qin, J., B.P. Rosen et al. Arsenic detoxification and evolution of trimethylarsine gas by a microbial arsenite S-adenosylmethionine methyltransferase. *Proceedings of the National Academy of Sciences of the United States of America* 103(7), 2006: 2075–2080.

Rahman, M.M., J.C. Ng et al. Chronic exposure of arsenic via drinking water and its adverse health impacts on humans. *Environmental Geochemistry and Health* 31(1), 2009: 189–200.

Ravenscroft, P., J.M. McArthur, and B.A. Hoque. Geochemical and palaeohydrological controls on pollution of groundwater by arsenic. *Arsenic Exposure and Health Effects IV*, 2001: 53–77.

Ravenscroft, P., W.G. Burgess et al. Arsenic in groundwater of the Bengal Basin, Bangladesh: Distribution, field relations, and hydrogeological setting. *Hydrogeology Journal* 13(5–6), 2005: 727–751.

Rhine, E.D., S.M. Ni Chadhain, G.J. Zylstra, and L.Y. Young. The arsenite oxidase genes (aroAB) in novel chemoautotrophic arsenite oxidizers. *Biochemical and Biophysical Research Communications* 354(3), 2007: 662–667.

Rosen, B.P. Families of arsenic transporters. *Trends in Microbiology* 7(5), 1999: 207–212.

Rosen, B.P. Biochemistry of arsenic detoxification. *FEBS Letters* 529(1), 2002: 86–92.

Rosen, B.P., and Z. Liu. Transport pathways for arsenic and selenium: A minireview. *Environment International* 35(3), 2009: 512–515.

Rosen, B.P., A.A. Ajees, and T.R. McDermott. Life and death with arsenic. *Bioessays* 33(5), 2011: 350–357.

Rowland, D.A. Polya et al. Characterisation of organic matter in a shallow, reducing, arsenic-rich aquifer, West Bengal. *Organic Geochemistry* 37(9), 2006: 1101–1114.

Rowland, H.A.L., A.G. Gault et al. Geochemistry of aquifer sediments and arsenic-rich groundwaters from Kandal Province, Cambodia. *Applied Geochemistry* 23(11), 2008: 3029–3046.

Salmassi, T.M, K. Venkateswaren et al. Oxidation of arsenite by *Agrobacterium albertimagni*, AOL15, sp. nov., isolated from Hot Creek, California. *Geomicrobiology Journal* 19(1), 2002: 53–66.

Salmassi, T.M., J.J. Walker et al. Community and cultivation analysis of arsenite oxidizing biofilms at Hot Creek. *Environmental Microbiology* 8(1), 2006: 50–59.

Saltikov, C.W., and D.K. Newman. Genetic identification of a respiratory arsenate reductase. *Proceedings of the National Academy of Sciences of the United States of America* 100, 2003: 10983–10988.

Saltikov, C.W., and B.H. Olson. Homology of *Escherichia coli* R773 arsA, arsB, and arsC genes in arsenic-resistant bacteria isolated from raw sewage and arsenic-enriched creek waters. *Applied and Environmental Microbiology* 68(1), 2002: 280–288.

Saltikov, C.W., R.A. Wildman, and D.K. Newman. Expression dynamics of arsenic respiration and detoxification in *Shewanella* sp. strain ANA-3. *Journal of Bacteriology* 187(21), 2005: 7390–7396.

Sanders, C.R., M. Kuroda et al. Antimonite is accumulated by the glycerol facilitator GlpFin *Escherichia coli*. *Journal of Bacteriology* 179, 1997: 3365–3367.

Santini, J.M., I.C.A. Streimann et al. *Bacillus macyae* sp. nov., an arsenate-respiring bacterium isolated from an Australian gold mine. *International Journal of Systematic and Evolutionary Microbiology* 54(6), 2004: 2241–2244.

Santini, J.M., L.I. Sly et al. A new chemolithoautotrophic arsenite-oxidizing bacterium isolated from a gold mine: Phylogenetic, physiological, and preliminary biochemical studies. *Applied and Environmental Microbiology* 66(1), 2000: 92–97.

Sardiwal, S., J.M. Santini et al. Characterization of a two-component signal transduction system that controls arsenite oxidation in the chemolithoautotroph NT-26. *FEMS Microbiology Letters* 313(1), 2010: 20–28.

Sarkar, A., SK. Kazy, and P. Sar. Characterization of arsenic resistant bacteria from arsenic rich groundwater of West Bengal, India. *Ecotoxicology* 22(2), 2013: 363–376.

Scanlon, B.R., J.P. Nicot et al. Elevated naturally occurring arsenic in a semiarid oxidizing system, Southern High Plains aquifer, Texas, USA. *Applied Geochemistry* 24(11), 2009: 2061–2071.

Sharma, A.K., J.C. Tjell et al. Review of arsenic contamination, exposure through water and food and low cost mitigation options for rural areas. *Applied Geochemistry* 41, 2014: 11–33.

Sheik, C.S., T.W. Mitchell et al. Exposure of soil microbial communities to chromium and arsenic alters their diversity and structure. *PloS One* 7(6), 2012: e40059.

Silver, S., and L.T. Phung. Genes and enzymes involved in bacterial oxidation and reduction of inorganic arsenic. *Applied and Environmental Microbiology* 71(2), 2005: 599–608.

Singh, A.K. Arsenic contamination in groundwater of North Eastern India. In *11th National Symposium on Hydrology with Focal Theme on Water Quality. Roorkee, Proceeding*, 2004, 255–262.

Singh, S., S.H. Kang et al. Systematic engineering of phytochelatin synthesis and arsenic transport for enhanced arsenic accumulation in *E. coli. Biotechnology and Bioengineering* 105(4), 2010: 780–785.

Slaughter, D.C., R.E. Macur et al. Inhibition of microbial arsenate reduction by phosphate. *Microbiological Research* 167(3), 2012: 151–156.

Slyemi, D., and V. Bonnefoy. How prokaryotes deal with arsenic. *Environmental Microbiology Reports* 4(6), 2012: 571–586.

Smedley, P.L., and D.G. Kinniburgh. A review of the source, behaviour and distribution of arsenic in natural waters. *Applied Geochemistry* 17(5) 2002: 517–568.

Smedley, P., and D.G. Kinniburgh. *Arsenic in Groundwater and the Environment.* Springer, Netherlands, 2013.

Smith, A.H., C. Hopenhayn-Rich et al. Cancer risks from arsenic in drinking water. *Environmental Health Perspectives* 97, 1992: 259.

Song, W.F., Q. Deng et al. Arsenite oxidation characteristics and molecular identification of arsenic-oxidizing bacteria isolated from soil. *Applied Mechanics and Materials* 188, 2012: 313–318.

Sorokin, D.Y., T.P. Tourova et al. *Desulfuribacillus alkaliarsenatis* gen. nov. sp. nov., a deep-lineage, obligately anaerobic, dissimilatory sulfur and arsenate-reducing, haloalkaliphilic representative of the order Bacillales from soda lakes. *Extremophiles* 16(4), 2012: 597–605.

Sriyosachati, S., and C.D. Cox. Siderophore-mediated iron acquisition from transferrin by *Pseudomonas aeruginosa. Infection and Immunity* 52(3), 1986: 885–891.

Stolz, D.J.E., J.S. Blum et al. *Sulfurospirillum barnesii* sp. nov. and *Sulfurospirillum arsenophilum* sp. nov., new members of the *Sulfurospirillum* clade of the epsilon *Proteobacteria. International Journal of Systematic Bacteriology* 49, 1999: 1177–1180.

Stolz, J.F., P. Basu et al. Arsenic and selenium in microbial metabolism. *Annual Review of Microbiology* 60, 2006: 107–130.

Stolz, J.F., P. Basu, and R.S. Oremland. Microbial arsenic metabolism: New twists on an old poison. *Microbe* 5(2), 2010: 5, 53–59.

Strnad, H., J. Ridl et al. Complete genome sequence of the haloaromatic acid-degrading bacterium *Achromobacter xylosoxidans* A8. *Journal of Bacteriology* 193(3), 2011: 791–792.

Stüben, D., Z. Berner, D. Chandrasekharam, and J. Karmakar. Arsenic enrichment in groundwater of West Bengal, India: Geochemical evidence for mobilization of As under reducing conditions. *Applied Geochemistry* 18(9), 2003: 1417–1434.

Stumm, W., and J.J. Morgan. *Aquatic Chemistry,* John Wiley & Sons, Inc, New York. 1995.

Sultana, M., C. Härtig et al. Bacterial communities in Bangladesh aquifers differing in aqueous arsenic concentration. *Geomicrobiology Journal* 28(3), 2011: 198–211.

Sun, S.L., B. Li et al. Current situation of endemic Asosis in China. *Environmental Science*, 8, 2001: 425–434.

Sutton, N.B., G.M. van der Kraan et al. Characterization of geochemical constituents and bacterial populations associated with As mobilization in deep and shallow tube wells in Bangladesh. *Water Research* 43(6), 2009: 1720–1730.

Swartz, C.H., N.K. Blute et al. Mobility of arsenic in a Bangladesh aquifer: Inferences from geochemical profiles, leaching data, and mineralogical characterization. *Geochimica et Cosmochimica Acta* 68, 22, 2004: 4539–4557.

Takaya, N., M.A.B. Catalan-Sakairi et al. Aerobic denitrifying bacteria that produce low levels of nitrous oxide. *Applied and Environmental Microbiology* 69(6), 2003: 3152–3157.

Takeuchi, M., H. Kawahata et al. Arsenic resistance and removal by marine and non-marine bacteria. *Journal of Biotechnology* 127(3), 2007: 434–442.

Tandukar, P.B., A.B. Mukherjee et al. Preliminary assessment of arsenic contamination in groundwater in Nepal. In Arsenic in the Asia-Pacific region: Managing arsenic for our future. *Proceedings of the International Conference on Arsenic in the Asia-Pacific Region*, Adelaide, South Australia, November 21–23, 2001: 103–105.

Vala, A.K., B.P. Dave, and H.C. Dube. Chemical characterization and quantification of siderophores produced by marine and terrestrial aspergilli. *Canadian Journal of Microbiology* 52(6), 2006: 603–607.

van Lis, R., W. Nitschke et al. Arsenics as bioenergetic substrates. *Biochimica et Biophysica Acta (BBA)—Bioenergetics* 1827(2), 2013: 176–188.

Vaxevanidou, K., N. Papassiopi et al. Removal of heavy metals and arsenic from contaminated soils using bioremediation and chelant extraction techniques. *Chemosphere* 70(8), 2008: 1329–1337.

Villadangos, A.F., K. Van Belle et al. *Corynebacterium glutamicum* survives arsenic stress with arsenate reductases coupled to two distinct redox mechanisms. *Molecular Microbiology* 82(4), 2011: 998–1014.

Villegas-Torres, M.F., O.C. Bedoya-Reina et al. Horizontal *arsC* gene transfer among microorganisms isolated from arsenic polluted soil. *International Biodeterioration & Biodegradation* 65(1), 2011: 147–152.

Vorontsov, I.I., G. Minasov et al. Crystal structure of an apo form of *Shigella flexneri* ArsH protein with an NADPH-dependent FMN reductase activity. *Protein Science* 16(11), 2007: 2483–2490.

Wang, H.B., Z.H. Ye, W.S. Shu, W.C. Li, M.H. Wong, and C.Y. Lan. Arsenic uptake and accumulation in fern species growing at arsenic-contaminated sites of southern China: Field surveys. *International Journal of Phytoremediation* 8(1), 2006: 1–11.

Weeger, W., D. Lievremont et al. Oxidation of arsenite to arsenate by a bacterium isolated from an aquatic environment. *Biometals* 12(2), 1999: 141–149.

Welch, A.H., and M.S. Lico. Factors controlling As and U in shallow ground water, southern Carson Desert, Nevada. *Applied Geochemistry* 13(4), 1998: 521–539.

Welch, A.H., D.B. Westjohn et al. Arsenic in ground water of the United States: Occurrence and geochemistry. *Ground Water* 38(4), 2000: 589–604.

World Bank Policy Report. 2005. Towards a more effective operational response: Arsenic contamination of groundwater in South and East Asian Countries. Vol I and II.

Wu, J.D., G. Zhuang et al. *Bacillus* sp. SXB and Pantoea sp. IMH, aerobic As (V)-reducing bacteria isolated from arsenic-contaminated soil. *Journal of Applied Microbiology* 114(3), 2013: 713–721.

Yamamura, S., K. Watanabe et al. Effect of antibiotics on redox transformations of arsenic and diversity of arsenite-oxidizing bacteria in sediment microbial communities. *Environmental Science & Technology* 2013.

Yamamura, S., M. Watanabe et al. Removal of arsenic from contaminated soils by microbial reduction of arsenate and quinone. *Environmental Science & Technology* 42, 16, 2008: 6154–6159.

Yang, H.-C., J. Cheng et al. Novel pathway for arsenic detoxification in the legume symbiont *Sinorhizobium meliloti*. *Journal of Bacteriology* 187(20), 2005: 6991–6997.

Yang, S.T., G. Wang et al. Effectiveness of applying arsenate reducing bacteria to enhance arsenic removal from polluted soils by *Pteris vittata* L. *International Journal of Phytoremediation* 14(1), 2012: 89–99.

Yang, T., M.-L. Chen et al. Iron (III) modification of *Bacillus subtilis* membranes provides record sorption capacity for arsenic and endows unusual selectivity for As (V). *Environmental Science & Technology* 46(4), 2012: 2251–2256.

Ye, J., H.-C. Yang et al. Crystal structure of the flavoprotein ArsH from *Sinorhizobium meliloti*. *FEBS Letters* 581(21), 2007: 3996–4000.

Yin, X.-X., J. Chen et al. Biotransformation and volatilization of arsenic by three photosynthetic cyanobacteria. *Plant Physiology* 156(3), 2011: 1631–1638.

Zaldivar, R., and W.H. Wetterstrand. Nitrate nitrogen levels in drinking water of urban areas with high- and low-risk populations for stomach cancer: An environmental epidemiology study. *Zeitschrift für Krebsforschung und Klinische Onkologie* 92(3), 1978: 227–234.

Zargar, K., S. Hoeft et al. Identification of a novel arsenite oxidase gene, arxA, in the haloalkaliphilic, arsenite-oxidizing bacterium *Alkalilimnicola ehrlichii* strain MLHE-1. *Journal of Bacteriology* 192(14), 2010: 3755–3762.

Zhang, X., Y. Jia et al. Bacterial reduction and release of adsorbed arsenate on Fe (III)-, Al- and coprecipitated Fe (III)/Al-hydroxides. *Journal of Environmental Sciences* 24(3), 2012: 440–448.

20

Microbial Degradation of Toxic Organic Compounds of Waste and Allied Contaminants

Yu. L. Gurevich, V. P. Ladygina and M. I. Teremova

CONTENTS

20.1 Introduction

Consumption of mineral resources increases year after year, and the associated growth of wastes necessitates the development and use of more advanced retrofit and detoxification technologies. Traditional wastewater treatment and solid detoxification technologies are based on non-biological treatment methods. Alternative approaches are based on the use of microorganisms to decompose contaminants (Marrot et al. 2006). Wastewater biotreatment and detoxification of toxic waste processes employed for the purpose with microorganisms and their communities with protozoa and other organisms may be more eco-friendly (Senthilvelan et al. 2014; Whiteley and Bailey 2000). The degradation of contaminants, including recalcitrant and toxic compounds, by microorganisms is at the core of natural mineralisation of organic compounds in the biosphere. In such processes, the probability of accumulating dangerous by-products is almost negligible.

Man-made biotreatment of organic wastes essentially differs from natural mineralisation processes. This is connected with very high concentrations of compounds subject to disposal with the necessity to process wastes in much less time. Besides, differences in waste bioprocessing are due to their peculiar and complex composition specific for each production. For example, petrochemical and coke-chemical plants produce large volumes of wastewaters containing a broad spectrum of hydrocarbons, including mono- and polycyclic aromatic compounds, cyanides and ammoniacal and other compounds (Jena et al. 2005). The volume of wastewater in synthetic resin production is much smaller, but it contains critically high concentrations of phenol, formaldehyde and methanol (Urtiaga et al. 2009). Municipal wastewaters differ in terms of very large volumes and relatively low concentrations of recalcitrants and toxic organic substances compared to that of industrial wastewaters (Fatone et al. 2011).

The mentioned features of liquid wastes make it possible to outline three types of waste treatment research and development approaches:

1. Oxidation of organic contaminants with critical concentrations
2. Performance of microbiological processes under high hydraulic load
3. Treatment of multi-component wastewaters

The main research trends and solutions of these problems (i.e., wastewater biotreatment from priority contaminants—phenol, polycyclic aromatic and some other compounds) are considered here.

20.2 Batch and Continuous Processes of Growth of Microbial Cultures

Wastes are bioprocessed in bioengineering plants of different designs and process solutions. The most investigated and widely used processes are conducted in closed (non-flow) plants for batch and continuous cultivation of microorganisms. Quantitative description of these processes is based on concepts of substrate consumption kinetics and microorganism growth (Baily and Ollis 1986; Pirt 1975). The balance model of microorganism growth in a batch culture has the form

$$dX/dt = -YdS/dt = \mu(S)X \qquad (20.1)$$

where X is the concentration of the microorganism biomass, S is the concentration of the substrate, μ is the specific growth rate, Y is the growth yield coefficient and t is the time.

Modern concepts of microorganism growth regularities are based on studies of continuous cultivation processes in continuously stirred-tank reactors (chemostat). Growth of microorganisms in continuous culture is described by

$$dX/dt = (\mu(S) - D)X \qquad (20.2)$$

$$dS/dt = D(So - S) - \mu(S)X/Y \qquad (20.3)$$

where So is the initial substrate concentration in the medium and D is the dilution rate. The concentration of the microorganism biomass in steady state of the culture is determined as

$$X = Y(So - S) \qquad (20.4)$$

In batch and continuous processes including their modifications, the dynamics of biomass and substrate concentration and their steady-state values are determined by specific form of μ and Y parameters, which have dependence on ambient conditions. They reflect the biological characteristics of growth and utilisation of microorganism substrates.

20.3 Kinetics and Stoichiometry of Microbial Growth

20.3.1 Monod and Haldane-Type Kinetic Models

Laboratory- and full-scale wastewater treatment processes create conditions under which the growth rate of microbial populations is well simulated by the Monod equation (Baily and Ollis 1986; Kovárova-Kovar and Egli 1998; Pirt 1975; Reardon et al. 2000); its form is similar to that of Michaelis–Menten equation (Cornish-Bowden 1976).

$$\mu = \mu_m S/(k_s + S) \qquad (20.5)$$

where μ_m is the maximum specific growth rate and k_s is the half-saturation coefficient. In addition to that, there are frequent situations when the specific growth rate decreases at elevated concentrations of the substrate (Arutchelvan et al. 2006; Hill and Robinson 1975; Kumar et al. 2005; Stoilova et al. 2006). In these cases, it is common to use the Haldane equation:

$$\mu = \mu_m S/(k_s + S + S^2/k_i) \tag{20.6}$$

where k_i is the substrate inhibition coefficient.

Equations 20.5 and 20.6 sufficiently approximate experimental data produced in processes with pure and mixed bacterial cultures (Bajaj et al. 2009; Dey and Mukherjee 2010; Hill and Robinson 1975; Yang and Humphrey 1975), activated sludge (Buitron et al. 1998; Henze et al. 1997; Marrot 2006). They are also used to describe processes of contaminant biodegradation in biofilm reactors and reactors with aerobic granules (Basheer and Farooqi 2012; Tay et al. 2004).

The nature of inhibition may be different—excessive substrate and related to it is the accumulation of "unfavourable" intermediate oxidation products, and toxicity of the substrate as it is, or of related substances in media with complex composition, and so on. The mechanisms of the environment affect the constants μ_m, k_s and k_i, and therefore the form of expression $\mu(S)$ may be different (Heijnen and Romein 1995; Kovarova-Kovar and Egli 1998; Yang and Humphrey 1975). At the same time, the dependence of $\mu(S)$ has only two specific forms—they are curves with saturation or maximum.

20.3.2 Stoichiometry of Microbial Growth

The value of coefficient Y_s observed in experiments depends on the growth rate of the microorganisms. This is well reflected by the equation (Bailey and Ollis 1986; Pirt 1975) of the form:

$$1/Y = 1/Y_{max} + m/\mu \tag{20.7}$$

Here, the maximum theoretical value of the coefficients of substrate-to-biomass yield Y_{max} and specific maintenance rate m describe the consumption of the substrate to synthesise the biomass of the microorganism and maintain cells in vital condition, respectively. Substrate consumption to maintain (m/μ) are minimal at maximum growth rate of microorganisms (μ_m) and increase when it decreases.

The value of the growth yield coefficient $Y(\mu)$ characterises material and energy balance (Bailey and Ollis 1986; Minkevich 2005) and the specific growth rate defines dynamic properties of the microorganism culture. It is possible using relation $Y(\mu)$ to define the value $\mu(S)$ on-line (Minkevich et al. 2000; Ponomarev and Gurevich 1977). In this case there is no need to determine the coefficients K and K2 and the possibility of on-line control of continuous cultures appears.

20.3.3 Alternative Models

Commonly, interrelations between microorganism growth and substrate consumption are described by Monod and Haldane models. However, given the qualitative agreement with experimental data, deviation from calculated values is frequently observed. This invites research for more accurate alternative models. A review with an extensive and in-depth analysis of microorganism growth kinetics studies was made by Kovárova-Kovar and Egli (1998). Noting stagnation in the development of theory and experimental difficulties, the authors outlined topical for the actual practice routes for in-depth research. These routes are the growth of microorganisms on a mixture of substrates (homo- and heterogeneous) at low concentrations of substrates (specific for highly toxic and recalcitrant compounds in wastewaters, solid wastes and for natural media), multi-variable limitation of growth rate, microorganism adaptation, formation of mixed cultures and consortia. Since then, the methods of microorganism cultivation and waste bioprocessing methods have advanced as described below. The theoretical basis for them was the same as the equations of microorganism growth kinetics (20.5, 20.6).

No physical substantiation of coefficients in the kinetic Monod model has been found (Liu 2007). This is also true for the Haldane equation and other alternative models. Strictly speaking, the analogy with the kinetics of biochemical reactions is only a formal substantiation of the adequacy to real performance mechanisms (Baily and Ollis 1986; Cornish-Bowden 1976; Trigueros et al. 2010). The problem is not the lack of rigorous and unambiguous substantiations. Retrospective analysis of theoretical and experimental studies shows that under the tremendous variety of environmental conditions, the coefficients of kinetic equations of microorganism growth cannot be constants (Jannasch and Egli 1993; Kovárová-Kovar and Egli 1998).

The challenge of quantitative descriptions of microbiological processes is also associated with the problem of coefficient evaluation (even in the case of simple Monod and Haldane models). So, for example, for the bacterial monoculture values, μ_m and k_s can vary within very wide limits (Kovarova-Kovar and Egli 1998). This is connected with their sensitivity to methods of evaluating concentrations of microorganisms and substrate (using batch or continuous culture, wash-out data) (Agarry et al. 2009; Yang and Humphrey 1975) and methods of fitting the chosen model (Cornish-Bowden 1976). In batch cultures, the concentration and activity of microorganisms, concentration of growth-limiting substrate, other nutrition elements, and metabolites are in a non-equilibrium state. The microorganism attains equilibrium in continuous cultures, but in this case to collect data is time intensive and is technically more difficult. In all said methods, the microorganism growth dynamics and, accordingly, the decrease of the substrate depend on the history of the culture. Values μ_m, k_s and k_i are greatly affected by the intensity of mixing, medium composition and other factors of the medium (Kovárova-Kovar and Egli 1998).

Earlier reports (Pawlowsky et al. 1973; Yang and Humphrey 1975) and later work (Agarry et al. 2009; Bajaj et al. 2009; Basak et al. 2014; Marrot et al. 2006; Saravaran et al. 2008) show that the growth of bacteria on phenol and other contaminating compounds is described by substrate inhibition equations of various forms. When compared, the choice of the model to better approximate experimental data is generally based on statistical analysis (Agarry et al. 2009; Khorasani et al. 2013; Trigueros et al. 2010; Yang and Humphrey 1975).

It is unlikely that the coefficients and models of the growth of monoculture found under laboratory conditions remain invariable when functioning as a part of real cultures in large-scale processes. However, in the media with specific environmental conditions, the Monod and Haldane models and their modifications sufficiently approximate the dependence of $\mu(S)$ for monoculture of the microorganisms (Heijnen and Romein 1995). They also work in the processes with mixed cultures, activated sludge, biofilms and granules in the biotreatment of simple and multi-component wastewaters. On the one hand, this proves their formalism, and on the other, they take into account key properties of microorganism cultures necessary for engineering calculations and to design biotreatment facilities. Among them is the determination of growth-rate-limiting factors and dependence of residual concentration of oxidisable compounds on hydraulic loads and input concentrations.

20.3.4 Steady State of Continuous Processes

In a homogeneous culture continuously stirred tank reactor (CSTR), as the dilution rate (equal to hydraulic load) increases, the residence time of water in the reactors $(T = 1/D)$ decreases, and so does the time when the microorganisms contact with oxidisable compounds. According to kinetics equations (20.5), this increases their residual substrate concentration and decreases concentration of biomass (Fig. 1, line 1). This is also observed when the growth rate is limited by oxygen (Figure 20.1, line 2). However, in the range of flows above point a, incomplete substrate oxidation products accumulate (not shown). The form of $X(D)$ can change in this manner with an increase of initial concentration of the substrate (Figure 20.2). For cultures of a single species (microorganism), each of the given forms of $X(D)$ can be a particular case observed under different cultivation conditions. Here, curve 2 (Figure 20.2) is an experimental dependence $X(D)$ normalised with coefficient equal to the S_{01}/S_{02} ratio.

In non-homogeneous processes, the carry-over of microorganisms from the reactors can be reduced by wall growth, retention of floccules, granules or biomass recycling. As a result, as the hydraulic load increases, the time of residence and contact of microorganisms in the reactors does not decrease and may even increase. So, the non-homogeneous process opens new ways to increase the output and integrated wastewater treatment.

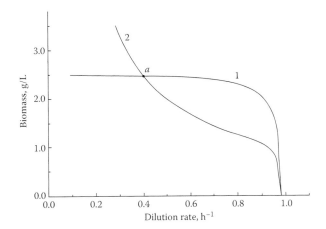

FIGURE 20.1
Typical dependence of biomass concentrations (1, 2) on flow rate in chemostat (1—growth rate limited by organic substrate, 2—by oxygen).

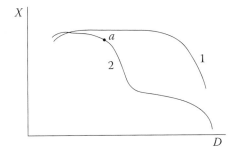

FIGURE 20.2
Characteristic form of dependence $X(D)$ in chemostat on concentration of substrate S_0. Curve 2 in the figure is an experimental dependence $X(D)$ normalized with coefficient equal to S_{01}/S_{02} ratio.

20.4 Modified Wastewater Biotreatment Methods

To increase the output and purification efficiency of wastewaters and to detoxify the solid wastes, different modifications of continuous and batch processes are under development. All of them are targeted to provide conditions to increase microorganism concentrations and flow rate (hydraulic load). In a continuous process to increase the concentration of microorganisms or of activated sludge, provisions are frequently made to return a part of the biomass from biotreated wastewater (effluent) to the influent (Henze et al. 1997; Villadsen 1999). The microorganisms aggregate with flocculation and more densely packed granules are formed (Basha et al. 2010; Gao et al. 2011; Jiang et al. 2006a; Maszenan et al. 2011), which are relatively more easily retained in bioreactors. This method increases the concentration of active microorganisms in bioreactors and the gross rate of substrate oxidation. Under the conditions of fluctuating hydraulic load, contaminant concentration, temperature, pH and other factors, which are observed in actual practice, the stability margin of wastewater treatment processes increases (Ajbar 2001; Galluzzo and Cosenza 2009; Villadsen 1999).

Another approach to increase biomass concentration and intensity of microbiological processes is based on the ability of microorganisms to grow on the surface of solid materials (Daugulis et al. 2011). In a flow apparatus, as opposed to suspended cells, the fixed microorganisms are not carried over with the medium flow. As a result, even with small inner surface area/volume ratios (S/V) with flow rate higher than critical μ_m, the culture is not washed out. Detached and surface-associated microorganisms remain in the liquid phase to continue oxidising the substrate. For example, in the work of Molin and Nilsson (1985) with an increase in the surface area/volume ratios from 0.5 to 5.5 cm^2/mL^-, the maximum oxidation rate of phenol was attained at a flow rate of 0.45 and 1.4 h^{-1}, respectively. The critical dilution rate value was found to be 0.4 h^{-1}. With an increase of the bacteria fixed in the biofilm, the maximum phenol oxidation output of the reactor more than tripled.

The wall growth capability of the microorganism is the basis of various bioreactor designs and processes. The reactors are added fixed or free-floating loads (supports) of inert material—fixed bed and mobile (fluidised, turbulent) beds. This creates conditions for the growth of biofilms and the biomass of attached bacteria. The mechanisms of microbial attachment to the inert surfaces can be different; however, irrespective of this, the total concentration of microorganisms in the apparatus increases (Jena et al. 2005).

Conditions of transfer of the substrate and of extracellular metabolic products in cell aggregates essentially change compared to the diffusion of substances in water. As the diffusion flow of nutrient elements from the aqueous phase to the cells in biofilms decreases (Beyenal et al. 1997; Stewart and Franklin 2008; Tay et al. 2004), so does the rate of transfer of extracellular metabolic products from the biofilms into the aqueous phase (Stewart and Franklin 2008). Using the data of several authors and of their own experiments as the base, Stewart (2003) assumed the diffusion coefficient in biofilms to decrease by 40% on the average for light gases (oxygen, carbon dioxide and methane) and by 75% for organic compounds. A similar conclusion is made by Beyenal et al. (1997) about the diffusion of oxygen and phenol into biofilms.

Similar changes in the availability of free and aggregated microorganisms take place in activated sludge floccules and aerobic granules. Diffusion flows of substances depend on the thickness of biofilms, size of floccules or granules (Pérez et al. 2005; Satoh et al. 2003; Zhang and Tay 2014), the structure and density of aggregates (Asadia et al. 2013). These physical parameters of cell aggregates vary with aging to change the activity of aggregated bacteria. As the biofilms, floccules, granules and a part of aggregated cells in the total biomass in the reactors grow, the positive effect increases to attain maximum and then drastically drops with complete disintegration of the aggregates (Zhang and Tay 2014; Zheng et al. 2006). The growth of biofilms and granules is time-limited because after the aggregates exceed a certain size, the diffusion flows of substances decrease and the activity of microorganisms drops drastically. This creates certain difficulties to operate and use such processes and apparatus in full-scale operations.

Decrease of the diffusion transfer results in the protection of microorganisms in the inner space of biofilms, floccules and granules from the effect of toxic substrates (Jiang et al. 2006b; Maszenan et al. 2011; Stewart and Franklin 2008). Microbial aggregates also form gradients of intermediate concentrations and cellular metabolites. As a result, areas with certain medium conditions localise to occupy cell sub-populations with corresponding physiological and functional properties (Stewart and Franklin 2008). Multi-species structures of biofilms, granules and activated sludge floccules form on the same principle (Whiteley and Bailey 2000). Co-aggregation of microorganisms with different properties can result in positive cooperative effects in the degradation of toxic compounds (Adav and Lee 2008; Jiang et al. 2006b) in combination with one reactor of nitrification and denitrification processes (Satoh et al. 2003).

The condition of the formation of biofilms, granules and aggregates with prescribed composition of the matrix, density of cells, porosity and so on is a great field of study covered by special reviews. Alongside this, the nature of the dependence of the growth of microorganisms is taken as predicted by Equations 20.5, 20.6 and 20.7 (Cruickshank et al. 2000; Pérez et al. 2005). A qualitative description of the processes is becoming more complex because it takes into account diffusion of substrates and space structure of biofilms, floccules and aggregates.

So far, the processes with granulation of activated sludge and selective strains of bacteria decomposers, membrane bioreactors and biofilm reactors have no full-scale design solutions (Daugulis et al. 2011; Gao et al. 2011). Still they can be used to treat small-scale wastes or as a component of large-scale processes; the latter calls for special investigation of process kinetics and design solutions.

A fundamentally different approach to the mechanism of increasing the treatment efficiency of wastewater containing toxic and volatile organic compounds is implemented in two-phase bioreactors. It is based on partitioning the aqueous phase of free and aggregated microorganisms and water-immiscible organic solvent or solid polymer phases. The latter (organic phase) accumulates toxic and volatile compounds, which diffuse into the aqueous phase as they are utilised by bacteria. Partitioning in the aqueous phase maintains sub-inhibition concentration of toxic substrates. Advantages of reactors with two-phase configuration are demonstrated by examples of phenol biodegradation with high concentration in the medium (Daugulis et al. 2011; Poleo and Daugulis 2013). However, this largely applies to moderately soluble volatile and recalcitrant organic compounds, which are, at the same time, very toxic (Muñoz et al. 2012).

The time until the treated wastewater is in contact with bacteria is longest in the membrane-type bioreactors (Fatone et al. 2011; Lesjean et al. 2009; Marrot et al. 2006). Going through the membrane is the liquid phase only; the solid phase including the microorganisms can be completely retained in the substrate oxidation zone. Residual concentration of substrates decreases by Equation 20.5 or other expressions for the dependence of $\mu(S)$. At the same time, under the minimum specific growth rate, the consumption to maintain and/or for lysis or cryptic growth of the cells increases. As a result, the excessive biomass to be buried is smaller. The membrane bioreactors are efficient to degrade recalcitrant compounds (Fatone et al. 2011).

20.5 Experimental Data and Interpretation of Model Parameters

20.5.1 Interpretation of Parameters of Growth Kinetics

The formalism of kinetic equations of microbial growth allows one to interpret μ_m, k_s and k_i as estimates of maximum specific growth rate, affinity and inhibition constants only in specific cases and with considerable restraints. This applies to the dependence of the growth of the microorganism on substrate transport systems and their non-zero consumption at zero growth rate (concepts of its threshold concentration, S_{min}) (Jannasch and Egli 1993; Kovárova-Kovar and Egli 1998).

The factors determining the growth rate of a microorganism on toxic substrates are related to their biochemical and genetic characteristics. Consider, for example, aromatic hydrocarbons. Catabolism of these compounds is attended by the formation of catechol and protocatechuate followed by cleaving of the benzene ring via *ortho-* (intradiol) or *meta-* (extradiol) pathways, to form various substrates of the tricarboxylic acid cycle (Cao et al. 2008; Feist and Hegeman 1969; Harwood and Parales 1996; Loh and Chua 2002). It is known that enzyme induction of intradiol or extradiol cleaving depends on the primary substrate and bacterial strain (Feist and Hegeman 1969; Harwood and Parales 1996). For example, when utilising phenol, *Pseudomonas putida* (NCIB 10015) and *Alcaligenes eutrophus* bacteria induced the *meta*-pathway, while benzoate induces the *ortho*-pathway (Murray and Williams 1974). On benzole and salicylic acids, the bacteria *P. putida* induced the *ortho*-pathway and on naphthalene the *meta*-pathway (Feist and Hegeman 1969; Hamzah and Al-Baharna 1994). Similar evidence is reported by numerous authors. Meanwhile, it was noted that in the utilisation of phenol by cultures of *Aspergillus fumigatus* and *Planococcus* sp. (Hupert-Kocurek et al. 2012; Jones et al. 1995), of benzoate cultures of *Pseudomonas cepacia* ATCC 29351 (Hamzah and Al-Baharna 1994; Nakazawa and Yokota 1973), *P. putida* P8 (Cao et al. 2008) and catechol by the culture of *P. putida* NCIB 10015 (Murray and Williams 1974), these divergent paths may be induced simultaneously. The latter work showed that in batch culture, the ratio of activities of enzymes of *ortho-* and *meta-*cleaving (1.2 – and 2.3 – dioxygenases) of catechol was 0.29, while in the fed-batch culture, this ratio was reverse –26. In periodic cultures of *P. putida* 451 and activated sludge utilising benzoate and phenol by the *ortho-*pathway, when the initial concentration of the substrates increased above a certain threshold, the *meta-*pathway was also induced at the same time (Lim et al. 2013; Loh and Chua 2002). It follows that induction of divergent pathways ambiguously depends on substrates and properties of the disruptor strains.

The concentration of key enzyme inducers can control the feeding mode. For instance, the diffusion of the substrate and oxygen, spatial inhomogeneity of activated sludge floccules, biofilms and granules and the complex composition of the wastewaters result in the formation of ecological microniches and a greater diversity of microorganism growth scenarios. The coexistence of bacteria with different metabolism types was observed by Jiang et al. (2006b). Some 10 isolates of phenol-oxidising bacteria isolated by the authors from aerobic granules demonstrated four metabolism types. Out of this, two strains of oxidised phenol by the *meta-*pathway, three strains by the *ortho-*pathway, three strains realised both pathways and one strain exhibited weak activity of enzymes of both pathways. Transition from the dominant *meta-*pathway to the *ortho-*pathway of phenol utilisation in continuous culture of *P. putida* was observed when it was introduced into the culture of protozoa *Colpoda steinii* (Gurevich et al. 1995; Manukovski et al. 1991).

A multi-enzyme system with alternative metabolism pathways should have kinetically complex functioning type. However, a vast majority of investigations analyse concentration dependencies, intermediates of the decomposition of various aromatic compounds, activity of catabolism enzymes and genetic characteristics of the microorganisms irrespective of the existence (demonstration) of the multiplicity of steady states in the multi-enzyme microorganism system. Populations, mixed cultures and consortia can choose one of the alternative pathways or concurrently all genetically and biochemically available ones, but with a certain ratio of activities. The driving force here is the energy efficiency of the substrate used. The ability to simultaneously induce divergent pathways expands competitive advantages of microorganisms providing a high growth rate and utilisation of mixed substrates (Watanabe et al. 2002; Zhang et al. 2004). The energy efficiency of growth on aromatic hydrocarbons depends on the catabolism path (Minkevich 2005).

20.5.2 Interpretation of Stoichiometric Parameters

For microorganisms with constant cell composition, the coefficient of substrate consumption to synthesise the biomass Y_{max} should probably be constant, provided other conditions of the environment are invariable. In experiments, the maximum value is observed at the growth rate close to μ_m. Actually, when the growth rate is inhibited or limited, the cell composition changes to some extent. By the concepts of material and energy balance of microorganism's growth (Erickson et al. 1978; Minkevich 2005; Pirt 1975), the coefficients m (specific maintenance rate) substrate consumption to maintain depend on the specific growth rate and environmental conditions. It should be noted that an absolute number of experiments exhibited linear dependence to $1/Y$. When fitting Equation 20.7 to the experimental data coefficient, m may turn out to be constant or equal to zero; the latter does not mean that the growth on the carbon substrate are strictly constant or absent.

During the process of utilising substrates that are the sources of carbon and energy, extracellular products of microbial metabolism are generally formed. A part of the substrate's energy is spent for thermogenesis to decrease the yield coefficient. In the case of growth at inhibiting concentrations of the substrate, the observed coefficient Y is considerably smaller. For example, in a continuous culture of phenol-oxidising bacteria, the coefficients of Equation 20.7 are $Y_{max} = 0.9$ mg/mg and $m = 0.24$ mg/mg h in the limitation range $(S < \sqrt{k_s k_i})$ and $Y_{max} = 0.59$ mg/mg and $m = 0.28$ mg/mg h in the inhibition range at $S > \sqrt{k_s k_i}$ (Jones et al. 1973). The variation of the yield coefficient is shown graphically in Figure 20.3. From Figure 20.3, it is evident that in the growth rate limitation and inhibition ranges, the culture of bacteria exhibits independent and disjoint dependences $1/Y(1/D)$. The form

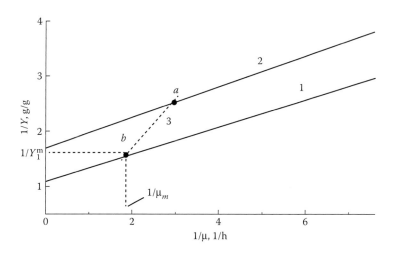

FIGURE 20.3
Dependence of the yield coefficient $(1/Y)$ on flow rate $(1/D)$ in the culture of phenol-oxidizing bacteria with limited (1) and inhibited (2) growth rate. (Adapted from Jones G.L., F. Jansen and A.J. McKay. 1973. *J. Gen. Microbiol.* 74: 139–148.) 3—hypothetical variation of $1/Y$ in 'limitation-inhibition' transition region. See explanations in the text.

of $\mu(s)$ is a continuous function and has a point common for both ranges. A question arises—why is the variation of the growth rate with the substrate concentration continuous while the substrate flow into the cells changes step-wise in the range of $S = \sqrt{k_s k_i}$? Probably, in this range, there exists a special mode of substrate metabolism (dash line in Figure 20.3), where coefficient m increases substantially. A vivid example of existence of a common point of dependences $1/Y$, produced in growth rate limitation/inhibition is given by Minkevich et al. (2000).

Another version assumes switching from one metabolism type to another. We would consider the experimental examples in favour of switching mechanisms (bistability) hereafter.

20.6 Bistability of Microbial Processes

20.6.1 Mechanism of Bistability under Formation of Partial Oxidation Products

The high efficiency of carbon substrates to synthesise biomass and numerous biologically active substances is found in fed-batch processes. Biotechnologies of this type are described in monographs by Baily and Ollis (1986) and Pirt (1975) and numerous dedicated investigations. Theoretically, controlled feeding of the substrate can produce biomass of very high concentrations. Within the framework of the Monod and Haldane models, the efficiency of the carbon source is explained by reduced consumption to maintain and eliminate substrate inhibition (Yamane and Hirano 1977). Experimentally high biomass concentrations were attained on the media with glucose (144 g/L for Escherichia coli) (Landwall and Holme 1977), methanol (133 g/L) (Kuraishi et al. 1979) and other substrates. Usually, at high biomass concentration, the coefficient of yield Y decreased with concurrent build-up of metabolite products. For example, in glucose utilisation, these are lactate, acetate, pyruvate and so on (Landwall and Holme 1977). A decrease of coefficient of yield Y notwithstanding, the controlled feeding produces high biomass concentrations.

On the other hand, in continuous cultivation when the carbon substrate concentration exceeds a certain threshold, the biomass concentration drops drastically (Figure 20.2, curve 2) (Fiechter 1975). In this situation, the quotient of substrate utilisation to synthesise the biomass decreases too. This variation of the type of dependence $X(D)$ (Figure 20.2, curve 2) can be considered as an indication of existence of steady states, which the Monod model does not predict.

Figure 20.1 (curve 2) shows that in the range of dilution rates beyond the point a, the biomass concentration drops relative to that predicted for preset concentration S_0 (Monod model), which can be caused by the microorganism growth rate limited by oxygen. In this case, dependence $X(D)$ takes the form of curve 2 in Figure 20.1, and the concentration of the biomass decreases by value ΔX. Stoichiometrically, the synthesis of biomass requires less substrate and the excess formed is used to synthesise partial oxidation products.

Summarising the performance types of growing microorganism cultures, dependence of $X(D)$ in Figure 20.1 can be considered as an example of existence of bistable states of great importance for wastewater processing. In this connection, Nagirny et al. (1977) analysed the growth of facultative aerobes in carbon-source-limited chemostat. Glucose was taken as an example of a substrate under this condition.

It is common knowledge that in cells of facultative anaerobes growing on glucose, the energy is generated at two stages: anaerobic (glycolysis) and aerobic oxidation in the cytochrome chain to CO_2; the specific time of increase of the carrying capacity of these stages differs considerably. Activity of glycolysis enzymes is induced within the time of about 1 s (Newsholme and Start 1973). To increase the capacity of the respiratory chains, it is necessary to synthesise additional cytochromes and this can take tens of minutes.

At the initial period of a batch or continuous culture or in different cases when the glucose concentration increases drastically, the glycolysis enzymes are activated. Glucose consumption attains a steady state within about 1 min or less. The carrying capacity of the respiratory chain may be insufficient, if the drastically increased load exceeds a certain threshold, L. To increase the respiratory chain, it is necessary to synthesise additional cytochromes. As the specific time of protein synthesis is much longer, the carrying capacity of the respiratory chain increases slowly. This means that when the load drastically increases above threshold L, the culture will build up partial oxidation products (Crabtree effect). The relatively high concentration of glucose inhibits their consumption (catabolic repression). With glucose concentration decreasing, the cells will use partial oxidation products as a secondary source of energy. The threshold L value is variable; it depends on the condition of the respiratory chain before the pulse increase of load on it.

The increasing concentration of microorganisms growing under the condition when the culture does not build up partial oxidation products results in the distribution of the growing load on the population between them. The specific load on a cell decreases at that point. The positive effect of a population increase is similar to that of the increase in the carrying capacity of the respirator chain. However, it must be kept in mind that the order of the specific time of the respiratory chain capacity increase for the population is in the order of microorganism generation time.

This is a general scenario of glucose utilisation to grow microorganisms, taking into account dependence of a specific growth rate only on a single source of energy and carbon, and on the initial and secondary sources.

Specific form of equations $\mu(S)$ and $\mu(S,P)$ is of no concern at all. Thus, the microorganism growth can follow three scenarios:

1. Growth without the formation of partial oxidation products, that is, high efficiency of substrate utilisation to synthesise the biomass
2. Growth with build-up of partial oxidation products
3. Growth on initial substrate and concurrently forming products

To illustrate and detail the given generalised analysis, Nagirny et al. (1977) considered the Monod (scenario 1) and Haldane (scenario 2 and 3) models. As a dynamic system, a culture of microorganisms can be in three states in addition to the trivial state at $X = 0$ and $S = 0$, respectively. According to the qualitative theory of dynamic systems (Andronov et al. 1966), a chemostat culture can be observed to have only two states out of three. One state corresponds to the Monod model and curve 1 and the second state to curve 2 in Figure 20.4. Both states are observed in experiments. The key observation is that state 2 occurs not in response to increase of concentration of S_0, but due to a drastic increase in the load and disbalance of substrate flows through glycolysis and respiratory chain. It follows that the dependence of $X(D)$ of type 1 (Figure 20.2) can be produced at high concentrations of S_0. For batch cultures, this is proven by numerous results of fed-batch processes (Landwall and Holme 1977; Whiffin et al. 2004).

In a continuous culture, the high efficiency state (Figure 20.2, curve 1) is *meta*-stable. For example, when the hydraulic load, substrate concentration, temperature and other factors vary for short periods, the culture can leave it and pass into the low-efficiency state (Figure 20.2, curve 2). After the disturbing factors normalise, it does not return. The condition necessary to maintain the high efficiency state is to prevent an over-flow metabolism disturbing the process.

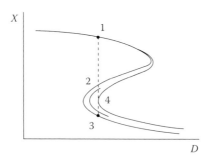

FIGURE 20.4
Dependence of steady states of continuous culture of microorganisms on flow rate in the wall growth chemostat with degradation of the toxic substrate (1, 3—steady and 2—unsteady states of the culture).

Conclusions of theoretical analysis were verified in a continuous culture of *E. coli* on a medium mineral with glucose. In experiments, with a steady concentration of *E. coli* B, the biomass increased from 5 to 40 g/L with glucose concentration from 10 to 40 g/L. In the culture of *E. coli* MRE 600, the biomass concentration was 30 g/L at $S_0 = 60$ g/L. The flow rate in experiments was set at 0.5 h^{-1}. The processes were stable for cultivation time up to 7 days.

Rearrangements in states of microbial cultures are regulation of wastewater bioprocessing processes at the cell level, population level and at the level of living organisms' communities. It comprises variations of the activity of metabolic pathways and their induction. The population level is implemented by an increase of microorganism concentration, regulation structures of phenotypical and genetic inhomogeneous populations. These levels demonstrate the example considered above. The third level consists of forming the species structure of communities, transfer of plasmids—this takes place in granules, biofilms, and activated sludge. At each of the levels and in their interaction, the conditions for occurrence of bistability of different biotechnological processes can exist.

20.6.2 Bistability under Biodegradation of Toxic Compounds

The peculiarity of the continuous processes of toxic compound biodegradation is that, in complete mixing flow reactors, steady operation of the culture is possible only in the range of dilution rates substantially below the maximum specific growth rate of microorganisms. The instability range was predicted in several theoretical studies (Hill and Robinson 1975). The entire space of states (X, S) is divided into two sub-domains—stability and instability—of the continuous processes. For real processes, this means that they can run at limited hydraulic load and concentration of oxidised compounds only. Natural oscillations of the composition of processed effluents, temperature, aeration and so on, can result in wash-out of the microorganisms.

In an explicit form, steady states of microbial cultures under conditions of the growth rate inhibited by the substrate were obtained in laboratory processes with controlled dilution rate by feedback (Bril'kov et al. 1980; Minkevich 2005). The quasi-steady state is achieved in the two-step process (Jones et al. 1973).

Actually, large-scale microbial processes are run in inhomogeneous media—with given biofilms, granules and activated sludge floccules. In inhomogeneous processes, toxic compounds can degrade within the limited range of substrate concentrations and hydraulic load. The work of Pawlowsky et al. (1973) showed the existence of the bistable mode of biodegradation of toxic substrate by example of phenol in biofilm processes. The culture that forms a biofilm on the inner surface of the bioreactor has two steady states of toxic substrate degradation in a flow process of chemostat type—with a high or low concentration of microorganisms (X) (Figure 20.4). In state 1 (upper branch of dependence $X(D)$ in Figure 20.2), the residual concentration of the substrate is low, and in state 2 (lower branch), it is high. Accordingly, a respective change in the purification efficiency of water from toxic contamination would result, which can be high or low. In the state of effective oxidation of toxic substrate (point 1 in Figure 20.4), the process is sensitive to external disturbances (fluctuations of temperature, input substrate concentration, etc.). According to the model, this state is meta-stable, as under the impact of the disturbances the process turns into stable state 3 (Figure 20.4). Even after the conditions are restored, the process does not return to state 1. A positive effect of the biofilms is that the flow rate (hydraulic load) for the steady processes can be higher than μ_{max}.

Similar results have been obtained for continuous oxidation of a mixture of dissimilar substrates (phenol and glucose) in the bioreactor with microorganism recycling (Ajbar 2001; González et al. 2001). Substrates S_1 and S_2 are assumed to interact with the mechanism of non-competitive inhibition of bacterial growth rate. The model of growth of aerobic culture with granular-supported biofilm and double-limiting substrate kinetics with phenol exhibition and oxygen limitation was considered in the study of Olivieri et al. (2011). These models predict bistability of the process but do not take into account the increase in growth rate and biomass attached cells with increasing concentrations of the inhibitory substrate. As a result, the state of culture with low degradation efficiency of carbon substrate remains stable.

Gurevich and Ladygina (1986, 1991) assumed a continuous culture of bacteria with active biofilm to be capable of both increasing and decreasing the mass of attached cells. Growth kinetics of free and attached bacteria is shown in Figure 20.5. A dependence of the growth rate of bacteria inside the biofilm is shown by curve 2. Its shift relative to curve 1 is due to the concentration

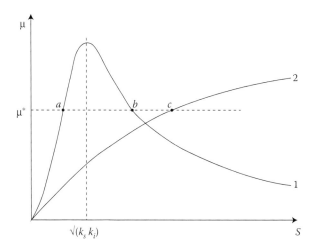

FIGURE 20.5
Dependence of specific growth rate of microorganism population on toxic substrate concentration (1—free and 2—attached cells).

of the substrate in the culture volume being higher than in the biofilm. At concentrations of S in the medium being higher than $\sqrt{k_s k_i}$, the effective concentration in the biofilm limits the growth of attached cells. As a result, for a certain given dilution rate and accordingly, for growth rate μ^*, the 'substrate—free cells—attached cells' system can have three states (simple version). Experiments with continuous cultures of free bacteria are known to have both weak and considerable growth of the biofilms. Considering this and according to the qualitative theory of dynamic systems (Andronov 1966), the system under consideration has two steady states (points 1 and 3 in Figure 20.5, the same in Figure 20.4) and one non-steady state (point 2). In the range of values $S > \sqrt{k_s k_i}$, the biomass concentration decreases and substrate concentration increases. But if the substrate concentration is more, then the film biomass grows at a greater rate. As a result, point 2 (Figure 20.4) shifts up. When point 2 becomes the tangent point of curve $X(D)$, the process turns into state 1 unevenly. Thus, under the conditions of active growth of the biofilm, the state of low substrate oxidation efficiency becomes *meta*-stable, and that of high efficiency is the stable state.

This model has been developed to explain observations for spontaneous restoration of the continuous culture of phenol-oxidising bacteria *P. putida* after drastic decrease of the volume and, accordingly, an increase of the flow rate from 0.09 to 0.6 h^{-1}. As a result, residual concentration of 0.2 mg/L phenol increased to 300 mg/L ($S_0 = 1$ g/L). The restoration mode was as follows: in the course of 2 h, the flow stopped, then the feed of the medium resumed. Biofilm growth was observed visually. The operating volume of the culture was restored after 10 h of medium feed. From this moment, the culture was continuous. After 8 h, the process reached its previous state, that is, the entire period of restoration took 18 h. Qualitatively, the model was verified in repeated cycles of disturbance and restoration of the continuous culture.

20.6.3 Bistability of Mixed Substrate Utilisation

Mixed cultures of microorganisms–decomposers and multi-component wastewaters are the reality of bioprocessing. Diversity of metabolism pathways of complex substrates and species composition of microflora would result in multiple steady states of bioprocessing processes.

The isolation and description of monocultures and simple mixed microorganism cultures oxidising different substrates are studied in numerous investigations. The challenge of selecting models and evaluation of coefficients in situations with mixed cultures and multi-component substrates is very sophisticated. Qualitative characteristics of the processes are realistic and useful in this sense. Usually, the conclusion is the potential possibility of using an isolated strain and adequate description of the processes of a known or proposed model. However, it should be kept in mind that such conclusions are applicable for particular cases and the probability of them occurring in real (large-scale) processes is very small.

Oxidation kinetics of substrate mixtures were studied in several very interesting works, which showed mutual influence of homologous compounds, including that of hydrocarbons on their oxidation. Potentially, this can be the reason of non-trivial growth of microorganisms on mixed substrates (Kovárova-Kovar and Egli 1998). Illustrative are the results reported by Bailey and Ollis (1986) and Imanaka et al. (1972). The authors found and described in detail the trigger mechanism of the induction of α-galactosidase synthesis in mould on a media containing glucose and galactose.

Wastewaters in the coke and by-product process are a multi-substrate medium for the bacterial community of treatment facilities. They contain phenols, naphthalenes, polycyclic aromatic hydrocarbons, cyanides, rhodanides and other compounds. Dominant by mass are phenol and associated naphthalene. It is a common knowledge that their oxidation is mutually dependent. Specifically, in a mixed bacterial culture (batch process), naphthalene reduces the oxidation rate of phenol. At the same time it does not oxidise until the phenol concentration is below some threshold (Meyer et al. 1984). The same interrelation was observed by Abuhamed et al. (2004) in culture *P. putida* under degradation of benzene, toluene and phenols.

Mutual dependence of phenol and naphthalene oxidation and its effect on the biotreatment of industrial phenol water was observed in several studies (Gurevich et al. 1989, 1995; Manukovski et al. 1991). In a laboratory experiment, a continuous culture of *Pseudomonas* sp. isolated from the aero-tank steadily grew on a medium with a naphthalene content of 2–4 g/L. The biomass concentration was 1–2 g/L by dry weight (Manukovski et al. 1991). Evidently, in the operating aero-tanks, the naphthalene-oxidising bacteria did not realise their potential, as the ratio of phenol- and naphthalene-oxidising species was not optimum in the spontaneously formed community.

As a dynamic system, the ratio of coexisting species in a community can be different; this includes even those that can be more efficient for the wastewaters of specific composition. On this basis, the stage circuit was made to include a short-term mode of building up naphthalene-oxidising bacteria. After they were involved in the continuous process, the bacterial community changed into a new steady state. The community considerably increased the population of bacteria oxidising naphthalene and phenanthrene. The transition period in aero-tanks was seen to take 3 weeks. Upon completion of the transition period, the treatment process maintained stability with naphthalene concentration increasing up to 900 mg/L. With the naphthalene content and other parameters fluctuating, residual concentration of phenol in the effluent decreased by 25%–40% and the variation coefficient –2.8 times. These results are based on the bio-treatment that was conducted and closely monitored for 2.5 years (Gurevich et al. 1989).

It should be noted that the transition into the state of a more efficient treatment of wastewater was the result of specific impact on the process. Spontaneous fluctuations of the process conditions during the previous period (about 10 years) did not result in such an effect of rearrangement in the composition of bacterial community. This demonstrates feasibility of forming a mixed culture with effective composition on a mixture of substrates. Similar to the examples described above, positive results can be realised, taking into account the dynamic properties and the phenomenon of bistability of the microbiological process.

20.7 Conclusion

Theoretical bases to develop and operate waste bioprocessing operations are the concepts based on the dependence of growth rate and concentration of substrate in the medium. These concepts can be understood at the initial period of investigating the growth kinetics of the microorganism. For microbiological processes that go with the participation of multi-species communities in multi-component wastewater and in a very wide range of concentrations of oxidizable substances, strict substantiation currently used kinetic equations cannot be found. In the experimental studies, mostly the basic batch culture methods are followed. Under this condition, they do not allow maintaining microorganisms and media in equilibrium states. Therefore, conclusions from numerous studies of growth and degradation kinetics of different toxic compounds take particular form. For real processes, they can be used with greater limitations only. Meanwhile, the observed dynamics of growth of microbial cultures and degradation of carbon compounds adequately characterises their interconnection. This explains the vitality of regularities earlier established by Monod and Haldane models and their modifications. The questions about regularities of contaminants' biodegradation at low concentrations, adaptation and coexistence of different microorganism in different media, including growth in aggregated states, remain open (Kovarova-Kovar and Egli 1998).

Biotechnological processes are dynamic systems specified by the phenomenon of multiplicity of steady states and bistability. Lack of information about their existence denies the use of the potential of microorganisms. Studies in this direction are only a few, even though there are examples of finding them and using in actual practice; the latter gives ground for its promise to consider them for exploitation.

References

Abuhamed T., E. Bayraktar, T. Mehmetoğlu and Ü. Mehmetoğlu. 2004. Kinetics model for growth of *Pseudomonas putida* F1 during benzene, toluene and phenol biodegradation. *Process Biochem.* (39): 983–988.

Adav S.S. and D.-J. Lee. 2008. Physiological characterization and interactions of isolates in phenol-degrading aerobic granules. *Appl. Microbiol. Biotechnol.* 78: 899–905.

Agarry S.E., O.K. Audu and B.O. Solomon. 2009. Substrate inhibition kinetics of phenol degradation by *P. fluorescens* from steady state and wash out data. *Int. J. Environ. Sci. Tech.* 6(3): 443–450.

Ajbar A. 2001. Stability analysis of the biodegradation of mixed wastes in a continuous bioreactor with cell recycle. *Water Res.* 35(5): 1201–1208.

Andronov A.A., A.A. Vitt and S.E. Khaikin. 1966. *Theory of Oscillators. (International Series of Monographs in Physics, Vol. 4)*. Oxford-London-Edinburgh-New York-Toronto-Paris-Frankfurt: Pergamon Press.

Arutchelvan V., V. Kanakasabai, R. Elangovan, S. Nagarajan and V. Muralikrishnan. 2006. Kinetics of high strength phenol degradation using *Bacillus brevis*. *J. Haz. Mat.* B129: 216–222.

Asadia A., A.A.L. Zinatizadeha, S. Sumathib, N. Rezaiec and S. Kiani. 2013. A comparative study on performance of two aerobic sequencing batch reactors with flocculated and granulated sludge treating an industrial estate wastewater: Process analysis and modeling. *IJE Trans. B Appl.* 26(2): 105–116.

Bailey J.E. and D.F. Ollis. 1986. *Biochemical Engineering Fundamentals*. 2nd Edn. New York: McGraw-Hill.

Bajaj M., C. Gallert and J. Winter. 2009. Phenol degradation kinetics of an aerobic mixed culture. *Biochem. Eng. J.* 46: 205–209.

Basak B., B. Bhunia, S. Dutta, S. Chakraborty, A. Dey. 2014. Kinetics of phenol biodegradation at high concentration by a metabolically versatile isolated yeast *Candida tropicalis* PHB5. *Environ. Sci. Pollut. Res. Int.* 21(2): 1444–1454.

Basha K.M., A. Rajendran and V. Thangavelu. 2010. Recent advances in the biodegradation of phenol: A review. *Asian L. Exp. Biol. Sci.* 1(2): 219–234.

Basheer F. and I.H. Farooqi. 2012. Biodegradation of *p*-cresol by aerobic granules in sequencing batch reactor. *J. Environ. Sci.* 24(11): 2012–2018.

Beyenal H., S. Seker and A. Tanyolaçç. 1997. Diffusion coefficients of phenol and oxygen in a biofilm of *Pseudomonas putida*. *AIChE J.* 43(1): 243–250.

Bril'kov A.V., N.S. Pechurkin and V.V. Litvinov. 1980. Substrate inhibition and limitation of growth of *Candida tropicalis* with phenol in continuous chemostat and pH-stat culture. *Microbiology.* 49(3): 466–472. (In Russian.)

Buitron G., A. Gonzalez and L.M. Lopez-Marin. 1998. Biodegradation of phenolic compounds by an acclimated activated sludge and isolated bacteria. *Water Sci. Technol.* 37(4): 371–378.

Cao B., A. Geng and K.C. Loh. 2008. Induction of *ortho*- and *meta*-cleavage pathways in *Pseudomonas* in biodegradation of high benzoate concentration: MS identification of catabolic enzymes. *Appl. Microbiol. Biotechnol.* 81: 99–107.

Cornish-Bowden A. 1976. *Principles of Enzyme Kinetics*. London–Boston: Butterworths.

Cruickshank S.M., A.J. Daugulis and P.J. McLellan. 2000. Dynamic modeling and optimal fed-batch feeding strategies for a two-phase partitioning bioreactor. *Biotechnol. Bioeng.* 67(2): 224–233.

Daugulis A.J., M.C. Tomei and B. Guieysse. 2011. Overcoming substrate inhibition during biological treatment of mono-aromatics: Recent advances in bioprocess design. *Appl. Microbiol. Biotechnol.* 90: 1589–1608.

Dey S. and S. Mukherjee. 2010. Performance and kinetic evaluation of phenol biodegradation by mixed microbial culture in a batch reactor. *Int. J. Water Resour. Environ. Eng.* 2(3): 40–49.

Erickson L.E., I.G. Minkevich and V.K. Eroshin. 1978. Application of mass and energy and energy balance regularities in fermentation. *Biotechnol. Bioeng.* 20: 1595–1621.

Fatone F., S. Di Fabio, D. Bolzonella and F. Cecchi. 2011. Fate of aromatic hydrocarbons in Italian municipal wastewater systems: An overview of wastewater treatment using conventional activated-sludge processes (CASP) and membrane bioreactors (MBRs). *Water Res.* 45: 93–104.

Feist C.F. and G.D. Hegeman. 1969. Phenol and benzoate metabolism by *Pseudomonas putida*: Regulation of tangential pathways. *J. Bacteriol.* 100(2): 869–877.

Fiechter A. 1975. Continuous cultivation of yeasts. *Meth. Cell. Biol.* 11: 97–130.

Gao D., L. Liu, H. Liang and W.M. Wu. 2011. Aerobic granular sludge: Characterization, mechanism of granulation and application to wastewater treatment. *Crit. Rev. Biotechnol.* 31(2): 137–52.

Galluzzo M. and B. Cosenza. 2009. Control of the biodegradation of mixed wastes in a continuous bioreactor by a type-2 fuzzy logic controller. *Comput. Chem. Eng.* 33(9): 1475–1483.

González G., M.G. Herrera, M.T. García and M.M. Peña. 2001. Biodegradation of phenol in a continuous process: Comparative study of stirred tank and fluidized-bed bioreactors. *Bioresour. Technol.* 76(3): 245–251.

Gurevich Yu.L. and V.P. Ladygina. 1986. Formation of biofilm and stability of toxic compounds biodegradation in continuous culture systems. In: *Problems of Ecological Monitoring and Ecosystem Modelling*, Ed.: Yu. A. Izrael. Leningrad: Gidrometeoizdat, 9, 22–30. (In Russian.)

Gurevich Y.L. and V.P. Ladygina. 1991. Effect of protozoa on bacterial degradation of aromatic hydrocarbons. *Stud. Environ. Sci.* 42, (Environ. Biotechnol.): 147–153.

Gurevich Y.L., M.I. Teremova, V.A. Chimarov, V.S. Shved and A.L. Shtein. 1989. Increase of efficiency of biological treatment of phenolic water. *Coke Chem. USSR.* 1: 51–52. (In Russian.)

Gurevich Yu.L., V.P. Ladygina and M.I. Teremova. 1995. Degradation of technogenic flows of organic matter by communities of microorganisms and ciliates. *Izv. Acad. Nauk. Biol.* 2: 226–230. (In Russian.)

Hamzah R.V. and B.S. Al-Baharna. 1994. Catechol ring-cleavage in *Pseudomonas cepacia*: The simultaneous induction of *ortho* and *meta* pathways. *Appl. Microbiol. Biotechnol.* 41(2): 250.

Harwood C.S. and R.E. Parales. 1996. The β-ketoadipate pathway and the biology of self-identify. *Annu. Rev. Microbiol.* 50:553–590.

Heijnen J.J. and B. Romein. 1995. Derivation of kinetic equations for growth on single substrate based on general properties of a simple metabolic network. *Biotechnol. Prog.* 11: 712–716.

Henze M., P. Harremoes, J. la Cour Jansen and E. Arvin. 1997. *Wastewater Treatment: Biological and Chemical Processes.* 2nd Edn. Berlin: Springer-Verlag.

Hill G.A. and C.W. Robinson. 1975. Substrate inhibition kinetics: Phenol degradation by *Pseudomonas putida*. *Biotechnol. Bioeng.* 17: 599–615.

Hupert-Kocurek K., U. Guzik and D. Wojcieszyńska. 2012. Characterization of catechol 2,3-dioxygenase from *Planococcus* sp. strain S5 induced by high phenol concentration. *Acta Biochim. Pol.* 59(3): 345–351.

Imanaka T., T. Kaieda, K. Sato and H. Taguchi. 1972. Optimization of α-galactosidase production by mold. 1. α-Galactosidase production on batch and continuous culture and a kinetic model for enzyme production. *J. Ferment. Technol.* 50(9): 633–646.

Jannasch H. W. and T. Egli. 1993. Microbial growth kinetics: A historical perspective. *Antonie Leeuwenhoek*. 63(3): 213–224.

Jena H.M., G.K. Roy and B.C. Meikap. 2005. Comparative study of immobilized cell bioreactors for industrial wastewater treatment. In: *WMCI-2005*, 1st and 2nd Oct – 2005. Rourkela: NIT.

Jiang H.-L., J.-H. Tay, A.M. Maszenan and S.T.-L. Tay. 2006a. Enhanced phenol biodegradation and aerobic granulation by two coaggregating bacterial strains. *Environ. Sci. Technol.* 40(19): 6137–6142.

Jiang H.-L., S.T.-L. Tay, A.M. Maszenan and J.-H. Tay. 2006b. Physiological traits of bacterial strains isolated from phenol-degrading aerobic granules. *FEMS Microbiol. Ecol.* 57(2): 182–191.

Jones G.L., F. Jansen and A.J. McKay. 1973. Substrate inhibition of the growth of bacterium NCIB 8250 by phenol. *J. Gen. Microbiol.* 74: 139–148.

Jones K.H., P.W. Trudgill and D.J. Hopper. 1995. Evidence of two pathways for the metabolism of phenol by *Aspergillus fumigatus*. *Arch. Microbiol.* 163: 176–181.

Khorasani A.C., M. Mashreghi and S. Yaghmaei. 2013. Study on biodegradation of mazut by newly isolated strain *Enterobacter cloacae* BBRC10061: Improving and kinetic investigation. *Iranian J. Environ. Health Sci. Eng.* 10: 2.

Kovárová-Kovar K. and T. Egli. 1998. Growth kinetics of suspended microbial cells: From single-substrate-controlled growth to mixed-substrate kinetics. *Microbiol. Mol. Biol. Rev.* 62(3): 646–666.

Kumar A., S. Kumar and S. Kumar. 2005. Biodegradation kinetics of phenol and catechol using *Pseudomonas putida* MTCC 1194. *Biochem. Eng. J.* 22: 151–159.

Kuraishi M., J. Terao, H. Ohkouchi, N. Matsuda and J. Nagai. 1979. SCP-process development with methanol as substrate. Microbiology applied to biotechnology. In: *Proceedings of the XII International Congress Microbial*. München, September 1978. Dechema monogr. 83(1704–1723): 111–124.

Landwall P. and T. Holme. 1977. Removal of inhibitors of bacterial growth by dialysis culture. *J. Gen. Microbiol.* 103(2): 345–352.

Lesjean B., V. Ferre, E. Vonghia and H. Moeslang. 2009. Market and design considerations of the 37 larger MBR plants in Europe. *Desalin. Water Treat.* 6(1–3): 227–233.

Lim J.W., C.E. Seng, P.E. Lim, S.L. Ng, K.C. Tan and S.L. Kew. 2013. Response of low-strength phenol-acclimated activated sludge to shock loading of high phenol concentrations. *Water SA*. 39(5): 695–700.

Liu Y. 2007. Overview of some theoretical approaches for derivation of the Monod equation. *Appl. Microbiol. Biotechnol.* 73: 1241–1250.

Loh K.C. and S.S. Chua. 2002. *Ortho* pathway of benzoate degradation in *Pseudomonas putida:* Induction of *meta* pathway at high substrate concentrations. *Enzyme Microb. Technol.* 30(5): 620–626.

Manukovski N.S., M.I. Teremova, Y.L. Gurevich and I.M. Pan'kova. 1991. Phenol and naphthalene degradation by mixed culture of microorganisms. *Stud. Environ. Sci.* 42, (Environ. Biotechnol.): 155–163.

Marrot B., A. Barrios-Martinez, P. Moulin and N. Roche. 2006. Biodegradation of high phenol concentration by activated sludge in an immersed membrane bioreactor. *Biochem. Eng. J.* 30: 174–183.

Maszenan A.M., Y. Liu and W.J. Ng. 2011. Bioremediation of wastewaters with recalcitrant organic compounds and metals by aerobic granules. *Biotechnol. Adv.* 29: 111–123.

Meyer J.S., M.D. Marcus and H.L. Bergman. 1984. Inhibitory interactions of aromatic organics during microbial degradation. *Environ. Toxicol. Chem.* 3: 583–587.

Minkevich I.G. 2005. *Material and Energy Balance and Kinetics of Microorganism Growth.* Moscow-Izhevsk: Institute of Computer Science, Scientific Publishing Centre "Regular and Chaotic Dynamics".

Minkevich I.G., S.V. Andreev and V.K. Eroshin. 2000. The effect of two inhibiting substrates on growth kinetics and cell maintenance of the yeast *Candida valida. Process. Biochem.* 36: 209–217.

Molin G. and I. Nilsson. 1985. Degradation of phenol by *Pseudomonas putida* ATCC 11172 in continuous culture at different ratios of biofilm surface to culture volume. *Appl. Environ. Microbiol.* 50(4): 946–950.

Muñoz R., A.J. Daugulis, M. Hernández and G. Quijano. 2012. Recent advances in two-phase partitioning bioreactors for the treatment of volatile organic compounds. *Biotechnol. Adv.* 30: 1707–1720.

Murray K. and P.A. Williams. 1974. Role of catechol and the methylcatechols as inducers of aromatic metabolism in *Pseudomonas putida. J. Bacteriol.* 117(3): 1153–1157.

Nagirny S.V., T.R. Khlebopros and Yu.L. Gurevich. 1977. Model of growth of the microorganism batch culture depending on inhibition with metabolism products. *Microbiol. J.* 34(4): 484–485. (In Ukrainian.)

Nakazawa T. and T. Yokota. 1973. Benzoate metabolism in *Pseudomonas putida* (*arvilla*) mt-2: Demonstration of two benzoate pathways. *J. Bacteriol.* (115): 262–267.

Newsholme E.A. and C. Start. 1973. *Regulation in Metabolism.* London: John Wiley & Sons.

Olivieri G., M.E. Russo, A. Marzocchella and P. Salatino. 2011. Modelling of an aerobic biofilm reactor with double-limiting substrate kinetics: Bifurcational and dynamical analysis. *Biotechnol. Prog.* 27(6): 1599–1613.

Pawlowsky U., J.A. Howell and C.T. Chi. 1973. Mixed culture biooxidation of phenol. III: Existence of multiple steady states in continuous culture with wall growth. *Biotechnol. Bioeng.* 15: 905–916.

Pérez J., C. Picioreanu and M. van Loosdrecht. 2005. Modeling biofilm and floc diffusion processes based on analytical solution of reaction-diffusion equations. *Water Res.* 39: 1311–1323.

Pirt S.J. 1975. *Principles of Microbe and Cell Cultivation.* Oxford: Blackwell Scientific Publishers.

Poleo E.E. and A.J. Daugulis. 2013. Simultaneous biodegradation of volatile and toxic contaminant mixtures by solid–liquid two-phase partitioning bioreactors. *J. Hazard. Mater.* (254–255): 206–213.

Ponomarev P.I. and Yu.L. Gurevich. 1977. Characteristics of continuous culture of hydrogen-reducing bacteria under conditions of the limitation of gases. *Microbiology* 46(1): 22–28. (In Russian.)

Reardon K.F., D.C. Mosteller and J.D.B. Rogers. 2000. Biodegradation kinetics of benzene, toluene and phenol as single and mixed substrates for *Pseudomonas putida* Fl. *Biotechnol. Bioeng.* 69(4): 385–400.

Saravanan P., K. Pakshirajan and P. Saha. 2008. Growth kinetics of an indigenous mixed microbial consortium during phenol degradation in a batch reactor. *Bioresour. Technol.* 99: 205–209.

Satoh H., Y. Nakamura, H. Ono and S. Okabe. 2003. Effect of oxygen concentration on nitrification and denitrification in single activated sludge flocs. *Biotechnol. Bioeng.* 83(5): 604–607.

Senthilvelan T., J. Kanagaraj, R.C. Panda and A.B. Mandal. 2014. Biodegradation of phenol by mixed microbial culture: An eco-friendly approach for the pollution reduction. *Clean. Techn. Environ. Policy.* 16: 113–126.

Stewart P.S. 2003. Diffusion in biofilms. *J. Bacteriol.* 185(5): 1485–1491.

Stewart P.S. and M.J. Franklin. 2008. Physiological heterogeneity in biofilms. *Nat. Rev. Microbiol.* 6(3):199–210.

Stoilova I., A. Krastanov, V. Stanchev, D. Daniel, M. Gerginova and Z. Alexieva. 2006. Biodegradation of high amounts of phenol, catechol, 2,4-dichlorophenol and 2,6-dimethoxyphenol by *Aspergillus awamori* cells. *Enzyme Microb. Technol.* 39(5): 1036–1041.

Tay S.T.L., H.L. Jiang and J.H. Tay. 2004. Functional analysis of microbial community in phenol-degrading aerobic granules cultivated in SBR. *Water Sci. Technol.* 50(10): 229–234.

Trigueros D.E.G., A.N. Módenes and M.A.S.S. Ravagnani. 2010. Biodegradation kinetics of benzene and toluene as single and mixed substrate: Estimation of biokinetics parameters by applying particle swarm optimization. *Latin Am. Appl. Res.* 40(3): 219–226.

Urtiaga A., R. Gutiérrez and I. Ortiz. 2009. Phenol recovery from phenolic resin manufacturing: Viability of the emulsion pertraction technology. *Desalination.* 245(1–3): 444–450.

Villadsen J. 1999. On the optimal design and control of a biodegradation process with substrate inhibition kinetics. *Ind. Eng. Chem. Res.* 38(3): 660–666.

Watanabe K., H. Futamata and S. Harayama. 2002. Understanding the diversity in catabolic potential of microorganisms for the development of bioremediation strategies. *Antonie van Leeuwenhoek* 81: 655–663.

Whiffin V.S., M.J. Cooney and R. Cord-Ruwisch. 2004. Online detection of feed demand in high V.S. cell density cultures of *Escherichia coli* by measurement of changes in dissolved oxygen transients in complex media. *Biotechnol. Bioeng.* 85(4): 422–433.

Whiteley A.S. and M.J. Bailey. 2000. Bacterial community structure and physiological state within an industrial phenol bioremediation system. *Appl. Environ. Microbiol.* 66(6): 2400–2407.

Yamane T. and Hirano S. 1977. Semibatch culture of microorganisms with constant fed of substrate—A mathematical simulation. *J. Ferment Technol.* 55(2): 156–165.

Yang R.D. and A.E. Humphrey. 1975. Dynamic and steady state studies of phenol biodegradation in pure and mixed cultures. *Biotechnol. Bioeng.* 17: 1211–1235.

Zhang X., P. Gao, Q. Chao, L. Wang, E. Senior and L. Zha. 2004. Microdiversity of phenol hydroxylase genes among phenol-degrading isolates of *Alcaligenes* sp. from an activated sludge system. *FEMS Microbiol. Lett.* 237(2): 369–375.

Zhang Y. and J.H. Tay. 2014. Rate limiting factors in trichloroethylene co-metabolic degradation by phenol-grown aerobic granules. *Biodegradation* 25(2): 227–237.

Zheng Y.-M., H.-Q. Yu, S.-J. Liu and X.-Z. Liu. 2006. Formation and instability of aerobic granules under high organic loading conditions. *Chemosphere* 63: 1791–1800.

Index

Printed and bound by CPI Group (UK) Ltd, Croydon, CR0 4YY

22/10/2024

01777647-0007